Constants and Conversion Factors

Planck's constant	$h = 6.626 \times 10^{-34}$ J·s $= 4.136 \times 10^{-15}$ eV·s
hbar	$\hbar = h/2\pi = 1.055 \times 10^{-34}$ J·s $= 6.582 \times 10^{-16}$ eV·s
Speed of light	$c = 2.998 \times 10^8$ m/s
Elementary charge	$e = 1.602 \times 10^{-19}$ C
Fine-structure constant	$\alpha = e^2/4\pi\epsilon_0\hbar c = 7.297 \times 10^{-3} = 1/137.036$
Boltzmann constant	$k_B = 1.381 \times 10^{-23}$ J/K $= 8.617 \times 10^{-5}$ eV/K
Avogadro constant	$N_A = 6.022 \times 10^{23}$ particles/mole
Electron mass	$m_e = 9.109 \times 10^{-31}$ kg $= 0.5110$ MeV/c^2
Proton mass	$m_p = 1.673 \times 10^{-27}$ kg $= 938.3$ MeV/c^2
Neutron mass	$m_n = 1.675 \times 10^{-27}$ kg $= 939.6$ MeV/c^2
Bohr radius	$a_0 = 4\pi\epsilon_0\hbar^2/m_e e^2 = 0.5292 \times 10^{-10}$ m
Rydberg energy	$hcR_\infty = m_e c^2 \alpha^2/2 = 13.61$ eV
Bohr magneton	$\mu_B = e\hbar/2m_e = 5.788 \times 10^{-5}$ eV/T

1 keV $= 10^3$ eV   1 MeV $= 10^6$ eV   1 GeV $= 10^9$ eV   1 TeV $= 10^{12}$ eV

1 μm $= 10^{-6}$ m   1 nm $= 10^{-9}$ m   1 pm $= 10^{-12}$ m   1 fm $= 10^{-15}$ m

1 eV $= 1.602 \times 10^{-19}$ J   1 Å $= 0.1$ nm $= 10^{-10}$ m

Constants and Conversion Factors

Quantum Physics

Quantum Physics

A Fundamental Approach to Modern Physics

John S. Townsend
HARVEY MUDD COLLEGE

Library
Quest University Canada
3200 University Boulevard
Squamish, BC V8B 0N8

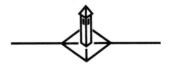

University Science Books
Mill Valley, California

University Science Books
www.uscibooks.com

Production Manager: Christine Taylor
Manuscript Editor: Lee Young
Design: Yvonne Tsang at Wilsted & Taylor
Illustrator: LM Graphics
Compositor: Macmillan Publishing Solutions
Cover Design: Genette Itoko McGrew
Printer & Binder: Edwards Brothers, Inc.

This book is printed on acid-free paper.

Copyright ©2010 by University Science Books

Reproduction or translation of any part of this work beyond that permitted by Section 107 or 108 of the 1976 United States Copyright Act without the permission of the copyright owner is unlawful. Requests for permission or further information should be addressed to the Permissions Department, University Science Books.

Library of Congress Cataloging-in-Publication Data

Townsend, John S.
　Quantum physics: a fundamental approach to modern physics / John S. Townsend.
　　p. cm.
　Includes bibliographical references and index.
　ISBN 978-1-891389-62-7 (alk. paper)
　1. Quantum theory—Textbooks. 2. Physics—Textbooks. I. Title.
QC174.12.T694 2010
530.12—dc22
　　　　　　　　　　　　　　2009022678

Printed in the United States of America
10 9 8 7 6 5 4 3 2

Contents in Brief

CHAPTER 1   Light   1

CHAPTER 2   Wave Mechanics   51

CHAPTER 3   The Time-Independent Schrödinger Equation   89

CHAPTER 4   One-Dimensional Potentials   113

CHAPTER 5   Principles of Quantum Mechanics   153

CHAPTER 6   Quantum Mechanics in Three Dimensions   177

CHAPTER 7   Identical Particles   211

CHAPTER 8   Solid-State Physics   257

CHAPTER 9   Nuclear Physics   277

CHAPTER 10   Particle Physics   317

APPENDIX A   Special Relativity   367

APPENDIX B   Power-Series Solutions   393

Contents

Preface xi

CHAPTER 1 **Light** 1

1.1 Waves 1
1.2 Interference and Diffraction: The Wave Nature of Light 4
1.3 Photons: The Particle Nature of Light 7
1.4 Probability and the Quantum Nature of Light 16
1.5 Interference with Single Photons 23
1.6 The Double-Slit Experiment 32
1.7 Diffraction Gratings 36
1.8 The Principle of Least Time 38
1.9 Summary 43

CHAPTER 2 **Wave Mechanics** 51

2.1 Atom Interferometry 52
2.2 Crystal Diffraction 58
2.3 The Schrödinger Equation 62
2.4 The Physical Significance of the Wave Function 64
2.5 Conservation of Probability 66
2.6 Wave Packets and the Heisenberg Uncertainty Principle 67
2.7 Phase and Group Velocity 72
2.8 Expectation Values and Uncertainty 76
2.9 Ehrenfest's Theorem 79
2.10 Summary 82

CHAPTER 3	**The Time-Independent Schrödinger Equation** 89
3.1	Separation of Variables 89
3.2	The Particle in a Box 91
3.3	Statistical Interpretation of Quantum Mechanics 98
3.4	The Energy Operator: Eigenvalues and Eigenfunctions 103
3.5	Summary 108

CHAPTER 4	**One-Dimensional Potentials** 113
4.1	The Finite Square Well 113
4.2	Qualitative Features 119
4.3	The Simple Harmonic Oscillator 123
4.4	The Dirac Delta Function Potential 130
4.5	The Double Well and Molecular Binding 132
4.6	Scattering and the Step Potential 135
4.7	Tunneling and the Square Barrier 140
4.8	Summary 145

CHAPTER 5	**Principles of Quantum Mechanics** 153
5.1	The Parity Operator 153
5.2	Observables and Hermitian Operators 155
5.3	Commuting Operators 160
5.4	Noncommuting Operators and Uncertainty Relations 162
5.5	Time Development 165
5.6	EPR, Schrödinger's Cat, and All That 169
5.7	Summary 173

CHAPTER 6	**Quantum Mechanics in Three Dimensions** 177
6.1	The Three-Dimensional Box 177
6.2	Orbital Angular Momentum 179
6.3	The Hydrogen Atom 187
6.4	The Zeeman Effect 194
6.5	Intrinsic Spin 196
6.6	Summary 204

CHAPTER 7	**Identical Particles** 211
7.1	Multiparticle Systems 211
7.2	Identical Particles in Quantum Mechanics 212
7.3	Multielectron Atoms 215
7.4	The Fermi Energy 221

7.5 Quantum Statistics 226
7.6 Cavity Radiation 235
7.7 Bose–Einstein Condensation 242
7.8 Lasers 247
7.9 Summary 251

CHAPTER 8 Solid-State Physics 257

8.1 The Band Structure of Solids 257
8.2 Electrical Properties of Solids 263
8.3 The Silicon Revolution 268
8.4 Superconductivity 273
8.5 Summary 275

CHAPTER 9 Nuclear Physics 277

9.1 Nuclear Notation and Properties 278
9.2 The Curve of Binding Energy 281
9.3 Radioactivity 290
9.4 Nuclear Fission 298
9.5 Nuclear Fusion 303
9.6 Nuclear Weapons: History and Physics 306
9.7 Summary 312

CHAPTER 10 Particle Physics 317

10.1 Quantum Electrodynamics 317
10.2 Elementary Particles 326
10.3 Hadrons 330
10.4 Quantum Chromodynamics 334
10.5 Quantum Flavor Dynamics 340
10.6 Mixing Angles 343
10.7 Symmetries and Conservation Laws 351
10.8 The Standard Model 358
10.9 Summary 363

APPENDIX A Special Relativity 367

A.1 The Relativity Principle 367
A.2 The Postulates of Special Relativity 368
A.3 The Lorentz Transformation 378
A.4 Four-vectors 382
A.5 Momentum and Energy 383

APPENDIX B **Power-Series Solutions** 393

    B.1   The Simple Harmonic Oscillator   393

    B.2   Orbital Angular Momentum   395

    B.3   The Hydrogen Atom   397

Answers to Odd-Numbered Problems   399

Index   405

Preface

As an undergraduate, my first course in quantum mechanics was in the spring semester of my senior year. This was, frankly, too little, too late. As a sophomore, I had taken a fairly standard course in modern physics. Despite the fact that I thought the professor teaching the course did a good job, I was not pleased with the content. The course seemed to be a summary of phenomenology, without giving me any understanding of the underlying physics. To a budding physicist, this was not a satisfying experience.

Today, I am confident we do better by our students, at least in the upper-division physics curriculum. This confidence is inspired by the quality and nature of the quantum mechanics textbooks that we use there, one of which (*A Modern Approach to Quantum Mechanics*) I am proud to have authored. At the introductory level the changes have been, in my judgment, less clear cut. I believe that students deserve a serious introduction to quantum mechanics, comparable to the introduction they receive to the subjects of mechanics and electromagnetism. Moreover, with the appropriate grounding in quantum mechanics, it is possible to give students real understanding and insight into an array of topics that often fall under the rubric of modern physics. Students can see in quantum mechanics a common thread that ties these topics together into a coherent picture of how the world works, a picture that gives students confidence that quantum mechanics itself really works, too.

While I have used the term "modern physics" to describe the material typically taught in an introductory course, I believe this term has reached the end of its useful life, at least in the way it is commonly used. Most if not all modern physics textbooks follow an historical ordering of the material, with, in order of appearance, Planck, Einstein, Rutherford, Bohr, and Schrödinger among the key characters in the story. Now I enjoy the history as much as anyone, and I try to weave it into the text in a natural way. But I don't think following the historical ordering so closely makes a lot of sense. After all, a story that starts in the early 1900s does not sound modern to students learning physics in the twenty-first century. Moreover, times have changed. We now have the advantage of truly modern experiments, such as single-photon and single-atom interferometry experiments, that have replaced the thought experiments that characterized much of the early discussions of quantum mechanics. So why not start with real experiments, which is what physics is really based on, after all.

Chapter 1 focuses on the quantum nature of light. While this chapter does include discussion of the photoelectric effect (the key to understanding the operation of a photodetector) and Compton scattering, the single-photon anticoincidence and interference experiments carried out by Alain Aspect and coworkers in the late 1980s are at the center of this chapter. Understanding the results of these experiments leads us to the concept of a probability amplitude and to the rules for multiplying and, in particular, adding these probability amplitudes when there are multiple paths that a photon can take between the source and the detector. This is really the sum-over-paths approach to quantum mechanics pioneered by Richard Feynman. One of the advantages of this approach is that students can see right away how quantum mechanics can explain everyday phenomena such as the the law of reflection, Snell's law, and a diffraction grating (in, say, the reflection of light from a CD) as straightforward extensions of the sum-over-paths approach from a few paths to many paths (leading naturally to Fermat's principle of least time). Although the approach that I follow in Chapter 1 is not the same as that given by Feynman in his short series of lectures titled *QED*, it is inspired by these lectures.

Chapter 2 starts with the double-slit experiment, a topic that was discussed in Chapter 1 as an illustration of the sum-over-paths approach to quantum mechanics. But in Chapter 2, the key experiment is a double-slit experiment with helium atoms carried out by Jürgen Mlynek's group in the 1990s. This beautiful experiment really brings home to students the strangeness of quantum mechanics. Since the sum-over-paths approach is not as useful for determining the behavior of particles such as electrons when they travel microscopic distances, Chapter 2 moves naturally toward the Schrödinger equation. This wave equation plays a similar role for nonrelativistic particles to that played by the wave equation for light in Chapter 1. Other topics in this chapter include wave packets, phase and group velocities, expectation values and uncertainty, and Ehrenfest's equations.

Chapter 3 and Chapter 4 are all about solving the Schrödinger equation for a variety of one-dimensional potentials. The centerpiece of Chapter 3 is the particle in a box, a great laboratory for seeing many quantum effects. Chapter 4 includes discussion of the finite square well, the harmonic oscillator, and the Dirac delta function potential, both as a simple model for an atom and, more interestingly, as a double well that can be used to capture the key features of molecular binding. Chapter 4 also includes a discussion of scattering (and tunneling) in one-dimensional quantum mechanics. One relatively novel feature of Chapter 3 at this level is the use of the particle in a box to illustrate the key features of the energy eigenvalue equation, including the principles of superposition and completeness and the way these principles are utilized to calculate the probability of events for a wave function that is not an energy eigenfunction. These ideas are generalized in Chapter 5 to the more general class of Hermitian operators corresponding to observables. Here the role that commuting and, in particular, noncommuting operators and uncertainty relations play in quantum mechanics is emphasized.

Chapter 6 extends the discussion of quantum mechanics to three-dimensional systems. Because the particle in a three-dimensional box, the orbital angular momentum eigenvalue problem, and the hydrogen atom can be attacked by the technique of separation of variables, these systems have much in common with the treatment of one-dimensional potentials from the earlier chapters. I make an effort to keep the mathematical level accessible to students. Some of the details, such as solving the hydrogen atom by a power-series solution, are left to an appendix for the interested reader. But the simple, direct way in which the eigenvalue problem for the z component of the orbital angular momentum, for example, quantizes the eigenvalues to integral multiples of $\hbar$ is a very important and natural extension of the techniques introduced in the earlier chapters. And the fact that the z component of the intrinsic spin angular momentum of an electron takes on only half-integral multiples of $\hbar$ tells us that intrinsic spin, real angular momentum that it is, is not connected to wave functions or to anything physically rotating, as is the case for orbital angular momentum.

Given the profound impact that the intrinsic spin of identical particles plays in multi-particle systems in quantum mechanics, one can argue that intrinsic spin is perhaps the single most important attribute of a particle. After introducing the exchange operator and seeing how the Pauli principle arises from basic quantum mechanics, Chapter 7 goes on to examine systems with identical fermions (multielectron atoms, electrons in a solid, and white dwarf and neutron stars) and identical bosons (cavity radiation, Bose–Einstein condensation, and lasers). Again, the focus is on showing how the behavior of these systems follows directly from quantum mechanics.

The remaining three chapters of the book are devoted to applications as well. But here too the focus is restricted so as to avoid the peril of too much phenomenology and too little explanation of the underlying physics. Chapter 8 is devoted to crystalline solids, namely solids for which the periodic nature of the potential energy leads to an energy band structure, allowing us to understand the electrical properties of metals, insulators, and semiconductors. The role that this band structure of semiconductors plays in the electronics/computer revolution in which we are all participating is emphasized. Chapter 9 is devoted to nuclear physics, focusing in one form or another on the all-important "curve of binding energy," including its impact on radioactivity, nuclear fission, and nuclear fusion. This chapter concludes with a discussion of the history and physics of nuclear weapons, a subject with which humanity is still struggling to deal. Finally, Chapter 10 is devoted to particle physics. One of the benefits of starting this book with a sum-over-paths approach to quantum mechanics is the natural way it leads into a description of relativistic quantum mechanics. Although it is not possible to explain particle physics at the same level of completeness as the earlier topics, it is important for students to see the role that quantum mechanics plays in the interactions of the particles in the Standard Model, where the probability amplitudes can be represented graphically by Feynman diagrams. This chapter concludes with a discussion of the close connection between symmetries and conservation laws and the fundamental role symmetry plays in determining the detailed nature of the interactions in quantum electrodynamics and quantum chromodynamics.

Although most of this book focuses on nonrelativistic quantum mechanics, the first chapter with its discussion of the quantum mechanics of light, the ninth chapter with its discussion of the curve of binding energy, and the last chapter with its discussion of Feynman diagrams presume some knowledge of special relativity. Consequently, a discussion of the basics of special relativity is included in an appendix for the benefit of students who have not had a previous exposure to this subject.

A comprehensive solutions manual for the instructor is available from the publisher, upon request of the instructor.

It is a pleasure to acknowledge the people who have helped me during the writing of this book. I have benefited greatly from the assistance of my colleagues at Harvey Mudd College. Tom Helliwell, Theresa Lynn, Dan Petersen, and Patti Sparks read portions of the manuscript and gave me helpful feedback. Tom Donnelly and Peter Saeta read the entire manuscript and gave me many valuable comments. In addition to giving me feedback on Chapter 10, Vatche Sahakian has been unfailingly patient and helpful with my LaTeX questions. For the better part of the past decade, the introductory physics course that is taken by all students at HMC in their first semester has combined an introduction to special relativity (see Appendix A) and an introduction to quantum mechanics, as detailed in Chapter 1. At Swarthmore College, where I completed a preliminary version of the text while on sabbatical, Frank Moscatelli provided thoughtful comments on the whole manuscript and Eric Jensen helped me fine-tune the astrophysics sections. My Swarthmore College Physics 14 students were a great audience for field testing this book. I very much appreciate the feedback and encouragement that I received from them. Jeff Dunham of Middlebury College was kind enough to give me detailed feedback on the entire manuscript. In addition to Harvey Mudd College and Swarthmore College, a number of institutions, including Bryn Mawr College, California State University (San Bernardino and San Marcos), Grinnell College, Haverford College, Pomona College, Rice University, and Ursinus College, have used a prepublication version of the text. I want to thank Michael Schulz, Michael Burin, Mark Schneider, Dwight Whitaker, Paul Padley, and Tom Carroll for their input based on their experiences teaching with the book. I also want to thank the Mellon Foundation for its support, Bob Wolf for assistance with the book's subtitle, Lee Young for helpful copyediting, Christine Taylor and her staff at Wilsted & Taylor Publishing Services for producing the book, Jane Ellis of University Science Books for all her efforts in overseeing the publishing process, and my publisher, Bruce Armbruster, for his support of this project.

Please do not hesitate to contact me if you find errors or have suggestions that might improve the book.

John S. Townsend
Department of Physics
Harvey Mudd College
Claremont, CA 91711
townsend@hmc.edu

Quantum Physics

CHAPTER 1

Light

Albert Einstein once remarked, "For the rest of my life, I will reflect on what light is." The greatest physicists of all time, including Newton and Maxwell as well as Einstein, have worried about the nature of light. And they certainly have not agreed with each other. Even now, there is probably not a clear consensus, with many physicists using terms such as wave–particle duality to express their perplexity. In this chapter we will examine the quantum nature of light, including the deep role that interference plays and its implications for understanding all the phenomena of optics.

1.1 Waves

There are a couple of ways to express a harmonic wave, that is, a wave with a single frequency (and wavelength). Since we will focus initially on light, we take our standard wave for this section to be an electromagnetic plane wave propagating in the x direction, for which the electric field can be expressed as

$$\mathcal{E} = \mathcal{E}_0 \cos(kx - \omega t) \tag{1.1}$$

where $\mathcal{E}_0$ is the amplitude of the wave. Light waves are transverse waves; that is the electric field oscillates perpendicular to the direction of propagation, like waves on a string, as indicated in Fig. 1.1. Thus (1.1) could be the component of the electric field

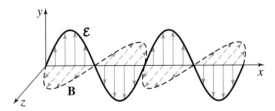

Figure 1.1 A plane polarized electromagnetic wave traveling in the x direction consists of oscillating electric and magnetic fields that are perpendicular to the direction of propagation of the wave.

along, say, the y or z directions. Figure 1.2a shows a plot of this wave at a particular time, say $t = 0$. The wave (1.1) does not have a definite position. Rather it extends infinitely far in the $+x$ and $-x$ directions, repeating itself over and over again with a spatial period λ, called the **wavelength**. The wave number k is related to the wavelength λ by

$$k = \frac{2\pi}{\lambda} \tag{1.2}$$

since when x changes by one wavelength, the cosine goes through one complete cycle and the argument of the cosine function changes by 2π when $k\lambda = 2\pi$. Similarly, if you sit at a particular position x, say $x = 0$, and watch the wave oscillate in time, the electric field will oscillate with a temporal **period** T, which is related to the angular frequency ω by

$$\omega = \frac{2\pi}{T} = 2\pi \nu \tag{1.3}$$

where in the last step we have introduced the ordinary frequency $\nu = 1/T$. Thus the units of k are m^{-1} and the units of ν are s^{-1}, or hertz.

We can, of course, also write the electric field as

$$\mathcal{E} = \mathcal{E}_0 \sin(kx - \omega t) \tag{1.4}$$

as shown in Fig. 1.2b. This form for the electric field looks essentially the same as the one in (1.1), it just has a different phase. If we were to shift our origin in Fig. 1.2b to the right by one-quarter wavelength, Fig. 1.2b and Fig. 1.2a would be the same. Thus making the shift $x \to x + \lambda/4$ in (1.4) leads to

$$\mathcal{E}_0 \sin[k(x + \lambda/4) - \omega t] = \mathcal{E}_0 \sin(kx - \omega t + \pi/2) = \mathcal{E}_0 \cos(kx - \omega t) \tag{1.5}$$

which is just (1.1). We say that there is a phase difference of $\pi/2$ between the waves (1.1) and (1.4). Clearly a more general way to write an electromagnetic wave is to express the electric field as

$$\mathcal{E} = \mathcal{E}_0 \cos(kx - \omega t + \phi) \tag{1.6}$$

where ϕ is an arbitrary phase that ranges between 0 and 2π. Note that a phase change of π, corresponding to a shift by one-half wavelength, leads to a wave that is the negative of the initial wave.

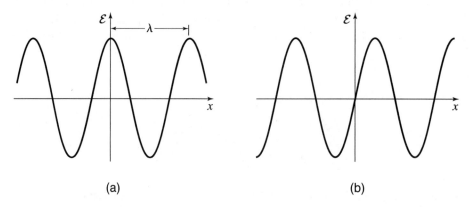

Figure 1.2 A wave of definite wavelength. In (a) the wave (1.1) and in (b) the wave (1.4) are shown at $t = 0$.

There is an additional way to write a wave that we will find to be extremely useful. We make use of the famous Euler identity

$$e^{i\theta} = \cos\theta + i\sin\theta \tag{1.7}$$

where $i = \sqrt{-1}$, to combine (1.1) and (1.4) together in the form of a complex wave

$$\mathcal{E}_0 e^{i(kx-\omega t)} = \mathcal{E}_0 [\cos(kx - \omega t) + i\sin(kx - \omega t)] \tag{1.8}$$

Expressing a wave in this complex form is often done for convenience in classical physics, where it is understood that the physical solution is the real part of (1.8). In quantum physics, on the other hand, using this complex form turns out to be essential. We will assume this to be the case in our discussion of the behavior of light in the rest of this chapter, and in Chapter 2 we will see explicitly how this complex form arises in nonrelativistic quantum mechanics when we examine the Schrödinger equation.

According to Maxwell's equations, the equations that govern the electromagnetic field, the electric field in free space (in vacuum, with no charges or currents present) satisfies the wave equation

$$\frac{\partial^2 \mathcal{E}}{\partial x^2} - \frac{1}{c^2}\frac{\partial^2 \mathcal{E}}{\partial t^2} = 0 \tag{1.9}$$

where c is the speed of light in vacuum. If we substitute $\mathcal{E} = \mathcal{E}_0 \cos(kx - \omega t)$ into this equation, we find

$$\left(-k^2 + \frac{\omega^2}{c^2}\right)\mathcal{E}_0 \cos(kx - \omega t) = 0 \tag{1.10}$$

Thus $\mathcal{E} = \mathcal{E}_0 \cos(kx - \omega t)$ will be a solution provided that

$$\omega = kc \tag{1.11}$$

Using (1.2) and (1.3), we see that (1.11) is equivalent to the familiar relation

$$\lambda \nu = c \tag{1.12}$$

It is important to note that when we talk about light, we are not necessarily talking about visible light. Rather, as Fig. 1.3 indicates, electromagnetic waves include waves of all wavelengths and frequencies, ranging from very long wavelength radio waves to very short wavelength gamma rays.

In a transparent medium such as glass that is characterized by an index of refraction n, the wave equation becomes

$$\frac{\partial^2 \mathcal{E}}{\partial x^2} - \frac{n^2}{c^2}\frac{\partial^2 \mathcal{E}}{\partial t^2} = 0 \tag{1.13}$$

which by repeating the steps corresponding to (1.10), (1.11), and (1.12) leads to

$$\lambda \nu = \frac{c}{n} \tag{1.14}$$

consistent with the speed of light in the medium being c/n. When light propagates from vacuum into a medium such as glass, the frequency ν of the light stays the same, but the wavelength of the light shifts from λ to λ/n, so as to satisfy (1.14).

Chapter 1: Light

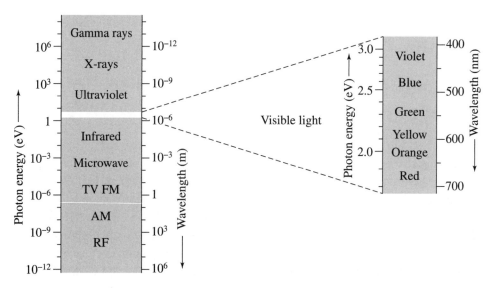

Figure 1.3 The electromagnetic spectrum.

1.2 Interference and Diffraction: The Wave Nature of Light

Figure 1.4a shows a schematic diagram (side view) of a curved glass surface resting on a flat glass plate, while Fig. 1.4b shows (top view) a series of concentric light and dark rings that are seen in the reflected light when the system is illuminated from above with monochromatic light, light with a single wavelength. These rings are known as Newton's rings. The light and dark regions are called interference fringes. Maxima in the intensity occur when the crests of the two waves, one coming from reflection from the glass–air interface at B and one from reflection from the air–glass interface at C (see Fig. 1.4a), coincide. Minima in the intensity, where the intensity is essentially zero, occur when the crest from one reflected wave coincides with the trough from the other reflected wave, giving cancellation. See Fig. 1.5. The phase difference between the two interfering waves varies with the thickness of the air gap between the two pieces of glass, since the light

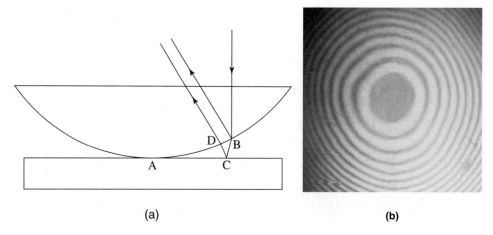

Figure 1.4 A spherical glass surface adjacent to a flat glass surface used to produce an interference pattern when illuminated from above with monochromatic light. Permission is granted to reproduce this 2007 image (Fig. 1.4b) by Winnie Summer under the terms of the GNU Free Documentation License.

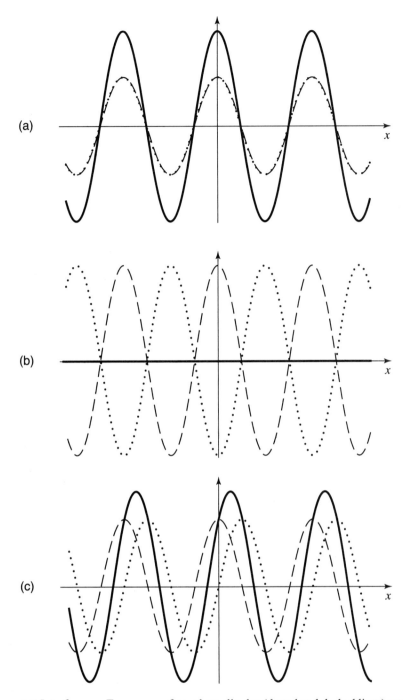

Figure 1.5 Interference: Two waves of equal amplitudes (dotted and dashed lines) are added together. In (a) the waves are in phase and interfere constructively, producing a wave (solid line) with an amplitude twice as large as that of the individual waves. In (b) the waves are π out of phase and interfere destructively (the solid line coincides with the x axis). In (c) the waves are $\pi/2$ out of phase and produce a resultant amplitude that is $\sqrt{2}$ times the amplitude of the individual waves. The intensity of a light wave is proportional to the square of the amplitude. Thus the resultant intensity of the light in (a) in which the two waves are in phase is four times as large as the intensity of the individual waves and twice as large as the resultant intensity in (c) in which the two waves are $\pi/2$ out of phase.

reflected from the air–glass interface at C travels an extra distance d from B to C to D relative to light reflected from the glass–air interface at B. In general, from the form of the electric field in, say, (1.5) we see that an extra distance d corresponds to a phase difference $kd = 2\pi d/\lambda$. When the extra path difference is 0, $\lambda/2$, λ, $3\lambda/2$, and so on, we see either destructive or constructive interference.

Note that the center region (in the vicinity of A in Fig. 1.4a) is dark. Right at the point A we might expect no reflection since one could argue that the light is simply traversing a solid piece of glass. But outside this point of contact there is a thin air gap with thickness smaller than the wavelength of the light which is also dark. This dark region arises from destructive interference because there is a phase shift of π when light is reflected at the air–flat glass interface (C in Fig. 1.4a) and no phase shift when light is reflected the the air–glass interface (B in Fig. 1.4a). In general, light reflected by a material for which the index of refraction is greater than that of the material in which the light is traveling undergoes a phase change of π. Thus the wave from the bottom interface is π out of phase with the wave from the top interface and the two cancel, or interfere destructively, when the thickness of the air film is very small compared to λ. This phase change of π upon reflection of an electromagnetic wave is reminiscent of the phase change that occurs when waves on a string are reflected at a discontinuity in the mass density of the string, as illustrated in Fig. 1.6.[1]

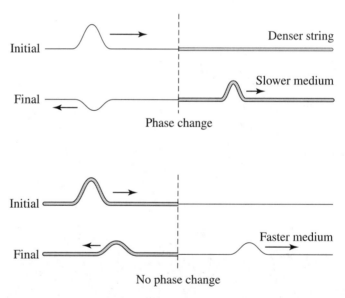

Figure 1.6 Reflection of waves from a discontinuity in a rope is similar to the reflection of electromagnetic waves from an interface where the index of refraction changes abruptly. The reflected wave undergoes a phase change of 180° when the reflection occurs at a junction where the transmitted wave slows down. When the transmitted wave speeds up at the junction, there is no phase change upon reflection.

---

[1] For an electromagnetic wave traveling in a transparent dielectric medium with index of refraction n_1 incident normally on an interface with a medium with an index of refraction n_2, the amplitude $\mathcal{E}_r$ of the reflected wave is related to the amplitude $\mathcal{E}_0$ of the incident wave by $\mathcal{E}_r = -\mathcal{E}_0(n_2 - n_1)/(n_2 + n_1)$. Thus for $n_2 > n_1$, there is a change of sign in $\mathcal{E}_r$ relative to $\mathcal{E}_0$, corresponding to a phase change of π since $e^{i\pi} = -1$.

Interference is the signature for wave behavior in classical physics. Thus, it is somewhat ironic that the interference pattern in Fig. 1.4 is referred to as Newton's rings, since Isaac Newton himself believed that light was composed of particles, or corpuscles as he called them. Newton noted that if light were indeed a wave it would "bend into the shadow," just as sound does in passing through an open door, and Newton had apparently not observed this. This bending of waves is called **diffraction** (really just interference with another name, as we will now show). Diffraction is readily observed for an appropriate aperture size (typically much smaller than a door!) with a monochromatic light source such as a laser, which of course did not exist in Newton's time. One common strategy for understanding diffraction for light passing through a slit of width a is to consider each point on the slit to be a source of secondary waves, or Huygens wavelets as they are often called. If we divide up the slit into two halves, as illustrated in Fig. 1.7, then for each point in the top half of the slit, there is a corresponding point a distance $a/2$ below it in the bottom half. If the path difference $(a/2)\sin\theta$ between the rays from each pair of points on the slit to a point P in the detection plane is $\lambda/2$, then there will be destructive interference between each pair and the total electric field at the point P will vanish, since a path difference of $\lambda/2$ leads to a phase difference between the two waves of π. Thus there will be a minimum in the intensity when $(a/2)\sin\theta = \lambda/2$ or

$$a\sin\theta = \lambda \tag{1.15}$$

We can go further in this vein and divide up the slit into four equal zones. In this case the path difference between a ray from a point in the top zone and one $a/4$ below it in the adjacent zone is $(a/4)\sin\theta$. Again there will be destructive interference between these two electric fields when the path difference is $\lambda/2$, that is, $(a/4)\sin\theta = \lambda/2$ or

$$a\sin\theta = 2\lambda \tag{1.16}$$

This cancellation will occur for each pair of rays originating from the top zone and the one $a/4$ below it. A similar cancellation will occur between every pair of rays in the third and fourth zones as well, leading to a complete interference minimum when the condition (1.16) is met. In general, there are minima whenever

$$a\sin\theta = \pm\lambda, \pm 2\lambda, \pm 3\lambda, \ldots \text{ (diffraction minima)} \tag{1.17}$$

Figure 1.8 shows a single-slit diffraction pattern and Fig. 1.9 shows plots of the relative intensity for slits of varying widths. For visible light, that is for λ in the 400–700 nm range, you need a very narrow slit, one that is only a couple of microns wide, in order for the first minimum to occur at, say, 15°. In comparison, note that a fine human hair has a diameter of 100 microns.[2]

1.3 Photons: The Particle Nature of Light

Newton's rings (Fig. 1.4) and single-slit diffraction (Fig. 1.8) provide convincing evidence of the wave nature of light. These photographs are taken with high levels of light illumination. What happens when a picture is taken with low-intensity light? Figure 1.10 shows such a picture (albeit not one involving interference). At first only

Figure 1.7 A single slit is represented as a large number of point sources. In the figure the slit is divided into halves. At the first minimum of the diffraction pattern the waves from the source at the top of the slit are 180° out of phase with the waves from the source at the middle of the slit and thus cancel, as do the waves from all the other pairs of sources in the slit. The rays shown emanating from the slit meet in the detection plane, which is where the interference pattern is observed. Because the detection plane is presumed to be quite distant from the slit, these rays are effectively parallel when viewed in the region near the slit.

---

[2] 1 micron $= 10^{-6}$ m. Thus 100 microns $= 10^{-4}$ m $= 0.1$ mm.

Figure 1.8 Diffraction pattern of laser light from a single slit. (Courtesy J. Newman)

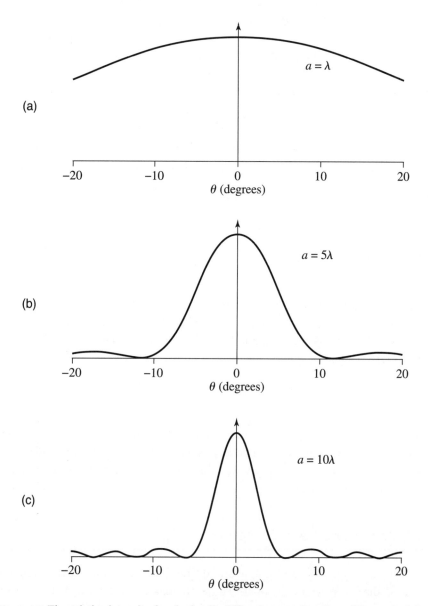

Figure 1.9 The relative intensity for single-slit diffraction as a function of the angle θ for slits of varying widths. The narrower the slit, the wider is the central maximum. The precise form of the intensity distribution is determined in Problem 1.33.

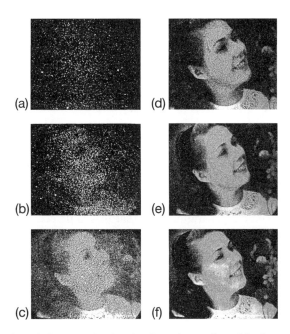

Figure 1.10 A series of photographs showing how the quality of the image improves with an increasing number of photons. The approximate numbers of photons involved in each exposure are (a) 3×10^3, (b) 1.2×10^4, (c) 9.3×10^4, (d) 7.6×10^5, (e) 3.6×10^6, and (f) 2.8×10^7. From A. Rose, *J. Opt. Soc. Am.* **43**, 715 (1953).

a seemingly random collection of bright spots is observed. Only when a sufficiently large number of spots has been generated does the image become clear. Each of these spots on the photographic film marks a location where a small chemical reaction has taken place, namely, a single particle of light has interacted with an ion in a grain of silver halide crystal in the photographic emulsion. A typical grain may be a micron or so in size and may contain 10 billion silver atoms. When the photographic emulsion is developed, all the silver atoms in a grain will be deposited as metallic silver if just a few of them have been activated in the original exposure, yielding an amplification of 10 billion.

Figure 1.10 presents a very different picture of how an image develops than has been suggested by our discussion of interference and diffraction in the previous section. In classical wave theory, making the source of the light dim, by reducing the strength of the electric field, can make the light intensity arbitrarily small. As the source gets dimmer and dimmer, the intensity decreases uniformly everywhere, which should make a photographic image progressively fainter and fainter. But the image would always be there in its entirety. This is at odds with what we see in Fig. 1.10.

The Photoelectric Effect

Despite the evidence for the wave nature of light, it was Albert Einstein in one of his famous 1905 papers, on the photoelectric effect, who first suggested that the electromagnetic field is composed of quanta, "bundles of energy" that are now called photons.[3] Einstein, inspired by the work of the physicist Max Planck in analyzing cavity radiation (a subject we will discuss in Section 7.6), proposed that a photon has energy

$$E = h\nu \qquad (1.18)$$

where ν is the frequency of the light. The constant h is a fundamental constant called **Planck's constant**. Its value is 6.63×10^{-34} J·s.

In the photoelectric effect, a photon strikes a clean metal surface, ejecting an electron from the metal, as illustrated in Fig. 1.11. This electron is typically referred to as a photoelectron. A schematic diagram of the apparatus that is used to study the photoelectric effect is shown in Fig. 1.12a. If the anode is at a positive electric potential difference relative to the cathode, then electrons emitted from the cathode will be accelerated toward and collected by the anode, generating a current that is detected by the galvanometer G in the circuit. If the intensity of the light striking the cathode is increased, so too is the maximum value (the saturation value) of the photocurrent, as shown in Fig. 1.12b. But if the polarity of the anode relative to the cathode is reversed, then electrons ejected from the cathode are repelled from the anode. Only those photoelectrons with sufficient kinetic energy will be able to overcome the retarding potential and reach the anode. When the retarding potential reaches a certain critical value, called the **stopping potential**, the current ceases. The magnitude of the stopping potential is independent of the intensity of the light, as is indicated in Fig. 1.12b. Thus the maximum kinetic energy of the photoelectrons is the charge of the electron ($e = 1.6 \times 10^{-19}$ C) times the magnitude of the the stopping potential measured in volts. For example, if the stopping potential were equal to 1 V, the maximum kinetic energy would be 1.6×10^{-19} C·V $= 1.6 \times 10^{-19}$ J, which is, by definition, 1 electron volt (1 eV).

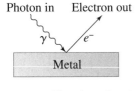

Figure 1.11 The photoelectric effect.

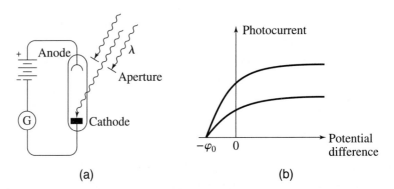

Figure 1.12 (a) A schematic diagram of the apparatus used to investigate the photoelectric effect. (b) The photoelectric current as a function of the potential difference (measured in volts) between the anode and the cathode for two different light intensities. Some of the photoelectrons emitted from the cathode have sufficient energy to reach the anode even when the anode is at negative polarity relative to the cathode. When the retarding potential exceeds φ_0, the current is zero. The saturation value of the photocurrent is proportional to the light intensity but the stopping potential φ_0 is independent of the intensity of the light.

---

[3] Within a seven-month period in 1905 Einstein submitted five articles for publication in the *Annalen der Physik*. In addition to the article on the photoelectric effect, which many people point to as the first bold step toward a quantum theory, there were two articles that laid out the foundations of special relativity and two articles on Brownian motion that provided convincing direct evidence of the atomic nature of matter. Hence 1905 is often referred to as a miraculous year, at least by physicists.

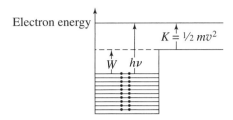

Figure 1.13 Electrons in a metal are bound in a potential energy well. The solid lines with dots indicate energy levels that contain electrons, generally two to each energy state. The work function W is the minimum amount of energy required to remove an electron from the metal. The maximum kinetic energy K of an electron ejected from the metal by a photon is the energy $h\nu$ of the photon minus the work function W.

In Einstein's analysis of the photoelectric effect, the energy of the incident photon is imparted to an electron in the metal. But before the electron can emerge from the metal it must overcome the forces that bind it within the metal. Thus the maximum kinetic energy K of an electron ejected from a metal by a photon is given by

$$K = h\nu - W \tag{1.19}$$

where W is the **work function** of the metal, the minimum energy necessary to extract an electron, as indicated in Fig. 1.13. Since the energy of a photon decreases linearly with decreasing frequency, there is a threshold frequency ν_0 for the ejection of photoelectrons from a metal that is determined by setting $K = 0$:

$$h\nu_0 = hc/\lambda_0 = W \tag{1.20}$$

since at that frequency the electrons would have no kinetic energy. For sodium, a typical alkali metal, the work function is 2.3 eV, which means that $\nu_0 = 5.5 \times 10^{14}$ Hz, or $\lambda_0 = 540$ nm, corresponding to green light. Therefore, *visible* light with $\lambda < \lambda_0$, for which it is easy to determine the wavelength, has sufficient energy to eject an electron from the metal. The maximum kinetic energy imparted to the electrons is then determined by measuring the stopping potential, permitting a test of (1.19). This is the reason Robert Millikan chose to use alkali metals to test Einstein's prediction. And testing was certainly called for, since (1.19) is in complete disagreement with the classical wave picture that we used in Section 1.1. In this wave picture, the kinetic energy of the electron should be proportional to the intensity of light (the energy per unit area per unit time) incident on the photocathode. This intensity is determined from the time average of the square of the electric field $\mathcal{E}$. Thus the intensity is proportional to $\mathcal{E}_0^2$, the square of the amplitude of the electric field, and is completely independent of the frequency of the light. Figure 1.14 shows Millikan's results for sodium, confirming Einstein's prediction.[4]

---

[4]Millikan did not believe Einstein's prediction and initially set out to disprove it. One of the nasty features of alkali metals is that they react very strongly with oxygen and therefore oxidize readily. Millikan had to take unusual precautions to keep the photocathode from being contaminated. In particular, he was forced to build a miniature "machine shop in vacuo." For example, he developed an ingenious technique of shaving the surface of the metal in a vacuum with a magnetically operated knife. In the end, after a long series of careful experiments, Millikan was forced to acknowledge the correctness of Einstein's photoelectric effect formulas "despite their unreasonableness." Millikan received the Nobel Prize in 1923 primarily for his work on the photoelectric effect, as did Einstein in 1921.

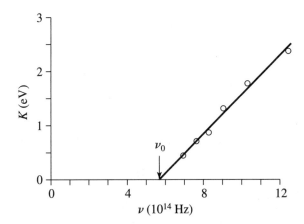

Figure 1.14 The maximum kinetic energy K of the photoelectrons vs. frequency for sodium. The data are from R. A. Millikan, *Phys. Rev.* **7**, 355 (1916). The threshold frequency ($\nu_0 = 5.6 \times 10^{14}$ Hz) has been adjusted to allow for the contact potential between the anode and the cathode (see Section 8.2). Millikan obtained a value for h, Planck's constant, from the slope of the straight line.

EXAMPLE 1.1 Determine the energy of a photon emitted by a helium–neon laser, for which $\lambda = 633$ nm. How many photons are emitted per second if the output power is 5.0 milliwatts?

SOLUTION From (1.12) and (1.18), the energy of a photon is

$$E = h\frac{c}{\lambda} = (6.63 \times 10^{-34} \text{ J·s})\frac{3.00 \times 10^8 \text{ m/s}}{6.33 \times 10^{-7} \text{ m}} = 3.14 \times 10^{-19} \text{ J}$$

which is a very small energy, reflecting the small size of Planck's constant, at least in macroscopic units. Put another way, the output power of the laser, which is 5 milliwatts, corresponds to the emission of

$$\frac{5.0 \times 10^{-3} \text{ J/s}}{3.14 \times 10^{-19} \text{ J/photon}} = 1.6 \times 10^{16} \text{ photons/s}$$

This illustrates why we don't notice the discrete, quantum nature of the electromagnetic field in everyday life. The number of photons is so large that adding or subtracting one here or there does not make a noticeable difference. A more appropriate energy scale for atomic processes is the electron volt. In electron volts, the energy of a photon emitted by a helium–neon laser is

$$E = \frac{3.14 \times 10^{-19} \text{ J}}{1.6 \times 10^{-19} \text{ J/eV}} = 2.0 \text{ eV}$$

EXAMPLE 1.2 The work function of cesium is 1.9 eV, the smallest work function of any pure metal. What is the threshold frequency ν_0 for the ejection of photoelectrons from cesium? What is the corresponding wavelength?

SOLUTION

$$W = h\nu_0$$

$$\nu_0 = W/h = \frac{(1.9 \text{ eV})(1.60 \times 10^{-19} \text{ J/eV})}{6.63 \times 10^{-34} \text{ J·s}} = 4.6 \times 10^{14} \text{ Hz}$$

Thus

$$\lambda_0 = \frac{c}{\nu_0} = 6.5 \times 10^{-7} \text{ m} = 650 \text{ nm}$$

namely red light.

Compton Scattering

Ironically, despite the beautiful confirmation that Millikan's results provided of Einstein's photoelectric effect formula (1.19), Millikan himself was not convinced of the particle nature of light until 1923. In that year, Arthur H. Compton carried out a scattering experiment with X-rays on a graphite target. X-rays are short wavelength electromagnetic radiation. In classical theory, these electromagnetic waves cause electrons in the carbon atom to oscillate up and down with the same frequency as the incident wave and therefore emit radiation at this frequency, since the electrons are being accelerated by the electromagnetic field of the incident wave. Figure 1.15 shows the experimental setup as well as Compton's results. Although the intensity of the incident radiation has a peak at one wavelength, the intensity of the scattered radiation has two peaks, one at the same wavelength as the incident radiation and one at a longer wavelength. The magnitude of this longer wavelength varies with the scattering angle θ.

There is a simple relationship between the wavelength and the momentum of a photon that follows from combining the results of two of Einstein's 1905 papers. The photoelectric effect tells us that $E = h\nu = hc/\lambda$ and, as noted in Section A.5, the special theory of relativity tells us that $E = pc$, a special case of the more general result $E = \sqrt{p^2c^2 + m^2c^4}$ for $m = 0$. Since $hc/\lambda = pc$, we find

$$p = \frac{h}{\lambda} \tag{1.21}$$

as the relationship that connects the momentum of the photon to its wavelength.[5] In a situation in which light is scattered from an electron at rest, transferring some kinetic energy to the electron, the wavelength of the scattered photon must be longer than that of the incident photon since the scattered photon has less energy than the incident one. Thus the particle nature of light leads to a result that is quite different from that predicted from a classical wave approach.

It is straightforward to analyze this scattering using conservation of energy and linear momentum, just as one would do for a collision between billiard balls (although the billiard balls are not usually relativistic). If we call the wavelength of the incident

---

[5] A rationale for taking the photon mass to be zero is given in Section 2.3.

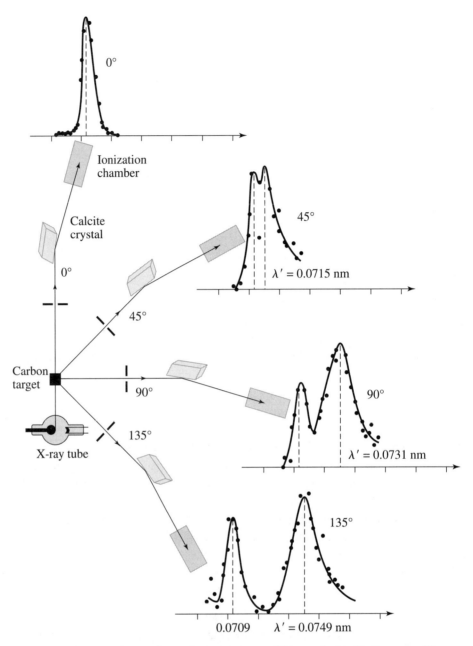

Figure 1.15 Compton's experimental arrangement and his results [A. H. Compton, *Phys. Rev.* **21**, 483; **22**, 409 (1923)]. The data show the intensity of the scattered radiation from a graphite target as a function of wavelength with a monochromatic source for four different scattering angles, namely $\theta = 0$, $\theta = 45°$, $\theta = 90°$, and $\theta = 135°$. The wavelength of the scattered radiation is measured using crystal diffraction (see Section 2.2) and the intensity is measured by determining the amount of ionization this radiation produces.

photon λ (and therefore its momentum $p = h/\lambda$) and that of the scattered photon λ' (and its momentum $p' = h/\lambda'$), as indicated in Fig. 1.16, then conservation of momentum requires

$$\mathbf{p} = \mathbf{p}' + \mathbf{p}_e \tag{1.22}$$

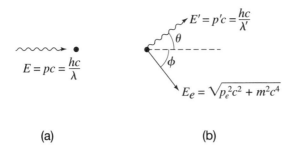

Figure 1.16 A photon is incident on an electron initially at rest in (a). The photon is scattered at angle θ and the electron recoils at angle ϕ, as shown in (b).

where $\mathbf{p}_e$ is the momentum of the scattered electron. Momentum is of course a vector and therefore each of its components must be separately conserved. If we denote by θ the angle the scattered photon's momentum makes with that of the incident photon (see Fig. 1.17), then using the law of cosines, for example, we see that

$$p_e^2 = p^2 + p'^2 - 2pp' \cos\theta \qquad (1.23)$$

Here the magnitude of $\mathbf{p}_e$ is denoted by p_e, the magnitude of $\mathbf{p}$ is denoted by p, and the magnitude of $\mathbf{p}'$ is denoted by p'. Conservation of energy requires

$$E + mc^2 = E' + E_e \qquad (1.24)$$

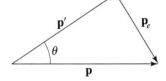

Figure 1.17 Conservation of linear momentum in Compton scattering.

or expressed in terms of the momenta

$$pc + mc^2 = p'c + \sqrt{p_e^2 c^2 + m^2 c^4} \qquad (1.25)$$

We now use (1.23) to eliminate p_e from (1.25). First move the $p'c$ term in (1.25) to the left side of the equation and square the result. Then substituting (1.23) into this equation, we obtain

$$p^2 c^2 + p'^2 c^2 - 2pp' c^2 + 2(p - p')mc^3 + m^2 c^4 = p^2 c^2 + p'^2 c^2 - 2pp' c^2 \cos\theta + m^2 c^4 \qquad (1.26)$$

which reduces to

$$pp'(1 - \cos\theta) = (p - p')mc \qquad (1.27)$$

Expressed in terms of the wavelength, we finally obtain the **Compton formula**

$$\lambda' - \lambda = \frac{h}{mc}(1 - \cos\theta) \qquad (1.28)$$

Compare this result with Compton's original data in Fig. 1.15. Notice that the scale of the shift in wavelength is determined by the length h/mc, which is called the Compton wavelength. For an electron as the target particle, $h/mc = 0.0024$ nm $= 2.4$ pm. You can now see why the effect was first observed with X-rays. For visible light, for which λ is 400–700 nm, this shift in wavelength is a very small fraction of the wavelength itself and is not as easily observed. Compton's X-rays, on the other hand, had a wavelength of 0.07 nm, so the shift in wavelength was a significant fraction of the wavelength itself.[6]

EXAMPLE 1.3 What is the energy of a photon with a wavelength equal to the Compton wavelength h/mc, where m is the mass of the electron? Suppose a photon with this energy makes a collision with a free electron initially at rest. What is the energy of the final photon if the scattering angle is 90°? How much kinetic energy is transferred to the electron?

SOLUTION The energy of the incident photon is

$$E = \frac{hc}{\lambda} = mc^2 = 0.511 \text{ MeV}$$

From (1.28), the wavelength of the scattered photon for $\theta = 90°$ is

$$\lambda' = \lambda + \frac{h}{mc} = 2\frac{h}{mc}$$

Thus the energy of the scattered photon is

$$E' = \frac{hc}{\lambda'} = \frac{1}{2}mc^2 = 0.256 \text{ MeV}$$

and the electron gains kinetic energy

$$K = E - E' = \frac{1}{2}mc^2 = 0.256 \text{ MeV}$$

in the collision.

1.4 Probability and the Quantum Nature of Light

Example 1.1 shows that a source of light with even a modest power output emits a very large number of photons. In this section and the next we will discuss experiments in which it is necessary to detect photons one at a time. It turns out to be pretty straightforward to "see" individual photons, although not quite directly with the human eye, which requires roughly 100 photons per second to produce a signal in the brain. (See Problem 1.5 to get a sense of the amazing sensitivity of the human eye.) A photomultiplier, on the other hand, can detect single photons. Figure 1.18 shows a schematic diagram of how it operates. A photon is first absorbed by the photocathode, an electrode coated with an alkali metal, ejecting an electron via the photoelectric effect. In the photomultiplier, this photoelectron is accelerated toward a second electrode at a higher electric potential. When this electron

---

[6]The data in Fig. 1.15 do show scattering occurring at the same wavelength as the incident radiation as well as at the shifted wavelength given by (1.28). How are we to understand this nominally classical result? The target carbon atom consists of a relatively heavy nucleus, at least in comparison with the mass of an electron, as well as the six electrons that are bound to the nucleus. If the incident photon interacts with the nucleus instead of a lightly bound electron, for example, the atom as a whole may recoil in the collision. In this case the mass m in (1.28) is the mass of the atom instead of the mass of the electron and the corresponding shift in wavelength is much reduced relative to that in scattering from an essentially free electron.

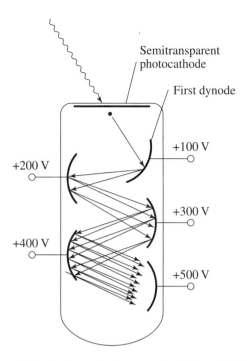

Figure 1.18 A schematic diagram of a photomultiplier.

strikes the electrode, it knocks out several (four or so) additional, or secondary, electrons, which are then accelerated toward a third electrode, which is at a still higher potential. Upon impact, these electrons knock out additional electrons, which are then accelerated toward an additional electrode, and so on. A typical photomultiplier may have as many as 14 or 15 different electrodes, or dynodes as they are called. Thus the initial photoelectron generates an avalanche of perhaps ten million electrons in a high-gain photomultiplier. These electrons form a current pulse, which can be amplified, generating an electrical signal that enables us to detect the arrival of a single photon.[7]

The photoelectric effect and Compton scattering provide indirect evidence for the particle nature of light, albeit evidence that is in striking disagreement with the predictions of classical physics. Direct evidence for the particle nature of light was, in fact, not obtained until quite recently. A. Aspect, P. Grangier, and G. Roger used the cascade radiative decay of an excited state of calcium (see Fig. 1.19) to generate single photons

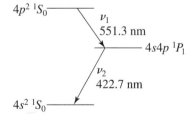

Figure 1.19 An energy-level diagram for calcium. The spectroscopic notation used to label the energy levels is explained in Section 7.3.

---

[7]A photodiode, a solid-state device constructed from the semiconductor silicon, can achieve a significantly higher efficiency than a photomultiplier for detecting single photons.

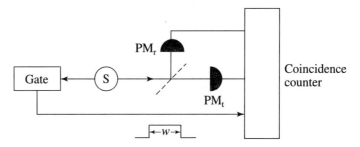

Figure 1.20 A schematic diagram of the anticoincidence experiment of Grangier et al. The source S emits two photons in a cascade decay. Detection of the first photon is used to generate an electrical pulse which activates the two photomultipliers.

in an experiment they carried out in 1986.[8] Two photons with frequencies ν_1 and ν_2 are emitted in this cascade decay. The first photon emitted is called the trigger photon, since if it is detected by the gate photodetector, shown to the left in Fig. 1.20, then an electronic signal is sent to the circuit containing the two photomultipliers PM_t and PM_r, turning on these photodetectors for a period $2\tau_s$, where $\tau_s = 4.7$ ns is the lifetime of the intermediate state shown in Fig. 1.19. The probability of detecting the photon with frequency ν_2 with either photodetector PM_t or PM_r from this particular cascade decay is much greater than that of observing a photon with this frequency from the decay of any other calcium atom in the source, essentially because the detectors are synchronized with the emission of the first photon and are active for such a short period of time.

Situated between the single-photon source and the photodetectors PM_t and PM_r is a beam splitter. The beam splitter is simply a device that reflects half the light incident upon it and transmits the other half. If light is incident on a sheet of glass at normal incidence, for example, 96% of the light is transmitted and 4% is reflected (which is why we make windows out of glass). If we coat the glass with a thin film of metal, it becomes a mirror that (ideally) reflects 100% of the light. If the surface is only lightly coated (silvered), we can create a half-silvered mirror that reflects 50% of the light and transmits 50% of the light incident upon it, say at an angle of $45°$.[9] Thus if we were to shine the 5 mW laser from Example 1.1 that emits 1.6×10^{16} photons/second on the beam splitter, we would find on average 8×10^{15} photons/second are reflected and 8×10^{15} photons/second are transmitted. This is an enormous number of photons, such that we would be hard pressed to notice the discrete particle nature of light.

But what happens if one photon at a time is incident on the beam splitter? Perhaps the photon splits in two, with half a photon being transmitted and half reflected. But if the photon were to split, then by energy conservation the reflected and the transmitted photon would have less energy than the incident photon and hence a longer wavelength. But this is not what is occurring since the reflected and the transmitted beams have the same color as the incident light. What Aspect and company find is that a count is obtained by either PM_t or PM_r *but not both simultaneously*. A count in PM_t never coincides with a count in PM_r, which is what we expect if light is composed of particles. Our everyday experience

---

[8]P. Grangier, G. Roger, and A. Aspect, *Europhys. Lett.* **1**, 173 (1986).
[9]A modern beam splitter may have a multilayer dielectric as the coating instead of a metal.

with a particle is that it is *either* "here" *or* "there." Waves, on the other hand, are spread out over an extended region of space and may be *both* "here" *and* "there." Consequently, the experiment is referred to as an anticoincidence experiment and the results with single photons are in complete accord with the fact that light is indeed composed of particles.

One may well ask then which of the two photomultipliers detects the photon with frequency ν_2. The answer is simple: it is not possible to predict whether the photon will be detected by PM_t or PM_r. In fact, the best we can do is give the probability for detection. The way we calculate this probability is to determine first a complex number z, called the **probability amplitude**. This terminology is reminiscent of the amplitude $\mathcal{E}_0$ of the complex form for the electromagnetic wave (1.8).

A general complex number z may be expressed as

$$z = x + iy \tag{1.29}$$

where x and y are real and $i = \sqrt{-1}$. In fact, all fundamental processes in nature can be described using a probability amplitude or a sum of amplitudes. The **complex conjugate** of the complex number z is defined to be

$$z^* = x - iy \tag{1.30}$$

Therefore the rule for obtaining the complex conjugate is to let $i \to -i$ everywhere. When a probability amplitude is multiplied by its complex conjugate, the result is a positive, real number:

$$z^*z = (x - iy)(x + iy) = x^2 + y^2 \tag{1.31}$$

which we postulate is the probability of detecting a photon, provided z is properly chosen. This is the **first major principle of quantum mechanics**:

$$\text{The probability of an event} = z^*z \tag{1.32}$$

There is an alternative way to represent a complex number that will be especially useful to us. Figure 1.21 shows the location of the complex number $z = x + iy$ in the complex plane. If we choose polar coordinates r and ϕ to label this point instead of the real and imaginary parts x and y, then

$$z = x + iy = r\cos\phi + ir\sin\phi = re^{i\phi} \tag{1.33}$$

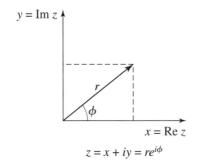

Figure 1.21 A complex number z is defined by giving its real and imaginary parts, x and y, respectively, or its magnitude r and phase ϕ.

where r is called the modulus, or magnitude, of the complex number and ϕ its phase. In the last step we have used the Euler identity (1.7). Notice that x, y, r, and ϕ are all real numbers. In terms of the magnitude and phase, the complex conjugate z^* is given by

$$z^* = r\cos\phi - ir\sin\phi = r(\cos\phi - i\sin\phi) = re^{-i\phi} \tag{1.34}$$

and the product z^*z involves only the magnitude squared:

$$z^*z = re^{-i\phi}re^{i\phi} = r^2 \tag{1.35}$$

which is the same as (1.31) since of course $r^2 = x^2 + y^2$. We will often use the somewhat more compact notation

$$|z|^2 = z^*z \tag{1.36}$$

In our discussion of interference of electromagnetic waves in Section 1.2 two aspects of the wave that we made use of were its amplitude and its phase. In classical physics the intensity of a light wave with amplitude $\mathcal{E}_0$ is proportional to $\mathcal{E}_0^2$. In quantum physics the probability of detecting a photon with probability amplitude $re^{i\phi}$ is equal to r^2, the magnitude squared of the probability amplitude. In general, the magnitude and phase of the probability amplitude are determined from first principles by solving Maxwell's equations. In free space, these equations can be reduced to the wave equation (1.9). Just as the spatial dependence of an electromagnetic wave in free space can be written as e^{ikx}, the phase ϕ of the corresponding probability amplitude for a photon with wavelength λ to travel a distance x is given by kx, where $k = 2\pi/\lambda$. We are interested in the spatial dependence of the amplitude since in the remainder of this chapter we will be analyzing situations in which there are a variety of distances that a photon may travel between the source and the detector.

Let's start with an example for which we know what the probability must be. Suppose the source of the light is a laser that is placed a distance x directly in front of a photomultiplier. If we assume we can ignore the spreading of the light beam as it exits the laser because the distance x is not too large, that is the photomultiplier is quite near the laser, then there should be a 100% chance of detecting a photon, presuming the photomultiplier is 100% efficient. If we express the probability amplitude in the form $z = re^{i\phi}$, then the fact that $z^*z = 1$ tells us that r must be one but it doesn't tell us anything about the phase ϕ. Since we are assuming that the region between the laser and the photomultiplier is empty space, the phase of the amplitude for a photon to travel a distance x is kx. But a phase that is proportional to the distance x traveled by the photon has an interesting consequence. Notice that if we break up the distance x into segments of length x_1 and x_2 where $x = x_1 + x_2$, then the phase factor e^{ikx} can be written as

$$e^{ikx} = e^{ik(x_1+x_2)} = e^{ikx_1}e^{ikx_2} \tag{1.37}$$

where the last step is an elementary consequence of the exponential function. Thus the phase factor e^{ikx} for traveling a distance x is the product of the phase factors e^{ikx_1} and e^{ikx_2} for traveling distances x_1 and x_2.

Although this is just one example, it suggests a **second major principle of quantum mechanics**: To determine the probability amplitude for a process that can be viewed as taking place in a series of steps—say a photon propagating a distance x_1 and then propagating an additional distance x_2—we *multiply* the probability amplitudes for each

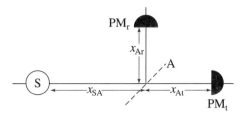

Figure 1.22 Distances traveled by the photon in the anticoincidence experiment.

of these steps:

$$z = z_a z_b \ldots \tag{1.38}$$

In classical physics, using complex exponentials for waves, for example, is a notational convenience. The physics is determined by the real part of these complex-valued functions. But using real valued functions such as $\cos kx_1$ and $\cos kx_2$ for the probability amplitudes would mean that the probability amplitudes would not obey this product rule. As we will see, the product rule and complex numbers are necessary for a complete description of the real world in which we live.

Let's now analyze the anticoincidence experiment shown in Fig. 1.20. If we call x_{SA} the distance the photon travels between the source and the beam splitter and x_{At} the distance the photon travels between the beam splitter and the photomultiplier PM_t, as indicated in Fig. 1.22, then the probability amplitude for a photon leaving the source to be detected by PM_t is given by

$$z_t = e^{ikx_{SA}} \left(\frac{1}{\sqrt{2}}\right) e^{ikx_{At}} = \frac{1}{\sqrt{2}} e^{ikd_t} \tag{1.39}$$

where we have labeled by d_t the total distance traveled between the source and the photodetector PM_t, that is, $d_t = x_{SA} + x_{At}$. Note that we have inserted a factor of $1/\sqrt{2}$ in the middle of the amplitude (1.39) to account for the fact that there is a 50% probability that the photon is transmitted at the beam splitter. Thus if we use the probability amplitude (1.39) to calculate the probability that a photon is detected by PM_t, we obtain

$$z_t^* z_t = \left(\frac{1}{\sqrt{2}} e^{-ikd_t}\right) \left(\frac{1}{\sqrt{2}} e^{ikd_t}\right) = \frac{1}{2} \tag{1.40}$$

Similarly, we can write the probability amplitude for a photon to travel from the source to photodetector PM_r as

$$z_r = e^{ikx_{SA}} \left(\frac{e^{i\pi}}{\sqrt{2}}\right) e^{ikx_{Ar}} = -\frac{1}{\sqrt{2}} e^{ikd_r} \tag{1.41}$$

where d_r is the total distance traveled between the source and the detector, that is the sum of the distance x_{SA} from the source to the beam splitter and the distance x_{Ar} between the beam splitter and the photodetector PM_r. Notice in (1.41) there is a minus sign in the overall amplitude indicative of the fact that there is a phase change of π upon reflection from the beam splitter, as was the case for the reflection of light from the air–glass interface in our analysis of Newton's rings in Section 1.2. Recall that $e^{i\pi} = -1$. Here too we obtain a 50% probability that a photon is detected:

$$z_r^* z_r = \left(-\frac{1}{\sqrt{2}} e^{-ikd_r}\right)\left(-\frac{1}{\sqrt{2}} e^{ikd_r}\right) = \frac{1}{2} \quad (1.42)$$

Of course, if the photon isn't detected by PM_t, it will be detected by PM_r, assuming our photodetectors are 100% efficient. Thus it must also be true that

$$z_t^* z_t + z_r^* z_r = 1 \quad (1.43)$$

Finally, notice how in this anticoincidence experiment the value of the phase for the amplitudes (1.39) and (1.41) cancels out in calculating the probabilities (1.40) and (1.42), respectively. In the next section, we will examine an experiment in which the phases of the probability amplitudes matter a great deal.

You may find it distressing that the best we can do is determine a probability for an event to occur. When we flip a coin, for example, we often say that there is a 50% chance that we get a head and a 50% chance that we get a tail. But we believe that if we have enough information about the underlying physics, such as how much force and how much torque is exerted on the coin when we flip it, we can determine how many flips the coin will make before, say, it hits the ground and whether it will land with a head up or not. Thus the flipping of the coin is not inherently probabilistic. But when it comes to predicting whether a single photon is transmitted or reflected by a beam splitter, the best we can do is give a probabilistic answer. Not everyone finds this situation satisfying. Einstein for one certainly did not, hence his famous dictum "God does not play dice with the Universe." To which Niels Bohr replied, "Who are you to tell God what to do?"

EXAMPLE 1.4 Suppose the beam splitter shown in Fig. 1.22 transmits one quarter of the light and reflects three quarters of the light of wavelength λ incident on it. What are the probability amplitudes for detecting the photon with photomultiplier PM_t and photomultiplier PM_r?

SOLUTION The probability amplitudes that photomultipliers PM_t and PM_r detect a photon are given by

$$z_t = e^{ikx_{SA}} \left(\sqrt{\frac{1}{4}}\right) e^{ikx_{At}} = \frac{1}{2} e^{ikd_t}$$

and

$$z_r = e^{ikx_{SA}} \left(e^{i\pi} \sqrt{\frac{3}{4}}\right) e^{ikx_{Ar}} = -\frac{\sqrt{3}}{2} e^{ikd_r}$$

respectively, where d_t is the total distance traveled between the source S and PM_t and d_r is the total distance traveled between the source S and PM_r. Consequently

$$z_t^* z_t = \frac{1}{4} \quad \text{and} \quad z_r^* z_r = \frac{3}{4}$$

as desired. Thus if the helium–neon laser in Example 1.1 that emits 1.6×10^{16} photons/s is the light source, then on average 0.4×10^{16} photons/s would be transmitted and 1.2×10^{16} photons/s would be reflected. But for a single photon all we can say is that there is a 25% chance the photon will be transmitted and a 75% chance it will be reflected.

EXAMPLE 1.5 A photon passes through a sheet of polarizing material. The polarization axis of the sheet is vertical. What is a probability amplitude that could be used to determine the probability that this photon will then pass through a polarizer with a polarization axis that makes an angle θ with the vertical?

SOLUTION We know from classical physics (the law of Malus) that the intensity of a beam of light passing through the second polarizer is reduced by a factor of $\cos^2 \theta$. Thus a probability amplitude that would work would be $z = \cos \theta$. If $\theta = 0$, there is a 100% probability of transmission (assuming ideal polarizers), and if $\theta = 90°$ (the transmission axes are perpendicular), there is no chance that the photon will be transmitted.

1.5 Interference with Single Photons

In 1989 A. Aspect, P. Grangier, and G. Roger followed up on their photon anticoincidence experiment by performing an experiment in which single photons were incident on a Mach–Zehnder interferometer.[10] The results of this experiment vividly illustrate what may well be one of the most disturbing aspects of quantum mechanics. Figure 1.23 shows a schematic diagram of the experiment. The first mirror, at A, is a half-silvered mirror that acts as a beam splitter. Fifty percent of the light is transmitted to mirror B while the other 50% is reflected to mirror C. Assume mirrors B and C reflect 100% of the light. Finally, mirror D is again a beam splitter. No matter whether light arrives from B or C, 50% is sent to one or the other photomultiplier, PM_1 or PM_2.

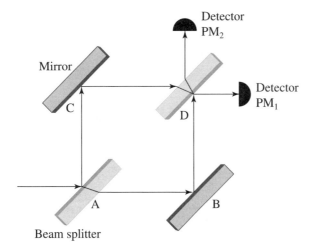

Figure 1.23 A diagram of a Mach–Zehnder interferometer. Photomultipliers PM_1 and PM_2 are used to detect a photon incident on beam splitter A from the left.

---

[10] A. Aspect, P. Grangier, and G. Roger, *J. Optics* (*Paris*) **20**, 119 (1989).

To analyze what is going to happen, start by imagining yourself to be a single photon traversing the interferometer. At each beam splitter you toss a coin. If the coin lands tails up (T), the photon is transmitted. If the coin lands heads up (H), the photon heads off in a different direction, that is, it is reflected. Thus a T at the first beam splitter and an H at the second beam splitter will cause the photon to be detected by PM_1. So will an H at the first beam splitter and a T at the second beam splitter. On the other hand, two H's or two T's will cause the photon to be detected by PM_2. Since HT, TH, HH, and TT occur with equal probability, there should be a 50% chance that the photon is detected by PM_1 and a 50% chance that the photon is detected by PM_2. Note that these predictions are independent of the distances between the beam splitters and the mirrors.

In the experiment Aspect et al. vary the distance traveled by the photon along one of the paths, say path ABD, by varying the position of mirror B in small steps each of size $\lambda/25$, where λ is the wavelength of the photon being detected by, say, PM_1. If you look at the data in the top row in Fig. 1.24, you see the number of photons detected if they count

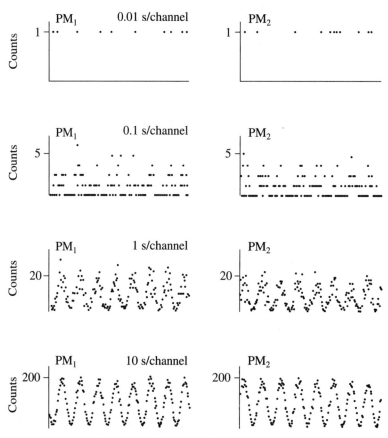

Figure 1.24 Development of the interference fringes one photon at a time. The vertical axis shows the number of photons detected in photomultipliers PM_1 and PM_2 for a particular position of the mirrors in the interferometer while the horizontal axis shows the path difference between the different arms of the interferometer measured in intervals of $\lambda/25$, where λ is the wavelength of the photon in the interferometer. For each fixed position of the mirrors photons are counted for time intervals ranging from 0.01 s to 10 s. The photomultipliers are active for 10 ns intervals following detection of the first photon emitted in the cascade decay. The data are from A. Aspect, P. Grangier, and G. Roger, *J. Optics (Paris)* **20**, 119 (1989), as are the data in Fig. 1.25.

for 0.01 s. Notice that in this time interval, they may detect a photon with photomultiplier PM_1, but they never detect two or more photons. This is consistent with there being at most one photon in the interferometer. Similarly, PM_2 detects at most one photon, and there are no coincidences in counts between PM_1 and PM_2 for a particular position of mirror B, in accord with our discussion in the previous section on the particle nature of light.

If you look at the results in the second row of Fig. 1.24 of counting for a 0.1 s interval for each position of mirror B, you see more counts but no clear pattern is evident yet. But look at the bottom row, where the counts are taken for 10 s for each position of mirror B, and you see a striking pattern. In some positions of mirror B essentially no photons are detected by PM_1. At those positions, many photons, up to 200, are counted by PM_2. But if mirror B is shifted slightly, say $\lambda/2$ to the right or to the left, this situation is reversed, with 200 photons counted by PM_1 and essentially none by PM_2. Or look at Fig. 1.25, which shows the results of counting for longer periods of time and moving mirror B by increments of $\lambda/50$. The data show a classic interference pattern, which is in marked disagreement with the predictions that we made by tossing coins.

We now face a conundrum. Interference is thought to be a characteristic of waves, but light is composed of particles. To account for the interference of the particles, we need an additional principle of quantum mechanics. To see what this principle may be, let's start by trying to calculate the probability that a photon is detected by PM_1. Let's call z_{ABD} the probability amplitude for a photon to reach PM_1 along path ABD. Our discussion in the previous section suggests that we should multiply the amplitudes for each leg of the path, that is,

$$z_{ABD} = \frac{1}{\sqrt{2}}(-1)\left(\frac{-1}{\sqrt{2}}\right)e^{ikd_1} = \frac{1}{2}e^{ikd_1} \tag{1.44}$$

where the first factor of $1/\sqrt{2}$ is the amplitude for transmission at the first beam splitter at A, the minus sign comes from the phase change upon reflection from the mirror at B, the second factor of $-1/\sqrt{2}$ coming from reflection from the second beam splitter at D with an associated phase change, and finally the factor of e^{ikd_1} arising from the

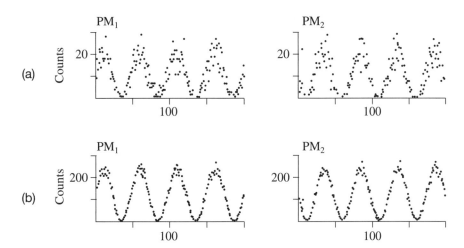

Figure 1.25 The location of the movable mirror in the Mach–Zehnder interferometer is varied in increments of $\lambda/50$. The photons are counted for 1 s for each position of the mirror in (a) and for 15 s in (b).

total distance d_1 that the photon follows from the source to PM$_1$ along the path ABD. Note that $z^*_{\text{ABD}} z_{\text{ABD}} = 1/4$, which is the probability of obtaining TH in the coin-tossing scenario.

There is, of course, a second route a photon can take to reach PM$_1$, namely along path ACD. The corresponding probability amplitude is

$$z_{\text{ACD}} = \left(\frac{-1}{\sqrt{2}}\right)(-1)\left(\frac{1}{\sqrt{2}}\right) e^{ikd_2} = \frac{1}{2} e^{ikd_2} \tag{1.45}$$

where the first factor of $-1/\sqrt{2}$ arises from the probability amplitude for reflection from the first beam splitter at A, the middle minus sign is the phase change associated with reflection from the fixed mirror at C, and the last factor of $1/\sqrt{2}$ arises from the probability amplitude for transmission at the second beam splitter at D. Here we have assumed that the total distance traveled by the photon in traversing the path ACD between the source and the photodetector is d_2. Note that $z^*_{\text{ACD}} z_{\text{ACD}} = 1/4$, the probability of HT in the coin-tossing scenario.

If we were to add the probabilities that a photon takes path ABD to reach the detector and a photon takes path ACD to reach the detector, we would obtain a 50% probability that the photon is detected by PM$_1$ independent of the position of mirror B. Even though we might think that a photon has to take one path or the other, this is at odds with the data. *Adding the probabilities does not work.* Instead, since there are no detectors that tell us which path the photon takes, we must write the probability amplitude for a photon to be detected by PM$_1$ as

$$z_{\text{PM}_1} = z_{\text{ABD}} + z_{\text{ACD}} \tag{1.46}$$

where the probability amplitudes z_{ABD} and z_{ACD} are given by (1.44) and (1.45), respectively. This is a special case of the **third major principle of quantum mechanics**: If there are multiple ways that an event can occur (in this case the multiple paths the light can take), we *add the amplitudes* for each of these ways:

$$z = z_1 + z_2 + \cdots \tag{1.47}$$

For the Mach–Zehnder interferometer example, (1.46) becomes

$$z_{\text{PM}_1} = \frac{1}{2} e^{ikd_1} + \frac{1}{2} e^{ikd_2} \tag{1.48}$$

The probability is then given by

$$\begin{aligned} z^*_{\text{PM}_1} z_{\text{PM}_1} &= \left(\frac{1}{2} e^{-ikd_1} + \frac{1}{2} e^{-ikd_2}\right)\left(\frac{1}{2} e^{ikd_1} + \frac{1}{2} e^{ikd_2}\right) \\ &= \frac{1}{4} + \frac{1}{4} + \frac{1}{4} e^{ik(d_1-d_2)} + \frac{1}{4} e^{-ik(d_1-d_2)} \\ &= \frac{1}{2} + \frac{1}{2} \cos[k(d_1-d_2)] = \cos^2\left[\frac{k(d_1-d_2)}{2}\right] \end{aligned} \tag{1.49}$$

where in the last step we have made use of the half-angle formula

$$\cos\frac{\theta}{2} = \sqrt{\frac{1+\cos\theta}{2}} \tag{1.50}$$

Since z_{PM_1} is the sum of two terms, $z_{ABD} = e^{ikd_1}/2$ and $z_{ACD} = e^{ikd_2}/2$, when z_{PM_1} is multiplied by its complex conjugate, there are two terms that arise in the multiplication, namely $z^*_{ABD}z_{ACD}$ and $z^*_{ACD}z_{ABD}$, which oscillate as $d_1 - d_2$ varies. These "cross terms," the third and fourth terms in the middle line of (1.49), are responsible for the interference effects. The overall probability (1.49) is in agreement with the experimental results.[11]

Note that if $d_1 - d_2 = 0, \pm\lambda, \pm 2\lambda, \ldots$, then $k(d_1 - d_2)/2 = 0, \pm\pi, \pm 2\pi, \ldots$ and the probability (1.49) is 1; that is, 100% of the photons incident on the interferometer will be detected by PM_1. On the other hand, if $d_1 - d_2 = \pm\lambda/2, \pm 3\lambda/2, \ldots$, then $k(d_1 - d_2)/2 = \pm\pi/2, \pm 3\pi/2, \ldots$ and the probability the photon is detected by PM_1 vanishes. But the detection probability is not solely 0 or 1, of course. For example, if $d_1 - d_2 = \pm\lambda/3, \pm 2\lambda/3, \ldots$, then $k(d_1 - d_2)/2 = \pm\pi/3, \pm 2\pi/3, \ldots$ and the probability the photon is detected by PM_1 is 1/4. As Example 1.6 shows, if the incident photon is not detected by PM_1, then it is detected by PM_2, assuming the photomultipliers are 100% efficient in detecting photons.

The data shown in Fig. 1.24 and Fig. 1.25 are in agreement with the result (1.49) (as well as the corresponding result for detection of a photon by PM_2 that is obtained in Example 1.6). We clearly see interference fringes, but only after we have detected a sufficiently large number of photons. Nonetheless, one can say that the interference is taking place for individual photons, since only one photon is present in the interferometer at any one time. Why is this state of affairs so troubling? As mentioned in the previous section, Aspect et al. have looked at the anticorrelations that exist when a single photon is incident on a beam splitter, as in Fig. 1.20. The photon is detected by either one photomultiplier or the other. The photon does not split at the beam splitter; it goes one way or the other. But by adding mirrors B and C and a second beam splitter D, Aspect et al. have created a situation in which both paths ABD and ACD matter for each photon. How can this be? How does the photon know that both paths are present once it passes the first beam splitter? What then is interfering?

To make matters perhaps stranger still, Aspect's group has recently carried out what the physicist John Wheeler called a **delayed-choice experiment**. In this experiment single photons were incident on a Mach–Zehnder interferometer with a 48-m path length (or, equivalently, a 160-ns time of flight) between the two beam splitters. The decision to include or exclude the second beam splitter was effectively made after the photon had entered the interferometer. Therefore as the photon passes the first beam splitter, it "cannot know" which experiment is being performed, the anticoincidence experiment described in the previous section or the single-photon interference experiment described in this section. Figure 1.26a shows the complementary interference fringes observed by the two photodetectors when the second beam splitter is present, while Fig. 1.26b shows the output from the photodetectors when the second beam splitter is removed, in which case we know which path the photon took between the source and the detector. In the latter case the characteristic anticorrelations in the count rate in the two detectors are observed, with 50% of the counts in one photodetector and 50% in the other.

---

[11]Of course if you toss a coin for which there is a 50% chance of obtaining a head, there is nothing inherently wrong with obtaining four heads in the first four tosses. It is just not highly probable. As the number of tosses increases, the distribution of heads and tails will approach the 50–50 mix that we expect based on the underlying probabilities. Similarly, the probability distribution (1.49) becomes most apparent as the number of photons counted increases.

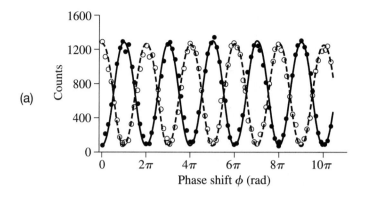
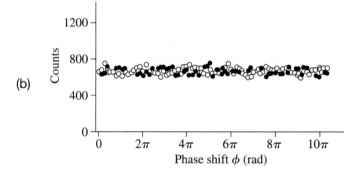

Figure 1.26 Results of a single-photon delayed-choice experiment with a Mach–Zehnder interferometer. In (a) both beam splitters are present and the path difference between the two arms of the interferometer is varied by tilting the second beam splitter. In (b) the second beam splitter has been effectively removed from the interferometer. Consequently, there are two paths between the light source and each photodetector in (a), while in (b) there is a single path between the light source and each photodetector. The decision to include or exclude the second beam splitter was made in such a way that the photon could not "know" whether the second beam splitter was present or not as it entered the interferometer. Adapted from V. Jacques, E. Wu, F. Grosshans, F. Treussart, P. Grangier, A. Aspect, and J.-F. Roch, *Science* **315**, 966 (2007). The solid circles indicate data from detector 1 and the open circles data from detector 2.

In summary, quantum mechanics tells us simply that if we set up a situation where there are multiple paths that a particle can follow between the source and the detector, then we must include a *probability amplitude* for each path. When we calculate the probability by multiplying this sum of amplitudes by its complex conjugate, the cross terms generate the interference that we observe. *We cannot say that the particle follows one path or the other.* This is the nature of the quantum world in which we live.[12]

---

[12] The issues raised so far may in part explain why a whole issue of *Optics and Photonics Trends* (Vol. 3, No. 1, 2003) was devoted to "The Nature of Light: What is a Photon?" The articles include two humorous quotes from Roy Glauber, who shared the Nobel prize in physics in 2005 for his contributions to the field of quantum optics: "I don't know anything about photons, but I know one when I see one." and "A photon is what a photodetector detects."

EXAMPLE 1.6 Determine the probability that a photon incident on the Mach–Zehnder interferometer shown in Fig. 1.23 is detected by photomultiplier PM_2. Show that the probability that a photon is detected by either photomultiplier PM_1 or photomultiplier PM_2 is one.

SOLUTION To calculate the probability that a photon is detected by PM_2, we must add the amplitudes for the photon to reach PM_2 along path ABD and along path ACD. The combined amplitude is given by

$$z_{PM_2} = \frac{1}{\sqrt{2}}(-1)\left(\frac{1}{\sqrt{2}}\right)e^{ikd_1} + \left(\frac{-1}{\sqrt{2}}\right)(-1)\left(\frac{1}{\sqrt{2}}\right)e^{ikd_2}$$

$$= -\frac{1}{2}e^{ikd_1} + \frac{1}{2}e^{ikd_2}$$

where in the first term d_1 is the distance traveled between the source and the photodetector along path ABD and in the second term d_2 is the distance traveled along path ACD. The second term is the tricky one. There are two phase changes (two minus signs) in this amplitude. The first arises from the reflection at the beam splitter at A and the second in reflection from the mirror at C. At first you might expect a third minus sign in reflection from the beam splitter at D into PM_2. But in this last reflection, light first travels through the glass of this second beam splitter before reaching the interface with air, which is where the reflection occurs. See Fig. 1.23. Thus in keeping with our discussion of phase changes in reflection in Section 1.2, there is *no* phase change for this reflection since the light approaches the interface from a medium with the larger index of refraction.[13] The probability corresponding to this combined amplitude is

$$z^*_{PM_2}z_{PM_2} = \left(-\frac{1}{2}e^{-ikd_1} + \frac{1}{2}e^{-ikd_2}\right)\left(-\frac{1}{2}e^{ikd_1} + \frac{1}{2}e^{ikd_2}\right)$$

$$= \frac{1}{4} + \frac{1}{4} - \frac{1}{4}e^{ik(d_1-d_2)} - \frac{1}{4}e^{-ik(d_1-d_2)}$$

$$= \frac{1}{2} - \frac{1}{2}\cos[k(d_1-d_2)] = \sin^2\left[\frac{k(d_1-d_2)}{2}\right]$$

where in the last step we have made use of the half-angle formula

$$\sin\frac{\theta}{2} = \sqrt{\frac{1-\cos\theta}{2}}$$

Adding the probability that the photon is detected by PM_1 from (1.49) to the probability that it is detected by PM_2, we find

$$z^*_{PM_1}z_{PM_1} + z^*_{PM_2}z_{PM_2} = \cos^2\left[\frac{k(d_1-d_2)}{2}\right] + \sin^2\left[\frac{k(d_1-d_2)}{2}\right] = 1$$

assuming of course that the photomultipliers are 100% efficient detectors.

---

[13]This discussion of phase change upon reflection at the beam splitter at D may prompt you to worry about reflection from the front interface of this beam splitter when it is approached on path ACD. As Example 1.8 shows, reflection from this interface can be minimized by coating the front surface with a thin film of, say, magnesium fluoride.

Note: It is worth emphasizing that the probabilities for the photon to be detected by PM$_1$ and PM$_2$ depend on the *difference* in the distances d_1 and d_2 traveled by the photon from the source to each photodetector along the paths ABD and ACD. Since on path ABD and on path ACD the photon travels the same distance from the source to the beam splitter at A, for example, this common distance cancels out when calculating $d_1 - d_2$. Hence the distances that really matter are the distances within the interferometer itself. Thus it is not imperative that photomultiplier PM$_2$ be positioned exactly the same distance from the second beam splitter as is PM$_1$, since this extra distance is the same for both paths that contribute to the amplitude for detecting a photon with either photomultiplier and thus cancels out when calculating the difference in path length.

Perhaps this is a good place to admit that our discussion of the calculation of the probability amplitudes has been somewhat oversimplified. When light is transmitted through the glass of a beam splitter, say of thickness d, on path ABD, the phase of the amplitude advances by $\phi = k'd$, where $k' = 2\pi/\lambda' = 2\pi n/\lambda$ and n is the index of refraction of the glass. That is, we need to account for the fact that the wavelength of the photon in the glass is $\lambda' = \lambda/n$. Thus the phase of the amplitude advances more rapidly in glass than it does in vacuum (or air). One might say that we have been treating the beam splitter as if it were arbitrarily thin, since we did not take this change of phase into account. But the real reason that we were able to ignore the finite thickness of the beam splitter is that the photon is presumed to traverse the same thickness of glass on path ABD or path ACD to reach detector PM$_1$ (or detector PM$_2$ for that matter) and therefore this phase change cancels out when we calculate the probability of detecting a photon, just as the common phase change due to travel on the distance between the source and the beam splitter at A cancels out in determining a probability such as (1.48). Example 1.8 discusses the phenomenon of thin-film interference, a phenomenon where the change of phase in traversing a medium other than air matters a great deal.

EXAMPLE 1.7 Suppose that the probability amplitude for a photon to be detected by one of the photomultipliers in the Mach–Zehnder interferometer is given by

$$z = e^{i\chi}(z_1 + z_2)$$

where χ is a real number and z_1 and z_2 are the probability amplitudes for the photon to take paths ABD and ACD, respectively. The quantity $e^{i\chi}$ is referred to as an **overall phase factor** (and χ is called the overall phase). Show the probability of the photon being detected is independent of χ.

SOLUTION The probability is given by

$$z^*z = e^{-i\chi}(z_1^* + z_2^*)e^{i\chi}(z_1 + z_2)$$
$$= (z_1^* + z_2^*)(z_1 + z_2)$$

since $e^{-i\chi}e^{i\chi} = 1$. Thus the probability is independent of χ. However, as we saw in the preceding example, the *phase difference* $\phi = k(d_1 - d_2)$ between the amplitudes z_1 and z_2, what is called the **relative phase**, matters greatly in determining the value of the probability.

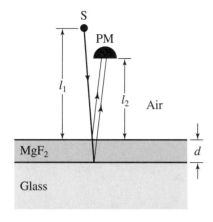 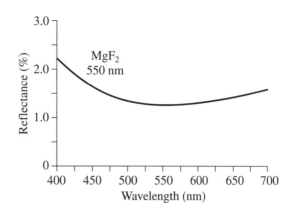

Figure 1.27 Reflection at normal incidence from a thin film of magnesium fluoride.

EXAMPLE 1.8 Glass lenses are often coated with a thin film of magnesium fluoride to reduce unwanted reflection of light. The index of refraction of MgF$_2$ is $n = 1.38$, which is intermediate between the index of refraction of air ($n_{\text{air}} = 1.00$) and that of glass ($n_{\text{glass}} = 1.52$). As shown in Fig. 1.27, a single-photon source of light S that emits photons of wavelength λ is located a distance l_1 above the film and a photomultiplier is located a distance l_2 above the film so that light reaches the detector by reflecting off the film at normal incidence. The magnitude of the probability amplitude for light to travel from the source to the detector via reflection from the top surface of the film is r; via reflection from the bottom surface it is fr, where f is a fraction less than one. Take the thickness of the film to be d.

(a) What is the probability amplitude for detecting a photon via reflection from the top surface of the film?

(b) What is the probability amplitude for detecting a photon via reflection from the bottom surface of the film?

(c) Calculate the probability that a photon is detected by the photomultiplier. Assume that the additional amplitudes that result from multiple reflections between the top and bottom surfaces of the film can be neglected in your calculation.

(d) Determine a value for d in terms of λ and n that minimizes the probability of detection. Obtain a numerical value for the minimum probability by taking $r = 0.16$ and $f = 0.30$ for $\lambda = 550$ nm. Compare your value with data for the reflectance of a thin film of magnesium fluoride in Fig. 1.27.

SOLUTION

(a) $z_t = -e^{ikl_1} r e^{ikl_2} = -r e^{ik(l_1+l_2)}$

where $k = 2\pi/\lambda$.

(b) $z_b = -e^{ikl_1} e^{ik'd} r f e^{ik'd} e^{ikl_2} = -rf e^{ik(l_1+l_2)} e^{2ik'd}$

where $k' = 2\pi/\lambda' = 2\pi n/\lambda$. The overall minus sign in each of these amplitudes results from the fact that light reflected by a material for which the index of refraction is greater than that of the material in which the light is traveling undergoes a phase change of π, as noted in the discussion of Newton's rings in Section 1.2.

(c)
$$z = z_t + z_b = -re^{ik(l_1+l_2)}\left(1 + fe^{2ik'd}\right)$$
$$z^*z = r^2\left(1 + fe^{-2ik'd}\right)\left(1 + fe^{2ik'd}\right)$$
$$= r^2\left(1 + f^2 + fe^{2ik'd} + fe^{-2ik'd}\right)$$
$$= r^2\left(1 + f^2 + 2f\cos 2k'd\right)$$

Thus the probability is given by

$$z^*z = r^2\left(1 + f^2 + 2f\cos\frac{4\pi nd}{\lambda}\right)$$

Notice how the distances l_1 and l_2 that are common to both amplitudes z_t and z_b cancel out in the determination of the probability. (d) The minimum probability occurs when the cosine equals -1. One possibility is that the argument of the cosine is equal to π. In this case

$$\frac{4\pi nd}{\lambda} = \pi$$

or

$$d = \frac{\lambda}{4n}$$

At a minimum

$$z^*z = r^2\left(1 + f^2 - 2f\right) = r^2(1-f)^2$$

Substituting in the values $r = 0.16$ and $f = 0.30$

$$z^*z = (0.16)^2(0.70)^2 = 0.0125$$

or a 1.25% probability of reflection, which compares quite favorably with the minimum value for the reflectance in Fig. 1.27.

In summary, note that the probability of reflection is *not* the sum of r^2, the probability of reflection from the top surface, and $r^2 f^2$, the probability of reflection from the bottom surface. The interference between the probability amplitudes for a photon to be reflected from the top surface of the film and for a photon to be reflected from the bottom surface of the film can be tuned by adjusting the thickness of the film for a specific wavelength to minimize the overall probability of reflection for a photon of that wavelength.

1.6 The Double-Slit Experiment

To gain more experience using this formalism of quantum mechanics, let's next look at a classic interference experiment, the double-slit experiment. This experiment was first carried out by Thomas Young in 1802, but not of course with single photons. In the experiment, monochromatic light of wavelength λ is incident on a screen with two narrow slits, as indicated in Fig. 1.28. Typically light is detected on a distant screen. A modern version of this experiment is easily carried out with a laser as the source of light. With a laser the number of photons per second incident on the screen with the slits is very large, as is illustrated by Example 1.1. While the experiment is not usually carried out one photon at a time, we will presume this experiment is done with a single-photon source

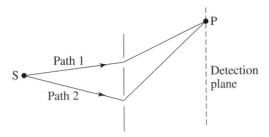

Figure 1.28 A schematic diagram of the double-slit experiment indicating two paths that light can follow from the source S to the detector located at point P.

and analyze it accordingly. We can suppose we have an array of small photomultipliers distributed throughout the detection plane.

We see the potential for interference immediately, since there are two paths that a photon can take to reach a photodetector centered at a point P in the detection plane. The probability amplitude for a photon to be detected by the photodetector at P is given by

$$z_P = re^{ikd_1} + re^{ikd_2} \tag{1.51}$$

where d_1 is the distance between the source of light and the detector for a path that traverses slit 1 and d_2 is the distance between the source and the detector for a path that traverses slit 2. As in our general discussion of complex numbers in Section 1.4, we are denoting the magnitude of each amplitude by r. Strictly, these two amplitudes would have slightly different magnitudes as well as different phases, but the differences in the magnitudes can be safely neglected when the distance between the screen containing the slits and the detection plane is large, much larger than the spacing d between the slits. If we were to close one of the slits, say slit 2, the amplitude would be reduced to $z_P = re^{ikd_1}$ and the probability of detecting a photon would be $z_P^* z_P = r^2$. The magnitude of this probability would depend on variables such as the actual size of the photomultiplier and how far the detecting plane is from the slits. We will leave this probability as a parameter that could be determined empirically by doing the experiment. In order to keep things as simple as possible, we will assume that each of the slits is very narrow, with a width substantially less than the wavelength of the light, so that the probability would roughly be uniform throughout the detection plane with a single slit open. However, when we open the second slit, there is a striking change in the probability of detecting a photon.

Let's assume, as is often the case, that the distance L between the screen containing the slits and the detection plane is sufficiently large that we can consider the rays drawn from the slits to the point P to be essentially parallel, as indicated in Fig. 1.29. Then

$$d_2 \cong d_1 + d\sin\theta \tag{1.52}$$

and we can rewrite (1.51) as

$$z_P = re^{ikd_1}(1 + e^{ikd\sin\theta}) = re^{ikd_1}(1 + e^{i\phi}) \tag{1.53}$$

where we have introduced $\phi = kd\sin\theta$ as the phase difference that arises for a photon taking path 2 relative to path 1. The probability that the photon is detected at the photomultiplier centered at point P is given by

$$\begin{aligned} z_P^* z_P &= re^{-ikd_1}(1 + e^{-i\phi})re^{ikd_1}(1 + e^{i\phi}) \\ &= r^2(1 + 1 + e^{i\phi} + e^{-i\phi}) \\ &= 2r^2(1 + \cos\phi) = 4r^2\cos^2(\phi/2) \end{aligned} \tag{1.54}$$

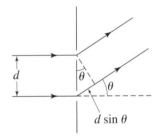

Figure 1.29 A close-up view of the two paths taken between the source and the detector in the double-slit experiment. The two rays shown are essentially parallel and the path difference between them is $d\sin\theta$, corresponding to a phase difference of $\phi = kd\sin\theta$.

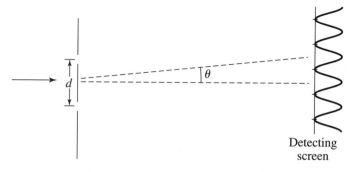

Figure 1.30 The intensity pattern for the double-slit experiment. Note that the slits are assumed to be sufficiently narrow that the probability distribution would be uniform if one of the slits were closed.

This probability distribution is indicated in Fig. 1.30. Notice that there are maxima when

$$\phi = 0, \pm 2\pi, \pm 4\pi, \ldots \qquad (1.55)$$

which means the path difference is an integral number of wavelengths:

$$d \sin \theta = 0, \pm \lambda, \pm 2\lambda, \ldots \qquad (1.56)$$

At the angles θ satisfying (1.56), we get constructive interference with a probability of detecting a photon that is four times as large as we would get if a single slit were open. There are minima when

$$\phi = \pm \pi, \pm 3\pi, \pm 5\pi, \ldots \qquad (1.57)$$

corresponding to a path difference

$$d \sin \theta = \pm \lambda/2, \pm 3\lambda/2, \pm 5\lambda/2, \ldots \qquad (1.58)$$

Thus opening the second slit has strikingly rearranged the likelihood of where a photon will strike the detector.[14]

It is helpful to have a geometric picture of how complex amplitudes add to give these results. We can represent a complex number as an arrow in the complex plane. In case of the double-slit experiment, we wish to add two complex numbers together, each of which is an arrow of length r but with different phases. The first complex number has a phase $\alpha = kd_1$ and the second has a phase $\alpha + \phi$, as shown in Fig. 1.31a. When we add complex numbers together, we add the real parts together and the imaginary parts together to determine the real and the imaginary part of the resultant complex number. This is the same rule for adding two-dimensional vectors together, where we add the x components of the vectors to obtain the x component of the resultant vector and add

---

[14] Young initally suggested the double-slit experiment to provide conclusive evidence that light, like sound, is a wave. Probably because of Newton's influence, the prevailing view was that light was composed of particles. To suggest that a screen uniformly illuminated by a single narrow slit could develop regions with less illumination (dark fringes) by opening a second slit was counterintuitive at that time. Now, we face a larger challenge, namely coming to grips with interference when we know that light is composed of particles after all. In meeting this challenge we have been forced to redefine our concept of a particle to be an object whose behavior is governed by probability amplitudes in the way we have outlined in the preceding sections.

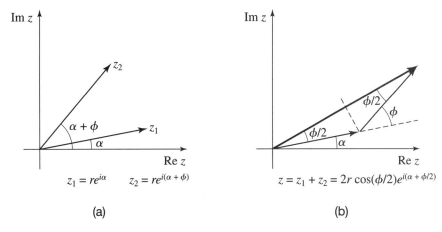

Figure 1.31 (a) Two complex numbers with the same magnitude r but a phase difference ϕ. (b) The addition of these complex numbers, which yields a complex number whose magnitude is $2r\cos(\phi/2)$.

the y components to obtain the y component of the resultant vector. In the same way that we add vectors by placing them head to tail, we can add these complex numbers together. For the special case where the two complex numbers have the same magnitudes, we generate an isosceles triangle for which it is easy to do the addition geometrically, leading to

$$z_P = 2r\cos(\phi/2)\, e^{i(\alpha+\phi/2)} \tag{1.59}$$

See Fig. 1.31b. Of course, if we calculate the probability of detecting a photon using this result, we obtain

$$z_P^* z_P = 4r^2 \cos^2(\phi/2) \tag{1.60}$$

as before. As we saw in the previous sections, here too the overall phase $\alpha + \phi/2$ of the resultant complex number z_P cancels out when determining the probability. As we will see in the next two sections, this geometric approach to adding the amplitudes turns out to be especially useful when the number of paths a photon can take between the source and the detector increases.

EXAMPLE 1.9 In a double-slit demonstration carried out with a He–Ne laser ($\lambda = 633$ nm), take the distance L between the slits and the screen to be 10 m and the separation of the slits d to be 0.1 mm $= 100$ μm. Determine the separation x on the screen between the central maximum and the next adjacent maximum.

SOLUTION The central maximum occurs at $\theta = 0$. The adjacent maximum (referred to as the first maximum) occurs when

$$\sin\theta = \lambda/d = 633 \times 10^{-9}\text{ m}/10^{-4}\text{ m} = 633 \times 10^{-5}$$

This is a very small angle and thus it is quite appropriate to use the approximation $\sin\theta \cong \tan\theta$. The separation between central maximum and the first maximum on the screen is then given by

$$x = L\tan\theta \cong L\sin\theta = L\lambda/d = 6.33 \times 10^{-2}\text{ m} = 6.33\text{ cm}$$

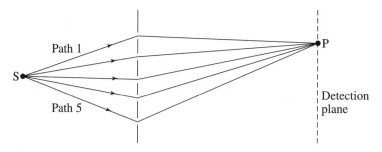

Figure 1.32 There are five paths that light can take between the source S and the detector at P when the screen has five narrow slits.

1.7 Diffraction Gratings

It is easy to generalize the results of Section 1.6 to a situation in which N slits, all equally spaced, are open. Such a device is called a diffraction grating. Such gratings have a great deal of utility. To be specific, let's consider the case of five slits, in which case there are five paths between the source and the detector, as indicated in Fig. 1.32. As for the double slit, when the distance between the screen containing the slits and the detection plane is large, the rays for each path from the screen to the detector are effectively parallel, as shown in Fig. 1.33. The difference in path length between adjacent paths is $d \sin\theta$, corresponding to a phase difference $\phi = kd \sin\theta$.

Following the procedures of the previous section, we can write the probability amplitude as

$$z_P = re^{ikd_1}(1 + e^{i\phi} + e^{2i\phi} + e^{3i\phi} + e^{4i\phi}) \tag{1.61}$$

where again $\phi = kd \sin\theta$. As before, we will get maxima when $\phi = 0, \pm 2\pi, \pm 4\pi, \ldots$, corresponding to a path difference between adjacent slits that is an integral number of wavelengths. At these maxima, $z_P = 5re^{ikd_1}$ and therefore the probability of detecting a photon by a photomultiplier centered on one of these maxima is $z_P^* z_P = 25r^2$, or twenty five times as large as the probability if only a single slit is open. It must of course be true that the probability of detecting a photon at locations other than the maxima is correspondingly reduced. It is not difficult to see why. If we go back to the double-slit example, we see that the condition for the first minimum is that the two complex amplitudes are π, or $180°$, out of phase with each other, as shown in Fig. 1.34a. In the case of five slits, the first minimum is reached when the phase difference between adjacent slits is $2\pi/5$, or $72°$, since in this case the five complex amplitudes will sum to zero, as illustrated in Fig. 1.34b. But since $\phi = kd \sin\theta$, the change in angle θ necessary to reach a minimum is smaller than it was for two slits, meaning that the maxima are correspondingly sharper. The maxima grow increasingly sharp as the number N of slits increases since the first minimum is reached when the phase difference is $2\pi/N$. See Fig. 1.35. This is one of the key features that makes a grating so useful, since it is possible to measure the angles at which the maxima occur very accurately when the number of slits is large and therefore to determine λ accurately.

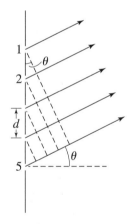

Figure 1.33 A close-up view of a five-slit grating. The phase difference between adjacent slits is $\phi = kd \sin\theta$.

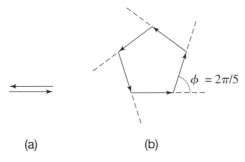

Figure 1.34 (a) How two complex amplitudes of equal magnitude cancel when the phase difference between them is π. (b) Five amplitudes cancel (sum to zero) when the phase difference between each is $2\pi/5$.

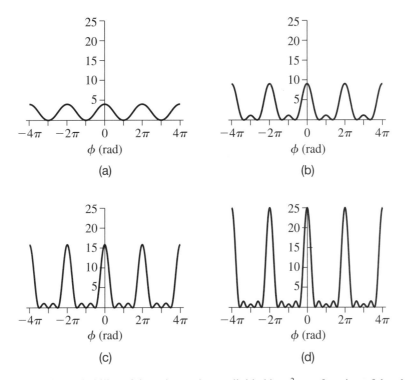

Figure 1.35 The probability of detecting a photon divided by r^2 as a function of the phase ϕ for (a) 2 slits, (b) 3 slits, (c) 4 slits, and (d) 5 slits.

EXAMPLE 1.10 Consider a grating composed of four very narrow slits each separated by a distance d. (a) What is the probability that a photon strikes a detector centered at the central maximum if the probability that a photon is counted by this detector with a single slit open is r^2? (b) What is the probability that a photon is counted at the first minimum of this four-slit grating if the bottom two slits are closed?

SOLUTION (a) At the central maximum ($\phi = 0$), all the amplitudes are in phase. Hence

$$z = re^{ikd}(1 + e^{i\phi} + e^{2i\phi} + e^{3i\phi}) = 4re^{ikd}$$

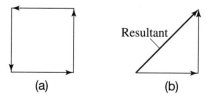

Figure 1.36 The amplitudes from a four-slit grating at (a) the first minimum and (b) the first minimum of the four-slit grating but with the bottom two slits closed.

Thus

$$z^*z = 16r^2$$

(*b*) At the first minimum, the four probability amplitudes sum to zero as shown in the Fig. 1.36a. With two slits blocked, two amplitudes are removed, as indicated in Fig. 1.36b. Hence

$$z^*z = \left(\sqrt{2}r\right)^2 = 2r^2$$

where $z = z_1 + z_2$. Thus the probability of detecting a photon is one eighth as large as it is at a maximum of the four-slit grating.

1.8 The Principle of Least Time

It is intriguing to think about the extension from two slits to five slits to an infinite number of slits, to the limit that there is no obstructing screen at all between the source of light and the detector. Clearly, in this latter case, instead of (1.51) for two slits or (1.61) for five slits, we would end up with

$$z_P = \sum_j r e^{ikd_j} \tag{1.62}$$

where the sum is over all possible paths between the source and the detector. As an illustration, we will show how this sum-over-paths approach to quantum mechanics can be used to derive the law of reflection from a mirror, namely, that the angle of incidence equals the angle of reflection.[15]

If we position a monochromatic source of light at S some distance above a mirror and locate the photodetector at point P, we can calculate the probability of detecting a photon at P by summing the probability amplitudes for all paths that light can take in traveling between the source and the detector. A block is inserted between S and P so that the paths that matter most are ones that reflect from the mirror, that is, if we were observing the source at point P, we would see it primarily via reflection from the mirror, as shown in Fig. 1.37. For convenience, we choose S and P to be the same distance above the mirror. Some of the paths that we should include in calculating the probability amplitude for a

---

[15]This example follows the discussion of R. P. Feynman, *QED: The Strange Theory of Light and Matter*, Princeton University Press, Princeton, NJ, 1985, pp. 36–49. Without mathematics, Feynman goes on to discuss how the principles of quantum electrodynamics (QED) can be used to understand other optical phenomena such as refraction and the focusing of light by a lens.

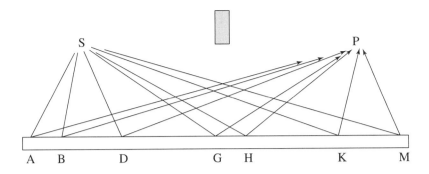

Figure 1.37 A light source S is observed at P via reflection from a mirror. A block is placed on the straight line between S and P so that light reaches P primarily by reflection from the mirror. The figures in this section are adapted from R. P. Feynman, *QED: The Strange Theory of Light and Matter*, Princeton University Press, Princeton, NJ, 1985.

photon to be detected at P are also shown. At first, this may seem somewhat crazy, since you may think you know how light reflects from a mirror and it isn't, for example, by taking path A. But be patient. Look at Fig. 1.38. Below each of the paths is an arrow representing the complex amplitude for light to take that path. Recall that the direction of the arrow is determined by the phase angle kd_i, where d_i is the distance traveled from S to P for that particular path.[16] This distance is also plotted in Fig. 1.38. Strictly speaking, the magnitude of the probability amplitudes (the length of the arrows) for the paths in which the light travels the longer distances should be somewhat less than those for which light travels a shorter distance.[17] Apart from this, we could say that the magnitude of the probability amplitudes are the same for light to be reflected from every point on the mirror. Unlike for our analysis of the Mach–Zehnder interferometer or the double-slit experiment in which there were just two paths involved, adding up these amplitudes for reflection from a mirror is challenging because there are so many paths. And of course, we have shown only a small fraction of the possible paths in Fig. 1.38. Nonetheless, adding the subset of the amplitudes shown in this figure illustrates the key idea: the contribution of the paths that are away from the middle of the mirror tend to cancel as they are added together while the paths in the vicinity of the middle tend to have similar phases and add constructively.

Why is this? What is special about the path in the middle? If we examine the plot of the total distance traveled, we see that the path to the middle of the mirror is the path with

---

[16]Thus a horizontal arrow pointing to the right corresponds to a phase of 0 or an integer multiple of 2π.

[17]You can see why if you consider a source with a fixed power output (say a light bulb) emitting light uniformly in all directions. The probability of a photon being detected a distance d from the source must fall off as $1/d^2$ (and hence the amplitude must vary as $1/d$) in order to maintain a fixed number of photons per unit time passing through an arbitrary spherical surface of radius d (and area $4\pi d^2$).

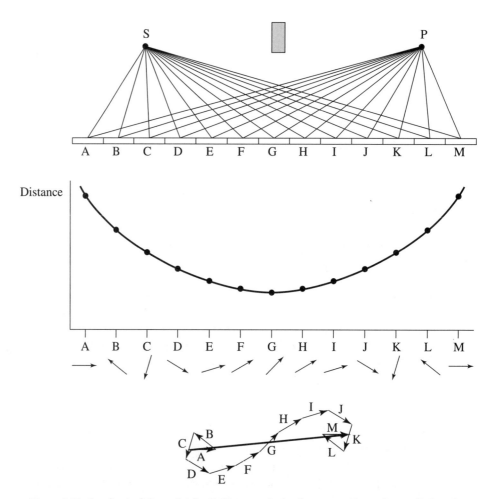

Figure 1.38 A subset of the paths for light to reach the detector at P are shown. Below the paths is a plot of the distance traveled for each path and the probability amplitude for each of these paths as well. Lastly, these probability amplitudes are summed. The long bold arrow is the resultant.

the shortest distance. Since this distance traveled is a minimum, the first derivative of the distance traveled vanishes in the middle. But if the first derivative vanishes, then the distance traveled changes most slowly in the vicinity of the path of minimum distance. But the phase of each amplitude is proportional to the distance traveled, so when the differences in the distance traveled in nearby paths are small, so are the differences in phase. Thus the amplitudes in the vicinity of the path of least distance have pretty much the same phase and pretty much point in the same direction in the complex plane. Therefore, the sum over paths is dominated by the paths in the vicinity of the path for which the distance is the shortest. But since the time required for a photon to traverse this path is the distance divided by the speed of light c, the paths that dominate are also the paths of least time.

What we see in this example is how the quantum mechanics of light can be used to derive **Fermat's principle of least time**. Notice that if light propagates into a medium other than the vacuum with index of refraction n, then since $\lambda \to \lambda/n$ the phase factor e^{ikx} becomes e^{iknx}, with $k = 2\pi/\lambda$. In this case, the amplitudes that add up coherently will be the ones in the neighborhood of the path for which the quantity nx is a minimum.

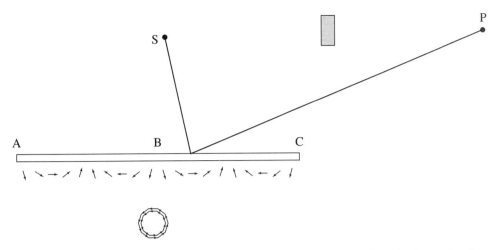

Figure 1.39 If only the end of the mirror is retained, the amplitudes for light to be detected at the point P effectively sum to zero. Only one of the paths between the source S and the point P is indicated in the figure.

This product of the regular path length times the index of refraction is commonly called the **optical path length**. Notice that the time light takes to travel a distance x in this medium is the distance x divided by the speed c/n, that is (nx/c), so again we are effectively minimizing the time when we minimize the optical path length. In short, light has complex probability amplitudes to take all paths, but the ones that matter in the summation are those in the vicinity of the path of least time (strictly an extremum of the time). Example 1.11 shows how we can use this principle to deduce the law of reflection.

Feynman gives an interesting twist to this discussion of reflection from a mirror. Suppose that we break off a piece of the mirror near one of the ends and discard the rest of the mirror. It now looks as if very little light will be reflected from the remaining piece, since it is impossible for light to reach the detector with an angle of incidence equal to the angle of reflection. Figure 1.39 shows how the amplitudes effectively cancel when added together. But if we take this piece of mirror and black out just the right portions of it, as shown in Fig. 1.40, seemingly making this piece appear to reflect even less light,

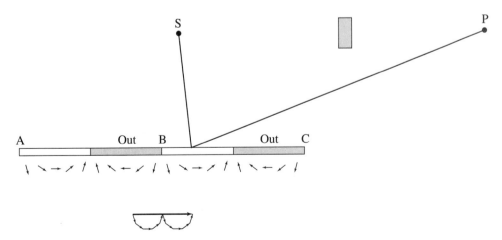

Figure 1.40 The probability that light is detected at P from the small piece of mirror in Fig. 1.39 can be significantly enhanced by making certain portions of the mirror less reflecting by blacking them out.

we can retain amplitudes that add up to a nonzero amplitude and remove those amplitudes that would cancel this nonzero amplitude. Given our discussion of the diffraction grating in the preceding section, you will probably recognize Fig. 1.40 as a reflection grating.

Finally, with this derivation of Fermat's principle in mind, let's go back to our earlier discussions. Why in our analysis of the double-slit experiment, for example, did we consider only the straight-line paths from the source S to the detector at point P? Why didn't we include all sorts of curvy paths? The correct approach is to include *all* paths, even the curvy ones. But as our discussion of reflection of light from a mirror illustrates, the paths that matter when we add up all the amplitudes will be the paths in close vicinity of the path of least time, that is, the straight-line path. The amplitudes for the other paths, including the curvy ones that deviate significantly from the path of least time, will cancel out in carrying out the sum over all paths.

EXAMPLE 1.11 Use Fermat's principle of least time to derive the law of reflection, namely, that the angle of incidence is equal to the angle of reflection.

SOLUTION For the most generality, assume that the source S and the detector P are at different elevations above the mirror, as indicated in Fig. 1.41. The time for light to travel from S to P is given by

$$t = \frac{1}{c}\left(\sqrt{a^2 + x^2} + \sqrt{(D-x)^2 + b^2}\right)$$

Minimizing the time by varying x, we find

$$\frac{dt}{dx} = \frac{1}{c}\left(\frac{x}{\sqrt{a^2+x^2}} - \frac{(D-x)}{\sqrt{(D-x)^2+b^2}}\right) = 0$$

or

$$\frac{x}{\sqrt{a^2+x^2}} = \frac{(D-x)}{\sqrt{(D-x)^2+b^2}}$$

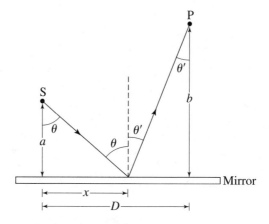

Figure 1.41 A general path between the source S and the detector at P in the reflection of light from a mirror.

In terms of the angles shown in the figure, this can be simply written as

$$\sin\theta = \sin\theta'$$

or $\theta = \theta'$, namely, the angle of incidence equals the angle of reflection. Problem 1.43 shows how Fermat's principle can be used to derive Snell's law ($n_1 \sin\theta_1 = n_2 \sin\theta_2$) for light refracted from a medium with index of refraction n_1 into a medium with index of refraction n_2.

1.9 Summary

Well, what are we to make of what we have seen so far? The original Young double-slit experiment helped to convince people that light was a wave. How else were they to understand what was causing the interference if it wasn't interference of two different parts of the wave? Before Young, the general consensus had been that light was composed of particles called corpuscles, because that was what Isaac Newton had believed. And despite what we often suppose, Young's experiment didn't change this view overnight. In fact, it wasn't until almost two decades later, when A. Fresnel calculated the amplitudes for reflection and refraction of waves incident on an interface between two media, that the tide started to turn to the wave theory of light.[18] And of course it wasn't until 1865, when James Clerk Maxwell modified Ampère's law with the addition of a "displacement" current term, that there was a theory that predicted the existence of electromagnetic waves.

Now we see that in some sense Newton was right all along. Light *is* composed of particles. The energy of the electromagnetic field is carried in discrete quanta called photons. In a monochromatic beam of light, each photon has energy $E = h\nu$, as suggested by Albert Einstein in his 1905 paper on the photoelectric effect. Given the smallness of Planck's constant h in conventional SI units, one should not be surprised that the number of photons in a beam of light in many situations is so large that the discrete particle nature of light is not apparent. However, recent experiments performed with single photons have clearly shown this particle nature of light. For example, if a single photon is incident on a beam splitter, the photon will be counted in only one of the two photomultipliers that might be triggered by this photon. There are no coincidences in the counting rate. When detected, a photon is either "here" or "there" but not both "here" and "there," as would be expected if light were a wave. Nonetheless, experiments carried out with single photons do show characteristic interference effects, provided there are multiple experimentally indistinguishable paths that photons can take between the source and the detector. To "explain" these phenomena, we need to modify our classical notions of a particle to include a new set of principles, the principles of quantum mechanics. So far, we have introduced three such principles:

1. The probability of detecting a particle is equal to z^*z, where z is called the probability amplitude and z^* is its complex conjugate.

---

[18] Ironically, Young apparently became discouraged with the slow acceptance of his results and gave up physics, devoting his time to deciphering the Egyptian hieroglyphics on the Rosetta Stone.

2. To determine the probability amplitude for a process that can be viewed as taking place in a series of steps, such as propagation of a photon from a light source to a beam splitter, transmission at the beam splitter, and propagation to a photodetector, we multiply the probability amplitudes for each of these steps:

$$z = z_a z_b \ldots$$

3. When there are multiple paths that a particle can take between the source and the detector, the probability amplitude for detecting the particle is the sum of the individual probability amplitudes for the particle to take each path:

$$z = z_1 + z_2 + \ldots$$

These individual amplitudes, each of which has a magnitude and a phase, can interfere with each other. Such interference in this sum-over-paths approach to quantum mechanics for light (for photons) can be used to derive such fundamental results of optics as the principle of least time.

The precise form for these probability amplitudes for light comes from solving Maxwell's equations in quantum electrodynamics (QED). We will return to a discussion of QED in Chapter 10. For now, we have stated without proof some simple rules that we can use to calculate the probability of detecting a photon, including the fact that the phase of the amplitude is equal to kd, where d is the distance traveled by the photon between the source and the detector and $k = 2\pi/\lambda$; that for propagation in a medium other than vacuum, such as glass, $\lambda \to \lambda/n$ where n is the index of refraction for that medium; and that there is a phase change of π in the amplitude for reflection when light is reflected by a material whose index of refraction is greater than that of the material in which the light is traveling. In the next chapter we will start to see how results such as these can be derived for the case of nonrelativistic quantum mechanics.

Problems

1.1. (a) Verify that $\mathcal{E} = \mathcal{E}_0 e^{i(kx-\omega t)}$ is a solution to the wave equation

$$\frac{\partial^2 \mathcal{E}}{\partial x^2} - \frac{1}{c^2} \frac{\partial^2 \mathcal{E}}{\partial t^2} = 0$$

provided $\omega = kc$. (b) The wave equation in a medium such as glass with index of refraction n is given by

$$\frac{\partial^2 \mathcal{E}}{\partial x^2} - \frac{n^2}{c^2} \frac{\partial^2 \mathcal{E}}{\partial t^2} = 0$$

What must be the relationship between the wavelength λ and the frequency ν in order that $\mathcal{E} = \mathcal{E}_0 e^{i(kx-\omega t)}$ is a solution?

1.2. What are the numerical values for k and ω for the electromagnetic wave

$$\mathcal{E} = \mathcal{E}_0 \sin[(9.72 \times 10^6 \text{ m}^{-1})x - (2.92 \times 10^{15} \text{ s}^{-1})t]$$

From these values determine the wavelength λ, the frequency ν, and evaluate $\lambda \nu$.

1.3. What must the width a of a slit be so that the first minimum of the single-slit diffraction pattern observed with light of wavelength $\lambda = 550$ nm occurs at an angle of $15°$? The diameter of a human hair is typically 100 μm. How does this compare in size with the width a?

1.4. A radio station broadcasts at a frequency $\nu = 91.5$ MHz with a total radiated power of $P = 20$ kW. (a) What is the wavelength λ of this radiation? (b) What is the energy of each photon in eV? How many photons are emitted each second? How many photons are emitted in each cycle? (c) A particular radio receiver requires 2.0 microwatts of radiation to provide intelligible reception. How many 91.5 MHz photons does this require per second? per cycle? (*d*) Do the answers to (*b*) and (*c*) indicate that the granularity of the electromagnetic radiation can be neglected in these circumstances?

1.5. The human eye can "see" with a signal of 100 photons per second. How far away can a 100-watt light bulb be seen by a dark-adapted eye? Assume the light bulb is in outer space, so that the light is not scattered by the atmosphere. Also assume that the bulb is monochromatic and radiates at a wavelength of 550 nm. Use a reasonable estimate for the diameter of the dark-adapted pupil of the eye.

1.6. An AM radio station broadcasts at 1000 kHz with an output power of 50,000 watts. Assuming the broadcast antenna is located 100 km away from the receiver and, for ease of calculation, the antenna radiates isotropically, estimate the number of photons per cubic meter at the receiver's location.

1.7. The output power of a diode laser in a DVD player is 50 milliwatts. How many photons per second strike the DVD? The wavelength $\lambda = 660$ nm.

1.8. A helium–neon laser ($\lambda = 633$ nm) used in a lecture demonstration of the double-slit experiment has an output power of 5 milliwatts. How many photons per second are emitted by the laser?

1.9. A beam of UV light of wavelength $\lambda = 197.0$ nm falls onto a metal cathode. The stopping potential needed to keep any electrons from reaching the anode is 2.08 V. (*a*) What is the work function W of the cathode surface, in eV? (*b*) What is the velocity v of the fastest electrons emitted from the cathode? *Note*: Since $K_{\max}/mc^2 \ll 1$, the nonrelativistic expression for the kinetic energy can be utilized here. (*c*) If Avogadro's number of photons strikes each square meter of the surface in one hour, what is the average intensity I of the beam, in units of W/m^2?

1.10. Use Millikan's data on the photoelectric effect (Fig. 1.14) to obtain a value for h, Planck's constant.

1.11. The work function of potassium is 2.26 eV. What is the maximum kinetic energy of electrons ejected from potassium by ultraviolet light of wavelength 200 nm?

1.12. The photoelectrons ejected in the photoelectric effect seem to appear instantaneously. In particular, no time delay has been observed, even with light of very low intensity. In classical physics, the energy of the electromagnetic wave is spread out uniformly over the surface of the metal. In this picture, calculate the amount of time it would require for 1 eV of energy (a typical binding energy) to be absorbed by an atom in a metal located 1 m away from a 1-watt bulb. Take the area of the atom to be 1Å^2. *Suggestion*: What fraction of the incident flux is absorbed by the atom?

1.13. The maximum kinetic energy of electrons ejected from sodium is 1.85 eV for radiation of 300 nm and 0.82 eV for radiation of 400 nm. Use this data to determine Planck's constant and the work function of sodium.

1.14. Show that a photon that strikes a free electron at rest cannot be absorbed:

$$\gamma + e^- \not\to e^-$$

Suggestion: Express the energy of the final-state electron in the form $\sqrt{p_e^2 c^2 + m^2 c^4}$ and see whether it is possible to conserve energy and momentum.

Such a reaction does take place in the photoelectric effect for an electron bound in an atom or in a solid. Why can absorption occur in this case?

1.15. Suppose a photon with a wavelength equal to the Compton wavelength makes a collision with a free electron initially at rest. What is the energy of the final photon if the scattering angle is 180°? How much kinetic energy is transferred to the electron?

1.16. It takes 3.1 eV to dissociate a AgBr molecule. What is the maximum wavelength of light required?

1.17. The energy required to break a chemical bond (as found in living tissue) is typically a few eV. Should we be worried about radiation damage from cell phones that operate in the 1 to 2 gigahertz range?

1.18. (*a*) Suppose that the probability amplitude for a photon to arrive at a detector is $1/(1 + i)$. What is the probability that the detector records a photon? (*b*) What is the probability of detecting a photon if the probability amplitude equals i? (*c*) Determine the probability of detecting a photon if the probability amplitude is

$$\frac{1}{1+i} + i$$

(d) Show that

$$\frac{1}{1+i} - i$$

is not a valid probability amplitude. *Suggestion*: What would be the probability of detecting a photon for this amplitude?

1.19. Express the complex number $z_1 = (\sqrt{3}+i)/2$ in the form $re^{i\phi}$. What about $z_2 = (1+\sqrt{3}i)/2$? If these complex numbers are the probability amplitudes for photons to be detected, what is the probability in each case?

1.20. Rewrite the following complex numbers in each of the forms $z = x+iy$ and $z = re^{i\phi}$, where x, y, r, and ϕ are real. (a) $(1+2i)^2$ (b) $1/(1+i)$ (c) $\sqrt{3-4i}$ (d) $e^{i\pi/4}$.

1.21. A certain photodetector can resolve the time of arrival of a photon to within 10^{-8} s. Two of these detectors are used in an anticoincidence experiment, as described in Section 1.4. (a) What is the maximum average rate of photon emission from the source if there is any hope of demonstrating anticoincidence? (b) Assuming the source emits light of wavelength 550 nm, what is the power output of the source?

1.22. In *QED: The Strange Theory of Light and Matter* Feynman states that the phase of the complex probability amplitude for photons makes about 36,000 revolutions per inch for red light. What wavelength is Feynman assuming in this calculation? Does this indeed correspond to red light? *Note*: 1 inch = 2.54 cm.

1.23. One photodetector is located in front of a thick piece of glass and another photodetector is located within the glass. At normal incidence, the glass reflects 4% of the light. A photon is incident on the glass, as indicated in Fig. 1.42. (a) What is the magnitude of the probability *amplitude* for reflection of the photon? (b) What is the magnitude of the probability amplitude for transmission of the photon?

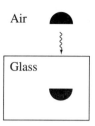

Figure 1.42 Partial reflection of light by a single surface of glass.

1.24. (a) Show that the probability of a photon of wavelength λ being reflected from a thin layer of glass of thickness d at normal incidence is given by

$$P = 0.16\sin^2(2\pi d/\lambda')$$

where λ' is the wavelength of light in glass, i.e., $\lambda' = \lambda/n$, where n is the index of refraction of glass. *Note*: In this calculation assume that the magnitude of the amplitude for reflection from the top or the bottom surface of the glass is 0.2 and that there is an additional phase change of π in the reflection from the top surface. Also assume that amplitudes that arise from multiple reflections between the top and bottom surfaces of the glass can be neglected in your calculation. Given the result of Problem 1.23, it is okay to approximate the magnitude of the amplitude for transmission as one. *Hint*: What extra distance does light travel in being reflected from the bottom surface relative to the top surface? (b) In *QED: The Strange Theory of Light and Matter* Feynman states that as the thickness of a thin layer of glass increases from zero thickness, the probability of reflection first reaches a value of 0.16 when the thickness of the layer of glass is 5 millionths of an inch. What index of refraction is being assumed? Take the wavelength of the light in air to be the same as you determined in Problem 1.22. (c) What is the minimum value of d necessary to produce zero reflection?

1.25. Suppose that a thin film of acetone (index of refraction $n = 1.25$) of thickness d is coating a thick plate of glass (index of refraction = 1.50). Take the magnitude of the amplitude for reflection of a photon from the top or the bottom surface of the acetone at normal incidence to be r and assume that there is an additional phase change of π in the reflection from the top *and* the bottom surface of the acetone, since at each of these surfaces light is passing from a medium with a lower index of refraction to one with a higher index of refraction. Calculate the probability that a photon of wavelength λ is reflected. Assume that amplitudes that involve multiple reflections at the bottom surface of the film can be neglected in your calculation. Express your answer in terms λ and r as well as the thickness d and the index of refraction n of the acetone. What is the minimum thickness of the coating necessary to produce zero reflection? *Note*: For the air–acetone and acetone–glass surfaces $r \cong 0.1$.

1.26. Assume that the first beam splitter at A in the Mach–Zehnder interferometer (Fig. 1.23) is a "third-silvered mirror," that is, a mirror that reflects one-third the light and transmits two-thirds. The two mirrors at B and C reflect 100% of the light, and the second beam splitter at D is a traditional half-silvered mirror that reflects one-half the light and transmits one-half. The probability of detecting a photon in either photomultiplier PM$_1$ or PM$_2$ varies with the position of the movable mirror, say mirror B. Determine the maximum probability and the minimum probability of obtaining a count in, say, PM$_1$. What is the visibility

$$V = \frac{P_{\max} - P_{\min}}{P_{\max} + P_{\min}}$$

of the interference fringes, where $P_{\max}$ and $P_{\min}$ are the maximum and minimum probabilities, respectively, that a photon is counted by the detector, as the position of the movable mirror varies? *Note*: In the experiment of Aspect et al. described in Section 1.5 the visibility of the fringes is 0.987 ± 0.005.

1.27. Figure 1.43 shows a Michelson interferometer with a movable mirror M_1, a fixed mirror M_2, and a beam splitter M_s, which is a half-silvered mirror that transmits one-half the light and reflects one-half the light incident upon it independent of the direction of the light. The source emits monochromatic light of wavelength λ. There are two paths that light can follow from the source to the detector, as indicated in the figure. Note that path 1 includes travel from the beam splitter M_s to the movable mirror M_1 and back to the beam splitter, while path 2 includes travel from the beam splitter M_s to the fixed mirror M_2 and back to the beam splitter. Assume the beam splitter introduces a phase change of π for light that follows path 1 from the source to the detector relative to light that follows path 2 from the source to the detector. Also assume the mirrors M_1 and M_2 reflect 100% of the light incident upon them and the

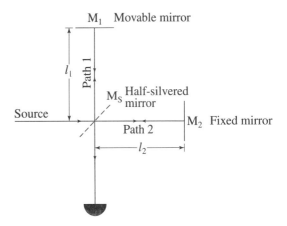

Figure 1.43 The Michelson interferometer.

photodetector (a photomultiplier) is 100% efficient as well. (*a*) Use the principles of quantum mechanics to determine the probability that a photon entering the interferometer is detected by the photodetector. Express your answer in terms of the lengths l_1, l_2, and λ. (*b*) Find an expression for l_1 in terms of l_2 and λ such that there is 100% probability that the photon is detected by the photodetector. (*c*) Suppose that the movable mirror is shifted upward by a distance $\lambda/6$ from the position(s) that you determined in part (*b*). Find the probability that the photon is detected at the photodetector in this case.

1.28. A beam of monochromatic light from a helium–neon laser ($\lambda = 633$ nm) illuminates a double slit. From there the light travels a distance $D = 10.0$ m to a screen. (*a*) If the distance between interference maxima on the screen is to be $\delta = 10.0$ mm, what should be the distance d between the two slits? (*b*) What would you see on the screen if a thin sheet of cellophane were placed over one of the slits so that there were 2.5 more wavelengths within the cellophane than within a layer of air of the same thickness? (Assume the interference maxima in question are at only a small angle with respect to the laser beam direction.)

1.29. Suppose that the two very narrow slits (widths $\ll \lambda$) in the double-slit experiment are not the same width and that the probability amplitude for a photon of wavelength λ to strike a photomultiplier centered at a particular point P in the detection plane that makes an angle θ with the horizontal from one of the slits is larger by a factor of $\sqrt{2}$ than for the other slit. Determine the visibility

$$V = \frac{P_{\max} - P_{\min}}{P_{\max} + P_{\min}}$$

of the interference fringes, where P_{max} is the maximum probability and P_{min} is the minimum probability that a photon is detected.

1.30. Suppose that a thin piece of glass were placed in front of the lower slit in a double-slit apparatus so that the amplitude for a photon of wavelength λ to reach that slit differs in phase by 180° with the amplitude to reach the top slit. (*a*) Describe in detail the interference pattern on the screen. At what angles will there be bright fringes? (*b*) What is the minimum thickness of glass required, assuming the index of refraction for the glass is n?

1.31. (*a*) A monochromatic light source S of wavelength λ is located to the left of an opaque screen with two very narrow slits of equal width. To the right of the screen in the detection plane is a photomultiplier at point P. The distance between S and P along the path 1 is d_1 and the distance between S and P along path 2 is d_2. Show that the probability of detecting the photon at P is given by

$$\text{Prob} = 2r^2 (1 + \cos \phi)$$

where r^2 is the probability that a photon strikes the photomultiplier with a single slit open. Obtain an expression for ϕ in terms of d_1, d_2, and λ. (*b*) Now suppose that thin pieces of partially silvered glass are placed in front of each of the slits in this double-slit experiment. Assume the glass covering the top slit transmits 1/2 the light incident upon it while the glass covering the bottom slit transmits 1/4 the light incident upon it. Determine the probability that a photon of wavelength λ hits the photomultiplier in this case. Express your answer in terms of r and ϕ.

1.32. Add the two complex numbers $z_1 = 1$ and $z_2 = e^{i\pi/3}$ by (*a*) adding the real and imaginary pieces together and (*b*) using geometry to "add the arrows" representing each of these complex numbers. Check that your results for the magnitude and phase of the complex number $z_1 + z_2$ agree.

1.33. A photon with wavelength λ is incident on a single slit of finite width a. Calculate the probability amplitude for the photon to strike a detector located at a distant point P at angle θ in the detection plane by integrating across the slit the infinitesimal probability amplitude

$$dz_{\text{P}} = r \frac{dx}{a} e^{ik(d_1 + x \sin \theta)}$$

for the photon to reach the point P by passing through dx, which is located a distance x below the top of the slit, as shown in Fig. 1.44. The distance d_1 is the distance traveled by the photon in reaching the point P from the top of the slit. Show that the probability of detecting the photon is given by

$$z_{\text{P}}^* z_{\text{P}} = r^2 \frac{\sin^2 \alpha}{\alpha^2}$$

where

$$\alpha = \frac{ka \sin \theta}{2}$$

Verify that minima in the probability occur at the angles given in (1.17). Evaluate the probability in the limit that $ka \ll 1$, that is, $a \ll \lambda$. Figure 1.9 shows how the probability varies with θ for $a = \lambda$, $a = 5\lambda$, and $a = 10\lambda$.

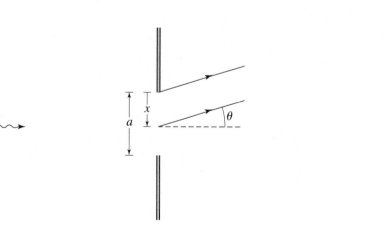

Figure 1.44 Single-slit diffraction.

1.34. Starting from first principles, show that the probability that a photon of wavelength λ hits a photomultiplier centered on a point P in the detection plane that makes an angle θ with the horizontal for a grating composed of three very narrow slits each separated by a distance d is given by

$$\text{Prob} = r^2(1 + 4\cos\phi + 4\cos^2\phi)$$

where r^2 is the probability that the photon would strike the photomultiplier with a single slit open and $\phi = kd\sin\theta = 2\pi d\sin\theta/\lambda$.

1.35. Determine the probability that a photon is detected at the location of the first minimum of a three-slit grating if the third slit is closed. Assume the magnitude of the probability amplitude due to each slit is r. *Suggestion*: Start by showing how the complex amplitudes from each slit add up to zero at the first minimum. What is the resulting amplitude if the third amplitude is eliminated?

1.36. Determine the probability that a photon is detected at the first minimum of a five-slit grating if the bottom three slits are closed. Assume the magnitude of the probability amplitude due to each slit is r. *Suggestion*: Start by showing how the complex probability amplitudes from each slit add up to zero at the first minimum.

1.37. Determine the probability that a photon is detected at the first minimum of a six-slit grating if the bottom two slits are closed. Assume the magnitude of the probability amplitude due to each slit is r. *Suggestion*: Start by showing how the complex probability amplitudes from each slit add up to zero at the first minimum.

1.38. For a grating with N equally spaced narrow slits, the amplitude for detecting a photon with a photomultiplier centered at point P in the detection plane is given by

$$z_P = re^{ikd_1}\left[1 + e^{i\phi} + e^{2i\phi} + \cdots + e^{i(N-1)\phi}\right]$$

Notice that each term in this series of terms can be obtained from the one preceeding it by multiplying by $e^{i\phi}$. Thus it is a geometric series that can be summed. Show that the probability of detecting a photon is given by

$$z_P^* z_P = r^2 \frac{\sin^2\frac{N\phi}{2}}{\sin^2\frac{\phi}{2}}$$

Verify that this result reduces to the double-slit result (1.60) for $N = 2$.

1.39. The yellow light from sodium consists of two wavelengths, 589.0 nm and 589.6 nm, known as the sodium doublet. In general, when incident upon a diffraction grating, light of two slightly different wavelengths, say λ and $\lambda + \Delta\lambda$, generates maxima satisfying $d\sin\theta = m\lambda$ and $d\sin(\theta + \Delta\theta) = m(\lambda + \Delta\lambda)$, respectively, where the integer m labels the order of the interference maximum. Show that the dispersion of the grating $\Delta\theta/\Delta\lambda$ is given by $m/d\cos\theta$. Assume $\Delta\theta \ll 1$. Thus the dispersion can be increased by reducing the separation d between the slits in the grating and/or working at higher order.

1.40. Light of wavelength λ is incident at an angle ψ to the normal on a transmission grating with spacing d between each slit, as shown in Fig. 1.45. At what angles θ to the normal will diffraction maxima be located?

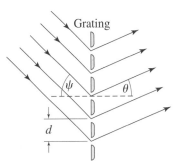

Figure 1.45 Light incident at angle ψ on a grating.

1.41. Figure 1.46 depicts Lloyd's mirror for the observation of interference upon a distant screen of monochromatic light of wavelength λ from a single narrow slit S. The slit is located a distance d above the mirror. Derive an expression in terms of λ and d for the angles θ at which bright bands appear on the screen due to constructive interference from light reaching P directly from the slit and light reaching P by reflection from the mirror. Simplify your expression as much as possible. *Note*: Experimentally, it is observed that a dark band (destructive interference) occurs as $\theta \to 0$. Why is this? *Suggestion*: Evaluate the path difference for the two paths shown in the figure for light to arrive at P. Assume the point P is sufficiently far away that the two rays to the point P can be taken to be parallel.

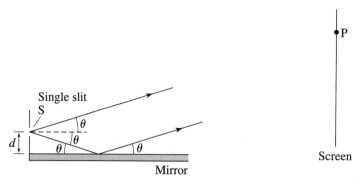

Figure 1.46 Lloyd's Mirror.

1.42. Explain carefully how the fact that photons have a complex probability amplitude to take all paths, say in reflection from a mirror, can be used to derive Fermat's principle, which is often referred to as the principle of least time, namely, that light follows the path of least time in traveling between two fixed points. Strictly, this principle should be referred to as the principle of extremum time since your discussion should show that light can be said to follow a path such that, compared with nearby paths, the time required is either a minimum or a maximum (or even remains unchanged).

1.43. Use the principle of least time to derive Snell's law, namely, $n_1 \sin \theta_1 = n_2 \sin \theta_2$ for light being refracted as it travels from a medium with index of refraction n_1 into a medium with index of refraction n_2. *Suggestion*: Follow a procedure similar to the one given in Example 1.11. Locate the source S in medium 1 and the point P in medium 2.

1.44. A candle is placed in a "room" made of reflecting surfaces. How should the "room" be shaped so that the observer sees the candle in any direction in which she looks? Explain your reasoning.

1.45. In classical physics, the probability of an event that can occur through a series of steps is the product of the probability of each step occurring. And if an event can occur in more than one way, the probability of the event is the sum of the probabilities for the different ways in which it can occur. How does this differ from the principles of quantum mechanics enunciated in this chapter?

CHAPTER 2

Wave Mechanics

The propagation of light is inherently a relativistic phenomenon that requires quantum field theory for a complete description. In Chapter 1, we applied the sum-over-paths approach to quantum mechanics that was developed by Richard Feynman in the 1940s to the propagation of light (photons) using (without proof) some results for the probability amplitudes that can be derived from quantum electrodynamics (QED). We saw how quantum mechanics allows us to understand the basis for the principle of least time for describing the propagation of light. In classical mechanics, there is a similar principle, the principle of least action, that can be used to describe the motion of particles. Just as the principle of least time implies that light in a vacuum travels in straight lines, the principle of least action leads to the result that a particle follows the path predicted by Newton's second law, $F = ma$. But how does the particle figure out how to take the path of least action? The answer is that there are probability amplitudes for a particle to take all paths and the paths that matter most in this sum over paths—at least for macroscopic particles—are those paths in the vicinity of the path of least action. Thus the underlying physics for particles, whether they be atoms or apples, is essentially the same as for photons, the particles of light. In this way we can say that quantum mechanics can be used to derive familiar results from classical mechanics such as $F = ma$. But we cannot derive quantum mechanics itself, since as far as we know it is the very foundation of the laws of nature.

While a sum-over-paths approach can be applied to particles on the microscopic as well as the macroscopic scale, this approach is not easy to implement when the distance scale over which these particles propagate is small. Figuring out the motion of an electron in an atom, for example, is difficult using this approach because so many paths contribute coherently to the sum. An alternative strategy that has a broad range of applicability to atomic and molecular physics, solid-state physics, and nuclear physics is to treat the particles nonrelativistically. This approach to quantum mechanics was developed in the 1920s by Erwin Schrödinger, Werner Heisenberg, Paul Dirac, and a number of other physicists. There are a number of variations on this approach, referred to as wave mechanics (Schrödinger), matrix mechanics (Heisenberg), or a more general operator approach (Dirac). We have chosen to focus on wave mechanics since it makes the most contact with the physics with which you are probably most familiar and is less abstract than the other variants.

The history of the discovery of the laws of quantum mechanics is itself quite interesting. We will not follow a strictly historical order, however. Rather we will start with a recent double-slit experiment carried out with helium atoms as the projectiles, an experiment that provides direct evidence for what is often termed the wave nature of particles. We will then introduce the Schrödinger equation, a wave equation for nonrelativistic particles.

2.1 Atom Interferometry

In our discussion of the interference of light, we focused on phenomena such as Young's double-slit experiment or the reflection of light from a mirror in which the distance traveled by the light between the source and the detector is a macroscopic distance. Similarly, we will start our discussion of wave mechanics by focusing on a double-slit experiment with atoms in which the distance traveled by the atoms is on the order of meters. In this case, just as for the Young double-slit experiment, the sum over paths is dominated by paths in the vicinity of two paths between the source and the detector, as shown in Fig. 2.1. Thus in this experiment we can observe interference effects of the probability amplitudes for atoms to take these two different paths.

Figure 2.2 gives a schematic representation of the experimental setup. Helium atoms are stored, as a gas under pressure, in a reservoir. An expansion of the gas through a nozzle produces an intense atomic beam. As the helium atoms travel toward entrance slit A, which serves to collimate the beam, they are bombarded by electrons that have been fired along the beam direction. As a result of these collisions, some of the helium atoms are in excited states that are metastable, that is, states with unusually long lifetimes. After this electron-impact excitation and collimation, the atoms pass (in a high vacuum) through microfabricated double-slit structure B in a thin gold foil. Figure 2.3 shows a scanning-electron microscope picture of the double-slit structure. Each slit has a width of 1 μm, that is, 1×10^{-6} m, or 1 micron, and the slits themselves are separated by a distance $d = 8$ μm. Note that the 100 μm scale shown in Fig. 2.3 is roughly the thickness of a human hair. These are *very* narrow slits, but a helium atom, whose linear size is roughly 10^{-4} times the size of the slit, has no trouble fitting through them. Finally, a secondary electron multiplier (SEM) serves as a detector in a plane located $L = 1.95$ m behind the double slit. An excited helium atom that strikes the SEM is very likely to be ionized; the SEM then generates an electronic pulse that can be amplified and counted, essentially allowing the measurement of single excited atoms. While roughly only one in a million of the helium atoms is in an excited state, the atoms that remain in their ground state are not detected because they are not ionized upon impact in the SEM.

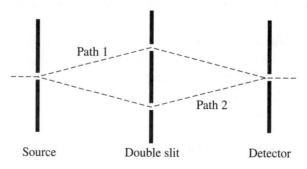

Figure 2.1 The double-slit experiment.

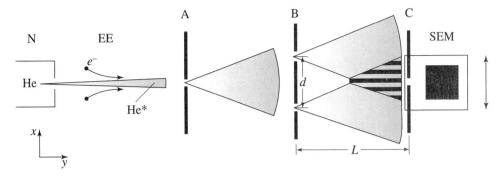

Figure 2.2 Schematic representation of the double-slit experiment with helium atoms, including a gas reservoir N, electron impact excitation EE, collimating entrance slit A, double slit B, the detection plane C, and a secondary electron multiplier (SEM).

Finally, measurement of the time of flight between electron excitation and the atom's impact permits the determination of the speed of each helium atom that is counted.

Now let's look at the results. Helium atoms are emitted from the reservoir with a range of speeds. Figure 2.4a shows the counts observed in the detection plane as a function of the lateral position x for those helium atoms with speeds greater than 30 km/s. At first glance, you may be relieved by these results. Isn't this what you expect when you fire particles at a screen with two slits in it? The particles travel in straight lines and, apart from those that perhaps bounce off the edges of the slits, should strike the detecting screen directly behind the openings formed by the slits. But look at Fig. 2.4b, which shows the counts for slower helium atoms, those with speeds of 2 km/s (actually between 2.1 and 2.2 km/s). First notice that the atoms land at many locations in the detection plane, not simply behind the slits as in Fig. 2.4a. In fact, Fig. 2.5 shows the way the counts accumulate over time. Here, we do not immediately see a pattern. And the beam intensity is quite low, with roughly one excited helium atom per second traversing the interferometer. It is only after many counts have been registered that the pattern—a double-slit interference pattern—becomes apparent (see Fig. 2.4b). But our description of interference [see (1.56), for example] requires a wavelength λ. This wavelength is called the **de Broglie wavelength**, since Louis de Broglie first suggested that a particle with momentum p possesses a wavelength λ given by

$$\lambda = \frac{h}{p} \tag{2.1}$$

where h is Planck's constant.[1] For helium atoms with a speed of 2.2 km/s, this means a de Broglie wavelength

$$\lambda = \frac{h}{mv} = \frac{6.63 \times 10^{-34} \text{ J} \cdot \text{s}}{(6.63 \times 10^{-27} \text{ kg})(2.2 \times 10^3 \text{ m/s})} = 45 \times 10^{-12} \text{ m} = 45 \text{ pm} \tag{2.2}$$

Figure 2.3 Scanning electron microscope picture of the double-slit structure. The horizontal supports are necessary to keep the slits rigidly positioned because the foil is very thin. This figure is reprinted with permission from O. Carnal and J. Mlynek, *Phys. Rev. Lett.* **66**, 2689 (1991). Copyright (1991) by the American Physical Society.

---

[1] In his 1924 PhD thesis de Broglie combined two results that follow from Einstein's 1905 papers: $E = h\nu = hc/\lambda$ (from the photoelectric effect paper) and $E = pc$ (from special relativity), results that we used in our discussion of Compton scattering in Section 1.3. Equating these two expressions for the energy yields $p = h/\lambda$ for light quanta. de Broglie suggested that a similar relationship should apply for all particles. Incidentally, the committee reviewing de Broglie's thesis didn't know what to make of these radical ideas, so they asked Einstein to review it. Einstein gave his approval and de Broglie got his PhD, as well as a Nobel Prize in physics five years later.

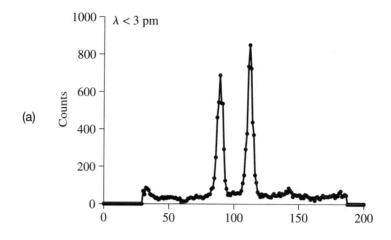

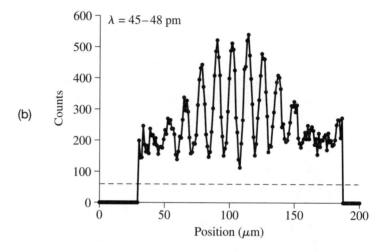

Figure 2.4 The number of helium atoms detected vs. the position x in the detection plane for atoms (a) with speeds greater than 30 km/s (and therefore $\lambda < 3$ pm) and (b) with speeds between 2.1 and 2.2 km/s (and therefore $\lambda \simeq 45$ pm). The dashed line in (b) shows the dark counts. This figure, together with Fig. 2.5, is from Ch. Kurtsiefer, T. Pfau, and J. Mlynek, private communication. See their article in *Nature* **386**, 150 (1997).

Just as in (1.56), the condition for an interference maximum in the intensity is given by the requirement that the difference in path length between the two paths shown in Fig. 2.1 is an integral number of wavelengths:

$$d \sin\theta = n\lambda \qquad n = 0, \pm 1, \pm 2, \ldots \qquad (2.3)$$

Notice that $\lambda/d = 5.6 \times 10^{-6}$, so the angles of deflection are very small and it is appropriate to make the approximation $\sin\theta \cong \tan\theta = x/L$, where x is the position of the maximum in the detection plane. Thus the distance between adjacent maxima is given by

$$x_{n+1} - x_n = \frac{L\lambda}{d} \qquad (2.4)$$

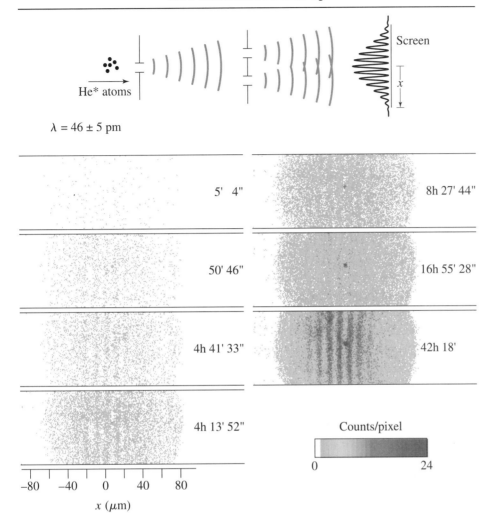

Figure 2.5 How the interference pattern shown in Fig. 2.4b builds up one atom at a time. The first data set is taken after 5 minutes of counting, while the last is taken after counts have accumulated for 42 hours. The "hotspot" in the data arises from an enhanced dark count due to an impurity in the microchannel plate detector.

Using the value of λ from (2.2), we obtain a separation between maxima of 11 μm, which is in good agreement with the observed separation.[2] (See Problem 2.3 and Problem 2.4 for some other examples.)

Given this new perspective, let's go back and reexamine the data for the fast helium atoms shown in Fig. 2.4a. Initially, we assumed it was natural for atoms to land directly behind the slits, but now we should ask: Is this result compatible with helium atoms having a wavelength? If slow helium atoms interfere, why not the fast ones as well? For interference to occur, the diffraction envelopes from each slit must overlap in space, as depicted in Fig. 2.2. Since the first zero of the single-slit diffraction envelope occurs

---

[2]Since $L/d \gg 1$, we can take the two rays drawn from the slits to a point in the detection plane to be essentially parallel, just as we did in the analysis of the Young double-slit experiment in Section 1.6. Also, from now on, we will use the symbol n to refer to an integer.

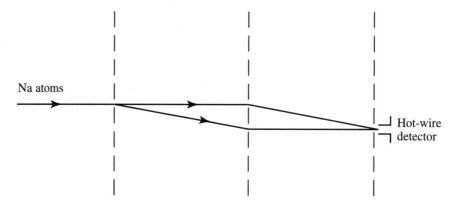

Figure 2.6 An interference experiment carried out with sodium atoms. Two 400-nm-period gratings serve as the beam splitters.

when $\sin\theta = \lambda/a$ [see (1.15) and Problem 1.33], where a is the width of the slit, the width of the central maximum in the detection plane is on the order of $(\lambda/a)L$, where L is the distance to the detecting screen. For the slower helium atoms, say those with a wavelength of 45 pm, this yields a width of 88 μm, which is substantially larger than the separation d between the slits. Thus the diffraction envelopes from the two slits overlap and interference occurs. For the faster helium atoms, those with a wavelength less than 3 pm, $(\lambda/a)L$ is less than 6 μm, which is less than the separation $d = 8$ μm, but larger than the width $a = 1$ μm of the slit. Thus although there is measurable diffraction of the helium atoms from each slit, there is insufficient diffraction for the diffraction envelopes from each slit to overlap substantially at the detection plane and hence produce interference.

Before concluding this section, there is an additional atom interferometry experiment that merits special note. This experiment, the initial results of which were published back to back in *Physical Review Letters* with the first results of the helium atom double-slit experiment, was carried out with sodium atoms. The interferometer consisted of three 400-nm-period gratings mounted 0.663 ± 0.003 m apart, as indicated in Fig. 2.6. The sodium atoms had a speed of 1 km/s, giving them a de Broglie wavelength of 16 pm. At the middle grating the width of the beams was 30 μm and they were separated by 27 μm. Thus just two paths, as shown in Fig. 2.6, overlap enough for interference to occur. Figure 2.7 shows the resulting interference pattern. Most interestingly, in a subsequent experiment the Pritchard group shined laser light through a port into the interferometer.[3] If the light scattered from the sodium atoms near the middle grating, it should have been possible to detect which of these two paths a sodium atom was taking if the experimentalists had chosen to observe the scattered laser light (although they didn't actually make this observation). In this case, the interference pattern disappeared. On the other hand, if the laser was positioned so that the light scattered from atoms close to the first grating, then it was impossible to resolve which of the two paths a sodium atom was taking and the interference pattern reappeared. This experiment therefore provided a beautiful confirmation of the fact that we should include multiple amplitudes in the

---

[3]M. S. Chapman, T. D. Hammond, A. Lenef, J. Schmiedmayer, R. R. Rubenstein, E. Smith, and D. E. Pritchard, *Phys. Rev. Lett.* **75**, 3783 (1995).

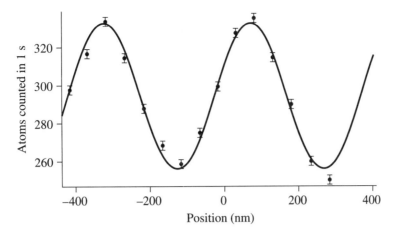

Figure 2.7 Sodium atom interference data adapted from D. W. Keith, C. R. Ekstrom, Q. A. Turchette, and D. E. Pritchard, *Phys. Rev. Lett.* **66**, 2693 (1991).

determination of the probability of detecting a particle only when it is impossible to know which of the paths is taken by that particle.

In conclusion, it is instructive to compare the data in Fig. 1.24 and Fig. 2.5. In the case of light, for which our everyday classical experience is with its wave nature, the particle nature of light is perhaps most surprising, although everyone has probably heard about the existence of photons. In the case of atoms, which we certainly think of as particles, it is the wave nature that forces us to revise our classical notions. In both cases, we are forced to try to understand how interference fringes—maxima and minima—can build up one particle, one photon or one atom, at a time. The similarities in the way we must deal with particles, whether they are photons or atoms, are striking. Clearly, we need a new set of laws, a new form of mechanics called **quantum mechanics**, not just for photons but for other particles as well.

EXAMPLE 2.1 What is the de Broglie wavelength of a tennis ball ($m = 0.05$ kg) moving at 10 m/s?

SOLUTION

$$\lambda = \frac{h}{p} = \frac{h}{mv} = \frac{6.63 \times 10^{-34} \text{ J} \cdot \text{s}}{(0.05 \text{ kg})(10 \text{ m/s})} = 1.33 \times 10^{-33} \text{ m}$$

a very small length.

EXAMPLE 2.2 Assume that the tennis ball of the previous example is thrown through a 1-m wide door at a speed of 10 m/s. Calculate the angular width of the central maximum. Would you be likely to observe the diffraction (bending) of tennis balls as they move through the door?

SOLUTION

$$\sin \theta = \frac{\lambda}{a} = 1.33 \times 10^{-33}$$

Therefore the angular width $2\theta = 2.7 \times 10^{-33}$ in this case.

These examples illustrate that it is the smallness of h, Planck's constant, on a macroscopic scale that prevents you from noticing this wave-like behavior for particles and protects your classical illusions that macroscopic particles are somehow different from what we are seeing on the microscopic level.

EXAMPLE 2.3 What is the de Broglie wavelength of an electron with a kinetic energy of 54 eV?

SOLUTION Since an electron with a kinetic energy of 54 eV is nonrelativistic, we can write the kinetic energy as

$$K = \frac{1}{2}mv^2 = \frac{p^2}{2m}$$

Thus

$$\lambda = \frac{h}{p} = \frac{h}{\sqrt{2mK}}$$

$$= \frac{6.63 \times 10^{-34} \text{ J} \cdot \text{s}}{\sqrt{2(9.11 \times 10^{-31} \text{ kg})(54 \text{ eV})(1.6 \times 10^{-19} \text{ J/eV})}} = 1.67 \times 10^{-10} \text{ m} = 1.67 \text{ Å}$$

Note that $1 \text{ Å} = 10^{-10} \text{ m} = 0.1 \text{ nm}$ is the characteristic size scale of atoms. Can you guess how such electrons might be utilized? See Problem 2.11.

2.2 Crystal Diffraction

Example 2.3 contains a very interesting result. Since an electron has such a small mass, the de Broglie wavelength of a 50-eV electron is not nearly as small as it is for a macroscopic particle such as the tennis ball in Example 2.1. Nonetheless, the electron's wavelength is short, equivalent to that of a photon in the X-ray regime. We know now, of course, that X-rays are short-wavelength photons.[4] In order to verify that X-rays were simply electromagnetic radiation, it was necessary to see some interference effects. But constructing a double-slit experiment or, better yet, a grating with a spacing between the slits on the order of angstroms is not easy. But it turns out that nature does this with regularity in the form of crystalline solids.

---

[4]The name X-ray derives from the fact that when X-rays were first discovered, by Wilhelm Röntgen in 1895, their nature was unclear. Röntgen happened to notice during experiments in which electrons with energy up to 50 keV struck the anode in an evacuated glass chamber that a fluorescent screen at the other end of the lab table started to glow faintly, even though the glass chamber was covered with black cardboard. The glow persisted even when a 1000-page book was placed between the tube and the fluorescent screen. Röntgen was startled to discover when he held a lead disk in front of the screen that not only did he see the shadow of the disk, as expected, but he also saw the shadows of the bones of his own fingers. He soon produced a more permanent record, taking an X-ray picture of his wife's hand that created a sensation. Within weeks X-rays were being used to help in the setting of broken bones. Röntgen received the first Nobel Prize in physics, in 1901.

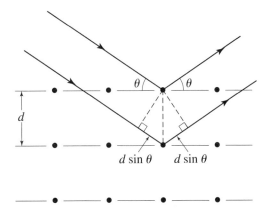

Figure 2.8 The atoms in a crystalline solid, one with a periodic structure, can be viewed as lying in a series of planes. The path difference for reflection from adjacent planes is $2d\sin\theta$.

X-rays incident on a periodic crystal such as NaCl scatter from each of the atoms in the crystal. In 1912 Max von Laue added up this scattered radiation and showed that very sharp maxima in the intensity would occur, in the same way that sharp maxima are generated when light is scattered by the simple grating discussed in Section 1.7. In the language of Chapter 1, the maxima occur when the probability amplitudes for scattering from each of the atoms are in phase. Shortly thereafter, in 1913, W. L. Bragg showed that the results of von Laue's calculation were the same as if the scattering took place from a series of planes, like a series of mirrors, that were generated by the individual atoms in the crystal. Figure 2.8 illustrates such a series of planes. Constructive interference occurs when the path difference for scattering from an adjacent plane is an integral number of wavelengths, that is,

$$2d\sin\theta = n\lambda \qquad n = 1, 2, 3, \ldots \qquad (2.5)$$

Equation (2.5) is often called the **Bragg relation**.[5] If the amplitudes are in phase for these adjacent planes, then the amplitudes will be in phase for all the planes, since the spacing between each of these planes is the same.

Figure 2.9a shows an X-ray diffraction pattern for rock salt, while Fig. 2.9b shows a similar picture taken with neutrons scattered from rock salt. The neutrons are scattered by the nuclei of the atoms, while the X-rays interact electromagnetically with the charged particles (nuclei and electrons) that constitute the atoms. Neutrons are typically thermalized (see Problem 2.14) so that their wavelengths are in the appropriate X-ray regime. This is another nice illustration of how particles behave in ways that classically we attribute only to waves.

---

[5]Unlike Snell's law where the incident angle is typically measured with respect to the normal to the surface, the angle θ in the Bragg relation is measured with respect to the horizontal, presumably because X-ray diffraction studies are often done with a grazing angle of incidence. There are numerous sets of planes that can be constructed from atoms in such a crystal array, so there are numerous angles for which the Bragg condition for constructive interference will be satisfied. It is possible to infer the underlying crystal structure from the nature of the X-ray diffraction pattern. W. L. Bragg and his father, W. H. Bragg, built an X-ray crystal spectrometer that could be used to analyze the structure of crystals, given a knowledge of λ. The Braggs shared the Nobel Prize in 1915.

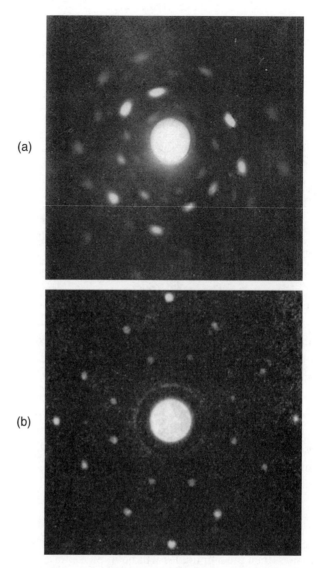

Figure 2.9 (a) X-ray diffraction and (b) neutron diffraction from a single sodium chloride crystal. The X-ray diffraction image, from W. L. Bragg, *Proc. Royal Soc.* **89**, 248 (1913), shows one of the very first attempts to use X-rays to analyze the structure of crystals. Modern X-ray crystal diffraction images are substantially sharper. The neutron diffraction image, reprinted with permission from E. O. Wollam, C. G. Shull, and M. C., Murray, *Phys. Rev.* **73**, 527 (1948), copyright (1948) by the American Physical Society, shows the first image of a single crystal taken with neutrons. Shull was awarded the Nobel Prize in physics in 1994.

You may have wondered (at least if you have been reading the footnotes) how de Broglie was able to win the Nobel Prize in 1929 just five years after making the unusual proposal that particles such as electrons have a wavelength. While the experiment on interference with helium atoms by Kurtsiefer et al. provides strong confirmation of de Broglie's idea, this particular experiment wasn't carried out until quite recently. And neutron scattering as a tool for crystal studies did not become available until the advent of a good source of neutrons, after World War II. The answer resides in a 1927 experiment that involved a bit of luck. C. J. Davisson and D. A. Germer, working at Bell Labs, were

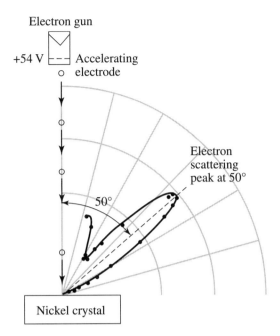

Figure 2.10 A schematic diagram of the Davisson–Germer apparatus and a plot of their scattering data for 54-eV electrons, for which $\lambda = 1.67$ Å. See Example 2.4. Adapted from Nobel Lectures, Physics 1922–1941, Elsevier, Amsterdam, 1965.

investigating what happens when 54-eV electrons were scattered from polycrystalline nickel, taking advantage of the relatively new vacuum technology. Their vacuum tube broke and the nickel became oxidized. In order to remove the oxide from the metal, Davisson and Germer heated the nickel. When the nickel cooled, it solidified as a single crystal. Subsequently, Davisson and Germer noticed a strong enhancement in the angular distribution of the scattered electrons, which turned out be at just the angle predicted by the Bragg condition, with a spacing between planes of 0.91 Å. See Example 2.4. Figure 2.10 shows their experimental setup as well as their data. Davisson shared the Nobel Prize with G. P. Thomson, who made a similar discovery independently in Scotland. Incidentally, G. P. Thomson was the son of J. J. Thomson, who is often credited with the discovery of the electron in 1897 and for which he, too, was awarded the Nobel Prize, in 1906. It is often said that the father discovered the electron as a particle while the son discovered its wave nature.

EXAMPLE 2.4 The 50° angle shown in Fig. 2.10 is the angle between the incoming and the outgoing ray in Fig. 2.8. Thus $2\theta = 180° - 50° = 130°$, or $\theta = 65°$. Determine the spacing between the planes of atoms in the nickel crystal.

SOLUTION Since an electron with a kinetic energy of 54 eV has a wavelength $\lambda = 1.67$ Å,

$$d = \frac{\lambda}{2 \sin \theta} = \frac{1.67 \text{ Å}}{1.81} = 0.92 \text{ Å}$$

in good agreement with the value $d = 0.91$ Å determined from X-ray diffraction.

2.3 The Schrödinger Equation

In 1926 in Zurich, after a physics colloquium by Erwin Schrödinger on de Broglie waves, Peter Debye made a remark to the effect that dealing with waves properly requires a wave equation. Schrödinger took this comment to heart, for he returned from a skiing vacation a few weeks later and announced at a subsequent colloquium, "My colleague Debye suggested that one should have a wave equation; well, I have found one!"[6] The Schrödinger equation is the equation of motion for nonrelativistic quantum mechanics. As we have noted, the Schrödinger equation cannot be derived from classical physics. Obtaining it involved some inspired guess work on Schrödinger's part. Rather than following his reasoning, we simply start with the equation itself in one dimension:

$$-\frac{\hbar^2}{2m}\frac{\partial^2 \Psi(x,t)}{\partial x^2} + V(x)\Psi(x,t) = i\hbar \frac{\partial \Psi(x,t)}{\partial t} \tag{2.6}$$

where m is the mass of the particle, $V(x)$ is the potential energy, and $\hbar$ (pronounced "h bar") is Planck's constant divided by 2π ($\hbar = h/2\pi$). $\Psi(x,t)$ is called the **wave function**. It is the probability amplitude for finding the particle at position x at time t. We will discuss how one solves this equation for $\Psi(x,t)$ in the next two chapters. For the remainder of this chapter we will give some plausibility arguments in support of the equation itself.

In this section, we will focus on the Schrödinger equation for a free particle, that is, a particle without forces acting on it. In this case, we can take $V(x) = 0$ and write the equation as

$$-\frac{\hbar^2}{2m}\frac{\partial^2 \Psi(x,t)}{\partial x^2} = i\hbar \frac{\partial \Psi(x,t)}{\partial t} \tag{2.7}$$

Equation (2.7) is the analogue of

$$\frac{\partial^2 \mathcal{E}}{\partial x^2} = \frac{1}{c^2}\frac{\partial^2 \mathcal{E}}{\partial t^2} \tag{2.8}$$

the wave equation for light that results from Maxwell's equations in free space. In Section 1.1 we commented that if we substitute a wave such as $\mathcal{E} = \mathcal{E}_0 e^{i(kx-\omega t)}$ into (2.8), we obtain the condition $\omega = kc$, where $\omega = 2\pi\nu$ and $k = 2\pi/\lambda$. We have seen that $E = h\nu$ and $p = h/\lambda$ for photons. If we write these expressions in terms of $\hbar$, we obtain

$$E = h\nu = \frac{h}{2\pi}2\pi\nu = \hbar\omega \tag{2.9}$$

and

$$p = \frac{h}{\lambda} = \frac{h}{2\pi}\frac{2\pi}{\lambda} = \hbar k \tag{2.10}$$

Notice that if we take the condition $\omega = kc$ and multiply both sides by $\hbar$, we obtain

$$\hbar\omega = \hbar k c \tag{2.11}$$

or

$$E = pc \tag{2.12}$$

---

[6]*Physics Today* **28** (12), 23 (1976).

which is just the relationship between energy and momentum for massless particles demanded by special relativity. Thus Maxwell's equations have naturally built into them that photons are massless quanta.

What then is built into the Schrödinger equation? Notice that if we substitute

$$\Psi(x,t) = Ae^{i(kx-\omega t)} \tag{2.13}$$

into (2.7), we obtain a solution provided

$$\frac{\hbar^2 k^2}{2m} Ae^{i(kx-\omega t)} = \hbar\omega Ae^{i(kx-\omega t)} \tag{2.14}$$

Thus for a nontrivial solution

$$\hbar\omega = \frac{\hbar^2 k^2}{2m} \tag{2.15}$$

Following de Broglie, if we presume that the relations

$$p = \frac{h}{\lambda} = \hbar k \tag{2.16}$$

and

$$E = h\nu = \hbar\omega \tag{2.17}$$

apply for *all* particles, not just for photons, then (2.15) requires

$$E = \frac{p^2}{2m} \tag{2.18}$$

which is the relationship between energy and momentum for a nonrelativistic free particle of mass m, since for a nonrelativistic particle $p = mv$ and $E = mv^2/2$.

Note that in order to obtain the relationship (2.15), which involves one factor of ω and two factors of k, we needed an equation that involves one time derivative and two space derivatives. But this means that waves of the form $\Psi(x,t) = A\cos(kx - \omega t)$ or $\Psi(x,t) = A\sin(kx - \omega t)$ are not solutions to the Schrödinger equation for a free particle, as you can readily verify by substituting them into the equation. The only simple wave solutions are in the form of complex exponentials given in (2.13). You can't just take the real part as you can for the electric field $\mathcal{E}$. The solutions to the Schrödinger equation for a free particle are inherently complex. Given the postulates of quantum mechanics that we introduced in Chapter 1, this may not be all that surprising. But it was definitely disturbing to the early practitioners of quantum mechanics, including Schrödinger himself. And of course, the Schrödinger equation itself explicitly involves $\hbar$ and $i = \sqrt{-1}$, unlike (2.8). Nobody had dared to write a physics equation that inherently involved complex numbers in this way before Schrödinger.

Because a helium atom, or an electron for that matter, has a rest mass, it is possible to restrict our focus to situations in which the particles are nonrelativistic. This is what the Schrödinger equation does. We will see that such nonrelativistic quantum mechanics has broad applicability in atomic, molecular, solid-state, and nuclear physics. In particle physics, on the other hand, relativistic considerations become quite important. At that point, our description must include at least some of the key features of quantum field theory. We will return to this issue in Chapter 10.

2.4 The Physical Significance of the Wave Function

What is the physical significance of $\Psi(x, t)$? In general, Ψ is a complex function, so it cannot correspond directly to something we can measure. As we have noted, we can also refer to the wave function as the probability amplitude of finding the particle at the position x at time t. You might therefore be tempted to identify $\Psi^*(x, t)\Psi(x, t) = |\Psi(x, t)|^2$ with the probability of finding the particle at x at time t. But this is not quite correct. Rather, the correct interpretation (often referred to as the Born interpretation since it was first suggested by Max Born) is that

$|\Psi(x, t)|^2 dx = $ the probability of finding the particle between x and $x + dx$ at time t if a measurement of the particle's position is carried out

Strictly, $|\Psi(x, t)|^2$ is the **probability density** since it must be integrated to determine the probability (like the charge density, which must be integrated to determine the electric charge), and has dimensions of probability per unit length, in one dimension. For example, it doesn't make physical sense to ask about the probability of a helium atom landing at a single position in the detecting plane in the experiment described in Section 2.1. Instead, if we make the width of the entrance aperture for our detector 2 microns and center the detector 4 microns above the line bisecting the two slits, then the probability that an atom lands in the detector is given by

$$\int_{x=3 \text{ microns}}^{x=5 \text{ microns}} |\Psi(x, t)|^2 dx$$

No matter how small we make the aperture to the detector, we cannot measure the probability that a helium atom lands at a single position out of the continuum of possible positions in the detection plane. Of course, if we make the aperture too small, helium atoms will be too big to fit through the aperture and we will not obtain any counts. We will then need to figure an alternative method of measuring the position of a helium atom to greater precision. (But our method might continue to work for electrons which, as far as we know, are point particles.)

Does this mean that before a measurement is carried out the helium atom does not have a definite position? The answer is yes. In quantum mechanics, you have to give up on the notion that a particle should be associated with a definite position. For if you thought the helium atom had a definite position just before it struck the detector, then you would presume it had a somewhat different definite position just in front of this position a moment earlier, and so on. You would effectively be presuming that the atom followed a definite trajectory to reach the detector. But if it followed a definite trajectory, what then is generating the interference pattern, which as we have seen requires multiple amplitudes (multiple paths) for a single particle to reach a particular point in the detection plane? This is of course a far cry from our classical experience and it explains why quantum mechanics was initially so controversial. In fact, Schrödinger himself was troubled by this probabilistic interpretation of the wave function.

Since the probability of finding the particle overall must be one, it is necessary that

$$\int_{-\infty}^{\infty} |\Psi(x, t)|^2 dx = 1 \qquad (2.19)$$

at all times t in order for this probabilistic interpretation of the wave function to be self-consistent. It is straightforward to satisfy this condition. For if $\Psi(x, t)$ is a solution to the Schrödinger equation, so is $N\Psi(x, t)$, where N is a (complex) constant. Thus we can always multiply $\Psi(x, t)$ by an appropriate constant so that (2.19) is satisfied. We call this **normalization of the wave function**.

Clearly, it is necessary that $\Psi \to 0$ as $|x| \to \infty$ in order for the integral in (2.19) to converge. The physically admissible wave functions are therefore said to be **square integrable**, where by the "square" we mean multiplying the complex wave function by its complex conjugate. A somewhat artificial but useful illustration is given in Example 2.5.

EXAMPLE 2.5 Normalize the wave function

$$\Psi(x) = \begin{cases} Nx(L-x) & 0 < x < L \\ 0 & \text{elsewhere} \end{cases}$$

depicted in Fig. 2.11.

SOLUTION

$$1 = \int_{-\infty}^{\infty} |\Psi|^2 dx = \int_0^L |N|^2 x^2 (L-x)^2 dx$$

$$= |N|^2 \int_0^L (x^2 L^2 - 2Lx^3 + x^4)\, dx$$

$$= |N|^2 L^5 \left(\frac{1}{3} - \frac{2}{4} + \frac{1}{5}\right) = \frac{|N|^2 L^5}{30}$$

Thus normalization of the wave function determines the value of N^*N. We might as well choose N to be real, namely

$$N = \sqrt{\frac{30}{L^5}}$$

but it is good to remember that other choices are possible as well. Note that $|N|^2 x^2(L-x)^2 dx$, the probability of finding the particle between x and $x + dx$, must be dimensionless. Since $x^2(L-x)^2 dx$ has the dimensions of length to the fifth power, $|N|^2$ must have the dimensions of inverse length to the fifth power.

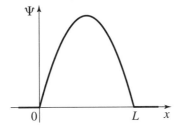

Figure 2.11 Wave function for Example 2.5.

2.5 Conservation of Probability

If we normalize the wave function at one time, will it stay normalized? Does (2.19) hold for all time? In short, is probability conserved? It is not difficult to verify that the Schrödinger equation assures conservation of probability. Let's start by calculating the time derivative of the probability density:

$$\frac{\partial |\Psi|^2}{\partial t} = \frac{\partial \Psi^* \Psi}{\partial t} = \Psi^* \frac{\partial \Psi}{\partial t} + \Psi \frac{\partial \Psi^*}{\partial t} \tag{2.20}$$

From the Schrödinger equation

$$\frac{\partial \Psi}{\partial t} = \frac{1}{i\hbar}\left(-\frac{\hbar^2}{2m}\frac{\partial^2 \Psi}{\partial x^2} + V(x)\Psi\right) \tag{2.21}$$

and, taking the complex conjugate of this equation,

$$\frac{\partial \Psi^*}{\partial t} = \frac{-1}{i\hbar}\left(-\frac{\hbar^2}{2m}\frac{\partial^2 \Psi^*}{\partial x^2} + V(x)\Psi^*\right) \tag{2.22}$$

where we have assumed the potential energy $V(x)$ is real. Thus

$$\frac{\partial |\Psi|^2}{\partial t} = \frac{i\hbar}{2m}\left(\Psi^* \frac{\partial^2 \Psi}{\partial x^2} - \Psi \frac{\partial^2 \Psi^*}{\partial x^2}\right)$$

$$= \frac{\partial}{\partial x}\left[\frac{i\hbar}{2m}\left(\Psi^* \frac{\partial \Psi}{\partial x} - \Psi \frac{\partial \Psi^*}{\partial x}\right)\right] = -\frac{\partial j_x}{\partial x} \tag{2.23}$$

where

$$j_x(x,t) = \frac{\hbar}{2mi}\left(\Psi^* \frac{\partial \Psi}{\partial x} - \Psi \frac{\partial \Psi^*}{\partial x}\right) \tag{2.24}$$

is referred to as the **probability current**.[7] Therefore

$$\frac{d}{dt}\int_{-\infty}^{\infty} |\Psi(x,t)|^2 dx = -\int_{-\infty}^{\infty} \frac{\partial j_x(x,t)}{\partial x} dx = -j_x(x,t)|_{-\infty}^{\infty} = 0 \tag{2.25}$$

where the last step follows from the fact that $\Psi(x,t) \to 0$ as $x \to \pm\infty$, for otherwise the wave function would not be normalizable. Thus the integral on the left-hand side of (2.25) is independent of time. If this integral is equal to one at a particular time, say $t = 0$, it will be equal to one for all time.

The reason for introducing the probability current j_x into our derivation is that it enables us to see that conservation of probability is not just a global phenomenon—the overall probability of finding the particle remains one—but a local phenomenon as

---

[7]In confirming (2.23), you may find it easiest to start with the second line, do the derivatives, and show that you obtain the first line. You may also recognize (2.23) as a one-dimensional version of the conservation of charge equation

$$\frac{\partial \rho}{\partial t} + \nabla \cdot \mathbf{j} = 0$$

where ρ is the electric charge density and $\mathbf{j}$ is the current density. This conservation of charge equation follows from Maxwell's equations just as (2.23) follows from the Schrödinger equation.

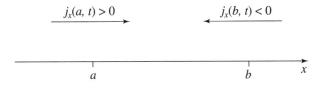

Figure 2.12 Conservation of probability: The probability current accounts for a change in the probability of finding a particle in [a, b].

well. To illustrate, instead of integrating the probability density over all positions, from $x = -\infty$ to $x = \infty$, integrate from $x = a$ to $x = b$. Then following the steps that lead to (2.25), we see that

$$\frac{d}{dt}\int_a^b |\Psi(x,t)|^2 dx = -j_x(x,t)\Big|_a^b = -j_x(b,t) + j_x(a,t) \quad (2.26)$$

Now there is no reason for $j_x(a, t)$ and $j_x(b, t)$ to vanish since a and b may be finite, and the probability of finding the particle in the region between $x = a$ and $x = b$ can change with time. But the reason for the change is clear; probability can flow into or out of the region at either endpoint (or at both). In particular, if $j_x(a, t)$, the probability flow per unit time at $x = a$, is positive, then the probability current contributes to an increase in the probability in the region to the right of $x = a$ (see Fig. 2.12), while if $j_x(b, t)$ is positive, this term contributes to a decrease of the probability in the region to the left of $x = b$. Or if $j_x(b, t)$ is negative, then the probability current is flowing toward the left at $x = b$, which contributes to increasing the probability in the region between $x = a$ and $x = b$. Thus if probability decreases in some region, it doesn't mysteriously appear somewhere else but rather flows continuously into or out of the region in question. Thus probability is said to be "locally conserved."

EXAMPLE 2.6 For the wave function $\Psi(x, t) = Ae^{i(kx-\omega t)}$ determine the probability current.

SOLUTION

$$j_x = \frac{\hbar}{2mi}\left(\Psi^*\frac{\partial \Psi}{\partial x} - \Psi\frac{\partial \Psi^*}{\partial x}\right)$$

$$= \frac{\hbar}{2mi}\left(A^*e^{-i(kx-\omega t)}ikAe^{i(kx-\omega t)} + Ae^{i(kx-\omega t)}ikA^*e^{-i(kx-\omega t)}\right) = \frac{\hbar k}{m}|A|^2$$

In this example, the probability current is the product of $\hbar k/m = p/m = v$ and the probability density $|A|^2$. Thus a large probability current can result from a large probability density and/or a large velocity for the particle.

2.6 Wave Packets and the Heisenberg Uncertainty Principle

Picture a wave with a single wavelength. The wave is *everywhere*. Now picture a particle with a single momentum. You tend to think of a particle as *somewhere*. But the momentum of the particle and the wavelength are related by $p = h/\lambda$. If you consider yourself to

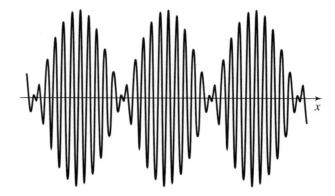

Figure 2.13 The sum of two sinusoidal waves with equal amplitudes but slightly different wavelengths.

be at rest as you read this book, then you would say your momentum is zero. Then your wave function should be one corresponding to an infinite wavelength, basically a constant wave function everywhere, although you certainly tend to think of yourself as having a well-defined position. How do we resolve this seeming contradiction? First note the wave function $\Psi(x,t) = Ae^{i(kx-\omega t)}$, a solution of the Schrödinger equation for a free particle with a single wavelength, does not approach zero as $|x| \to \infty$. Thus it is not a normalizable wave function and cannot correspond to a physically allowed state.[8] It is possible, however, to generate physically acceptable, normalizable solutions to the Schrödinger equation for a free particle by superposing solutions with different wavelengths.[9] Such a superposition produces a **wave packet**.

As we discussed in Chapter 1, when waves are in phase, they add constructively, and when waves are π out of phase, they interfere destructively. Here we want to superpose waves with different wavelengths to form a localized wave packet. To see the pattern, we start by adding two waves together each with the same amplitude A but slightly different wavelengths, one with wave number $k + \Delta k/2$ and the other with wave number $k - \Delta k/2$:

$$A \sin\left[(k - \Delta k/2)x - (\omega - \Delta\omega/2)t\right] + A \sin\left[(k + \Delta k/2)x - (\omega + \Delta\omega/2)t\right]$$
$$= 2A \sin(kx - \omega t) \cos\left[(\Delta k/2)x - (\Delta\omega/2)t\right] \quad (2.27)$$

where we have made use of the trigonometric identity

$$\sin\alpha + \sin\beta = 2\cos\frac{\alpha - \beta}{2}\sin\frac{\alpha + \beta}{2} \quad (2.28)$$

Figure 2.13 shows the result. Notice that we have a traveling wave $2A \sin(kx - \omega t)$ modulated by the factor $\cos\left[(\Delta k/2)x - (\Delta\omega/2)t\right]$, which is itself a traveling wave with a longer wavelength and lower (beat) frequency. We have chosen to add two sine waves together rather than the complex exponentials that are solutions to the Schrödinger equation, so we can graph the result. Or you can think of (2.27) as the result of adding the imaginary parts of the complex solutions. Problem 2.15 shows a similar effect for adding

---

[8] A wave function with a single momentum (a single wavelength) is, nonetheless, a very useful abstraction.

[9] Such a superposition is possible because the Schrödinger equation is a **linear differential equation**, which means that if ψ_1 and ψ_2 are two solutions to the Schrödinger equation, then so is $c_1\psi_1 + c_2\psi_2$, where c_1 and c_2 are arbitrary complex numbers. See Problem 2.16.

Figure 2.14 The addition of five sinusoidal waves with varying wavelengths can produce a more localized pulse which nonetheless repeats indefinitely.

the real parts. Figure 2.14 shows the result of adding five waves together. Here the "beats" are better articulated. If we add an infinite number of waves together with a continuous infinity of different wave numbers of the form

$$\Psi(x, t) = \int_{-\infty}^{\infty} A(k) e^{i(kx - \omega t)} dk \qquad (2.29)$$

we obtain a wave packet that doesn't repeat itself, as illustrated in Fig. 2.15a. You can suppose that each of the waves has a crest at the origin, at $x = 0$, as shown in Fig. 2.15b. Here the waves add constructively. But as we move away from the origin, the crests (and the troughs) of the different waves no longer coincide since the wavelength of each wave in the superposition is different. In this way, we can generate a region where the waves add together to give a large amplitude, while in other regions the waves effectively cancel. The more waves we introduce in the superposition, the more complete the cancellation can in principle be. Example 2.7 provides a nice illustration of how this superposition works.

You can probably foresee that the more waves with differing wavelengths that we add together the narrower we can make the region of space where the wave packet is nonzero. In fact, it is possible to show that the width Δx in position of the wave packet and the

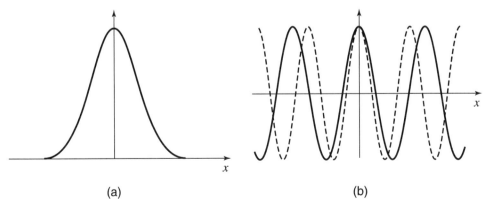

Figure 2.15 (a) A wave packet can be thought of as a superposition of a continuum of waves. Two of these waves, each with a maximum at the origin, are shown in (b). These two waves add constructively at the origin, but interfere destructively at some locations and constructively at other locations.

width Δk in wave numbers that are involved in the superposition satisfy the relation

$$\Delta x \, \Delta k \geq \frac{1}{2} \tag{2.30}$$

provided one uses a precise definition (see Section 2.8) of what we mean by the quantities Δx and Δk. This is a very general result from the physics of waves, one with which you probably already have some experience. For example, lightning produces a local, intense heating (and expansion) of the air in the immediate vicinity of a strike, namely, a narrow wave packet. This pressure pulse propagates outward from the strike as a sound wave. When it reaches your ear, you hear the many different wavelengths (or frequencies) in the thunder clap. This is quite different from the sound of an almost pure frequency from, say, a violin string, which would have a substantially longer wave train.

What makes (2.30) unusually interesting in quantum physics is the connection between the wavelength λ and the momentum of the particle. Since $p = \hbar k$, we can express the relation (2.30) in terms of the momentum p_x (the x component of the linear momentum, since we are considering one-dimensional motion along the x axis) as

$$\Delta x \, \Delta p_x \geq \frac{\hbar}{2} \tag{2.31}$$

which is the famous **Heisenberg uncertainty principle**.[10] It shows that a particle cannot have both a definite position and a definite momentum simultaneously, since the right-hand side of (2.31) is nonzero. If a particle were to have a well-defined momentum ($\Delta p_x \to 0$), the position of the particle would be very uncertain ($\Delta x \to \infty$). On the other hand, if the particle were to have a very well-defined position ($\Delta x \to 0$), then the particle would not have a definite momentum at all ($\Delta p_x \to \infty$). Of course, the scale for the product of these uncertainties in the position and the momentum is set by the size of Planck's constant. As Example 2.8 illustrates, these restrictions are not noticeable for macroscopic particles.

EXAMPLE 2.7 Determine the form of the wave packet

$$\Psi(x, 0) = \int_{-\infty}^{\infty} A(k) e^{ikx} \, dk$$

generated by the superposition with $A(k) = A$ for $k_0 - (\Delta k)/2 < k < k_0 + (\Delta k)/2$ and $A(k) = 0$ otherwise. Estimate the value of $\Delta x \, \Delta k$ for this wave packet.

SOLUTION

$$\Psi(x, 0) = \int_{k_0-(\Delta k)/2}^{k_0+(\Delta k)/2} A e^{ikx} \, dk$$

$$= A e^{ik_0 x} \frac{e^{ix\Delta k/2} - e^{-ix\Delta k/2}}{ix}$$

$$= 2A e^{ik_0 x} \frac{\sin(x\Delta k/2)}{x}$$

---

[10]Our "derivation" of the Heisenberg uncertainty relation follows directly from (2.30), which is a fundamental result of Fourier analysis. The main point, at this stage, is to get a sense of how the superposition and interference of waves can lead to this result. Using a different strategy, we will give a rigorous derivation of (2.31) in Chapter 5.

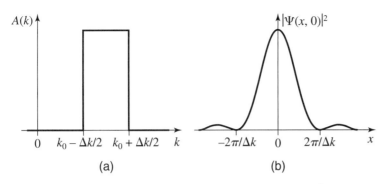

Figure 2.16 Graphs of (a) $A(k)$ and (b) $|\Psi(x, 0)|^2$.

Thus

$$|\Psi(x, 0)|^2 = 4|A|^2 \frac{\sin^2(x\Delta k/2)}{x^2}$$

Figure 2.16 shows graphs of $A(k)$ and $|\Psi(x, 0)|^2$. A quick estimate of the width of $A(k)$ is to say it is Δk. The width of the wave packet in position can be estimated from the distance between the first two nodes on either side of the central maximum. These nodes occur when $x\Delta k/2 = \pm\pi$, so $x = \pm 2\pi/\Delta k$. Setting $\Delta x = 4\pi/\Delta k$, we find $\Delta x \Delta k = 4\pi$. Thus our result is consistent with the general limit expressed in (2.30). The smaller the magnitude of Δk, the larger is the width Δx of the wave packet. Notice that the width of the wave packet depends solely on Δk and not on the value of k_0, which in momentum terms means that the width of the packet does not depend on the value of the central momentum $p_0 = \hbar k_0$.

EXAMPLE 2.8 Estimate the uncertainty in your velocity as you sit reading if the uncertainty in your position is 10^{-10} m.

SOLUTION From the Heisenberg uncertainty relation,

$$\Delta p_x \geq \frac{\hbar}{2\Delta x} = \frac{1.055 \times 10^{-34} \text{ J·s}}{2 \times 10^{-10} \text{ m}} = 5.3 \times 10^{-25} \text{ kg·m/s}$$

Since $p_x = mv_x$, you can obtain an estimate of the uncertainty in your velocity by dividing Δp_x by your mass. For example, if $m = 60$ kg, then

$$\Delta v \geq 8.8 \times 10^{-27} \text{ m/s}$$

which is such a small value we cannot in practice distinguish this spread in velocity from the classical notion that your speed is zero as you sit reading.

EXAMPLE 2.9 Estimate the uncertainty in the velocity of an electron confined to a region that is 10^{-10} m in size. This is a typical size scale for an atom.

SOLUTION Again from the Heisenberg uncertainty relation,

$$\Delta p_x \geq \frac{\hbar}{2\Delta x} = \frac{1.055 \times 10^{-34} \text{ J·s}}{2 \times 10^{-10} \text{ m}} = 5.3 \times 10^{-25} \text{ kg·m/s}$$

In this example the mass is the mass of an electron, namely, $m = 9.11 \times 10^{-31}$ kg. Thus

$$\Delta v = \frac{\Delta p_x}{m} \geq 5.8 \times 10^5 \text{ m/s}$$

The uncertainty of 580 km/s in the electron's velocity in an atom is certainly not negligible, showing the importance of the Heisenberg uncertainty relation on the microscopic scale.

2.7 Phase and Group Velocity

Superposing waves to make a wave packet resolves the problem of localizing a particle so that it has a reasonably well-defined position, with some uncertainty, but at the expense of introducing an uncertainty in the momentum of the particle as well. Superposition also solves another problem, namely, the speed of the wave and the speed of the particle do not seem to be equal. If you take a wave with a definite wavelength such as $Ae^{i(kx-\omega t)}$, you can express it in the form

$$Ae^{i(kx-\omega t)} = Ae^{i\{k[x-(\omega/k)t]\}} = Ae^{i[k(x-v_{\text{ph}}t)]} \tag{2.32}$$

where v_{ph} is called the **phase velocity** of the wave:

$$v_{\text{ph}} = \frac{\omega}{k} = \frac{2\pi \nu}{(2\pi/\lambda)} = \lambda \nu \tag{2.33}$$

The phase velocity is the speed at which a particular point on the wave, such as a crest, moves. For example, if you set $x = 0$ and $t = 0$ for the wave $A \cos(kx - \omega t)$ (the real part of $Ae^{i(kx-\omega t)}$), you are at the maximum of the wave. A time dt later, you have to move along the x axis to a point dx which satisfies $kdx - \omega dt = 0$ in order to keep up with the crest. (See Fig. 2.17.) This speed is thus given by $v_{\text{ph}} = dx/dt = \omega/k$. For light, $v_{\text{ph}} = c$. But for the Schrödinger equation, we see that the phase velocity for a nonrelativistic particle of mass m is given by

$$v_{\text{ph}} = \frac{\omega}{k} = \frac{\hbar \omega}{\hbar k} = \frac{E}{p} = \frac{mv^2/2}{mv} = \frac{v}{2} \tag{2.34}$$

That is, the phase velocity is one half the velocity of the particle.

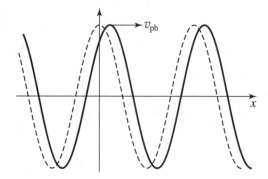

Figure 2.17 The dashed line shows a sinusoidal traveling wave at $t = 0$ and the solid line shows the same wave at a slightly later time.

At first, this seems quite troubling. From our discussion in the preceding section, however, we know that the location of the particle is determined by superposing waves so that they form a wave packet; it is the location of the packet that tells us roughly where the particle is located. But if you look back at (2.27), our expression for the superposition of two waves with slightly different wavelengths, you will see there are two velocities present. The term $\sin(kx - \omega t)$ has the characteristic phase velocity ω/k, while the factor $\cos[(\Delta k/2)x - (\Delta \omega/2)t]$ modulating the superposition travels with a different velocity, namely $\Delta \omega / \Delta k$ since

$$\cos[(\Delta k/2)x - (\Delta \omega/2)t] = \cos\{(\Delta k/2)[x - (\Delta \omega/\Delta k)t]\} \tag{2.35}$$

We will now show that this is a special case of a more general result: namely, for a superposition of an infinite number of waves, a continuum of waves, the wave packet itself moves with a velocity that is given by

$$v_g = \frac{d\omega}{dk} \tag{2.36}$$

where v_g is called the **group velocity**. Whereas the phase velocity is the speed of a particular point, such as the crest, of a wave with a single wavelength, the group velocity is the speed of a localized packet of waves that has been generated by superposing many waves together.

To establish (2.36) we assume that the wave packet is peaked near some central momentum p_0. That is, the amplitude $A(k)$ in the superposition

$$\Psi(x, t) = \int_{-\infty}^{\infty} A(k) e^{i(kx - \omega t)} dk \tag{2.37}$$

has a peak at wave vector k_0, where $p_0 = \hbar k_0$, as indicated in Fig. 2.18. Since most of the contribution to this integral comes in the vicinity of k_0 [because $A(k)$ is largest in this region], we can obtain an approximate expression for this integral by expanding ω in a Taylor series about $k = k_0$. In particular, if we retain the two leading terms in the Taylor series for ω, then

$$\omega \cong \omega_0 + \left(\frac{d\omega}{dk}\right)_{k=k_0} (k - k_0) \tag{2.38}$$

where ω_0 is the value of ω when $k = k_0$. Using the definition of the group velocity in (2.36)

$$\omega \cong \omega_0 + v_g(k - k_0) \tag{2.39}$$

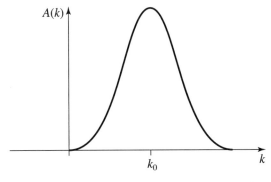

Figure 2.18 The amplitude $A(k)$ has a maximum at $k = k_0$.

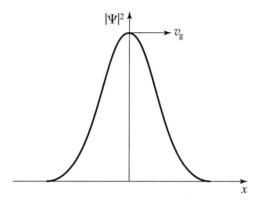

Figure 2.19 A superposition of traveling waves forms a wave packet which moves at speed v_g.

where it is now understood that $d\omega/dk$ is evaluated at k_0, the value of k at which $A(k)$ has its maximum. Thus

$$kx - \omega t \cong kx - \left[\omega_0 + v_g(k - k_0)\right]t \tag{2.40}$$

Substituting this result for the exponent in (2.37), we obtain

$$\Psi(x, t) \cong e^{i(k_0 x - \omega_0 t)} \int_{-\infty}^{\infty} A(k) e^{i(k - k_0)(x - v_g t)}\, dk \tag{2.41}$$

But the initial form for the wave packet is given by [put $t = 0$ in (2.37)]

$$\Psi(x, 0) = \int_{-\infty}^{\infty} A(k) e^{ikx}\, dk = e^{ik_0 x} \int_{-\infty}^{\infty} A(k) e^{i(k - k_0)x}\, dk \tag{2.42}$$

and therefore

$$\Psi(x, t) \cong e^{i[k_0 v_g - \omega_0]t} \Psi(x - v_g t, 0) \tag{2.43}$$

The term multiplying $\Psi(x - v_g t, 0)$, namely $e^{i(k_0 v_g - \omega_0)t}$, is an overall phase factor that cancels out when we calculate the probability density:

$$|\Psi(x, t)|^2 = |\Psi(x - v_g t, 0)|^2 \tag{2.44}$$

Thus we see that the wave packet moves with a speed v_g, as illustrated in Fig. 2.19. Since for the Schrödinger equation for a free particle

$$\omega = \frac{\hbar k^2}{2m} \tag{2.45}$$

the group velocity is

$$v_g = \frac{d}{dk}\left(\frac{\hbar k^2}{2m}\right) = \frac{\hbar k}{m} = \frac{p}{m} = v \tag{2.46}$$

Thus the wave packet moves at the speed of the particle. Another success for the Schrödinger equation![11]

---

[11] In general, different wave equations yield different ratios for the phase and group velocities. See Problem 2.24 and Problem 2.25 for water waves in deep and shallow water, respectively.

One final comment on this subject is probably in order. You may be wondering about the effect of the higher order terms that we have neglected in the Taylor-series expansion (2.38). One of the many things that distinguishes the Schrödinger equation from the wave equation (2.8) for light is that the relationship between ω and k is not simply a linear one for the Schrödinger equation since $\omega = \hbar k^2/2m$ for a free particle. Thus, in particular, the second-order terms in the Taylor-series expansion are nonzero. But this also means that the phase velocity depends on k since $\omega/k = \hbar k/2m$. Thus the very fine balancing of constructive and destructive interference that is required to localize the wave packet in Fig. 2.15 will be diminished as time increases since each wave in the superposition moves with a different speed. This causes the wave packet to spread out. However, the time for spreading depends sensitively on the mass of the particle and the size of the initial wave packet. The smaller the wave packet, for example, the larger the number of different momentum components in the wave function and hence the larger the variation in the phase velocities. See Problem 2.21. Given this discussion, you will not be surprised to learn that the general relationship between ω and k is referred to as a **dispersion relation**.

EXAMPLE 2.10 The relation between the wavelength λ and frequency ν for the propagation of electromagnetic waves through a wave guide (typically a hollow rectangular or cylindrical metal pipe) is given by

$$\lambda = \frac{c}{\sqrt{\nu^2 - \nu_0^2}}$$

What are the phase and group velocities of these waves? *Note*: For a wave guide the constant ν_0 is the minimum frequency for which the waves will propagate.

SOLUTION The relationship between the wavelength λ and the frequency ν can also be expressed in terms of the wave vector k and the angular frequency ω as

$$kc = \sqrt{\omega^2 - \omega_0^2}$$

or

$$\omega = \sqrt{(kc)^2 + \omega_0^2} = c\sqrt{k^2 + (\omega_0/c)^2}$$

where $\omega_0 = 2\pi\nu_0$. The phase velocity is given by

$$v_{\text{ph}} = \frac{\omega}{k} = \frac{c\sqrt{k^2 + (\omega_0/c)^2}}{k} = c\sqrt{1 + (\omega_0/kc)^2}$$

which is greater than c, while the group velocity is given by

$$v_g = \frac{d\omega}{dk} = c\frac{k}{\sqrt{k^2 + (\omega_0/c)^2}} = \frac{c}{\sqrt{1 + (\omega_0/kc)^2}}$$

which is less than c. Note that $v_{\text{ph}} v_g = c^2$.

A phase velocity that exceeds the speed of light may seem troubling at first. But it is the group velocity that determines how fast, say, information is transmitted by a localized wave packet. The phase velocity is simply the velocity of a particular point on a wave with a definite wavelength, a wave that extends throughout space.

2.8 Expectation Values and Uncertainty

We have seen that a particle in a physically allowed state does not have a definite position or a definite momentum. Consequently, there is an inherent uncertainty with respect to what value to assign for the particle's position and the particle's momentum. There is a straightforward way to quantify this uncertainty, a way that is quite similar to the way we calculate the standard deviation of, say, the results for an exam.

One of the first things everybody wants to know about an exam is: What's the average? We calculate the average, or the mean, by adding up the scores and dividing by the total number of people (say, 25) who took the exam. Equivalently, we can generate a distribution function in which we divide the number of people $N(n)$ who obtained a particular score n by the total number N in the class, which yields the fraction of the class that obtained that score:

$$P(n) = \frac{N(n)}{N} \tag{2.47}$$

$P(n)$ is like a probability distribution. In particular, it satisfies the condition

$$\sum_{n=0}^{100} P(n) = \sum_{n=0}^{100} \frac{N(n)}{N} = \frac{1}{N}\sum_{n=0}^{100} N(n) = 1 \tag{2.48}$$

The average score on the exam is

$$\langle n \rangle = \frac{\sum_{n=0}^{100} nN(n)}{N} = \sum_{n=0}^{100} nP(n) \tag{2.49}$$

Note that the average need not be one of the scores obtained on the exam, or even an integer for that matter.

The other piece of information often sought about an exam is the standard deviation σ, which is determined from

$$\sigma^2 = \langle (n - \langle n \rangle)^2 \rangle \tag{2.50}$$

that is, we subtract $\langle n \rangle$ from each score on the exam, square this result, and find its average for the class:

$$\sigma^2 = \frac{\sum_{n=0}^{100} (n - \langle n \rangle)^2 N(n)}{N}$$
$$= \sum_{n=0}^{100} (n - \langle n \rangle)^2 P(n) = \sum_{n=0}^{100} (n^2 - 2n\langle n \rangle + \langle n \rangle^2)P(n)$$
$$= \sum_{n=0}^{100} n^2 P(n) - 2\langle n \rangle \sum_{n=0}^{100} nP(n) + \langle n \rangle^2 \sum_{n=0}^{100} P(n)$$
$$= \langle n^2 \rangle - 2\langle n \rangle \langle n \rangle + \langle n \rangle^2 = \langle n^2 \rangle - \langle n \rangle^2 \tag{2.51}$$

This last result, namely

$$\sigma^2 = \langle n^2 \rangle - \langle n \rangle^2 \tag{2.52}$$

is worth remembering. It provides a fast route to calculating the standard deviation.

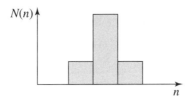

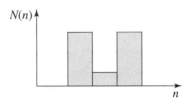

Figure 2.20 Histograms of the grade distribution for two different classes.

Figure 2.20 shows the distributions for two classes on the same exam. The average in each class is the same, but the spread is quite different in the two classes. One class is filled with mostly B students, while the other has a significant number of A and C students and a few B students. Consequently, the standard deviations for the two classes are quite different, much larger in the latter class reflecting the broad nature of the exam-score distribution. The limiting case of a narrow distribution is one in which every student in the class has the same score, in which case $\langle n^2 \rangle = \langle n \rangle^2$ and the standard deviation $\sigma = 0$.

The expressions we have been using for the average value and standard deviation can be taken over directly to continuous distributions. The average value of the position is given by [the analogue of (2.49)]

$$\langle x \rangle = \int_{-\infty}^{\infty} x |\Psi(x,t)|^2 dx \tag{2.53}$$

where we are presuming that the wave function is normalized [the analogue of (2.48)], that is,

$$\int_{-\infty}^{\infty} |\Psi(x,t)|^2 dx = 1 \tag{2.54}$$

which is equivalent to saying that the total probability of finding the particle somewhere is one. Average values, such as (2.53), are often referred to as **expectation values** in quantum mechanics. Potentially this is somewhat misleading, since $\langle x \rangle$ is not necessarily the value of the position we most expect to find if we measure the position of the particle once, but it is accepted usage nonetheless. In fact, as Fig. 2.21 shows, the particle may have no chance of actually being in the vicinity of $\langle x \rangle$. The expectation value is just the average value and, as for the exam scores, gives you only one piece of information about the probability distribution $|\Psi|^2$.

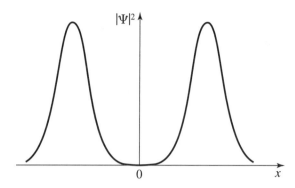

Figure 2.21 A plot of $\Psi^*\Psi$ for which $\langle x \rangle$ is zero but there is no chance of finding the particle in the vicinity of the origin if a measurement of the position of the particle is carried out.

As for an exam, it is often informative to ask about the standard deviation as well as the average. Since

$$\langle x^2 \rangle = \int_{-\infty}^{\infty} x^2 |\Psi(x,t)|^2 dx \tag{2.55}$$

the standard deviation in position is given by

$$(\Delta x)^2 = \langle (x - \langle x \rangle)^2 \rangle = \langle x^2 \rangle - \langle x \rangle^2$$
$$= \int_{-\infty}^{\infty} x^2 |\Psi(x,t)|^2 dx - \left(\int_{-\infty}^{\infty} x |\Psi(x,t)|^2 dx \right)^2 \tag{2.56}$$

where we have denoted the standard deviation by the symbol Δx. In quantum mechanics Δx is generally called the **uncertainty** in the particle's position. This is a reflection of the fact that a single particle whose wave function is Ψ does not have a definite position. This is a fundamentally different meaning from the one we assign to the quantity σ in our exam example. There each student had a definite score on the exam and σ is a measure of the spread in exam scores in the class. Here a single particle does not have a definite position and Δx is a measure of the uncertainty in that single particle's position. If the particle is charged, for example, we could determine the location of the particle by scattering light from it. Such measurements will inevitably change the particle's wave function. Therefore, testing expressions such as (2.53) and (2.56) requires measurements carried out on an ensemble of particles each of which is in the state Ψ. The larger the number of particles in the ensemble, the more closely the experimental results for the average position and the uncertainty in the position will approach (2.53) and (2.56).

Incidentally, (2.56) is the precise definition of Δx that appears in the Heisenberg uncertainty relation $\Delta x \Delta p_x \geq \hbar/2$. The momentum uncertainty is defined similarly, namely,

$$(\Delta p_x)^2 = \langle p_x^2 \rangle - \langle p_x \rangle^2 \tag{2.57}$$

In the next section, we will see a way to calculate $\langle p_x \rangle$ from the wave function.

EXAMPLE 2.11 Calculate $\langle x \rangle$, $\langle x^2 \rangle$, and Δx for the wave function of Example 2.5.

SOLUTION

$$\langle x \rangle = \int_{-\infty}^{\infty} x|\Psi|^2 dx = \int_0^L x \frac{30}{L^5} x^2 (L-x)^2 dx$$

$$= \frac{30}{L^5} \int_0^L (L^2 x^3 - 2Lx^4 + x^5) dx = 30L\left(\frac{1}{4} - \frac{2}{5} + \frac{1}{6}\right) = \frac{L}{2}$$

as could also be seen from the symmetry of the wave function.

$$\langle x^2 \rangle = \int_{-\infty}^{\infty} x^2 |\Psi|^2 dx = \int_0^L x^2 \frac{30}{L^5} x^2 (L-x)^2 dx$$

$$= \frac{30}{L^5} \int_0^L (L^2 x^4 - 2Lx^5 + x^6) dx = 30L^2 \left(\frac{1}{5} - \frac{2}{6} + \frac{1}{7}\right) = \frac{2L^2}{7}$$

Thus

$$\Delta x = \left(\langle x^2 \rangle - \langle x \rangle^2\right)^{1/2} = L\left(\frac{2}{7} - \frac{1}{4}\right)^{1/2} = \frac{L}{2\sqrt{7}}$$

Notice that while we could have determined $\langle x \rangle$ by inspection of the wave function, it takes a calculation to determine the precise value for Δx.

2.9 Ehrenfest's Theorem

The Schrödinger equation is the equation of motion in quantum mechanics. We can use it to determine how expectation values such as $\langle x \rangle$ vary with time. But if wave mechanics is to be consistent with classical physics in the appropriate limit (this is often referred to as the **correspondence principle**), then we should find that

$$\frac{d\langle x \rangle}{dt} = \frac{\langle p_x \rangle}{m} \tag{2.58}$$

Thus if we follow through on the evaluation of

$$\frac{d\langle x \rangle}{dt} = \frac{d}{dt} \int_{-\infty}^{\infty} x |\Psi(x,t)|^2 dx$$

$$= \int_{-\infty}^{\infty} x \frac{\partial \Psi^* \Psi}{\partial t} dx \tag{2.59}$$

we will find a way to calculate not only $\langle x \rangle$ but also $\langle p_x \rangle$ from $\Psi(x,t)$.

We start by substituting the expression (2.23) for the time derivative of the probability density into (2.59). We thus obtain

$$\frac{d\langle x \rangle}{dt} = \int_{-\infty}^{\infty} x \frac{\partial}{\partial x}\left[\frac{i\hbar}{2m}\left(\Psi^* \frac{\partial \Psi}{\partial x} - \Psi \frac{\partial \Psi^*}{\partial x}\right)\right] dx \tag{2.60}$$

This expression can be simplified by integration by parts:[12]

$$\frac{d\langle x \rangle}{dt} = -\int_{-\infty}^{\infty} \left[\frac{i\hbar}{2m}\left(\Psi^* \frac{\partial \Psi}{\partial x} - \Psi \frac{\partial \Psi^*}{\partial x}\right)\right] dx + \left[x \frac{i\hbar}{2m}\left(\Psi^* \frac{\partial \Psi}{\partial x} - \Psi \frac{\partial \Psi^*}{\partial x}\right)\right]\Big|_{-\infty}^{\infty} \quad (2.61)$$

The boundary term vanishes since Ψ goes to zero as $x \to \pm\infty$. If we do a second integration by parts on the second term in the integral [the $\Psi(\partial \Psi^*/\partial x)$ term], we find that

$$\frac{d\langle x \rangle}{dt} = \frac{1}{m} \int_{-\infty}^{\infty} \Psi^* \frac{\hbar}{i} \frac{\partial \Psi}{\partial x} dx \quad (2.62)$$

and therefore, comparing with (2.58),

$$\langle p_x \rangle = \int_{-\infty}^{\infty} \Psi^* \frac{\hbar}{i} \frac{\partial \Psi}{\partial x} dx \quad (2.63)$$

Equation (2.63) is a very strange-looking expression. After all, you may have been willing to buy into the idea that a single particle doesn't have a definite position, and that its wave function can be used to determine (through $|\Psi|^2 dx$) the probability that the particle is between x and $x + dx$. Given this, expressions such as (2.53) for the average value of x are quite understandable. But what about (2.63)? What sense does it make to calculate the average value of the momentum by taking the derivative of Ψ, multiplying by Ψ^*, and then integrating over all positions, with a factor of $\hbar/i$ thrown in for good measure? In Chapters 3 and 5 we will provide justification for why this strange process works.[13] For now, note that if we take the time derivative of the average value of the momentum, as given by (2.63), using the Schrödinger equation for the time derivative of the wave function, we find

$$\frac{d\langle p_x \rangle}{dt} = \left\langle -\frac{\partial V}{\partial x} \right\rangle \quad (2.64)$$

which looks like Newton's second law, since in classical physics $-\partial V/\partial x = F_x = dp_x/dt$, where F_x is the force acting on the particle. See Problem 2.32. In fact, (2.58)

---

[12]Since

$$\frac{d}{dx}(uv) = u\frac{dv}{dx} + v\frac{du}{dx}$$

it follows that

$$\int_a^b \frac{d}{dx}(uv)\,dx = \int_a^b u\frac{dv}{dx}\,dx + \int_a^b v\frac{du}{dx}\,dx$$

and therefore

$$\int_a^b u\frac{dv}{dx}\,dx = -\int_a^b v\frac{du}{dx}\,dx + uv\Big|_a^b$$

Thus under the integral sign we can switch the derivative operation from one term in the product to the other term, provided we remember to insert a minus sign and add a boundary contribution (evaluated at the end points).

[13]In addition

$$\langle p_x^2 \rangle = \int_{-\infty}^{\infty} \Psi^* \left(\frac{\hbar}{i}\right)^2 \frac{\partial^2 \Psi}{\partial x^2} dx = -\hbar^2 \int_{-\infty}^{\infty} \Psi^* \frac{\partial^2 \Psi}{\partial x^2} dx$$

so we can calculate, for example, the expectation value of p_x^2 directly from the wave function.

and (2.64) are often referred to as Ehrenfest's theorem, since P. Ehrenfest first established these results. At the very least, (2.64) suggests that we are on the right track in making the identification (2.63), for with it we see how results of wave mechanics can correspond with those of classical physics.

EXAMPLE 2.12 Calculate $\langle p_x \rangle$, $\langle p_x^2 \rangle$, and Δp_x for the wave function of Example 2.5.

SOLUTION

$$\langle p_x \rangle = \int_{-\infty}^{\infty} \Psi^* \frac{\hbar}{i} \frac{\partial \Psi}{\partial x} dx$$

$$= \frac{30}{L^5} \int_0^L x(L-x) \frac{\hbar}{i} (L - 2x) \, dx = 0$$

and

$$\langle p_x^2 \rangle = -\hbar^2 \int_{-\infty}^{\infty} \Psi^* \frac{\partial^2 \Psi}{\partial x^2} dx$$

$$= -\frac{30\hbar^2}{L^5} \int_0^L x(L-x)(-2) \, dx = \frac{10\hbar^2}{L^2}$$

Thus

$$\Delta p_x = \left(\langle p_x^2 \rangle - \langle p_x \rangle^2 \right)^{1/2} = \sqrt{10} \frac{\hbar}{L}$$

Note: if we use the value for

$$\Delta x = \left(\langle x^2 \rangle - \langle x \rangle^2 \right)^{1/2} = \frac{L}{2\sqrt{7}}$$

from Example 2.11, we see that

$$\Delta x \, \Delta p_x = \left(\frac{L}{2\sqrt{7}} \right) \left(\frac{\sqrt{10}\hbar}{L} \right) = \frac{1}{2} \sqrt{\frac{10}{7}} \hbar = 0.60 \hbar$$

which is consistent with the Heisenberg uncertainty principle $\Delta x \, \Delta p_x \geq \hbar/2$.

EXAMPLE 2.13 Show that $\langle p_x \rangle = 0$ if the spatial part of the wave function $\Psi(x)$ is real ($\Psi^* = \Psi$).

SOLUTION

$$\langle p_x \rangle = \int_{-\infty}^{\infty} \Psi^* \frac{\hbar}{i} \frac{\partial \Psi}{\partial x} dx = \frac{\hbar}{i} \int_{-\infty}^{\infty} \Psi \frac{\partial \Psi}{\partial x} dx$$

$$= \frac{\hbar}{2i} \int_{-\infty}^{\infty} \frac{\partial \Psi^2}{\partial x} dx = \frac{\hbar}{2i} \Psi^2 \Big|_{-\infty}^{\infty} = 0$$

where the last step follows since the wave function must vanish as $x \to \pm\infty$ if it is normalizable.

2.10 Summary

The fundamental equation of motion in nonrelativistic quantum mechanics is the Schrödinger equation:

$$-\frac{\hbar^2}{2m}\frac{\partial^2 \Psi(x,t)}{\partial x^2} + V(x)\Psi(x,t) = i\hbar\frac{\partial \Psi(x,t)}{\partial t} \quad (2.65)$$

where $\Psi(x,t)$ is called the wave function and

$$|\Psi(x,t)|^2 dx = \text{the probability of finding the particle between } x \text{ and } x+dx \text{ at time } t \text{ if a measurement of the particle's position is carried out} \quad (2.66)$$

provided that the wave function is appropriately normalized such that the total probability of finding the particle somewhere is one:

$$\int_{-\infty}^{\infty} |\Psi(x,t)|^2 dx = 1 \quad (2.67)$$

The probability density $|\Psi(x,t)|^2$ obeys a local conservation law of the form

$$\frac{\partial |\Psi|^2}{\partial t} + \frac{\partial j_x}{\partial x} = 0 \quad (2.68)$$

where

$$j_x(x,t) = \frac{\hbar}{2mi}\left(\Psi^*\frac{\partial \Psi}{\partial x} - \Psi\frac{\partial \Psi^*}{\partial x}\right) \quad (2.69)$$

is called the probability current. Integrating over a region of space, say from a to b, yields

$$\frac{d}{dt}\int_a^b |\Psi(x,t)|^2 dx = -j_x(x,t)\Big|_a^b = -j_x(b,t) + j_x(a,t) \quad (2.70)$$

Thus the probability in a region can change as the probability current flows into or out of that region. Since for a normalizable wave function $\Psi \to 0$ as $|x| \to \infty$, we are assured that the probability current vanishes as $|x| \to \infty$ and hence

$$\frac{d}{dt}\int_{-\infty}^{\infty} |\Psi(x,t)|^2 dx = 0 \quad (2.71)$$

Thus if (2.67) holds at one time, it holds at all times and probability is globally conserved as well.

Taking advantage of (2.66), we can calculate average values, or expectation values, of the position through

$$\langle x \rangle = \int_{-\infty}^{\infty} x|\Psi(x,t)|^2 dx = \int_{-\infty}^{\infty} \Psi^*(x,t)\, x\, \Psi(x,t)\, dx \quad (2.72)$$

The uncertainty Δx in the particle's position is determined from

$$(\Delta x)^2 = \langle (x - \langle x \rangle)^2 \rangle = \langle x^2 \rangle - \langle x \rangle^2$$
$$= \int_{-\infty}^{\infty} x^2 |\Psi(x,t)|^2 dx - \left(\int_{-\infty}^{\infty} x|\Psi(x,t)|^2 dx\right)^2 \quad (2.73)$$

Among the consequences of the Schrödinger equation are the results

$$\frac{d\langle x \rangle}{dt} = \frac{\langle p_x \rangle}{m} \quad \text{and} \quad \frac{d\langle p_x \rangle}{dt} = \left\langle -\frac{\partial V}{\partial x}\right\rangle \quad (2.74)$$

that is, the expectation values obey the principles of classical physics provided we make the identification

$$\langle p_x \rangle = \int_{-\infty}^{\infty} \Psi^* \frac{\hbar}{i} \frac{\partial \Psi}{\partial x} dx \qquad (2.75)$$

One of the important consequences of superposing waves with varying wavelengths is the Heisenberg uncertainty principle

$$\Delta x \Delta p_x \geq \frac{\hbar}{2} \qquad (2.76)$$

where

$$\Delta p_x = \left(\langle p_x^2 \rangle - \langle p_x \rangle^2 \right)^{1/2} \qquad (2.77)$$

Lastly, if we restrict our attention to the Schrödinger equation for a free particle, a solution with a particular wavelength (and frequency) is given by

$$\Psi(x, t) = A e^{i(kx - \omega t)} \qquad (2.78)$$

where $p = \hbar k = h/\lambda$ and $E = \hbar \omega = h\nu$. While (2.78) is not a normalizable wave function, a physically acceptable wave function can be constructed in the form of a wave packet, namely a superposition of waves with different wavelengths:

$$\Psi(x, t) = \int_{-\infty}^{\infty} A(k) e^{i(kx - \omega t)} dk \qquad (2.79)$$

We have seen that this packet of waves moves with a speed $v_g = d\omega/dk$ called the group velocity. For the Schrödinger equation for a free particle $\omega = \hbar k^2 / 2m$ and hence $v_g = \hbar k/m = p/m$.

Problems

2.1. What is the speed of helium atoms with a de Broglie wavelength of 1.03 Å?

2.2. In an early version of the double-slit experiment discussed in Section 2.1, Carnal and Mlynek used helium atoms (from a reservoir maintained at 295 K) that exit the nozzle after expansion with a wavelength $\lambda = 0.56$ Å. (a) What is the speed of these helium atoms? (b) The spacing d between the slits is 8 ± 0.6 μm and the distance between the slits and the detection plane is 0.64 m in their experiment. Calculate the spacing between maxima in the detection plane. The observed spacing is 4.5 ± 0.6 μm.

2.3. Our discussion of the helium-atom interferometer focused on the location of the interference fringes. In addition, the interference pattern is modulated by a broader single-slit diffraction envelope. Determine the width of this diffraction envelope, that is, calculate the distance in the detecting plane between the first nodes of the diffraction pattern on either side of the central maximum. How many interference fringes fit in this envelope for a wavelength of 45 pm? Compare your result with the data in Fig. 2.4.

2.4. In the sodium atom interferometry experiment described in Section 2.1, Keith et al. note that the gas leaves the nozzle with a speed $v = 10^3$ m/s, giving a de Broglie wavelength for the sodium atoms of 16 pm (1 pm = 10^{-12} m). Check their calculation.

2.5. In 1999 Anton Zeilinger's research group reported de Broglie wave interference of C_{60} molecules, or buckyballs, the most massive particles for which such interference has been observed. A 100-nm SiN_x grating with slits nominally 50 nm wide was utilized as the beam splitter and interference was observed 1.25 m behind the grating. Figure 2.22a shows the interference pattern produced by C_{60} molecules (open circles) and a fit using diffraction theory (solid line). Figure 2.22b shows the molecular beam profile without the grating in the path of the molecules. The experimentalists note that the most probable velocity of the molecules is 220 m/s. (*a*) What is the corresponding wavelength? The mass of a C_{60} molecule is 1.2×10^{-24} kg. (*b*) Determine the distance between the central maximum and the first maximum in the detection plane. Don't be put off by the lack of precise agreement between your result and the experimental results shown in Fig. 2.22. In the experiment there is a large spread in the velocities of the C_{60} molecules. Moreover, the molecular beam is itself quite broad and the grating used in the experiment had a significant variation in the widths of the slits. When these effects are taken into account, the agreement between theory and experiment is quite good.

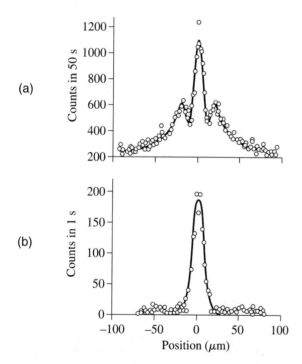

Figure 2.22 Interference of C_{60} molecules, adapted from M. Arndt et al., *Nature* **401**, 680 (1999).

2.6. Zeilinger et al., *Rev. Mod. Phys.* **60**, 1067 (1988) carried out a double-slit experiment with neutrons with a wavelength of 2 nm. (*a*) What was the speed of these neutrons? (*b*) The two slits were 22 μm and 23 μm wide and were separated by a distance $d = 104$ μm. The detection plane was located 5 m downstream from the two slits. Given this information, determine whether or not the single-slit diffraction envelopes from the two slits overlapped in the detection plane, generating an interference pattern.

2.7. What is the wavelength λ for an electron with a kinetic energy of 1 eV, 1 keV, and 1 MeV?

2.8. Through what potential difference must an electron be accelerated so that the electron's wavelength is 1 nm = 10^{-9} m? Repeat the calculation for $\lambda = 1$ pm = 10^{-12} m and $\lambda = 1$ fm = 10^{-15} m.

2.9. Suppose a lecture hall is evacuated and (Schrödinger) cats are projected with speed v at the two doors leading out of the lecture hall in a double-slit experiment. Assume that in order for interference fringes to be observed as the cats pile up against a distant wall the wavelength of each cat must be greater than 1 m. Estimate the maximum speed of each cat. If the distance between the front of the lecture hall to the wall is 30 m, how long will it take to carry out the experiment? Compare this time with the age of the universe, roughly 10^{10} yr.

2.10. Figure 2.23, which is not drawn to scale, shows a sketch of the intensity pattern on a screen located $D = 20$ cm from a single slit of width $a = 0.5$ nm when a monoenergetic beam of electrons is incident upon the slit. If the width of the central maximum is $w = 2$ cm on the screen, what is the kinetic energy (in eV) of the incident electrons?

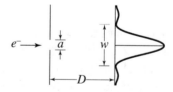

Figure 2.23 Single-slit diffraction.

2.11. An electron microscope takes advantage of the fact that the wavelength of sufficiently energetic electrons is much less than that of visible light. Because of diffraction, the resolving power of any optical instrument is proportional to the wavelength of the radiation used. Light with a wavelength of 0.1 nm is in the X-ray regime, a regime for which it is not possible to focus the radiation adequately to obtain clear images. Electrons, on the other hand, are charged and can be manipulated and focused with electric and magnetic fields. The technology of magnetic lenses does not permit us to reach the diffraction limit, but it is possible to achieve much better resolution and magnification than with an optical microscope. The smallest detail that can be resolved is roughly equal to $0.6\lambda/\sin\theta$, where λ is the wavelength used in forming the image, as shown in Fig. 2.24. Suppose we wish to "see" some of the details of a large molecule, so that a resolution of 0.3 nm is needed. (*a*) If an electron microscope is used, in which θ is typically 10^{-2} radians, what minimum kinetic energy for the electrons is needed? (*b*) If a photon microscope is used, in which θ can be nearly $90°$, what energy for the photons is needed? Which microscope seems more practical?

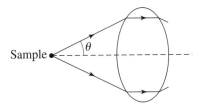

Figure 2.24 Aperture angle of a microscope.

2.12. The Stanford Linear Accelerator Center (SLAC) accelerates electrons to 50 GeV. What is the wavelength of an electron with this energy? How does it compare with the 10^{-15} m size scale of a proton? Is SLAC a good microscope for investigating the internal structure of the proton? *Suggestion*: For a particle with energy $E \gg mc^2$ it is okay to use the extreme relativistic approximation $E = pc$.

2.13. A thermal neutron is a neutron whose most probable kinetic energy is about 0.025 eV. What is the de Broglie wavelength of a thermal neutron? *Note*: Neutrons in a nuclear reactor are thermalized to enhance their probability of fissioning ^{235}U.

2.14. Neutron diffraction is a powerful tool for studying the structure of crystals, especially organic crystals containing hydrogen. Neutrons from a nuclear reactor are sent through a column of graphite, which "thermalizes" the neutrons, that is it slows them down so their average energy is the same as the average energy of the carbon atoms in the graphite. They then bounce off a crystal of known structure, off an unknown crystal, and into a neutron detector, as illustrated in Fig. 2.25. (*a*) If the known crystal has a lattice spacing of 1.5 Å, which angle(s) θ of incidence will give a strong reflection of neutrons with kinetic energy $K = (1/40)$ eV? Assume the lattice planes are parallel to the surface of the crystal. (*b*) If these neutrons reflect strongly off the unknown crystal at angle $\phi = 45°$, what is the lattice spacing of this crystal? (*c*) Explain why neutrons are thermalized to do these experiments.

2.15. Verify that $\Psi(x, t) = A\cos(kx - \omega t)$ and $\Psi(x, t) = A\sin(kx - \omega t)$ are *not* solutions to the Schrödinger equation for a free particle:

$$-\frac{\hbar^2}{2m}\frac{\partial^2 \Psi(x,t)}{\partial x^2} = i\hbar \frac{\partial \Psi(x,t)}{\partial t}$$

2.16. Show that $c_1\psi_1 + c_2\psi_2$ is a solution to the Schrödinger equation (2.6) provided ψ_1 and ψ_2 are solutions and c_1 and c_2 are arbitrary complex numbers.

2.17. Repeat the steps leading to (2.27) but start by superposing two cosines with slightly different wavelengths instead of two sines.

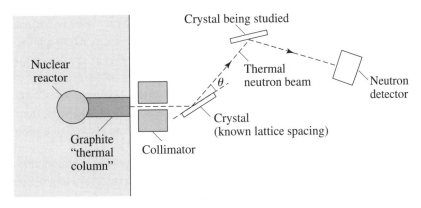

Figure 2.25 Neutron diffraction.

2.18. For the unnormalized wave function shown in Fig. 2.26, where is the particle most likely to be found if a measurement of the position of the particle is carried out? Where is the particle least likely to be found? Is the particle more likely to be found in a region in which $x > 0$ or $x < 0$?

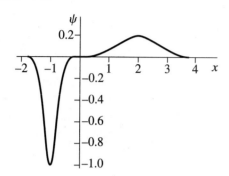

Figure 2.26 An unnormalized wave function.

2.19. For the wave function $\Psi = Ae^{ikx} + Be^{-ikx}$ evaluate the probability current

$$j_x = \frac{\hbar}{2mi}\left(\Psi^*\frac{\partial \Psi}{\partial x} - \Psi\frac{\partial \Psi^*}{\partial x}\right)$$

2.20. (a) Normalize the wave function

$$\psi(x) = \begin{cases} Ae^{\kappa x} & x \leq 0 \\ Ae^{-\kappa x} & x \geq 0 \end{cases}$$

Note: This wave function is the ground-state wave function for the Dirac delta function potential energy well to be discussed in Section 4.4. (b) What is the probability that the particle will be found within $1/\kappa$ of the origin if a measurement of its position is carried out?

2.21. (a) The Heisenberg uncertainty relation can be used to infer an uncertainty in the velocity of a particle, namely

$$\Delta v = \frac{\Delta p}{m} \sim \frac{\hbar}{2m\Delta x_0}$$

where Δx_0 is the uncertainty in the position of the particle at time $t = 0$. The additional spreading in time t in the uncertainty in the position of the particle is $\Delta x = \Delta v\, t$. Show that the time required for the additional position uncertainty due to this spreading to be equal to the uncertainty at $t = 0$ is given by

$$t \sim \frac{2m(\Delta x_0)^2}{\hbar}$$

(b) Evaluate the time t for a marble with an initial position uncertainty of a micron and for an electron with an initial position uncertainty of an angstrom.

2.22. Imagine that you played baseball in a parallel universe in which Planck's constant $h = 0.663$ J·s. What would be the uncertainty in position of a 0.15-kg baseball thrown at 30 m/s with an uncertainty in velocity of 1.0 m/s? What would it be like to catch the ball? Ignore the fact that size of atoms would be different in this parallel universe.

2.23. Lasers can now be designed to emit pulses of light smaller than 30 microns wide in their direction of motion. (a) Estimate the uncertainty in the momentum of a photon in such a pulse. (b) The momentum of a photon is $p = h/\lambda$. Estimate the uncertainty in the wavelength of a photon in the pulse, assuming a nominal wavelength of 800 nm.

2.24. The relationship between the frequency and the wavelength for ocean waves is given by

$$\nu = \left(\frac{g}{2\pi\lambda}\right)^{1/2}$$

where g is the acceleration due to gravity. Show that the group velocity is one half the phase velocity.

2.25. For surface tension waves in shallow water, the relation between the frequency ν and the wavelength λ is given by

$$\nu = \sqrt{\frac{2\pi T}{\rho\lambda^3}}$$

where T is the surface tension and ρ is the density. (a) Determine the phase and group velocities of these waves. (b) The surface tension is defined by the work W necessary to increase the surface area A of the liquid through $dW = T\,dA$. Although the surface tension is often thought of as a force since it arises from the attractive forces of the molecules within the liquid, it has different units. Use dimensional analysis to verify that the frequency, surface tension, density, and wavelength must be related as given in the problem statement.

2.26. When a pebble is tossed into a pond, a circular wave pulse propagates outward from the disturbance. In addition, surface ripples move inward through the circular disturbance. Explain this effect in terms of group and phase velocity, given that the phase velocity of the ripples is given by

$$v_{\rm ph} = \sqrt{\frac{2\pi T}{\lambda \rho}}$$

where T is the surface tension, ρ is the density of the liquid, and λ is the wavelength.

2.27. Prove that the group velocity of a wave packet is equal to the particle's velocity for a relativistic free particle. *Recall:* $E = \hbar \omega = \sqrt{p^2 c^2 + m^2 c^4}$.

2.28. In his doctoral thesis, de Broglie assumed that the photon has an extremely small mass m and travels at speeds less than (although very close to) c. It is possible to place strict upper limits on m. (*a*) Use the relativistic relation $E = \sqrt{p^2 c^2 + m^2 c^4}$ for particles with rest mass m to generate a dispersion relation (the relation between ω and k) that replaces $\omega = kc$, which we derived from Maxwell's equations. (*b*) Evaluate the group velocity v_g in terms of m, c, k, and $\hbar$. Find an approximate expression for $(c - v_g)/c$ in the case $mc^2 \ll pc$. de Broglie assumed that radio waves of wavelength 30 km travel with a speed at least 99% of the speed of visible light. Check de Broglie's calculation of a 10^{-47} kg limit for the photon mass. (*c*) A pretty good limit on the mass of the photon comes from the practically simultaneous arrival at earth of radio waves ($\lambda \approx 1$ m) and visible light from a flare star 20 light-years away. B. Lovell, F. L. Whipple, and L. H. Solomon, *Nature* **202**, 377 (1964), found the velocities are the same to 4 parts in 10^7, that is, $(c - v_g)/c < 4 \times 10^{-7}$ for these 1-m radio waves. What limits does this permit you to place on the photon rest mass?

2.29. (*a*) Show that $\Psi(x, t) = A e^{i(kx - \omega t)}$ is a solution to the Klein–Gordon equation

$$\frac{\partial^2 \Psi(x,t)}{\partial x^2} - \frac{1}{c^2}\frac{\partial^2 \Psi(x,t)}{\partial t^2} - \frac{m^2 c^2}{\hbar^2}\Psi(x,t) = 0$$

if

$$\omega = \sqrt{k^2 c^2 + (m^2 c^4/\hbar^2)}$$

The Klein–Gordon equation is a relativistic quantum mechanical wave equation for a free particle. (*b*) Determine the group velocity of a wave packet made of waves satisfying the Klein–Gordon equation. (*c*) The Klein–Gordon equation describes the motion of particles of mass m. Using your result from (*a*), show that

$$E = \sqrt{p^2 c^2 + m^2 c^4}$$

for these particles. (*d*) Show that the speed v of these particles is equal to the group velocity that you determined in (*b*).

2.30. Suppose the wave function for a particle is given by the symmetric "tent" wave function in Fig. 2.27

$$\Psi(x) = \begin{cases} \sqrt{\frac{12}{a^3}}\left(\frac{a}{2} - |x|\right) & |x| \leq a/2 \\ 0 & |x| > a/2 \end{cases}$$

Show that $\Psi(x)$ is properly normalized. What is $\langle x \rangle$ for the particle? Calculate the uncertainty Δx in the particle's position. *Note:* The wave function is an even function.

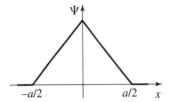

Figure 2.27 A "tent" wave function.

2.31. Normalize the wave function

$$\Psi(x) = \begin{cases} Nx^2(L-x) & 0 < x < L \\ 0 & \text{elsewhere} \end{cases}$$

What is $\langle x \rangle$ for this wave function?

2.32. In Chapter 5 we will see a quick route to deriving the second of Ehrenfest's equations, namely

$$\frac{d\langle p_x \rangle}{dt} = \left\langle -\frac{\partial V}{\partial x} \right\rangle$$

Alternatively, follow the procedure outlined in Section 2.9 by taking the time derivative of

$$\langle p_x \rangle = \int_{-\infty}^{\infty} \Psi^*(x,t)\frac{\hbar}{i}\frac{\partial \Psi(x,t)}{\partial x}\,dx$$

2.33. Evaluate $\langle x \rangle$, $\langle p_x \rangle$, Δx, Δp_x, and $\Delta x \Delta p_x$ for the normalized wave function

$$\Psi(x) = \begin{cases} \sqrt{\frac{2}{L}}\sin\frac{\pi x}{L} & 0 < x < L \\ 0 & \text{elsewhere} \end{cases}$$

In the next chapter we will see that this wave function is the ground-state wave function for a particle confined in the potential energy well

$$V(x) = \begin{cases} 0 & 0 < x < L \\ \infty & \text{elsewhere} \end{cases}$$

2.34. Assume $\psi(x)$ is an arbitrary normalized real function. Calculate $\langle p_x \rangle$ for the wave function $\Psi(x) = e^{ikx}\psi(x)$.

2.35. Determine $\langle x \rangle$ and Δx for the wave function in Problem 2.20.

2.36. You are dropping darts trying to hit a crack in the floor. To aim, you have the most precise equipment possible. Assuming each dart has mass m and is released a distance s above the floor, determine the root mean square distance by which on average you will miss the crack. Obtain a numerical value assuming reasonable values for m and s. Take $g = 9.8$ m/s^2.

CHAPTER 3

The Time-Independent Schrödinger Equation

Within wave mechanics, the Schrödinger equation in one dimension is a partial differential equation in x and t. Through a technique known as separation of variables, this equation can be reduced to two ordinary differential equations, one of which is called the time-independent Schrödinger equation or, as we will see toward the end of this chapter, the energy eigenvalue equation. We will focus in this chapter on solving this equation for a particle confined in an infinite potential well (the particle in a box). This example has much to teach us about quantum mechanics.

3.1 Separation of Variables

In Chapter 2 we solved the Schrödinger equation

$$-\frac{\hbar^2}{2m}\frac{\partial^2 \Psi(x,t)}{\partial x^2} + V(x)\Psi(x,t) = i\hbar\frac{\partial \Psi(x,t)}{\partial t} \qquad (3.1)$$

for a free particle. In this chapter we will begin to examine solutions when the potential energy is nonzero. If we were trying to solve the problem of the mass on a spring in quantum mechanics, for example, we would take $V(x) = Kx^2/2$, since according to Hooke's law $F_x = -dV/dx = -Kx$.[1]

When the potential energy $V(x)$ is independent of t, we can solve the Schrödinger equation by a technique known as separation of variables. We start by writing the wave function $\Psi(x, t)$ as

$$\Psi(x,t) = \psi(x)f(t) \qquad (3.2)$$

that is, as a product of two functions: $\psi(x)$, which is solely a function of x, and $f(t)$, which is solely a function of t. Then

$$\frac{\partial^2 \Psi(x,t)}{\partial x^2} = f(t)\frac{d^2\psi(x)}{dx^2} \qquad (3.3)$$

---

[1] We are calling the force constant K to avoid confusion with the wave vector k.

and

$$\frac{\partial \Psi(x,t)}{\partial t} = \psi(x)\frac{df(t)}{dt} \tag{3.4}$$

where we have replaced the partial derivatives on the left-hand side of (3.3) and (3.4) with ordinary derivatives on the right-hand side since the functions being differentiated are functions of a single variable. Therefore, when we substitute (3.2) into (3.1), we obtain

$$-\frac{\hbar^2}{2m}f(t)\frac{d^2\psi(x)}{dx^2} + V(x)\psi(x)f(t) = i\hbar\psi(x)\frac{df(t)}{dt} \tag{3.5}$$

If we now divide (3.5) by the wave function $\psi(x)f(t)$, we obtain

$$\frac{1}{\psi(x)}\left[-\frac{\hbar^2}{2m}\frac{d^2\psi(x)}{dx^2} + V(x)\psi(x)\right] = \frac{i\hbar}{f(t)}\frac{df(t)}{dt} \tag{3.6}$$

We now have all the x dependence on one side of the equation and all the t dependence on the other, hence the name **separation of variables**. Since x and t are entirely independent variables, which can be varied arbitrarily, the only way that (3.6) can be satisfied is for both sides of this equation to be equal to a constant, which we choose to call E:

$$\frac{1}{\psi(x)}\left[-\frac{\hbar^2}{2m}\frac{d^2\psi(x)}{dx^2} + V(x)\psi(x)\right] = \frac{i\hbar}{f(t)}\frac{df(t)}{dt} = E \tag{3.7}$$

Thus we have reduced the partial differential equation (3.1) to two ordinary differential equations:

$$\frac{df(t)}{dt} = \frac{-iE}{\hbar}f(t) \tag{3.8}$$

and

$$-\frac{\hbar^2}{2m}\frac{d^2\psi(x)}{dx^2} + V(x)\psi(x) = E\psi(x) \tag{3.9}$$

This latter equation is called the **time-independent Schrödinger equation**. This equation plays such a large role in solving (3.1), the time-dependent Schrödinger equation, that sometimes (3.9) is simply referred to as the Schrödinger equation. We will devote the remainder of this chapter and the next one as well to solving this equation, which requires specification of $V(x)$.

The first of these equations [(3.8)], on the other hand, is easy to solve, since the derivative of f is proportional to f itself, which is characteristic of an exponential function. The solution to (3.8) is simply

$$f(t) = f(0)e^{-iEt/\hbar} \tag{3.10}$$

as you can verify by substituting (3.10) into (3.8). We can also express (3.10) in the form

$$f(t) = f(0)e^{-i\omega t} \tag{3.11}$$

where the angular frequency ω of this periodic function is related to E by

$$E = \hbar\omega \tag{3.12}$$

This is the same relationship between the energy E of the particle and the angular frequency ω of the wave function that we saw in Chapter 2 for the free particle [see the discussion leading up to (2.18)]. Also note from (3.9) that the separation constant E must have the dimensions of energy, the same dimensions as $V(x)$. These are good reasons to have called the separation constant E. We generally absorb the constant $f(0)$ into ψ, which will eventually be normalized in any case. Thus a solution to the time-dependent Schrödinger equation is

$$\Psi(x, t) = \psi(x) e^{-iEt/\hbar} \tag{3.13}$$

where $\psi(x)$ is the solution (3.9). The wave function (3.13) is often referred to as a **stationary state** because the probability density

$$|\Psi(x, t)|^2 = \psi^*(x) e^{iEt/\hbar} \psi(x) e^{-iEt/\hbar} = |\psi(x)|^2 \tag{3.14}$$

is independent of time.

3.2 The Particle in a Box

Let's solve the time-independent Schrödinger equation for a specific example to make things more definite. In this section we will take the potential energy $V(x)$ to be

$$V(x) = \begin{cases} 0 & 0 < x < L \\ \infty & \text{elsewhere} \end{cases} \tag{3.15}$$

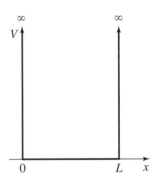

as shown in Fig. 3.1. Classically, since $F_x = -\partial V/\partial x$, there is no force on the particle for $0 < x < L$. At the edges of the well, where the potential energy rises abruptly to infinity, there are infinitely large forces that repel the particle. A particle moving in this potential energy well is often referred to as a particle in a box, although in this chapter it is strictly a one-dimensional box, with motion only along a line [like a bead on a (frictionless) wire that is free to move only in a limited region]. In the next chapter we will consider some more realistic potential energy wells, including a finite square well (for which $V = V_0$ outside the well), but we can learn a lot about quantum mechanics from the simple example of the infinite square well. And in fact we will see later that the infinite square well has a broader degree of applicability than you might at first expect. For now, we will use it as a laboratory in which we will see many of the fundamental features of quantum mechanics.

Figure 3.1 The potential energy for the particle in the box.

We start by solving the time-independent Schrödinger equation for $0 < x < L$, the region to which the particle is strictly confined:

$$-\frac{\hbar^2}{2m} \frac{d^2\psi}{dx^2} = E\psi \qquad 0 < x < L \tag{3.16}$$

since $V = 0$ in this region. It is convenient to introduce the parameter

$$k^2 = \frac{2mE}{\hbar^2} \tag{3.17}$$

in terms of which the equation becomes

$$\frac{d^2\psi}{dx^2} = -k^2\psi \qquad 0 < x < L \tag{3.18}$$

This differential equation, which is similar in form to the classical equation of motion for a mass on a spring, recurs repeatedly throughout our discussion of wave mechanics. Equation (3.18) shows that taking two derivatives of the wave function $\psi(x)$ yields the wave function multiplied by a negative constant (we are assuming that $E > 0$, so $k^2 > 0$ as well). This behavior is characteristic of the trigonometric functions $\sin kx$ and $\cos kx$:

$$\frac{d^2}{dx^2}\sin kx = \frac{d}{dx}k\cos kx = -k^2 \sin kx \tag{3.19}$$

and

$$\frac{d^2}{dx^2}\cos kx = -\frac{d}{dx}k\sin kx = -k^2 \cos kx \tag{3.20}$$

Thus the most general solution to the second-order differential equation (3.18) can be written as

$$\psi(x) = A\sin kx + B\cos kx \qquad 0 < x < L \tag{3.21}$$

where A and B are arbitrary constants. If this were the solution for all space (i.e., the particle were truly a free particle), then k would be equal to $2\pi/\lambda$ where λ is the wavelength. This is the rationale for setting $2mE/\hbar^2 = k^2$ in (3.18). Of course, for the particle confined in the infinite well, $\psi(x) = 0$ outside the region $0 < x < L$, since the potential energy rises abruptly to infinity there.

We now need to apply the appropriate boundary conditions to our solutions to the differential equation (3.16). Since the time-independent Schrödinger equation

$$-\frac{\hbar^2}{2m}\frac{d^2\psi(x)}{dx^2} + V(x)\psi(x) = E\psi(x) \tag{3.22}$$

is a second-order differential equation, the wave function $\psi(x)$ must be continuous everywhere. Physically, this means that there is no point in space where the likelihood of finding the particle is discontinuous, which is reassuring.[2] In particular, we must guarantee that $\psi(0) = \psi(L) = 0$:

$$\psi(0) = A\sin 0 + B\cos 0 = B = 0 \tag{3.23}$$

and

$$\psi(L) = A\sin kL = 0 \tag{3.24}$$

One strategy, albeit not an interesting one, is to set $A = 0$. But then since $\psi = 0$ everywhere, there is no particle in the box. Alternatively, we can set

$$kL = n\pi \qquad n = 1, 2, 3, \ldots \tag{3.25}$$

Notice that we must exclude $n = 0$, since in this case $k = 0$, which also implies $\psi = 0$ everywhere. We also exclude the solutions with negative n. Since $\sin(-kx) = -\sin kx$, a negative n just changes the overall phase of the wave function and thus does not lead to a linearly independent solution.

The allowed values of k can thus be labeled by the integer n:

$$k_n = \frac{n\pi}{L} \tag{3.26}$$

---

[2] In general, the first derivative must be continuous as well. The infinite square well is an exception, as the wave functions shown in Fig. 3.3 illustrate. We will discuss the boundary conditions in more detail in Chapter 4.

From (3.17), the energies are given by

$$E_n = \frac{\hbar^2 k_n^2}{2m} = \frac{n^2 \hbar^2 \pi^2}{2mL^2} \quad n = 1, 2, 3, \ldots \quad (3.27)$$

and the corresponding wave functions are

$$\psi_n(x) = A_n \sin \frac{n\pi x}{L} \quad 0 < x < L \quad (3.28)$$

The overall amplitude A_n is determined not from the boundary conditions but rather from the normalization requirement:

$$\begin{aligned}
\int_{-\infty}^{\infty} |\psi_n(x)|^2 dx &= \int_0^L |A_n|^2 \sin^2 \frac{n\pi x}{L} dx \\
&= |A_n|^2 \int_0^L \frac{1}{2}\left(1 - \cos \frac{2n\pi x}{L}\right) dx \\
&= \frac{|A_n|^2}{2} \left[x - \frac{L}{2n\pi} \sin \frac{2n\pi x}{L}\right]\bigg|_0^L \\
&= |A_n|^2 \frac{L}{2} = 1 \quad (3.29)
\end{aligned}$$

We take $A_n = \sqrt{2/L}$ (choosing the phase of the amplitude so that the amplitude is real, at least at $t = 0$) and therefore

$$\psi_n(x) = \begin{cases} \sqrt{\frac{2}{L}} \sin \frac{n\pi x}{L} & 0 < x < L \\ 0 & \text{elsewhere} \end{cases} \quad n = 1, 2, 3, \ldots \quad (3.30)$$

Discussion

A number of comments about the energies E_n and wave functions ψ_n are in order:

1. Classically, a particle confined in the box bounces back and forth between the walls without any constraint on its energy E. That is, you could imagine putting a particle in the box with any energy that you want. However, in quantum mechanics only certain energies of the particle are permitted. Notice how the boundary condition that the wave function ψ vanish at $x = 0$ and $x = L$ has led to the discrete values for the energy given in (3.27). We say the energies are **quantized**. The key role that the boundary conditions play in determining the allowed energies of the system will be evident in the one-dimensional potentials that we will examine in the next chapter. Figure 3.2 gives an energy-level diagram for the four lowest energies.

2. Figure 3.3a shows the wave functions for $n = 1$ through $n = 3$. These functions are reminiscent of the modes of vibration of a (violin) string. As for the string, the boundary conditions that there are nodes at each end for the particle in a box lead to certain allowed k values that result from the requirement that an integral number of half-wavelengths fit between the end points. But there are significant differences between the vibration of the string and a particle in a box. Only for the particle is it appropriate to identify the energy with the frequency ($E = \hbar\omega$). For the waves on a string, the energy of each mode increases with

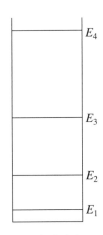

Figure 3.2 The infinite potential energy well with the lowest four allowed energies.

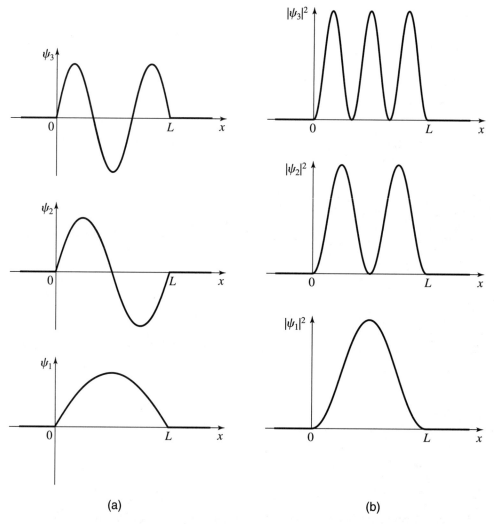

Figure 3.3 (a) The wave functions and (b) the probability densities for the three lowest energy states of the particle in a box.

the amplitude of the mode. For the particle in the box, on the other hand, the energy increases as the number of nodes of the wave function increases. The more the wave function oscillates—the higher its spatial frequency—the more energy the particle has. The energy of the particle is entirely independent of the amplitude of the wave function, which is fixed by normalization.

3. Interestingly, there is no $E = 0$ solution. You are encouraged to work out Problem 3.1 to confirm this. Classically, we would have found it easy to imagine putting a particle in a box with zero energy. Just put it at rest somewhere in the box. But that turns out to be impossible, since even the ground-state wave function has to oscillate so that it will have a node at each wall, leading to a nonzero value of the lowest energy (often called the zero-point energy),

$$E_1 = \frac{\hbar^2 \pi^2}{2mL^2} \quad (3.31)$$

We can also argue for this result from the uncertainty principle. If the particle were at rest, its momentum would be zero, with no uncertainty. But the

Heisenberg uncertainty principle implies that if $\Delta p_x = 0$, then Δx should be infinite. But Δx cannot be larger than L, since we know the particle is somewhere in the box. Thus Δp_x must be nonzero as well.[3] The smaller the box, the smaller the value of Δx, and hence the larger the value of Δp_x and the larger the energy, as (3.31) shows.

4. Figure 3.3b shows the probability density functions for the first three energy levels. Look, in particular, at the first excited state (at $|\psi_2(x)|^2$), the one corresponding to energy E_2. The probability density has two lumps with a node in the middle. If your picture of what is going on is that the particle is bouncing back and forth in the well with energy E_2 and $|\psi_2(x)|^2 dx$ is the probability of finding the particle between x and $x + dx$, then you should be troubled by the probability density for the first excited state. For if the particle bounces from one side of the well to the other, it must pass through the middle of the well. But there is negligible probability of finding the particle in the immediate neighborhood of $x = L/2$.

So what is wrong with this picture? It is based on a classical notion that a particle follows a definite path. In fact, remember that the full wave function for the particle in a state with energy E_2 is

$$\Psi_2(x, t) = \psi_2(x) e^{-iE_2 t/\hbar} \tag{3.32}$$

which is a stationary state. That is,

$$|\Psi_2(x, t)|^2 = |\psi_2(x) e^{-iE_2 t/\hbar}|^2 = |\psi_2(x)|^2 \tag{3.33}$$

is independent of time. Thus the particle isn't moving; it's not bouncing back and forth as your classical experience says it must. Rather, the particle simply does not have a definite position. For the wave function ψ_2, the particle has a split personality, an amplitude to be in the left half of the well and an amplitude to be in the right half. It just doesn't have an amplitude to be exactly in the middle.

Time Dependence

The notion that the allowed energy states are stationary states is not the whole story, fortunately, for otherwise no movement—no dynamics—would be possible. To illustrate how time dependence arises, let's consider the wave function

$$\Psi(x) = \frac{1}{\sqrt{2}} \psi_1 + \frac{1}{\sqrt{2}} \psi_2 \tag{3.34}$$

namely a superposition of the ground-state and first-excited-state wave functions. The wave function $\Psi(x)$ is properly normalized, although we will hold off demonstrating this

---

[3] In fact, a word of caution is in order here. You may tend to think as you examine the wave functions in Fig. 3.3a that in each one the particle has a definite wavelength (or at least half-wavelength). However, if the particle had a definite wavelength, its wave function would have to extend beyond the region $0 < x < L$. In essence, the wave functions for the particle in a box are wave trains that vanish outside the region $0 < x < L$. They have a finite Δx as well as a finite Δp_x. See Problem 2.33, for example.

until the next section. Time evolution is generated by including the appropriate factor of $e^{-iE_n t/\hbar}$ for each ψ_n, that is,

$$\Psi(x, t) = \frac{1}{\sqrt{2}} \Psi_1(x, t) + \frac{1}{\sqrt{2}} \Psi_2(x, t)$$

$$= \frac{e^{-iE_1 t/\hbar}}{\sqrt{2}} \psi_1 + \frac{e^{-iE_2 t/\hbar}}{\sqrt{2}} \psi_2$$

$$= e^{-iE_1 t/\hbar} \left[\frac{1}{\sqrt{2}} \psi_1 + \frac{e^{-i(E_2 - E_1)t/\hbar}}{\sqrt{2}} \psi_2 \right] \quad (3.35)$$

where in the last step we pulled the phase factor $e^{-iE_1 t/\hbar}$ out in front as an overall phase factor. The probability density is therefore

$$|\Psi(x, t)|^2 = \Psi^* \Psi$$

$$= e^{iE_1 t/\hbar} \left[\frac{1}{\sqrt{2}} \psi_1^* + \frac{e^{i(E_2 - E_1)t/\hbar}}{\sqrt{2}} \psi_2^* \right] e^{-iE_1 t/\hbar} \left[\frac{1}{\sqrt{2}} \psi_1 + \frac{e^{-i(E_2 - E_1)t/\hbar}}{\sqrt{2}} \psi_2 \right]$$

$$= \frac{1}{2} |\psi_1|^2 + \frac{1}{2} |\psi_2|^2 + \frac{1}{2} \psi_2^* \psi_1 e^{i(E_2 - E_1)t/\hbar} + \frac{1}{2} \psi_1^* \psi_2 e^{-i(E_2 - E_1)t/\hbar}$$

$$= \frac{1}{2} \psi_1^2 + \frac{1}{2} \psi_2^2 + \psi_2 \psi_1 \cos(E_2 - E_1)t/\hbar \quad (3.36)$$

where in the last step we have taken advantage of the fact ψ_1 and ψ_2 are real functions [look back at (3.30)]. Because of this, we could have written ψ_1^* as ψ_1, for example, in the second line, but it is good to get in the habit of writing the complex conjugate since in general the wave functions are complex functions. You can see in moving from the second line to the third line how the overall phase factor $e^{-iE_1 t/\hbar}$ cancels out. The relative phase factor $e^{-i(E_2 - E_1)t/\hbar}$ does *not* cancel out, however. It appears in the cross terms when you carry out the multiplication.[4] From the last line of (3.36) we see that probability density is periodic in time with period $T = h/(E_2 - E_1)$. Figure 3.4 shows plots of $|\Psi(x, t)|^2$ at two times, $t = 0$ and a half period later. The probability of finding the particle is oscillating back and forth in the well.

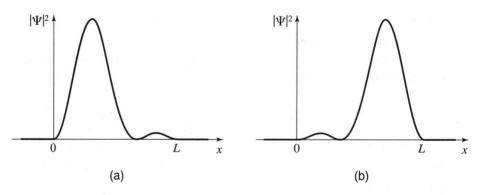

Figure 3.4 (a) The probability density $\Psi(x, t)^* \Psi(x, t)$ at $t = 0$ and (b) a half a period later at $t = \pi \hbar/(E_2 - E_1)$.

---

[4] Recall it was the cross terms that led to interference effects when we added the amplitudes for a single photon to take both paths in the Mach–Zehnder interferometer in Section 1.5.

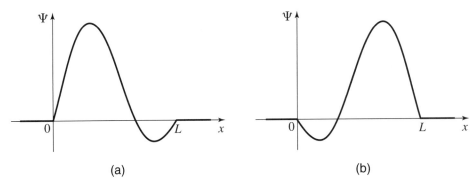

Figure 3.5 (a) The wave function (3.35) at $t=0$ and (b) a half a period later at $t = \pi\hbar/(E_2 - E_1)$, excluding the overall phase factor $e^{-iE_1 t/\hbar}$ in the last line of (3.35).

Figure 3.5 shows plots of $\Psi(x, t)$ at $t=0$ and when $(E_2 - E_1)t/\hbar = \pi$, ignoring the overall phase factor $e^{-iE_1 t/\hbar}$ in this latter case. At $t=0$ you can see how the wave functions add constructively for $x < L/2$, where ψ_1 and ψ_2 are both positive, while they interfere destructively for $x > L/2$, where ψ_1 is positive and ψ_2 is negative. On the other hand,

$$\Psi[x, t = \pi\hbar/(E_2 - E_1)] = e^{-iE_1 t/\hbar}\left[\frac{1}{\sqrt{2}}\psi_1 - \frac{1}{\sqrt{2}}\psi_2\right] \quad (3.37)$$

and here the wave functions interfere destructively for $x < L/2$ and add constructively for $x > L/2$. Thus the time dependence of the wave function arises from the interference of the wave functions $\Psi_1(x, t)$ and $\Psi_2(x, t)$ in the superposition.

EXAMPLE 3.1 What is the energy scale for electrons confined to a box whose size is 1 Å, a typical size for an atom? Repeat the calculation for protons or neutrons in a box whose size is 5 fm, the size of a medium-sized nucleus. Note: 1 Å = 10^{-10} m and 1 fm = 10^{-15} m.

SOLUTION The ground-state energy sets the energy scale for a particle in a box. For electrons in an atom ($m = 9.1 \times 10^{-31}$ kg, $L = 10^{-10}$ m)

$$E_1 = \frac{\hbar^2 \pi^2}{2mL^2} = 6.0 \times 10^{-18} \text{ J} = 38 \text{ eV}$$

while for protons or neutrons in a nucleus ($m = 1.67 \times 10^{-27}$ kg, $L = 5 \times 10^{-15}$ m)

$$E_1 = \frac{\hbar^2 \pi^2}{2mL^2} = 1.3 \times 10^{-12} \text{ J} = 8.2 \times 10^6 \text{ eV} = 8.2 \text{ MeV}$$

Notice how quantum mechanics determines the energy scale of the world we live in once the size of the box as well as the mass of the particle confined in the box are specified. Typically, the energy scale of atomic physics is electron volts while the energy scale of nuclear physics is millions of electron volts. In Chapter 6 we will see for the hydrogen atom how quantum mechanics determines the size of the box as well.

> **EXAMPLE 3.2** Suppose that the difference in energy between two allowed energy states in an atom is 2 eV. What is the wavelength of a photon that would be emitted (or absorbed) in a transition between these levels? In what portion of the electromagnetic spectrum does this photon reside?
>
> **SOLUTION**
>
> $$h\nu = hc/\lambda = 2 \text{ eV}$$
>
> Therefore $\lambda = 620 \times 10^{-9}$ m $= 620$ nm, which is orange light. Thus in a helium–neon laser, which emits light with wavelength 633 nm, the spacing between two energy levels in neon, which is the atom emitting the light, must be almost 2 eV.

3.3 Statistical Interpretation of Quantum Mechanics

The wave function (3.34) is a specific example of the more general superposition

$$\Psi(x) = c_1 \psi_1(x) + c_2 \psi_2(x) \tag{3.38}$$

where c_1 and c_2 are complex numbers. The full time-dependent wave function $\Psi(x, t)$ in (3.35) can also be cast in this form with the complex numbers that are time dependent:

$$c_1(t) = \frac{1}{\sqrt{2}} e^{-iE_1 t/\hbar} \quad \text{and} \quad c_2(t) = \frac{1}{\sqrt{2}} e^{-iE_2 t/\hbar} \tag{3.39}$$

An even more general superposition, involving in principle all of the ψ_n, can be written as

$$\Psi = \sum_{n=1}^{\infty} c_n \psi_n(x) \tag{3.40}$$

There is a very powerful analogy between (3.40) and the expansion of a vector **V** in terms of the unit vectors **i**, **j**, and **k**:

$$\mathbf{V} = V_x \mathbf{i} + V_y \mathbf{j} + V_z \mathbf{k} \tag{3.41}$$

Like the ψ_n, which have been normalized so that

$$\int_{-\infty}^{\infty} \psi_n^*(x) \psi_n(x) \, dx = 1 \tag{3.42}$$

the vectors **i**, **j**, and **k** are unit vectors that satisfy the condition

$$\mathbf{i} \cdot \mathbf{i} = \mathbf{j} \cdot \mathbf{j} = \mathbf{k} \cdot \mathbf{k} = 1 \tag{3.43}$$

Also, in the same way that **i**, **j**, and **k** form an orthogonal set in that

$$\mathbf{i} \cdot \mathbf{j} = \mathbf{i} \cdot \mathbf{k} = \mathbf{j} \cdot \mathbf{k} = 0 \tag{3.44}$$

there is an orthogonality condition that holds for the ψ_n, namely

$$\int_{-\infty}^{\infty} \psi_m^*(x) \psi_n(x) \, dx = 0 \qquad \text{for } m \neq n \tag{3.45}$$

This orthogonality condition is easily shown to be satisfied for the ψ_n that arise from solving the infinite square well:

$$\int_{-\infty}^{\infty} \psi_m^*(x)\psi_n(x)\,dx = \frac{2}{L}\int_0^L \sin\frac{m\pi x}{L}\sin\frac{n\pi x}{L}\,dx$$

$$= \frac{1}{L}\int_0^L \left[\cos\frac{(m-n)\pi x}{L} - \cos\frac{(m+n)\pi x}{L}\right]dx$$

$$= \left[\frac{1}{(m-n)\pi}\sin\frac{(m-n)\pi x}{L} - \frac{1}{(m+n)\pi}\sin\frac{(m+n)\pi x}{L}\right]\Bigg|_0^L$$

$$= \frac{\sin(m-n)\pi}{(m-n)\pi} - \frac{\sin(m+n)\pi}{(m+n)\pi} = 0 \qquad \text{for } m \neq n \quad (3.46)$$

where in the second line we have made use of the trigonometric identity

$$\frac{1}{2}[\cos(\alpha - \beta) - \cos(\alpha + \beta)] = \sin\alpha\sin\beta \tag{3.47}$$

A convenient way to capture all this information contained in (3.42) and (3.45) is with the aid of the Kronecker delta, which is defined by

$$\delta_{mn} = \begin{cases} 1 & m = n \\ 0 & m \neq n \end{cases} \tag{3.48}$$

Thus

$$\int_{-\infty}^{\infty} \psi_m^*(x)\psi_n(x)\,dx = \delta_{mn} \tag{3.49}$$

We say the wave functions ψ_n form an **orthonormal set**.

The ψ_n share another important property with the vectors **i**, **j**, and **k**, namely **completeness**. In the same way that **i**, **j**, and **k** span the space of vectors in that any ordinary vector (the force **F**, the acceleration **a**, the momentum **p**, etc.) can be expressed as a linear combination of them, the ψ_n form a complete set as well. Any wave function Ψ can be written as

$$\Psi = \sum_{n=1}^{\infty} c_n \psi_n(x) \tag{3.50}$$

Proving completeness is not as straightforward as proving orthogonality. We will assume completeness holds since, as we will see, it is an essential component of quantum mechanics. For the ψ_n for the infinite square well, the superposition takes the form

$$\Psi(x) = \sum_{n=1}^{\infty} c_n \sqrt{\frac{2}{L}}\sin\frac{n\pi x}{L} \qquad 0 < x < L \tag{3.51}$$

which you may recognize as a standard Fourier series.

Given the vector **V** and the unit vectors **i**, **j**, and **k**, we can determine the coefficients V_x, V_y, and V_z in (3.41) by taking the dot product of **V** with each unit vector. For example,

$$\mathbf{i} \cdot \mathbf{V} = V_x \mathbf{i} \cdot \mathbf{i} + V_y \mathbf{i} \cdot \mathbf{j} + V_z \mathbf{i} \cdot \mathbf{k} = V_x \tag{3.52}$$

where we have taken advantage of (3.43) and (3.44) in evaluating the dot products between the unit vectors. In the same way, we see that

$$\mathbf{j}\cdot\mathbf{V} = V_y \quad \text{and} \quad \mathbf{k}\cdot\mathbf{V} = V_z \tag{3.53}$$

Similarly, we can take advantage of (3.49) to determine the c_n in the expansion

$$\Psi(x) = \sum_{n=1}^{\infty} c_n \psi_n(x) \tag{3.54}$$

Simply multiply $\Psi(x)$ by one of the ψ_n^* and integrate, for

$$\int_{-\infty}^{\infty} \psi_n^*(x)\Psi(x)\,dx = \int_{-\infty}^{\infty} \psi_n^*(x) \sum_{m=1}^{\infty} c_m \psi_m(x)\,dx$$

$$= \sum_{m=1}^{\infty} c_m \int_{-\infty}^{\infty} \psi_n^*(x)\psi_m(x)\,dx$$

$$= \sum_{m=1}^{\infty} c_m \delta_{nm}$$

$$= c_n \tag{3.55}$$

Note that in obtaining the result (3.55) we have been careful to label the dummy index in the sum in this equation m so as to distinguish the dummy index from the particular value n for which we are multiplying by ψ_n^*.

As an interesting illustration of the completeness of the ψ_n, consider the wave function

$$\Psi(x) = \begin{cases} \sqrt{\frac{2}{L}} \sin \frac{\pi x}{L} & 0 < x < L \\ 0 & \text{elsewhere} \end{cases} \tag{3.56}$$

that is, a sinusoidal bump in the left side of a well of width $2L$. Figure 3.6 shows how the right-hand side of (3.54) approaches $\Psi(x)$ with an increasing number of terms in the superposition.[5]

The c_n play a very important role in quantum mechanics. If Ψ as well as the ψ_n are normalized, then

$$1 = \int_{-\infty}^{\infty} |\Psi|^2 dx$$

$$= \int_{-\infty}^{\infty} \left(\sum_{m=1}^{\infty} c_m^* \psi_m^* \right) \left(\sum_{n=1}^{\infty} c_n \psi_n \right) dx$$

$$= \sum_{m=1}^{\infty} c_m^* \sum_{n=1}^{\infty} c_n \int_{-\infty}^{\infty} \psi_m^* \psi_n \, dx$$

$$= \sum_{m=1}^{\infty} c_m^* \sum_{n=1}^{\infty} c_n \delta_{mn}$$

$$= \sum_{n=1}^{\infty} c_n^* c_n = \sum_{n=1}^{\infty} |c_n|^2 \tag{3.57}$$

---

[5]Problem 3.12 gives a mechanism for generating the wave function $\Psi(x)$ and Problem 3.13 gives the values of c_n for $n \neq 2$. As a warm up, try your hand at calculating c_2.

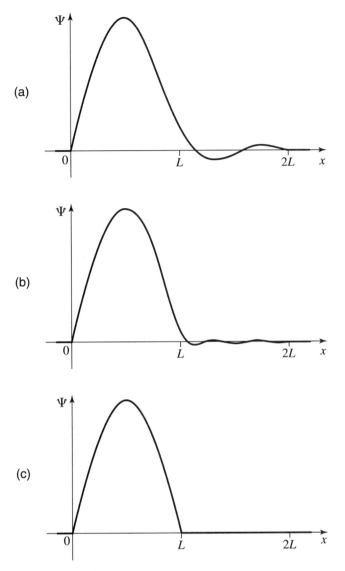

Figure 3.6 The sum $\Psi = \sum_{n=1}^{N} c_n \psi_n$ with (a) $N = 4$, (b) $N = 10$, and (c) $N = 50$ for the wave function (3.56).

where we have labeled by m and n the separate dummy indices that appear in the two independent sums that occur in the second line of this equation. Thus we see that

$$\sum_{n=1}^{\infty} |c_n|^2 = 1 \tag{3.58}$$

The natural (and correct) interpretation of this result is that

$$|c_n|^2 = P_n \tag{3.59}$$

is the probability of obtaining E_n if a measurement of the energy of a particle with wave function Ψ is carried out. Then (3.58) assures us that

$$\sum_{n=1}^{\infty} P_n = 1 \tag{3.60}$$

The only result of a measurement of the energy is one of the allowed energies E_n. We can take advantage of this result to calculate, for example, the expectation value of the energy for a particle in the state Ψ:

$$\langle E \rangle = \sum_{n=1}^{\infty} |c_n|^2 E_n \tag{3.61}$$

While you may be getting used to the idea that a particle doesn't have a definite position, you may find it harder to swallow the idea that it needn't have a definite energy either. If this is the case, you will just need to get over it.

EXAMPLE 3.3 Show that the wave function

$$\Psi(x,t) = \frac{e^{-iE_1 t/\hbar}}{\sqrt{2}} \psi_1(x) + \frac{e^{-iE_2 t/\hbar}}{\sqrt{2}} \psi_2(x)$$

is normalized. Calculate $\langle E \rangle$ and ΔE for this state.

SOLUTION A measurement of the energy yields E_1 with probability

$$P_1 = |c_1|^2 = \left| \frac{e^{-iE_1 t/\hbar}}{\sqrt{2}} \right|^2 = \frac{1}{2}$$

or E_2 with probability

$$P_2 = |c_2|^2 = \left| \frac{e^{-iE_2 t/\hbar}}{\sqrt{2}} \right|^2 = \frac{1}{2}$$

Since $\sum_{n=1}^{\infty} |c_n|^2 = 1$, $\Psi(x,t)$ is properly normalized. The expectation value of the energy for the state $\Psi(x,t)$ is

$$\langle E \rangle = \sum_{n=1}^{\infty} P_n E_n = \frac{1}{2} E_1 + \frac{1}{2} E_2 = \frac{1}{2}(E_1 + E_2)$$

independent of time. Also

$$\langle E^2 \rangle = \sum_{n=1}^{\infty} P_n E_n^2 = \frac{1}{2} E_1^2 + \frac{1}{2} E_2^2 = \frac{1}{2}(E_1^2 + E_2^2)$$

Thus the uncertainty in energy for a particle in state $\Psi(x,t)$ is

$$\Delta E = \left(\langle E^2 \rangle - \langle E \rangle^2 \right)^{1/2}$$

$$= \left[\frac{1}{2}(E_1^2 + E_2^2) - \frac{1}{4}(E_1 + E_2)^2 \right]^{1/2}$$

$$= \left[\frac{1}{4}(E_1^2 + E_2^2 - 2E_1 E_2) \right]^{1/2} = \frac{1}{2}(E_2 - E_1)$$

As a check, note that $\Delta E = 0$ if we set $E_2 = E_1$, for in this case there is no uncertainty in the energy of the particle.

EXAMPLE 3.4 The wave function for a particle in the box is given by

$$\Psi(x) = \begin{cases} \sqrt{\frac{30}{L^5}}x(L-x) & 0 < x < L \\ 0 & \text{elsewhere} \end{cases}$$

Example 2.5 shows that Ψ is normalized. A measurement of the energy of the particle is carried out. What is the probability of obtaining E_1, the ground-state energy for the particle in a box?

SOLUTION To determine the probability, evaluate c_1:

$$c_1 = \int_{-\infty}^{\infty} \psi_1^*(x)\Psi(x)\,dx = \int_0^L \sqrt{\frac{2}{L}}\sin\frac{\pi x}{L}\sqrt{\frac{30}{L^5}}x(L-x)\,dx$$

Making the change of variables $\pi x/L = y$,

$$c_1 = \frac{\sqrt{60}}{\pi^2}\int_0^\pi \sin y\left(y - \frac{y^2}{\pi}\right)dy$$

$$= \frac{\sqrt{60}}{\pi^2}\left(-y\cos y + \frac{y^2-2}{\pi}\cos y\right)\bigg|_0^\pi$$

$$= \frac{4\sqrt{60}}{\pi^3}$$

Thus the probability of obtaining E_1 is $|c_1|^2 = 960/\pi^6 = 0.9986$. The similarity between the ground-state wave function ψ_1 for the particle in the box and the parabolic wave function given in this example (see Fig. 2.11) is reflected in how close this probability is to one. Put another way, the probability of obtaining an energy other than the ground-state energy is $1 - 0.9986 = 0.0014$.

Caution: When c_1 is real and positive, it is easy to forget that the probability is given by $c_1^* c_1$ and not by c_1 itself. It may help to remember that $c_1(t) = c_1(0)e^{-iE_1 t/\hbar}$, so that even if $c_1(0)$ is real and positive, $c_1(t)$ is not.

3.4 The Energy Operator: Eigenvalues and Eigenfunctions

The time-independent Schrödinger equation

$$-\frac{\hbar^2}{2m}\frac{d^2\psi(x)}{dx^2} + V(x)\psi(x) = E\psi(x) \tag{3.62}$$

is often referred to as the energy eigenvalue equation. An eigenvalue equation is one in which an operator acting on a function yields a constant multiplying the function, that is, an equation of the form

$$A_{\text{op}}\psi_a = a\psi_a \tag{3.63}$$

where the constant a is called the **eigenvalue** and ψ_a is called the **eigenfunction** of the operator A_{op}. The subscript a on the wave function ψ_a indicates that it is the eigenfunction corresponding to the eigenvalue a. In general, a given operator may have (infinitely) many eigenvalues and corresponding eigenfunctions.

So what's an operator? Let's take some examples to get our bearings. In wave mechanics the position operator is

$$x_{\text{op}} = x \tag{3.64}$$

so that

$$x_{\text{op}} \psi(x) = x\, \psi(x) \tag{3.65}$$

The position operator just multiplies the wave function $\psi(x)$ by x, leading to a different function, namely $x\, \psi(x)$. As we will argue in the next few paragraphs, the momentum operator is

$$p_{x_{\text{op}}} = \frac{\hbar}{i} \frac{\partial}{\partial x} \tag{3.66}$$

Thus

$$p_{x_{\text{op}}} \psi(x) = \frac{\hbar}{i} \frac{\partial \psi(x)}{\partial x} \tag{3.67}$$

In general, this too is a different function, namely the derivative of the wave function $\psi(x)$ multiplied by the constant $\hbar/i$. Typically, in one-dimensional wave mechanics the operators are mixtures of derivatives with respect to x, multiplication by functions of x, such as $V(x)$, and multiplication by constants.

Now let's look at a momentum eigenfunction. Recall that $p = h/\lambda$. Thus a wave function with a particular momentum p is one with a particular wavelength λ. If we apply the momentum operator (3.66) to the wave function Ae^{ikx} (remember $k = 2\pi/\lambda$), where A is a constant, we obtain

$$p_{x_{\text{op}}} A e^{ikx} = \frac{\hbar}{i} \frac{\partial}{\partial x} A e^{ikx} = \hbar k A e^{ikx} \tag{3.68}$$

Notice that the operator moves through the constant A and differentiates e^{ikx}. The net effect in the end is to return the function Ae^{ikx} multiplied by the constant $\hbar k$, or h/λ, the momentum eigenvalue. Thus we can characterize Ae^{ikx} as a momentum eigenfunction with eigenvalue $p = \hbar k$. We express the momentum eigenvalue equation in the form

$$p_{x_{\text{op}}} \psi_p(x) = p\, \psi_p(x) \tag{3.69}$$

where

$$\psi_p(x) = A e^{ipx/\hbar} \tag{3.70}$$

With this background, we can now see why the time-independent Schrödinger equation is the energy eigenvalue equation. In nonrelativistic quantum mechanics the energy operator is the sum of the kinetic energy and potential energy operators:

$$E_{\text{op}} = \frac{(p_{x_{\text{op}}})^2}{2m} + V(x_{\text{op}}) \tag{3.71}$$

Using the forms (3.64) and (3.66) for the position and momentum operators, we see that the energy operator, which is generally referred to by the symbol H (for Hamiltonian), is given by

$$H \equiv E_{\text{op}} = \frac{1}{2m} \left(\frac{\hbar}{i} \frac{\partial}{\partial x} \right) \left(\frac{\hbar}{i} \frac{\partial}{\partial x} \right) + V(x) = -\frac{\hbar^2}{2m} \frac{\partial^2}{\partial x^2} + V(x) \tag{3.72}$$

Thus the time-independent Schrödinger equation is really just the eigenvalue equation

$$H\psi = E\psi \quad (3.73)$$

To make the correspondence between (3.73) and (3.63) more complete, we could put a subscript E on the the energy eigenfunction corresponding to the eigenvalue E, that is,

$$H\psi_E = E\psi_E \quad (3.74)$$

In fact we have done this already in dealing with the particle in a box in the previous section. There we saw that the discrete energies (eigenvalues) and corresponding wave functions (eigenfunctions) could be labeled by an integer n. See (3.27) and (3.30), respectively. Using this notation, the time-independent Schrödinger equation for the particle in a box becomes

$$H\psi_n = E_n\psi_n \quad (3.75)$$

Our identification of observables—things that are measurable such as energy, momentum, and position—with operators gives us an alternative way to calculate expectation values. Let's return to the particle in the box and take the normalized wave function to be a superposition of two energy eigenfunctions:

$$\Psi = c_1\psi_1 + c_2\psi_2 \quad (3.76)$$

In the previous section we saw that [see (3.61)]

$$\langle E \rangle = |c_1|^2 E_1 + |c_2|^2 E_2 \quad (3.77)$$

But now we can show that there is an alternative way to obtain $\langle E \rangle$. Note that the operator H is a linear operator, which means that

$$H\Psi = c_1 H\psi_1 + c_2 H\psi_2 \quad (3.78)$$

Since ψ_1 and ψ_2 are eigenfunctions of H with corresponding eigenvalues E_1 and E_2, respectively, we can write

$$H\Psi = c_1 H\psi_1 + c_2 H\psi_2 = c_1 E_1 \psi_1 + c_2 E_2 \psi_2 \quad (3.79)$$

Therefore

$$\begin{aligned}
\int_{-\infty}^{\infty} \Psi^* H \Psi \, dx &= \int_{-\infty}^{\infty} (c_1\psi_1 + c_2\psi_2)^* (c_1 E_1 \psi_1 + c_2 E_2 \psi_2) \, dx \\
&= |c_1|^2 E_1 \int_{-\infty}^{\infty} |\psi_1|^2 dx + |c_2|^2 E_2 \int_{-\infty}^{\infty} |\psi_2|^2 dx \\
&\quad + c_1^* E_2 c_2 \int_{-\infty}^{\infty} \psi_1^* \psi_2 \, dx + c_2^* E_1 c_1 \int_{-\infty}^{\infty} \psi_2^* \psi_1 \, dx \\
&= |c_1|^2 E_1 + |c_2|^2 E_2 \quad (3.80)
\end{aligned}$$

where in going from the second to the last line we have taken advantage of the orthonormality of the eigenfunctions [see (3.49)]. But the last line is just the expectation value of the energy for a particle in the state Ψ since $|c_1|^2$ and $|c_2|^2$ are the probabilities of obtaining E_1 and E_2, respectively.[6] Thus we see that

$$\langle E \rangle = \int_{-\infty}^{\infty} \Psi^* H \Psi \, dx \qquad (3.81)$$

Example 3.5 gives a nice illustration of (3.81).

As we will discuss more generally in Chapter 5, this operator approach for determining expectation values in which the operator is sandwiched between Ψ^* and Ψ and then the resulting function is integrated over all space works for observables other than the energy. We have seen two additional examples so far. The expectation value of the position is given by

$$\langle x \rangle = \int_{-\infty}^{\infty} x |\Psi(x,t)|^2 dx$$
$$= \int_{-\infty}^{\infty} \Psi^*(x,t) \, x \, \Psi(x,t) dx \qquad (3.82)$$

where in the second line we have simply positioned the factor of x in the integrand between Ψ^* and Ψ. We also saw in Section 2.9 that[7]

$$\langle p_x \rangle = \int_{-\infty}^{\infty} \Psi^* p_{x_{\text{op}}} \Psi \, dx = \int_{-\infty}^{\infty} \Psi^* \frac{\hbar}{i} \frac{\partial \Psi}{\partial x} dx \qquad (3.83)$$

EXAMPLE 3.5 Determine $\langle E \rangle$ for a particle in the box with wave function

$$\Psi(x) = \begin{cases} \sqrt{\frac{30}{L^5}} x(L-x) & 0 < x < L \\ 0 & \text{elsewhere} \end{cases}$$

SOLUTION

$$\langle E \rangle = \int_{-\infty}^{\infty} \Psi^* H \Psi \, dx = \int_0^L \sqrt{\frac{30}{L^5}} x(L-x) \left(-\frac{\hbar^2}{2m} \frac{d^2}{dx^2} \right) \sqrt{\frac{30}{L^5}} x(L-x) \, dx$$

$$= \frac{30}{L^5} \frac{\hbar^2}{m} \int_0^L x(L-x) \, dx = \frac{30}{L^5} \frac{\hbar^2}{m} L^3 \left(\frac{1}{2} - \frac{1}{3} \right) = \frac{5\hbar^2}{mL^2}$$

Note:

$$E_1 = \frac{\hbar^2 \pi^2}{2mL^2} = 4.93 \frac{\hbar^2}{mL^2}$$

Thus our result for $\langle E \rangle$ is consistent with a probability of 0.9985 of obtaining E_1 (see Example 3.4). The magnitude of E_1 is just slightly less than the magnitude

---

[6]It is straightforward to extend our derivation to the more general case where $\Psi = \sum c_n \psi_n$. In the next chapter, we will solve the time-independent Schrödinger equation for a variety of different potential energy functions $V(x)$. It is probably worth emphasizing at this point that the allowed wave functions for these systems include superpositions of energy eigenfunctions—superpositions that themselves are *not* energy eigenfunctions as long as the energies involved in the superposition are distinct, as (3.79) shows.

[7]The argument establishing (3.83) is a little more complicated than that leading to (3.81), since there is a continuum of momentum states and we need to express the superposition of momentum states as an integral rather than a sum.

of $\langle E \rangle$. Notice how easy it is to calculate $\langle E \rangle$ using (3.81).[8] Alternatively, we could calculate the probability $P_n = |c_n|^2$ for each energy eigenvalue E_n and then determine

$$\langle E \rangle = \sum_{n=1}^{\infty} |c_n|^2 E_n$$

In this example, using (3.81) is certainly the way to go.

EXAMPLE 3.6 The energy operator, the Hamiltonian, for a particle of mass m on a spring is given by

$$H = -\frac{\hbar^2}{2m} \frac{\partial^2}{\partial x^2} + \frac{1}{2} K x^2 = -\frac{\hbar^2}{2m} \frac{\partial^2}{\partial x^2} + \frac{1}{2} m \omega^2 x^2$$

where in the last step the Hamiltonian has been expressed in terms of the classical spring frequency $\omega = \sqrt{K/m}$. Determine the value of a so that the Gaussian function $\psi(x) = N e^{-ax^2}$ is an eigenfunction of H. What is the associated energy eigenvalue?

SOLUTION

$$\frac{\partial}{\partial x} e^{-ax^2} = -2ax e^{-ax^2}$$

and

$$\frac{\partial^2}{\partial x^2} e^{-ax^2} = \frac{\partial}{\partial x}\left(-2ax e^{-ax^2}\right) = -2a e^{-ax^2} + 4a^2 x^2 e^{-ax^2}$$

Thus

$$H\psi = \left(-\frac{\hbar^2}{2m} \frac{\partial^2}{\partial x^2} + \frac{1}{2} m\omega^2 x^2\right) N e^{-ax^2}$$

$$= \left[-\frac{\hbar^2}{2m}\left(-2a + 4a^2 x^2\right) + \frac{1}{2} m\omega^2 x^2\right] N e^{-ax^2}$$

$$= \left[\frac{\hbar^2 a}{m} + \left(-\frac{2\hbar^2 a^2}{m} + \frac{1}{2} m\omega^2\right) x^2\right] N e^{-ax^2} = E N e^{-ax^2}$$

where in the last step we have inserted the requirement that ψ be an energy eigenfunction with eigenvalue E. Since E must be a constant, the x^2 terms in the brackets on the left-hand side must cancel. Hence

$$a = \frac{m\omega}{2\hbar}$$

---

[8] In fact, the calculation outlined here is deceptively easy. Since the wave function $\Psi(x)$ has a discontinuous derivative at $x = 0$ and $x = L$, the second derivative of the wave function generates a Dirac delta function at $x = 0$ and at $x = L$. See Section 4.4 for a discussion of the Dirac delta function and Problem 4.21, where it is shown that the delta function is generated by taking the derivative of the step function. The Dirac delta functions that arise in taking the second derivative of $\Psi(x)$, however, do not contribute to $\langle E \rangle$ since the wave function $\Psi^*(x)$ vanishes at $x = 0$ and $x = L$.

and therefore

$$E = \frac{\hbar^2 a}{m} = \frac{1}{2}\hbar\omega$$

Note that we must choose the positive value for a so that the wave function is normalizable.

The harmonic oscillator is a physically important system and the corresponding energy eigenvalue equation is an unusually nice equation mathematically, too. The ground-state wave function is a Gaussian function, one of the nicest functions around. As we will see in Section 4.3, this wave function is the minimum uncertainty wave function, namely, one that satisfies $\Delta x \Delta p_x = \hbar/2$.

3.5 Summary

When the potential energy $V(x)$ is independent of time, a solution to the Schrödinger equation

$$-\frac{\hbar^2}{2m}\frac{\partial^2 \Psi(x,t)}{\partial x^2} + V(x)\Psi(x,t) = i\hbar\frac{\partial \Psi(x,t)}{\partial t} \tag{3.84}$$

can be obtained by expressing

$$\Psi(x,t) = \psi(x)e^{-iEt/\hbar} \tag{3.85}$$

where $\psi(x)$ obeys the time-independent Schrödinger equation

$$-\frac{\hbar^2}{2m}\frac{d^2\psi(x)}{dx^2} + V(x)\psi(x) = E\psi(x) \tag{3.86}$$

As an instructive example, we have solved this equation for a particle confined to an infinite potential energy well (a particle in a box) and found that the (normalized) solutions $\psi_n(x)$ are discrete, namely,

$$\psi_n(x) = \begin{cases} \sqrt{\frac{2}{L}} \sin \frac{n\pi x}{L} & 0 < x < L \\ 0 & \text{elsewhere} \end{cases} \quad n = 1, 2, 3, \ldots \tag{3.87}$$

with quantized energies

$$E_n = \frac{n^2\hbar^2\pi^2}{2mL^2} \quad n = 1, 2, 3, \ldots \tag{3.88}$$

These $\psi_n(x)$ are characterized as orthonormal in that they satisfy

$$\int_{-\infty}^{\infty} \psi_m^*(x)\psi_n(x)\,dx = \delta_{mn} \tag{3.89}$$

with the Kronecker delta defined by

$$\delta_{mn} = \begin{cases} 1 & m = n \\ 0 & m \neq n \end{cases} \tag{3.90}$$

These functions form a complete set, since any wave function can be expressed as the superposition

$$\Psi(x, t) = \sum_{n=1}^{\infty} c_n(t)\psi_n(x) \tag{3.91}$$

where

$$c_n(t) = c_n(0)e^{-iE_n t/\hbar} \tag{3.92}$$

in order that $\Psi(x, t)$ is a solution to the time-dependent Schrödinger equation. If $\Psi(x, t)$ is itself normalized as well, then it is straightforward to show [see (3.57)] that

$$\sum_{n=1}^{\infty} |c_n|^2 = 1 \tag{3.93}$$

which suggests that we identify

$$|c_n|^2 = P_n \tag{3.94}$$

where P_n is the probability of obtaining E_n if a measurement of the energy of a particle with wave function Ψ is carried out. Thus the only result of a measurement of the energy is one of the allowed energies E_n. With this identification, the expectation value of the energy for a particle in the state Ψ is given by

$$\langle E \rangle = \sum_{n=1}^{\infty} |c_n|^2 E_n \tag{3.95}$$

Given a particular wave function $\Psi(x)$, it is possible to use the orthonormality of the $\psi_n(x)$ to determine c_n:

$$c_n = \int_{-\infty}^{\infty} \psi_n^*(x)\Psi(x)\,dx \tag{3.96}$$

An alternative way to express the time-independent Schrödinger equation is in the form of an energy eigenvalue equation

$$H\psi = E\psi \tag{3.97}$$

where

$$H = \frac{(p_{x_{\text{op}}})^2}{2m} + V(x) \tag{3.98}$$

is the energy operator, that is the sum of the kinetic and potential energy operators, once we make the identification that

$$p_{x_{\text{op}}} = \frac{\hbar}{i}\frac{\partial}{\partial x} \tag{3.99}$$

and the constant E is the energy eigenvalue. Thus $\psi_n(x)$ is often referred to as an energy eigenfunction with corresponding energy eigenvalue E_n. It is then straightforward to verify that

$$\langle E \rangle = \sum_{n=1}^{\infty} |c_n|^2 E_n = \int_{-\infty}^{\infty} \Psi^* H \Psi \, dx \tag{3.100}$$

which shows that we can calculate the expectation value of the energy either by summing the probabilities of obtaining each value of the energy multiplied by that energy or by carrying out the integral over all space with the energy operator sandwiched between Ψ^* and Ψ. This starts to make more palatable the result that we saw at the end of the previous chapter, namely

$$\langle p_x \rangle = \int_{-\infty}^{\infty} \Psi^* p_{x_{\text{op}}} \Psi \, dx = \int_{-\infty}^{\infty} \Psi^* \frac{\hbar}{i} \frac{\partial \Psi}{\partial x} \, dx \qquad (3.101)$$

Problems

3.1. Show there is no solution to the time-independent Schrödinger equation for a particle in the infinite square well

$$V(x) = \begin{cases} 0 & 0 < x < L \\ \infty & \text{elsewhere} \end{cases}$$

for $E = 0$. *Suggestion*: Start with the differential equation for ψ within the well for $E = 0$:

$$-\frac{\hbar^2}{2m} \frac{d^2 \psi}{dx^2} = 0$$

What is the most general solution to this second-order differential equation? Show that the requirement that the wave function vanish at the boundaries of the well leads to $\psi = 0$.

3.2. Show there are no negative energy solutions for a particle confined in the potential energy well

$$V(x) = \begin{cases} 0 & 0 < x < L \\ \infty & \text{elsewhere} \end{cases}$$

3.3. The wave function for a particle in a box is

$$\Psi(x) = \begin{cases} N & 0 < x < L \\ 0 & \text{elsewhere} \end{cases}$$

where N is a constant. (*a*) Determine a value for N so that the wave function is appropriately normalized. *Note*: In reality, the wave function $\Psi(x)$ does not drop discontinuously to zero at the ends of the well. Assume the change in the wave function occurs over such a small distance that you can neglect this effect in your calculations. (*b*) Calculate the uncertainty Δx in the particle's position.

3.4. At time $t = 0$ the wave function for a particle in a box is given by

$$\Psi(x) = \sqrt{\frac{2}{3}} \psi_1(x) + \sqrt{\frac{1}{3}} \psi_2(x)$$

where $\psi_1(x)$ and $\psi_2(x)$ are the ground-state and first-excited-state wave functions with corresponding energies E_1 and E_2, respectively. What is $\Psi(x, t)$? What is the probability that a measurement of the energy yields the value E_1? What is $\langle E \rangle$? How would you go about testing these predictions?

3.5. The wave function for a particle in a box is given by

$$\Psi = \frac{i}{2} \psi_1 + \frac{\sqrt{3}}{2} \psi_2$$

where ψ_1 and ψ_2 are energy eigenfunctions with energy eigenvalues E_1 and E_2, respectively. What is the probability that a measurement of the energy yields the value E_1? What is the probability that a measurement of the energy yields the value E_2? What are $\langle E \rangle$ and ΔE?

3.6. At time $t = 0$ the normalized wave function for a particle of mass m in the one-dimensional infinite well

$$V(x) = \begin{cases} 0 & 0 < x < L \\ \infty & \text{elsewhere} \end{cases}$$

is given by

$$\Psi(x) = \begin{cases} \frac{1+i}{2} \sqrt{\frac{2}{L}} \sin \frac{\pi x}{L} + \frac{1}{\sqrt{2}} \sqrt{\frac{2}{L}} \sin \frac{2\pi x}{L} & 0 < x < L \\ 0 & \text{elsewhere} \end{cases}$$

(*a*) What is $\Psi(x, t)$? (*b*) What is the probability that a measurement of the energy at time t will yield the result $\hbar^2 \pi^2 / 2mL^2$? (*c*) What is $\langle E \rangle$ for the particle at time t? *Suggestion*: This result can be obtained by inspection. No integrals are required. (*d*) Is $\langle x \rangle$ time dependent? Justify your answer.

3.7. Solve the time-independent Schrödinger equation for the infinite square well centered at the origin, namely

$$V(x) = \begin{cases} 0 & -L/2 < x < L/2 \\ \infty & \text{elsewhere} \end{cases}$$

Determine the allowed energies and corresponding wave functions. Check that the allowed energies agree with (3.27) and the energy eigenfunctions can be obtained from (3.30) by the substitution $x \to x + L/2$.

3.8. Show that the probability of obtaining E_n for a particle in a box with wave function

$$\Psi(x) = \begin{cases} \sqrt{\frac{30}{L^5}} x(L-x) & 0 < x < L \\ 0 & \text{elsewhere} \end{cases}$$

is given by

$$|c_n|^2 = \frac{240}{n^6 \pi^6} [1 - (-1)^n]^2$$

Note: The probability of obtaining the ground-state energy E_1 is determined in Example 3.4.

3.9. A particle of mass m is confined in the potential well

$$V(x) = \begin{cases} 0 & 0 < x < L \\ \infty & \text{elsewhere} \end{cases}$$

(a) At time $t = 0$, the wave function for the particle is the one given in Problem 3.3. Calculate the probability that a measurement of the energy yields the value E_n, one of the allowed energies for a particle in the box. What are the numerical values for the probabilities of obtaining the ground-state energy E_1 and the first-excited-state energy E_2? *Note*: The energy eigenvalues and eigenfunctions are given in (3.27) and (3.30), respectively. (b) What is $\Psi(x, t)$? Is the particle in a stationary state? Explain why or why not.

3.10. A particle of mass m is confined in the potential energy well

$$V(x) = \begin{cases} 0 & |x| < a/2 \\ \infty & |x| > a/2 \end{cases}$$

(a) Show that

$$\psi_1(x) = \begin{cases} \sqrt{\frac{2}{a}} \cos \frac{\pi x}{a} & |x| \leq a/2 \\ 0 & |x| > a/2 \end{cases}$$

is an energy eigenfunction and determine its corresponding energy eigenvalue E_1. Sketch the wave function. *Note*: It is sufficient to verify that $\psi_1(x)$ satisfies the energy eigenvalue equation as well as the appropriate boundary conditions. Argue that the eigenvalue E_2 of the energy eigenfunction

$$\psi_2(x) = \begin{cases} \sqrt{\frac{2}{a}} \sin \frac{2\pi x}{a} & |x| \leq a/2 \\ 0 & |x| > a/2 \end{cases}$$

satisfies the relation $E_2 = 4E_1$. Sketch the wave function. (b) Suppose the wave function for a particle confined in this potential well is given by the symmetric "tent" wave function (see Problem 2.30)

$$\Psi(x) = \begin{cases} \sqrt{\frac{12}{a^3}} \left(\frac{a}{2} - |x| \right) & |x| \leq a/2 \\ 0 & |x| > a/2 \end{cases}$$

What is the probability that a measurement of the energy of the particle with wave function $\Psi(x)$ yields E_1, the ground-state energy? (c) What is the probability that a measurement of the energy of the particle with wave function $\Psi(x)$ yields E_2, the first-excited-state energy? Explain your answer. *Hint*: You can do the integral by inspection.

3.11. The wave function for a particle of mass m in a potential energy well for which $V = 0$ for $0 < x < L$ and $V = \infty$ elsewhere is given by

$$\Psi(x) = \begin{cases} \sqrt{\frac{105}{L^7}} x^2 (L-x) & 0 < x < L \\ 0 & \text{elsewhere} \end{cases}$$

What is the probability that a measurement of the energy of the particle yields the ground-state energy?

3.12. A particle of mass m is in the lowest energy (ground) state of the infinite potential energy well

$$V(x) = \begin{cases} 0 & 0 < x < L \\ \infty & \text{elsewhere} \end{cases}$$

At time $t = 0$ the wall located at $x = L$ is suddenly pulled back to a position at $x = 2L$. This change occurs so rapidly that instantaneously the wave function does not change. (a) Calculate the probability that a measurement of the energy will yield the ground-state energy of the new well. What is the probability that a measurement of the energy will yield the first excited energy of the new well? (b) Describe the procedure you would use to determine the time development of the system. Is the system in a stationary state?

3.13. Show that the probability that a measurement of the energy in Problem 3.12 yields one of the energy eigenstates of the new well other than one with $n = 2$ is given by

$$|c_n|^2 = \left[\frac{4\sqrt{2}\sin(n\pi/2)}{\pi(4-n^2)} \right]^2$$

3.14. Verify that

$$\psi(x) = Nxe^{-ax^2}$$

is an energy eigenfunction for the simple harmonic oscillator with energy eigenvalue $3\hbar\omega/2$ provided $a = m\omega/2\hbar$.

CHAPTER 4

One-Dimensional Potentials

In this chapter we will focus on solutions to the one-dimensional time-independent Schrödinger equation. We will see how the nature of the potential energy $V(x)$ determines the spectrum of allowed energies, whether these energies are discrete or continuous, and how they are spaced. We will extend our discussion beyond the particle in a box to the finite square well, the harmonic oscillator, and Dirac delta function potentials, as well as scattering from potential energy barriers and wells. These examples give us a lot of insight into how nature behaves on the microscopic level and they will prove to be very useful as we move toward applications of quantum mechanics in three-dimensional systems.

4.1 The Finite Square Well

The potential energy

$$V(x) = \begin{cases} 0 & |x| < a/2 \\ V_0 & |x| > a/2 \end{cases} \quad (4.1)$$

for the finite square well is depicted in Fig. 4.1. Although the finite square well with its straight edges does not seem especially "natural," such quantum wells are reasonable approximations to ones that can be made from certain semiconductors using molecular

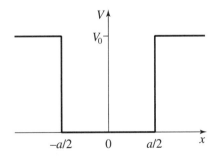

Figure 4.1 A finite square well potential.

114 Chapter 4: One-Dimensional Potentials

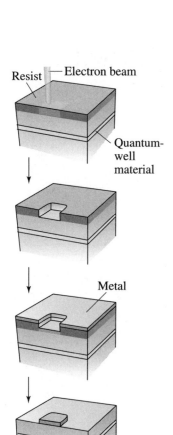

Quantum dot

beam epitaxy, as illustrated in Fig. 4.2. This technology has led to a new growth industry in "designer atoms" and other semiconductor devices. By adjusting the width and depth of the well, it is possible to adjust the spacing of the energy levels. In this way, for example, semiconductor lasers that emit in the green have been produced. In three dimensions, such a quantum well is often referred to as a quantum dot.

In this section we will restrict ourselves to finding solutions to the time-independent Schrödinger equation

$$-\frac{\hbar^2}{2m}\frac{d^2\psi}{dx^2} + V\psi = E\psi \tag{4.2}$$

for energies in the range $0 < E < V_0$. At the end of this section we will see how we recover the results for the infinite square well in the limit $V_0 \to \infty$.

Given the piecewise constant nature of this potential energy, it is straightforward to solve this equation inside and outside the potential energy well. Inside the well, where $V = 0$, the differential equation can be written as

$$\frac{d^2\psi}{dx^2} = -k^2\psi \qquad |x| < a/2 \tag{4.3}$$

where

$$k = \frac{\sqrt{2mE}}{\hbar} \tag{4.4}$$

This is, of course, the same differential equation that we solved in Section 3.2 for the particle in a box. We can write the most general solution as a combination of $\sin kx$ and $\cos kx$, as we did in (3.21), or we can use the complex exponentials

$$\psi(x) = Ae^{ikx} + Be^{-ikx} \qquad |x| < a/2 \tag{4.5}$$

where A and B are constants. Because we have chosen to place the origin of our coordinates at the center of the well, we will see that there is a set of solutions with cosine functions within the well as well as a set with sine functions. At this stage, using the complex exponentials is a handy way to include both cosine and sine functions.

Outside the well, the differential equation (4.2) becomes

$$\frac{d^2\psi}{dx^2} = \kappa^2\psi \qquad |x| > a/2 \tag{4.6}$$

where

$$\kappa = \frac{\sqrt{2m(V_0 - E)}}{\hbar} > 0 \tag{4.7}$$

since we are assuming that $0 < E < V_0$. The most general solution is given by

$$\psi(x) = Ce^{\kappa x} + De^{-\kappa x} \qquad |x| > a/2 \tag{4.8}$$

Figure 4.2 The figure illustrates one of the early strategies for making quantum dots. A semiconductor containing a quantum well material is coated with a polymer. This coating, known as a resist, is decomposed with an electron beam. After a metal layer is deposited on the resulting surface, a solvent removes the remaining resist. Reactive ions then etch the chip except where it is protected by metal, leaving a quantum dot. The role of the size of the wells in scattering light is beautifully illustrated in the colors of a variety of cadmium selenide crystallites synthesized by Michael Steigerwald of AT&T Bell Laboratories. See M. A. Reed, *Sci. Am.* January, 1993, p. 118.

This wave function involves real exponentials instead of complex exponentials since, according to (4.6), two derivatives of the wave function return the wave function multiplied by a *positive* constant. In order to satisfy the requirement that the wave function be normalizable—that the wave function not diverge as $|x| \to \infty$—we must discard the exponential solution that blows up as $x \to -\infty$ in the region for which $x < -a/2$ and the solution that blows up as $x \to +\infty$ in the region for which $x > a/2$. Thus

$$\psi(x) = Ce^{\kappa x} \qquad x < -a/2 \qquad (4.9)$$

and

$$\psi(x) = De^{-\kappa x} \qquad x > a/2 \qquad (4.10)$$

We now face the job of knitting together the solutions in the three different regions, namely $x < -a/2$, $-a/2 < x < a/2$, and $x > a/2$. Since the Schrödinger equation is a second-order differential equation, the wave function must be continuous everywhere, including at the boundaries between the various regions. Otherwise, the first derivative of the wave function would not exist, which would make it difficult to make sense of an equation that involves the second derivative. Rewriting (4.2) in the form

$$\frac{d^2\psi}{dx^2} = -\frac{2m(E-V)}{\hbar^2}\psi \qquad (4.11)$$

makes it clear that the first derivative of the wave function must be continuous as well, at least for any potential energy function for which V is finite. After all, if the right-hand side of (4.11) is finite, the left-hand side must be finite as well. And if the first derivative is not continuous, then the second derivative is infinite. This explains, by the way, why the wave functions for the particle in a box (see Fig. 3.3) have a discontinuous derivative at the boundaries of the box, since V jumps to infinity there.

Imposing the requirement that the wave function is continuous and has a continuous first derivative at $x = -a/2$ yields the conditions

$$Ce^{-\kappa a/2} = Ae^{-ika/2} + Be^{ika/2} \qquad (4.12)$$

$$\kappa Ce^{-\kappa a/2} = ik(Ae^{-ika/2} - Be^{ika/2}) \qquad (4.13)$$

and at $x = a/2$ yields the conditions

$$De^{-\kappa a/2} = Ae^{ika/2} + Be^{-ika/2} \qquad (4.14)$$

$$-\kappa De^{-\kappa a/2} = ik(Ae^{ika/2} - Be^{-ika/2}) \qquad (4.15)$$

We can eliminate the constant C from equations (4.12) and (4.13) by dividing these two equations, which yields

$$\frac{ik}{\kappa} = \frac{Ae^{-ika/2} + Be^{ika/2}}{Ae^{-ika/2} - Be^{ika/2}} \qquad (4.16)$$

Solving for the ratio A/B, we find

$$\frac{A}{B} = e^{ika}\left(\frac{\kappa + ik}{-\kappa + ik}\right) \qquad (4.17)$$

Similarly, we can eliminate the constant D from equations (4.14) and (4.15) by dividing these equations, leading to

$$-\frac{ik}{\kappa} = \frac{Ae^{ika/2} + Be^{-ika/2}}{Ae^{ika/2} - Be^{-ika/2}} \qquad (4.18)$$

and
$$\frac{A}{B} = e^{-ika}\left(\frac{\kappa - ik}{-\kappa - ik}\right) = e^{-ika}\left(\frac{-\kappa + ik}{\kappa + ik}\right) \quad (4.19)$$

[This last result can also be obtained from (4.17) by noting that we can go from (4.16) and (4.17) to (4.18) and (4.19) with the replacement $k \to -k$.] If we multiply (4.17) by (4.19), we obtain

$$\left(\frac{A}{B}\right)^2 = e^{ika}\left(\frac{\kappa + ik}{-\kappa + ik}\right)e^{-ika}\left(\frac{-\kappa + ik}{\kappa + ik}\right) = 1 \quad (4.20)$$

Thus the wave functions fall into two classes: those for which $A/B = 1$ ($A = B$) and those for which $A/B = -1$ ($A = -B$). This is an interesting result and suggestive in its simplicity.

Let's look separately at these two classes of functions. When $A = B$, then (4.12) and (4.14) show that $C = D$. Thus the wave function can be written as

$$\psi(x) = \begin{cases} Ce^{\kappa x} & x \leq -a/2 \\ 2A\cos kx & -a/2 \leq x \leq a/2 \\ Ce^{-\kappa x} & x \geq a/2 \end{cases} \quad (4.21)$$

where we have taken advantage of the identity

$$\cos kx = \frac{e^{ikx} + e^{-ikx}}{2} \quad (4.22)$$

to express the wave function in the region $|x| < a/2$ as a cosine function. Thus this wave function satisfies $\psi(x) = \psi(-x)$; it is an even function. Also when $A = B$, the condition (4.16) becomes

$$\frac{ik}{\kappa} = \frac{(e^{-ika/2} + e^{ika/2})/2}{i(e^{-ika/2} - e^{ika/2})/2i} = \frac{\cos(ka/2)}{-i\sin(ka/2)} \quad (4.23)$$

or

$$\tan(ka/2) = \frac{\kappa a/2}{ka/2} \quad (4.24)$$

If we define the dimensionless variables

$$\xi = ka/2 \quad \text{and} \quad \xi_0 = \frac{a}{\hbar}\sqrt{\frac{mV_0}{2}} \quad (4.25)$$

then this equation becomes

$$\tan \xi = \frac{\sqrt{\xi_0^2 - \xi^2}}{\xi} \quad (4.26)$$

Similarly, when $A = -B$, then (4.12) and (4.14) show that $C = -D$. The wave function can be written in the form

$$\psi(x) = \begin{cases} Ce^{\kappa x} & x \leq -a/2 \\ 2iA\sin kx & -a/2 \leq x \leq a/2 \\ -Ce^{-\kappa x} & x \geq a/2 \end{cases} \quad (4.27)$$

In this case we have taken advantage of the identity

$$\sin kx = \frac{e^{ikx} - e^{-ikx}}{2i} \quad (4.28)$$

to express the wave function in the region $|x| < a/2$ as a sine function.[1] This wave function satisfies $\psi(-x) = -\psi(x)$; it is an odd function. Now the condition (4.16) becomes

$$-\frac{ik}{\kappa} = \frac{e^{ika/2} - e^{-ika/2}}{e^{ika/2} + e^{-ika/2}} = \frac{i\sin(ka/2)}{\cos(ka/2)} \tag{4.29}$$

or

$$-\cot\xi = \frac{\sqrt{\xi_0^2 - \xi^2}}{\xi} \tag{4.30}$$

using the same definitions of ξ and ξ_0 as before.

Equations (4.26) and (4.30) are transcendental equations that determine the allowed values of the energy for the particle in the finite square well. Remember that the energy E is contained in the expression for $\xi = ka/2 = \sqrt{mE/2\hbar^2 a}$. Unlike an equation such as $\tan\xi = \infty$ for which we immediately know the solutions [$\xi = (2n+1)\pi/2$, where $n = 0, 1, 2\ldots$], it is not so obvious what is the solution to (4.26) or (4.30). One strategy for determining the allowed energies is to solve these equations numerically, on your calculator for example. Another strategy, which allows us to get an overview of what is happening is to solve the equations graphically. Figure 4.3a shows a plot of $\tan\xi$ and $\sqrt{\xi_0^2 - \xi^2}/\xi$. The intersections of these two curves determine the allowed energies for the even wave functions. The energy is determined by noting the value of ξ for which the intersection occurs. Notice that for ξ_0^2 small, there is only one intersection, that is, one allowed energy. As ξ_0 increases, the number of energy eigenvalues grows. Figure 4.3b shows a plot of $-\cot\xi$ and $\sqrt{\xi_0^2 - \xi^2}/\xi$. For the odd wave functions, there is no solution for $\xi_0 < \pi/2$ since $-\cot\xi$ is not positive unless $\xi > \pi/2$. Notice that as $\xi_0 \to \infty$, say for $V_0 \to \infty$, there will be an infinite number of allowed energies determined by the condition $\xi = n\pi/2$, with n a positive integer. This is precisely the result

$$E_n = \frac{n^2\hbar^2\pi^2}{2ma^2} \qquad n = 1, 2, 3\ldots \tag{4.31}$$

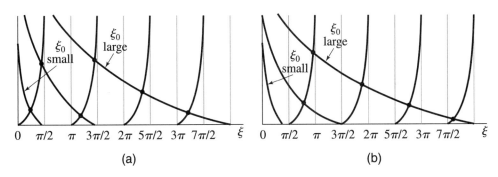

Figure 4.3 Graphs of (a) $\tan\xi$ and $\sqrt{\xi_0^2 - \xi^2}/\xi$ and (b) $-\cot\xi$ and $\sqrt{\xi_0^2 - \xi^2}/\xi$ for various values of ξ_0. The intersections correspond to allowed energies.

---

[1]From (4.12), we see that when $A = B$, $Ce^{-qa/2} = 2A\cos ka/2$, and when $A = -B$, $Ce^{-qa/2} = -i2A\sin ka/2$. Thus there is just one overall constant (either C or A, depending on which you choose to eliminate) that is then fixed by normalization. See Problem 4.7.

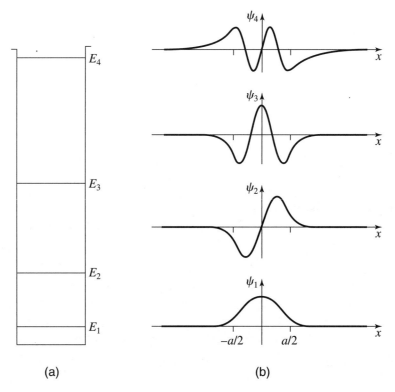

Figure 4.4 (a) A finite square well with four allowed energies and (b) the corresponding energy eigenfunctions.

that we obtained in Chapter 3 for an infinite well of width a. Interestingly, we would obtain the same result if we let the width a of our finite well become very large (holding V_0 fixed), since this too leads asymptotically to intersections at $\xi = n\pi/2$. As we will see in Chapter 8, a useful model of an electron in a metal is to treat it like a particle in a very large box, with the size a measured in, say, centimeters instead of tenths of nanometers, which is the characteristic length scale for an electron bound in an atom. Thus it is not unreasonable to use the results from the infinite square well in determining the properties of a metal in this free-electron model of a solid.

Figure 4.4 shows plots of the wave functions for a finite square well that possesses four allowed energy states. The ground state and the second excited state are even wave functions, while the first and third excited states are odd wave functions.[2] The larger the number of nodes, the higher the energy, just as for the infinite square well. Notice in particular how the wave functions penetrate into the region where $V > E$. The degree of penetration is determined by the magnitude of κ in the wave functions (4.21) and (4.27). The wave function falls to $1/e$ of its value at $x = \pm a/2$ in a distance $1/\kappa = \hbar/\sqrt{2m(V_0 - E)}$. Thus the wave functions with the smallest energy—those closest to the bottom of the energy well—penetrate less deeply into the classically disallowed region where $V > E$. Notice that as $V_0 \to \infty$, $1/\kappa \to 0$ independent of E, confirming our supposition in Section 3.2 that the wave functions for the particle in a box (the infinite potential energy well) are indeed zero outside the box.

---

[2] The fact that the energy eigenfunctions are either even or odd functions of x is not an accident. We will see why this happens in Section 5.3.

4.2 Qualitative Features

Before going on to solve the time-independent Schrödinger equation for other potential energy functions, it is instructive to examine more qualitatively the character of the solutions to this equation. First, a word about terminology. Consider a general potential energy such as the one shown in Fig. 4.5. For the particular energy E shown in the figure, there is a region $a < x < b$ where $E > V$ that is bounded by a region where $V > E$. The region $a < x < b$ thus forms a potential energy well. The points a and b are often called turning points. At these points the total energy E of the particle is equal to the potential energy, meaning that the particle has zero kinetic energy. In classical physics, we would say that at this point the particle has stopped and the force $F = -dV/dx$ accelerates the particle back into the region $a < x < b$. Thus the particle is bound in the potential energy well. Consequently, the allowed energy states of the Schrödinger equation in this energy regime are referred to as **bound states**. Although the particle's motion classically is strictly limited to the region $a \leq x \leq b$, we have seen in our discussion of the finite square well that the wave function does not stop at the boundaries. Consequently, there is a nonzero probability that a measurement of the position of the particle will yield a location for which $V > E$.

Solving the time-independent Schrödinger equation for a potential energy such as the one shown in Fig. 4.5 is not as straightforward as determining the allowed energies for the finite square well, where the potential energy is piece-wise constant. The Schrödinger equation

$$-\frac{\hbar^2}{2m} \frac{d^2\psi}{dx^2} + V(x)\psi = E\psi \tag{4.32}$$

for the region within the potential energy well can be put in the form

$$\frac{d^2\psi}{dx^2} = -k^2(x)\psi \tag{4.33}$$

where we have defined

$$k^2(x) = \frac{2m[E - V(x)]}{\hbar^2} \tag{4.34}$$

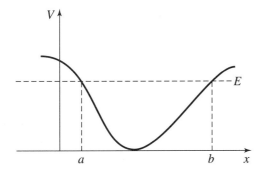

Figure 4.5 A particle with energy E (the dashed line) would classically be confined to move in the region between a and b in which $E > V$.

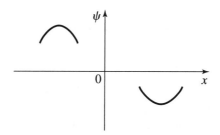

Figure 4.6 If $E > V$, the energy eigenfunctions are oscillatory in nature.

Note that $k^2(x)$ is positive in a region in which $E > V$. If V were a constant, the solution could be written as

$$\psi = A \sin kx + B \cos kx \qquad V = \text{constant with } E > V \qquad (4.35)$$

But we are not assuming here that $V(x)$ is a constant independent of x, so the solutions to (4.32) are not simply sines and cosines. Nonetheless, there are some strong similarities. Notice that since $k^2(x)$ is positive, the second derivative has the opposite sign to the wave function, assumed here to be real. Thus if the wave function is positive, the second derivative is negative, or concave down. Similarly, if the wave function is negative, the second derivative is positive, or concave up. Thus the function bends back toward the x axis independent of the sign of the wave function. Hence the wave function oscillates. See Fig. 4.6. And the bigger the difference between E and V, the larger the second derivative and the faster the oscillation. These are the same sorts of results that we saw inside the well for the finite square well in the previous section. But again, we are not assuming that the potential energy is constant as we were there.

Now let's examine a region in which $V > E$. In this case it is convenient to rewrite the Schrödinger equation in the form

$$\frac{d^2\psi}{dx^2} = \frac{2m[V(x) - E]}{\hbar^2}\psi \qquad (4.36)$$

Here we define

$$\kappa^2(x) = \frac{2m[V(x) - E]}{\hbar^2} \qquad (4.37)$$

so that

$$\frac{d^2\psi}{dx^2} = \kappa^2(x)\psi \qquad (4.38)$$

Because $V > E$, $\kappa^2(x)$ is positive. If $V(x)$ were a constant, then κ would be constant and the solution would be

$$\psi = Ce^{\kappa x} + De^{-\kappa x} \qquad V = \text{constant with } V > E \qquad (4.39)$$

with real exponentials. If κ is not constant but depends on x, we can characterize the solutions as exponential-like. For example, if the wave function is positive in a region in which $V > E$, the second derivative is positive and the function is concave up, bending away from the axis. Similarly, if the wave function is negative in a region in which $V > E$, the second derivative is negative as well. Thus the function is concave down, again bending away from the axis. See Fig. 4.7. Although wave mechanics is telling

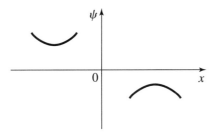

Figure 4.7 If $E < V$ the energy eigenfunctions are exponential-like, bending away from the axis.

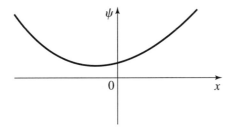

Figure 4.8 An energy eigenfunction with $V > E$ everywhere.

us that while there is a nonzero exponential-like wave function in a region for which $V > E$, there is no solution if $V > E$ everywhere. For if the wave function were to start out positive and bend away from the axis everywhere, it would eventually diverge, as indicated in Fig. 4.8.

How then do we end up with a square integrable wave function, one for which the integral of $|\psi|^2$ over all space is finite? There must be a region in which $E > V$ so that the wave function can bend back toward the axis. Let's examine the potential energy shown in Fig. 4.5 for which there is a bound state. Follow the wave function along in the positive x direction, starting at negative infinity. As was noted in the discussion of the finite square well, the wave function and its derivative should be continuous everywhere. Since the Schrödinger equation is a second-order differential equation, it has two linearly independent solutions. We choose the exponential-like solution that goes asymptotically to zero as $x \to -\infty$. As we move toward point a, the wave function remains positive and concave up since it is in a region in which $V > E$. At a, $E = V$ and, as (4.32) shows, the second derivative vanishes. This is, therefore, an inflection point. As the wave function moves into the region to the right of point a, a region in which $E > V$, the sign of the second derivative changes from positive to negative and the wave function changes from exponential-like to oscillatory. As we follow the wave function along, we eventually reach a region in which $V > E$ on the other side of the well, at point b. We are now searching for a wave function that asymptotically approaches zero for large positive x. The trick is to make it happen. Unless we are fortunate in our choice of E, when we join the wave function inside the well to the wave function outside the well with a continuous function and a continuous derivative, we find that both the increasing and the decreasing exponential-like solutions are present in the wave function and therefore the wave function diverges. For example, if we start with an energy that is much too small, then the wave function oscillates too slowly in the region in which $E > V$ and is positive with a positive slope, or first derivative, at the turning point. But if the wave

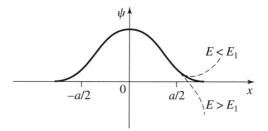

Figure 4.9 If the energy of the eigenfunction for the finite square well is too large or too small, the eigenfunction (shown with a dashed line) doesn't tie into an exponentially decreasing wave function as $x \to \infty$. There is one wave function (the solid line) for which the energy is just right.

function, the first derivative, and the second derivative are all positive to the right of the turning point, where $V > E$, the wave function must diverge as $x \to \infty$. To get the wave function to turn back toward the x axis would require negative curvature, which is not allowed for a positive wave function in this $V > E$ region. Even if the wave function oscillates sufficiently rapidly in the well so that the slope is negative at the turning point, the wave function will in general diverge as both exponential-like solutions are present, as illustrated in Fig. 4.9. But there does turn out to be one energy where the wave function is well behaved for large negative and positive x. This is the ground state.

Are there additional allowed energies? For a higher energy solution, the wave function oscillates even more rapidly within the well. It is indeed possible that higher energy states, with nodes within the well, exist. As we saw in the discussion of the finite square well, this depends on the depth and width of the potential energy well. In any case, we are assured that the allowed energies are quantized, provided the particle's energy is such that it would be classically confined within the well. On the other hand, if the particle were not confined, say by not having a region to the right where $V > E$, then the wave function oscillates along, all the way to infinity. This may not look promising either, since we are looking for square integrable solutions. But in this case there is no constraint of having to tie into a decreasing exponential-like solution at infinity. Thus there is a continuum of solutions that can be superposed to create a solution that dies off as x approaches infinity. This superposition is just the sort of superposition that we discussed in Section 2.6 when we were constructing wave packets. We will examine these continuum solutions when we discuss scattering in Section 4.6 and Section 4.7.

Finally, we can ask how the amplitude of the wave function varies in a region in which $E > V$ but V is not constant. Consider a potential energy well in which there is a step, as shown in Fig. 4.10. In a region of constant potential energy, say to the left of the step, we can write the wave function as[3]

$$\psi = A \sin(kx + \phi) \tag{4.40}$$

where

$$k = \frac{\sqrt{2m(E - V)}}{\hbar} \tag{4.41}$$

---

[3]Note that $A \sin(kx + \phi) = A \cos\phi \sin kx + A \sin\phi \cos kx$, so this is just another way to write the wave function as a combination of a sine and a cosine.

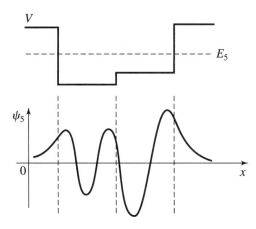

Figure 4.10 An energy eigenfunction for a potential energy well with a step illustrating how the amplitude of the wave function increases as the difference between the energy E and the potential energy $V(x)$ decreases.

Thus the derivative of the wave function is given by

$$\frac{d\psi}{dx} = kA\cos(kx + \phi) \tag{4.42}$$

and hence if we square this result, divide by k^2, and add it to the square of (4.40), we obtain

$$A = \sqrt{\psi^2 + \frac{1}{k^2}\left(\frac{d\psi}{dx}\right)^2} \tag{4.43}$$

Since ψ and $d\psi/dx$ are continuous everywhere, including at the step, the change in the amplitude at the step is dictated by what happens to k. Thus as k decreases in moving from the immediate left to the immediate right of the step, the amplitude A must increase, since ψ and $d\psi/dx$ don't change. Classically, the particle would slow down as it moves from the left to the right side of the well, since as the potential energy increases, the kinetic energy decreases, assuming a particular energy E. One is tempted to say that the amplitude increases as the particle "slows down" because the particle spends more time in the region in which it is going more slowly; hence the probability of finding it there should be larger. The trouble with this line of reasoning is that we are discussing the behavior of an energy eigenfunction and, as we saw in Section 3.1, such states are stationary states, with no time dependence for the probability density. Thus perhaps we should use these classical arguments as a helpful mnemonic, but not necessarily as the reason for this behavior of the amplitude.

4.3 The Simple Harmonic Oscillator

As we did for the particle in a box, we can solve the harmonic oscillator analytically in a completely closed form, determining the energy eigenvalues and eigenfunctions exactly. In nonrelativistic quantum mechanics, the harmonic oscillator plays a very important role, since any smooth potential energy in the vicinity of a minimum looks like a harmonic oscillator. Consider a potential energy $V(x)$ (see Fig. 4.11) that has a minimum at the

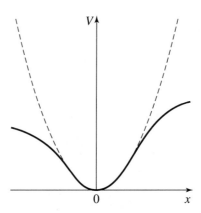

Figure 4.11 In the vicinity of a minimum the potential energy can be approximated as a parabola, namely the potential energy of the simple harmonic oscillator.

origin (we can always position the origin of our coordinates so it coincides with the minimum). If we take a Taylor-series expansion for $V(x)$, we obtain

$$V(x) = V(0) + \left(\frac{dV}{dx}\right)_{x=0} x + \frac{1}{2!}\left(\frac{d^2V}{dx^2}\right)_{x=0} x^2 + \cdots \qquad (4.44)$$

But since $x = 0$ is a minimum, the first derivative vanishes there and hence

$$V(x) = V(0) + \frac{1}{2}\left(\frac{d^2V}{dx^2}\right)_{x=0} x^2 + \cdots \qquad (4.45)$$

Thus, provided the amplitude of the oscillation is not too large, the potential energy is of the form

$$V(x) = \frac{1}{2}Kx^2 \qquad (4.46)$$

where we have called the constant $(d^2V/dx^2)_{x=0} = K$, the spring constant, and dropped the overall additive constant $V(0)$, since we are free to locate the zero of potential energy wherever we choose, such as $V(0) = 0$. In terms of the spring frequency ω,

$$V(x) = \frac{1}{2}m\omega^2 x^2 \qquad (4.47)$$

since $\omega^2 = K/m$.

To determine the energy eigenvalues and eigenfunctions of the harmonic oscillator in wave mechanics, we must solve the differential equation

$$-\frac{\hbar^2}{2m}\frac{d^2\psi}{dx^2} + \frac{1}{2}m\omega^2 x^2 \psi = E\psi \qquad (4.48)$$

Right away we see that this will be challenging since

$$\frac{d^2\psi}{dx^2} = \frac{2m}{\hbar^2}\left(\frac{1}{2}m\omega^2 x^2 - E\right)\psi \qquad (4.49)$$

and therefore two derivatives of the wave function do not yield a constant (whether it's positive or negative) multiplying the wave function. The x^2 dependence of the potential energy complicates matters considerably. Nonetheless, we can learn something about the wave functions by looking at the behavior for large $|x|$, sufficiently large that the x^2 term on the right-hand side of (4.49) dominates. Then

$$\frac{d^2\psi}{dx^2} \approx \frac{m^2\omega^2 x^2}{\hbar^2}\psi \qquad (4.50)$$

You can readily verify (see Problem 4.6) that an approximate solution for large $|x|$ is

$$\psi(x) = Ax^n e^{-m\omega x^2/2\hbar} + Bx^n e^{m\omega x^2/2\hbar} \qquad (4.51)$$

where n is an integer. Of course, we must discard the increasing exponential since we want a normalizable solution, not one that blows up as $|x| \to \infty$. If you look at Example 3.6, you will see that

$$\psi_0(x) = A_0 e^{-m\omega x^2/2\hbar} \qquad (4.52)$$

is in fact an exact energy eigenfunction for the harmonic oscillator. It satisfies the differential equation (4.48) for all x provided the energy E is set to

$$E_0 = \frac{1}{2}\hbar\omega \qquad (4.53)$$

This wave function must be the ground state, the lowest permitted energy state of the oscillator, by the way, since $\psi_0(x)$ has no nodes. This ground-state energy is often referred to as the zero-point energy of the harmonic oscillator.

We have labeled the wave function and the energy with a zero subscript since the allowed energies (the energy eigenvalues) for the harmonic oscillator are given by

$$E_n = \left(n + \frac{1}{2}\right)\hbar\omega \qquad n = 0, 1, 2, \ldots \qquad (4.54)$$

Figure 4.12 shows the energy spectrum. Following the procedure outlined in Example 3.6, you can readily verify, for example, that the first-excited-state wave function is given by

$$\psi_1(x) = A_1 x e^{-m\omega x^2/2\hbar} \qquad (4.55)$$

with corresponding energy

$$E_1 = \frac{3}{2}\hbar\omega \qquad (4.56)$$

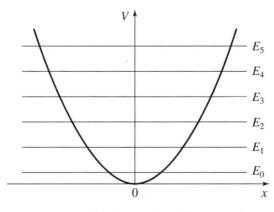

Figure 4.12 The energy spectrum of the harmonic oscillator superimposed on the potential energy $V(x) = m\omega^2 x^2/2$.

Figure 4.13 shows the wave functions $\psi_n(x)$ and the probability densities $|\psi_n(x)|^2$ for the six lowest energy states. In general, the energy eigenfunctions of the harmonic oscillator are of the form

$$\psi_n(x) = A_n H_n(\sqrt{m\omega/\hbar}\, x)\, e^{-m\omega x^2/2\hbar} \qquad (4.57)$$

where H_n is a polynomial in x of order n called a Hermite polynomial. If you are interested in seeing one way in which these wave functions and allowed values for the energy can be derived, check out Section B.1.

Notice how the wave functions in Fig. 4.13 oscillate in the regions in which $E > V$ and are exponentially damped in regions in which $V > E$, in accord with our discussion in Section 4.2. Compare the plots of the wave functions for the bound states of the finite square well in Fig. 4.4 with those of the harmonic oscillator. The similarities are striking, even though the functions themselves are quite different. One feature that is present for the harmonic oscillator that is not present for the finite square well is how the amplitude of the wave functions grows as the wave function approaches the classical turning points (the points at which $E = V$), since $E - V$ decreases as one moves away from the origin for the harmonic oscillator but not for the finite square well, again in accord with our general discussion at the end of Section 4.2. As for the particle in a box, the energies of the harmonic oscillator are all discrete since a particle is confined no matter how high its energy. Even though the allowed energies are labeled by an integer n in each case, the dependence on n is quite different. For the infinite square well the energies grow like n^2 while for the harmonic oscillator they are linear in n.[4] This illustrates how differences in the nature of the potential energy manifest themselves in the spectrum of the energy levels, providing us with a handle on the nature of the potential energy on a microscopic scale.[5]

In Example 4.1, at the end of this section, we determine the probability that a particle in the ground state will be found in a region in which $V > E$ if a measurement of the position of the particle is carried out. To answer this question, we need to normalize the ground-state wave function. The normalization requirement is

$$\int_{-\infty}^{\infty} |\psi_0(x)|^2 dx = 1 \qquad (4.58)$$

which translates into

$$|A_0^2| \int_{-\infty}^{\infty} e^{-m\omega x^2/\hbar} dx = 1 \qquad (4.59)$$

Integrals of the form

$$I(b) = \int_{-\infty}^{\infty} e^{-bx^2} dx \qquad (4.60)$$

---

[4]The harmonic oscillator plays an extremely important role in quantum field theory. The constant spacing of the energy levels gives us a clue as to why this might be so. Instead of thinking of E_n as the energy of the nth excited state of the harmonic oscillator, we might view it as the energy of n quanta, each with energy $\hbar\omega = h\nu$, apart from the zero-point energy.

[5]There is good reason to think that the potential energy of interaction between heavy quarks varies linearly with separation of the quarks at large distances. One piece of evidence comes from the spacing of energy levels of a two-body bound state such as charmonium, which was discovered in 1974. At that time, theorists who had been focusing on the intricacies of quantum field theory had to remind themselves how to solve the Schrödinger equation for potentials other than the ones that are discussed in most textbooks. See Problem 4.14.

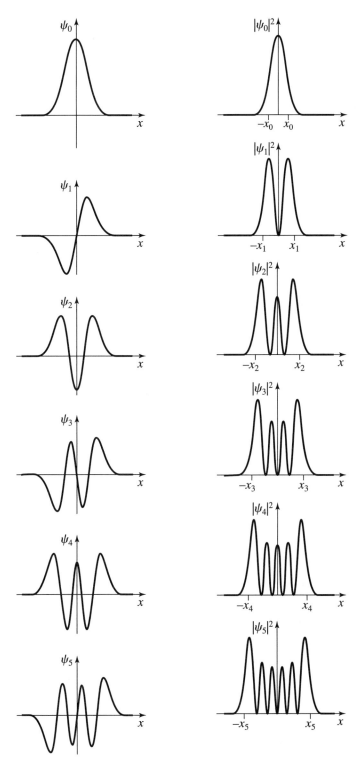

Figure 4.13 The energy eigenfunctions and probability densities for the first six energy states of the harmonic oscillator. The classical turning points x_n are determined by setting the potential energy equal to the total energy, that is $m\omega^2 x_n^2/2 = E_n$.

are referred to as Gaussian integrals and there are some cute tricks for evaluating integrals of this type.[6] Instead of trying to calculate $I(b)$, let's start with $I^2(b)$:

$$I^2(b) = \int_{-\infty}^{\infty} e^{-bx^2} dx \int_{-\infty}^{\infty} e^{-by^2} dy \qquad (4.61)$$

where we have suggestively called the dummy variable of integration in the second expression for $I(b)$ y instead of x. If we write this expression as a double integral, we see that

$$I^2(b) = \int_{-\infty}^{\infty} \int_{-\infty}^{\infty} e^{-b(x^2+y^2)} dx\, dy \qquad (4.62)$$

which cries out for a change of variables to polar coordinates:

$$I^2(b) = \int_0^{2\pi} \int_0^{\infty} e^{-br^2} r\, dr\, d\theta \qquad (4.63)$$

It is now straightforward to carry out the r integral, since we have an exact differential. And the θ integral is trivial, since there is no θ dependence in the integrand. Thus

$$I^2(b) = -\frac{\pi}{b} \int_0^{\infty} e^{-br^2}(-2br)\, dr = -\frac{\pi}{b} e^{-br^2} \Big|_0^{\infty} = \frac{\pi}{b} \qquad (4.64)$$

Therefore

$$\sqrt{\frac{b}{\pi}} \int_{-\infty}^{\infty} e^{-bx^2} dx = 1 \qquad (4.65)$$

and hence for the ground state of the harmonic oscillator, for which $b = m\omega/\hbar$, the normalized wave function is given by

$$\psi_0(x) = \left(\frac{m\omega}{\pi\hbar}\right)^{1/4} e^{-m\omega x^2/2\hbar} \qquad (4.66)$$

The zero-point energy and the Heisenberg uncertainty principle

The existence of a nonzero ground-state energy for the harmonic oscillator is both interesting and important. We can gain some insight into the underlying physics with the aid of the Heisenberg uncertainty principle. Suppose that we take the ground-state wave function ψ_0 and calculate $\langle E \rangle$:

$$\langle E \rangle = \int_{-\infty}^{\infty} \psi_0^* H \psi_0\, dx = \frac{\langle p_x^2 \rangle}{2m} + \frac{1}{2} m\omega^2 \langle x^2 \rangle \qquad (4.67)$$

But since $H\psi_0 = E_0 \psi_0$, $\langle E \rangle = E_0$. And because the ground-state wave function is an even, real function, $\langle x \rangle$ and $\langle p_x \rangle$ are both zero. Thus using $(\Delta x)^2 = \langle x^2 \rangle - \langle x \rangle^2$ and $(\Delta p_x)^2 = \langle p_x^2 \rangle - \langle p_x \rangle^2$, we obtain

$$E_0 = \frac{(\Delta p_x)^2}{2m} + \frac{1}{2} m\omega^2 (\Delta x)^2 \qquad (4.68)$$

---

[6]Once you can do an integral such as (4.60), then it is straightforward to evaluate integrals such as $\int_{-\infty}^{\infty} x^2 e^{-bx^2} dx$. See Problem 4.11.

But since $\Delta x \Delta p_x \geq \hbar/2$, we see that

$$E_0 \geq \frac{\hbar^2}{8m(\Delta x)^2} + \frac{1}{2}m\omega^2 (\Delta x)^2 \qquad (4.69)$$

Nature is trying to minimize this energy in the ground state. We can do this as well by taking the derivative of the right-hand side of (4.69) with respect to Δx and setting the result equal to zero:

$$-\frac{\hbar^2}{4m(\Delta x)^3} + m\omega^2 \Delta x = 0 \qquad (4.70)$$

which means

$$(\Delta x)^2 = \frac{\hbar}{2m\omega} \qquad (4.71)$$

Plugging this value for $(\Delta x)^2$ back into (4.69), we find

$$E_0 \geq \frac{1}{4}\hbar\omega + \frac{1}{4}\hbar\omega = \frac{1}{2}\hbar\omega \qquad (4.72)$$

Thus the smallest value for E_0 consistent with the Heisenberg uncertainty principle is $\hbar\omega/2$. But this is in fact the exact ground-state energy for the harmonic oscillator. Thus we infer that the Gaussian ground-state wave function (4.52) is the minimum uncertainty wave function. After all, we could have replaced the inequality in (4.69) with an equality if we had presumed from the beginning that we could set $\Delta x \Delta p_x = \hbar/2$.

Our analysis shows the profound difference between classical and quantum mechanics. In classical mechanics, the particle in the lowest energy state would just sit at the bottom of the potential well, at rest with the "spring" unstretched. But this would require we know both the position and the momentum simultaneously, in contradiction to the Heisenberg uncertainty principle. For the harmonic oscillator, the lowest energy state has equal mixtures for the expectation values of the kinetic and potential energies, as (4.72) shows. This procedure of using the Heisenberg uncertainty principle to estimate the ground-state energy worked especially well since not only did the expectation value of the kinetic energy involve $(\Delta p_x)^2$, but the expectation value of the potential energy comes out directly in terms of $(\Delta x)^2$. Only a potential energy that varies as x^2—only the harmonic oscillator—has this property. For other potential energies, you can still use the uncertainty principle to estimate the ground-state energy, but some approximations will be required. See Problem 4.14 and Problem 4.16.

EXAMPLE 4.1 What is the probability that a particle in the ground state of the harmonic oscillator will be found in a classically disallowed region if a measurement of the location of the particle is carried out? Before looking at the solution, test your quantum intuition by making a guess at the answer.

SOLUTION The locations of the classical turning points are determined by setting the potential energy equal to the total energy:

$$V(x_0) = \frac{1}{2}m\omega^2 x_0^2 = E = \frac{1}{2}\hbar\omega \quad \Rightarrow \quad x_0 = \pm\sqrt{\frac{\hbar}{m\omega}}$$

Thus the probability of finding the particle in a classically disallowed region (outside the turning points) is given by

$$\text{Prob} = \int_{-\infty}^{-\sqrt{\hbar/m\omega}} |\psi_0|^2 dx + \int_{\sqrt{\hbar/m\omega}}^{\infty} |\psi_0|^2 dx = 2\sqrt{\frac{m\omega}{\pi\hbar}} \int_{\sqrt{\hbar/m\omega}}^{\infty} e^{-m\omega x^2/\hbar} dx$$

where in the last step we have taken advantage of the fact that $|\psi_0|^2$ is an even function. To obtain a numerical value, it is convenient to convert this integral into a form in which we can use the tabulated values of the error function or a calculator. If we define $m\omega x^2/\hbar = y^2/2$, then

$$\text{Prob} = \frac{2}{\sqrt{2\pi}} \int_{\sqrt{2}}^{\infty} e^{-y^2/2} dy = 2\left(\frac{1}{\sqrt{2\pi}} \int_0^{\infty} e^{-y^2/2} dy - \frac{1}{\sqrt{2\pi}} \int_0^{\sqrt{2}} e^{-y^2/2} dy\right)$$

$$= 2\left(\frac{1}{2} - 0.42\right) = 0.16$$

Thus there is a 16% chance of finding the particle in a classically forbidden region when it is in the ground state.

EXAMPLE 4.2 Estimate the quantum number n for a macroscopic mass on a spring.

SOLUTION Assume we want $E = 1$ J and that $\omega = 1$ s^{-1}. Then since $\hbar = 1.055 \times 10^{-34}$ J·s, we need n to be roughly 10^{34}. Our everyday experience is with very highly excited harmonic oscillators, not with ones in the ground state. This may be the reason that there is a tendency to guess too low in Example 4.1.

4.4 The Dirac Delta Function Potential

The best way to think about a Dirac delta function (which isn't, strictly, a function) is as the limit of a sequence of functions, each with unit area, that progressively become narrower and consequently higher as some parameter is varied. A nice example from the previous section is to take the limit of the integrand in (4.65) as $b \to \infty$:

$$\delta(x) = \lim_{b \to \infty} \sqrt{\frac{b}{\pi}} e^{-bx^2} \qquad (4.73)$$

This function has been normalized so that

$$\int_{-\infty}^{\infty} \delta(x) \, dx = 1 \qquad (4.74)$$

and as $b \to \infty$ it satisfies the requirement that

$$\delta(x) = \begin{cases} 0 & x \neq 0 \\ \infty & x = 0 \end{cases} \qquad (4.75)$$

As we will see, in the endgame a delta function generally appears as part of the integrand within an integral. It makes doing integrals easy. For example,

$$\int_{-\infty}^{\infty} f(x)\delta(x) \, dx = f(0) \int_{-\infty}^{\infty} \delta(x) \, dx = f(0) \qquad (4.76)$$

where $f(x)$ is any well-behaved function. The first step in (4.76) follows since $\delta(x)$ is zero everywhere except $x = 0$. Thus we can replace $f(x)$ by $f(0)$ within the integrand and then pull the constant $f(0)$ outside the integral. In fact, (4.76) is sometimes taken as the defining relationship for a Dirac delta function.

In this section we will assume that the potential energy is zero everywhere except at the origin, and at the origin there is an infinitely deep potential well, as indicated in Fig. 4.14. Such a potential energy well can be expressed in terms of a Dirac delta function as[7]

$$\frac{2m}{\hbar^2} V(x) = -\frac{\alpha}{a} \delta(x) \tag{4.77}$$

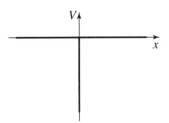

Figure 4.14 The Dirac delta function potential energy (4.77).

Notice that $2mV(x)/\hbar^2$ has the dimensions of $1/\text{length}^2$. Since a delta function has the dimensions of $1/\text{length}$ [see (4.74)], the constant multiplying the delta function must have the dimensions of $1/\text{length}$ as well. We have accounted for this length by introducing an arbitrary constant a as a length and we have inserted a dimensionless constant α as well. Varying α gives us a way to adjust the strength of the potential.

Next we investigate the boundary conditions on the wave function. Instead of the three regions ($x < -a/2$, $-a/2 < x < a/2$, and $x > a/2$) that we considered in solving the finite potential well, here there are only two regions ($x < 0$ and $x > 0$) for the delta function. Since the Schrödinger equation is a second-order differential equation, we must, as always, require the wave function to be continuous everywhere. But if we integrate the Schrödinger equation from just below to just above $x = 0$, we see that the first derivative of the wave function is not continuous at $x = 0$:

$$\int_{0^-}^{0^+} \frac{d^2\psi}{dx^2} dx = \left(\frac{d\psi}{dx}\right)_{0^+} - \left(\frac{d\psi}{dx}\right)_{0^-}$$

$$= \frac{2m}{\hbar^2} \int_{0^-}^{0^+} [V(x) - E]\psi(x)\, dx \tag{4.78}$$

Now

$$\frac{2mE}{\hbar^2} \int_{0^-}^{0^+} \psi(x)\, dx = 0 \tag{4.79}$$

since the width of the integration region is infinitesimal and the wave function itself is finite. On the other hand,

$$\frac{2m}{\hbar^2} \int_{0^-}^{0^+} V(x)\psi(x)\, dx = -\frac{\alpha}{a} \int_{0^-}^{0^+} \delta(x)\psi(x)\, dx = -\frac{\alpha}{a} \psi(0) \tag{4.80}$$

and therefore

$$\left(\frac{d\psi}{dx}\right)_{0^+} - \left(\frac{d\psi}{dx}\right)_{0^-} = -\frac{\alpha}{a} \psi(0) \tag{4.81}$$

Thus the first derivative is discontinuous at $x = 0$. Our derivation shows, by the way, why the first derivative is typically continuous, since integrals such as the one on the

---

[7]Since $V(x) = 0$ except at $x = 0$, the minus sign on the right-hand side shows that we are dealing with a potential energy well, as opposed to a potential energy barrier. In our example of the finite square well in Section 4.1, we could have defined the potential energy function with the zero of potential energy outside a potential energy well of depth $-V_0$. Then we might view the delta function potential well as resulting from taking the limit as $V_0 \to \infty$ as the width of the well goes to zero.

left-hand side of (4.80) will vanish just as (4.79) does if the potential energy $V(x)$ is not singular.

Let's finally turn our attention to solving the Schrödinger equation for a delta function potential energy well. For $x \neq 0$, the Schrödinger equation is simply

$$\frac{d^2\psi}{dx^2} = -\frac{2mE}{\hbar^2}\psi \tag{4.82}$$

Since we are searching for a bound state ($E < 0$), the solution to (4.82) is

$$\psi = Ae^{-\kappa x} + Be^{\kappa x} \tag{4.83}$$

where

$$\kappa = \frac{\sqrt{-2mE}}{\hbar} \tag{4.84}$$

As for the finite square well, the requirement that the wave function be normalizable as well as continuous at $x = 0$ dictates that

$$\psi(x) = \begin{cases} Ae^{\kappa x} & x \leq 0 \\ Ae^{-\kappa x} & x \geq 0 \end{cases} \tag{4.85}$$

Figure 4.15 shows a plot of the wave function. Clearly, the derivative is discontinuous at $x = 0$. Substituting this wave function into (4.81), we find

$$-2\kappa A = -\frac{\alpha}{a}A \tag{4.86}$$

or

$$\sqrt{\frac{-2mE}{\hbar^2}} = \frac{\alpha}{2a} \tag{4.87}$$

Thus the energy of the single bound state for a particle in the delta function well is

$$E = -\frac{\hbar^2 \alpha^2}{8ma^2} \tag{4.88}$$

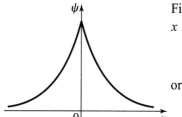

Figure 4.15 The wave function (4.85) for a particle bound in a Dirac delta function potential energy well.

That the delta function well has a single bound state is almost obvious, at least in retrospect. We saw in the discussion of the one-dimensional finite square well in Section 4.1 that there was always at least one bound state solution independent of the width and depth of the well. Moreover, if you look at the wave function in Fig. 4.15 for the Dirac delta function well, the requirement that the wave function be exponentially damped outside the well as well as continuous at the origin means there is no other wave function that we can conceivably sketch, up to an overall phase.

4.5 The Double Well and Molecular Binding

If we imagine that a finite square well is a simple model for the potential energy well confining an electron in an atom such as hydrogen, then a double well such as shown in Fig. 4.16 might be a model for an electron bound in the hydrogen molecule ion H_2^+. In this case, the two potential wells would be centered on each of the protons. By solving the Schrödinger equation for this simple system, we can gain an important insight into why molecules exist.

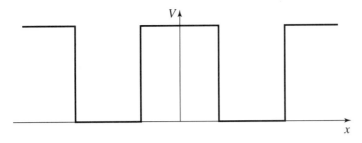

Figure 4.16 The potential energy of a double square well.

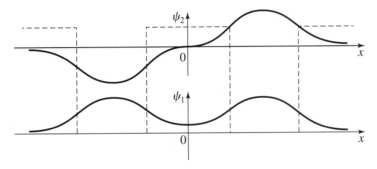

Figure 4.17 The ground state ψ_1 and the first excited state ψ_2 of a double square well potential. Note the exponential nature of the wave function in the region between the wells, a region for which $V > E$.

For a bound-state solution, the solutions to the Schrödinger equation must be oscillatory within the wells and involve exponential functions outside the wells. This eigenvalue problem is certainly more complex than the single finite square well since there are five regions for which we must solve the Schrödinger equation. Moreover, the requirement that the wave function and its first derivative be continuous at each of the boundaries leads to eight boundary conditions. We can simplify things a bit by noting that the potential energy in Fig. 4.16 is an even function $[V(-x) = V(x)]$ and, as was the case for the other systems that we have treated in this chapter, the ground-state wave function is also an even function. Figure 4.17 shows a sketch of what the wave function must look like for the ground state and the first excited state. You might at first have thought that the electron in H_2^+ would be attracted to one or the other of the two protons, but quantum mechanics gives quite a different picture. The electron is equally likely to be in both wells and, moreover, there is a significant chance of finding the electron in the region between the two wells in the ground state. This sharing of the electron between the wells is responsible for the increased binding that makes the molecule stable.

To see more concretely how this works, let's take advantage of the Dirac delta function potential to make a really simple model, namely one in which

$$\frac{2m}{\hbar^2}V(x) = -\frac{\alpha}{a}[\delta(x-a) + \delta(x+a)] \tag{4.89}$$

that is two delta function wells separated by a distance $2a$. The potential energy (4.89) is sketched in Fig. 4.18. The $E < 0$ solutions to the Schrödinger equation must satisfy

$$\frac{d^2\psi}{dx^2} = -\frac{2mE}{\hbar^2}\psi \tag{4.90}$$

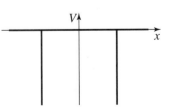

Figure 4.18 A double delta function well.

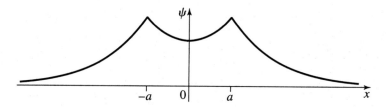

Figure 4.19 The wave function (4.92) for the ground state of a double delta function well.

Here again we define

$$\kappa = \frac{\sqrt{-2mE}}{\hbar} \qquad (4.91)$$

Note that since $E < 0$, κ is real and positive. While we must discard those exponential solutions that blow up as $|x| \to \infty$, in the region between the wells the solution involves both exponentials:

$$\psi(x) = \begin{cases} Ce^{\kappa x} & x < -a \\ A\cosh \kappa x & -a < x < a \\ Ce^{-\kappa x} & x > a \end{cases} \qquad (4.92)$$

where in the region $-a < x < a$ we have written the solution in terms of the combination of increasing and decreasing exponentials that is even, or symmetric, under $x \to -x$:

$$\cosh \kappa x = \frac{e^{\kappa x} + e^{-\kappa x}}{2} \qquad (4.93)$$

It is worth emphasizing that it is okay to have both the increasing and the decreasing exponentials in the region $-a < x < a$ since $|x| < a$ in this region. See Fig. 4.19.

Since the wave function is an even function, it is sufficient to apply the boundary conditions at $x = a$. Continuity of the wave function demands that

$$A\cosh \kappa a = Ce^{-\kappa a} \qquad (4.94)$$

and the condition on the discontinuity of the first derivative in this case becomes

$$\left(\frac{d\psi}{dx}\right)_{a+} - \left(\frac{d\psi}{dx}\right)_{a-} = -\frac{\alpha}{a}\psi(a) \qquad (4.95)$$

or

$$-\kappa Ce^{-\kappa a} - \kappa A \sinh \kappa a = -\frac{\alpha}{a}A\cosh \kappa a \qquad (4.96)$$

where

$$\sinh \kappa x = \frac{e^{\kappa x} - e^{-\kappa x}}{2} \qquad (4.97)$$

Substituting the continuity condition (4.94) into (4.96) and dividing by $\cosh \kappa a$, we find

$$\tanh \kappa a = \frac{\alpha}{\kappa a} - 1 \qquad (4.98)$$

where

$$\tanh \kappa a = \frac{\sinh \kappa a}{\cosh \kappa a} = \frac{e^{\kappa a} - e^{-\kappa a}}{e^{\kappa a} + e^{-\kappa a}} \qquad (4.99)$$

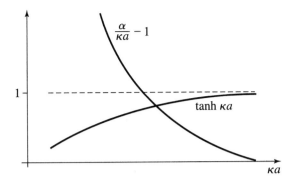

Figure 4.20 A graph of the left and right hand sides of (4.98).

Equation (4.98) is the eigenvalue condition that determines the allowed value of the energy. It is a transcendental equation, which can be solved graphically in a manner similar to the one we used in determining the allowed energies of the finite square well. Figure 4.20 shows a plot of $\tanh \kappa a$ and $\alpha/\kappa a - 1$. Since $\tanh \kappa a$ is always less than one, the intersection must occur when $\alpha/\kappa a - 1 < 1$, or $\alpha/\kappa a < 2$. Going back to the definition (4.91) of κ, we see this implies

$$E < -\frac{\hbar^2 \alpha^2}{8ma^2} \tag{4.100}$$

Comparing this result with the allowed energy $E = -\hbar^2\alpha^2/8ma^2$ for a single delta function well, we see that the "molecule" is more tightly bound than the "atom." This is good news, since otherwise there wouldn't be an energy advantage for molecules to form at all. And life as we know it would not exist without molecules.

What about the first excited state? For the two square wells, the first excited state wave function is an odd wave function, with a node at the origin, as shown in Fig. 4.17. The extra bending of the wave function required to reach this node is indicative of the higher energy of this wave function relative to the ground state.[8] The solutions are of course exponentials between the wells, but in this case involve a $\sinh \kappa x$ rather than a $\cosh \kappa x$. Just as the finite square well is not guaranteed to have an excited state, there are no guarantees that the double well has one either. For the delta function molecule, for example, it is necessary to chose the parameter α appropriately for an excited bound-state solution to exist. See Problem 4.18. In more realistic examples, nature makes these choices. The hydrogen molecule ion H_2^+, for example, does not possess an excited state.

4.6 Scattering and the Step Potential

We have seen that the allowed energies for the bound-state solutions to the Schrödinger equation are quantized, that is the energy spectrum takes on certain discrete values. In this section we will begin our examination of the solutions to the Schrödinger equation for which the particle is not confined. As we will show, in this case there is no constraining condition on the energy. There exists a continuum of allowed solutions. We can generate

---

[8] If the wells are far apart or the barrier between them is high, then the difference in bending between the wave function for the ground state and the first excited state will be small. See Problem 4.18.

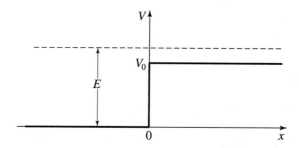

Figure 4.21 The step potential.

physically acceptable wave functions by superposing these solutions to form a wave packet. Since the wave packet includes solutions with different energies, the wave packet evolves in time, in a process that is often characterized as scattering.

As a first step, consider the potential energy

$$V(x) = \begin{cases} 0 & x < 0 \\ V_0 & x > 0 \end{cases} \quad (4.101)$$

Initially, we will assume the energy E is greater than V_0, as indicated in Fig. 4.21. For negative x the Schrödinger equation is

$$\frac{d^2\psi}{dx^2} = -\frac{2mE}{\hbar^2}\psi \qquad x < 0 \quad (4.102)$$

As before we define

$$k = \frac{\sqrt{2mE}}{\hbar} \quad (4.103)$$

and write the general solution in terms of complex exponentials as

$$\psi(x) = Ae^{ikx} + Be^{-ikx} \qquad x < 0 \quad (4.104)$$

As we will see, in scattering problems it is imperative to use complex exponentials instead of trigonometric functions so that we can identify the incoming and reflected waves in the solution. Similarly, the Schrödinger equation for positive x is

$$\frac{d^2\psi}{dx^2} = -\frac{2m(E - V_0)}{\hbar^2}\psi \qquad x > 0 \quad (4.105)$$

If we define

$$k_0 = \frac{\sqrt{2m(E - V_0)}}{\hbar} = \sqrt{k^2 - \frac{2mV_0}{\hbar^2}} \quad (4.106)$$

then since $E > V_0$, the solution is

$$\psi(x) = Ce^{ik_0 x} + De^{-ik_0 x} \qquad x > 0 \quad (4.107)$$

The requirement that the wave function be continuous and have a continuous first derivative yields the following two equations:

$$A + B = C + D \quad (4.108)$$

and

$$ik(A - B) = ik_0(C - D) \quad (4.109)$$

Thus we see that we have only two constraints on the four unknowns A, B, C, and D. We can satisfy these equations in many ways. For example, we are free to choose $D = 0$ (we will justify this choice momentarily), in which case

$$\psi(x) = Ce^{ik_0 x} \qquad x > 0 \tag{4.110}$$

We can easily solve the resulting equations

$$A + B = C \tag{4.111}$$

and

$$ik(A - B) = ik_0 C \tag{4.112}$$

for B and C in terms of A:

$$B = \frac{k - k_0}{k + k_0} A \tag{4.113}$$

$$C = \frac{2k}{k + k_0} A \tag{4.114}$$

Notice that we have obtained a solution for any value of k, that is for any value of the energy. Thus the energy eigenvalues are continuous, consistent with our discussion in Section 4.2, where we argued that the energies take on discrete values only when the particle is confined in a potential well.

You might at first think that we can determine the constant A by normalization of the wave function, but this wave function does not approach zero as $|x| \to \infty$ and is not therefore normalizable. But we can determine the probability of reflection and transmission with the aid of the probability current

$$j_x = \frac{\hbar}{2mi} \left(\psi^* \frac{\partial \psi}{\partial x} - \psi \frac{\partial \psi^*}{\partial x} \right) \tag{4.115}$$

which we introduced in Section 2.5 in our discussion of conservation of probability. For the wave function

$$\psi = \begin{cases} Ae^{ikx} + Be^{-ikx} & x < 0 \\ Ce^{ik_0 x} & x > 0 \end{cases} \tag{4.116}$$

the probability current is given by

$$j_x = \begin{cases} \frac{\hbar k}{m} (|A|^2 - |B|^2) & x < 0 \\ \frac{\hbar k_0}{m} |C|^2 & x > 0 \end{cases} \tag{4.117}$$

Here the positive term indicates a probability current in the positive x direction and the negative term indicates a probability current in the negative x direction. Each term in the probability current is equal to the probability density multiplied by the velocity of propagation ($\hbar k/m$ for $x < 0$, for example). You can now see why we were able to set $D = 0$ in (4.107). Such a term would lead to a probability current traveling in the negative x direction for $x > 0$. If the scattering experiment we are modeling involves particles incident on the step potential from the left, then there is no source of particles traveling toward the barrier from the right.

We can define a reflection coefficient R, a probability of reflection, as the ratio of the magnitudes of reflected current to the incident current:

$$R = \frac{j_{\text{ref}}}{j_{\text{inc}}} = \frac{(\hbar k/m)|B|^2}{(\hbar k/m)|A|^2} = \frac{|B|^2}{|A|^2} \qquad (4.118)$$

Similarly the transmission coefficient T is defined as the ratio of the transmitted probability current to the incident current:

$$T = \frac{j_{\text{trans}}}{j_{\text{inc}}} = \frac{(\hbar k_0/m)|C|^2}{(\hbar k/m)|A|^2} = \frac{k_0|C|^2}{k|A|^2} \qquad (4.119)$$

Note the transmission coefficient is not equal to $|C|^2/|A|^2$ since the particle moves more slowly in the $x > 0$ region than it does in the $x < 0$ region, and the probability current includes this effect.

If we substitute the values for B/A and C/A from (4.113) and (4.114), we find the probabilities of reflection and transmission are given by

$$R = \frac{(k - k_0)^2}{(k + k_0)^2} \qquad (4.120)$$

and

$$T = \frac{4kk_0}{(k + k_0)^2} \qquad (4.121)$$

respectively. It is comforting to check that

$$R + T = 1 \qquad (4.122)$$

consistent with conservation of probability.

It is worth emphasizing how strange these results are from the perspective of classical mechanics. A particle passing over the step potential would slow down without being reflected. But from the perspective of wave mechanics, the results seem natural. After all, a wave propagating on a rope would be reflected and transmitted if there is a discontinuity in the density (and consequently in the speed of propagation).

At the beginning of this section it was noted that the way to generate physically acceptable (i.e., normalizable) solutions from the continuum of allowed solutions to the time-independent Schrödinger equation is to superpose these solutions to form a wave packet just as we did for the free particle (see the discussion in Section 2.6). For the step potential we have been analyzing in this section, for $E > V_0$ this superposition can be written as

$$\Psi(x, t) = \begin{cases} \int_{-\infty}^{\infty} \left[A(k)e^{i(kx - Et/\hbar)} + B(k)e^{-i(kx + Et/\hbar)} \right] dk & x < 0 \\ \int_{-\infty}^{\infty} C(k)e^{i(k_0 x - Et/\hbar)} dk & x > 0 \end{cases} \qquad (4.123)$$

In this integral over k, we typically assume that the amplitude $A(k)$ is peaked at a particular value of k corresponding to an energy E that is greater than V_0. We have inserted the $e^{-iEt/\hbar}$ factor with the values of E given by (4.103). The $A(k)e^{i(kx - Et/\hbar)}$ term is the incident wave in the $x < 0$ region, a wave traveling in the $+x$ direction, while the $B(k)e^{-i(kx + Et/\hbar)}$ term is the reflected wave in the $x < 0$ region, a wave traveling in the $-x$ direction. As indicated in Fig. 4.22, the wave packet starts out to the far left of the potential and travels toward the potential, where it interacts, producing a reflected wave in $x < 0$ region and a transmitted wave in the $x > 0$ region, the $C(k)e^{i(k_0 x - Et/\hbar)}$ term. Given this picture, it is worth reiterating why we dropped the

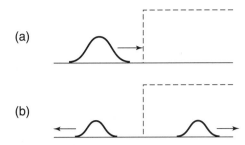

Figure 4.22 A wave packet initially incident on the step potential with $E > V_0$ from the left [in (a)] has an amplitude to be reflected and transmitted, as indicated in (b).

$De^{-ik_0 x}$ term in the solution (4.107) in the $x > 0$ region. When we include the time dependence, this term becomes $De^{-i(k_0 x + Et/\hbar)}$, which is a wave traveling in the $-x$ direction. If the wave packet is incident on the barrier from the left, there is nothing that produces a wave traveling from $+\infty$ *toward* the step in the $x > 0$ region. Such a term, however, could be the *incident* wave for a wave packet that approaches the barrier from the right. Clearly, the mathematics of solving the equation should permit such a situation. We just do not need such a term from the general solution to describe the situation outlined in Fig. 4.22.

Calculating the probability that a particle is reflected or transmitted is not easy using the wave packet approach. We would need to evaluate the integrals in (4.123) and then determine the area under the $|\Psi|^2$ curve for the reflected wave and the transmitted wave. Alternatively, if we assume the incident particle is nearly monoenergetic, then the wave packet will be quite broad and we can get a lot of information from the energy eigenfunctions alone. An energy eigenfunction is, of course, a stationary state. Analyzing scattering via such stationary states is the analogue of investigating the existence of submerged objects in a pond by sending out a steady flow of waves with a particular wavelength. Such waves might be generated by beating up and down for a long period of time in the water with a stick at a fixed frequency, generating a steady-state situation. These waves interact with the submerged object and this leads to waves traveling outward from the submerged object that are different from what would occur in the absence of the object. As an alternative to this steady state investigation, we could simply drop a rock in the pond, which would generate a wave packet that would scatter when it reaches the submerged object.

EXAMPLE 4.3 Determine the reflection coefficient for a particle incident on the step potential of Fig. 4.21 with energy E less than the height of the barrier, that is, $E < V_0$.

SOLUTION As before,

$$\psi(x) = Ae^{ikx} + Be^{-ikx} \qquad x < 0$$

For positive x the wave function is given by

$$\psi(x) = Ce^{-\kappa x} + De^{\kappa x} \qquad x > 0$$

where
$$\kappa = \frac{\sqrt{2m(V_0 - E)}}{\hbar}$$

Here we *must* set $D = 0$, discarding the rising exponential since it blows up as $x \to \infty$. Thus
$$\psi(x) = Ce^{-\kappa x} \qquad x > 0$$

Continuity of the wave function and its derivative at $x = 0$ lead to the equations
$$A + B = C$$
and
$$ik(A - B) = -\kappa C$$

Solving for B/A, we find
$$\frac{B}{A} = \frac{k - i\kappa}{k + i\kappa}$$

Consequently
$$R = \frac{B^*B}{A^*A} = \left(\frac{k + i\kappa}{k - i\kappa}\right)\left(\frac{k - i\kappa}{k + i\kappa}\right) = 1$$

Thus 100% of the particles incident with energy less than the height of the barrier are reflected. How does this square with the fact that the wave function is nonzero within the barrier? If we substitute $Ce^{-\kappa x}$ into the expression for the probability current (4.115), we see that $j_x = 0$. Thus the transmission coefficient T vanishes, consistent with conservation of probability, namely $R + T = 1$. In the next section, we will see that transmission through the barrier can occur, however, if the barrier is of finite width.

Finally, it is worth noting that there is a quick way to determine the reflection coefficient for $E < V_0$ from our result for $E > V_0$. Notice that we can move from the wave function $Ce^{ik_0 x}$ for $x > 0$ to the wave function $Ce^{-\kappa x}$ with the substitution $ik_0 \to -\kappa$. Consequently, we can obtain the result for B/A from the result (4.113) with the substitution $k_0 \to i\kappa$. While this strategy can also work for figuring out C/A when $E < V_0$, it does not mean that you can calculate T by making this substitution in (4.119) since the wave function is no longer a traveling wave in the region $x > 0$. So when in doubt, start with the probability current.

4.7 Tunneling and the Square Barrier

Let's now ask what happens when a particle with energy $0 < E < V_0$ is incident on the energy barrier
$$V(x) = \begin{cases} 0 & x < 0 \\ V_0 & 0 < x < a \\ 0 & x > a \end{cases} \tag{4.124}$$

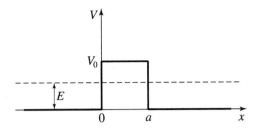

Figure 4.23 The square barrier.

pictured in Fig. 4.23. Classically, the probability of transmission should of course be zero but, as we will now show, quantum mechanics leads to a nonzero probability of transmission, or **tunneling**, as it is called.

For a particle incident on the barrier from the left, we can express the wave function as

$$\psi(x) = \begin{cases} Ae^{ikx} + Be^{-ikx} & x < 0 \\ Fe^{\kappa x} + Ge^{-\kappa x} & 0 < x < a \\ Ce^{ikx} & x > a \end{cases} \quad (4.125)$$

where, as in the previous section,

$$k = \frac{\sqrt{2mE}}{\hbar} \quad (4.126)$$

and

$$\kappa = \frac{\sqrt{2m(V_0 - E)}}{\hbar} \quad (4.127)$$

The important thing to note is that in contrast to the infinitely long barrier of the preceding section the wave function in the region $0 < x < a$ includes an increasing as well as a decreasing exponential. We are not compelled to discard the increasing exponential since the region of the barrier is finite in length. While the increasing exponential would blow up as $x \to \infty$, here x is never larger than a within the barrier.

The boundary conditions arising from continuity of the wave function are

$$A + B = F + G \quad \text{at} \quad x = 0 \quad (4.128)$$

and

$$Fe^{\kappa a} + Ge^{-\kappa a} = Ce^{ika} \quad \text{at} \quad x = a \quad (4.129)$$

and from continuity of the first derivative are

$$ik(A - B) = \kappa(F - G) \quad \text{at} \quad x = 0 \quad (4.130)$$

and

$$\kappa\left(Fe^{\kappa a} - Ge^{-\kappa a}\right) = ikCe^{ika} \quad \text{at} \quad x = a \quad (4.131)$$

Determining the transmission coefficient means solving for C in terms of A, which involves some algebra. See Problem 4.26. The ratio A/C is given by

$$\frac{A}{C} = \frac{e^{ika}}{2ik\kappa}\left[(k^2 - \kappa^2)\sinh\kappa a + 2ik\kappa \cosh\kappa a\right] \quad (4.132)$$

Thus the transmission coefficient is given by

$$T = \frac{j_{\text{trans}}}{j_{\text{inc}}} = \frac{(\hbar k/m)|C|^2}{(\hbar k/m)|A|^2} = \frac{|C|^2}{|A|^2} = \left[1 + \frac{(k^2+\kappa^2)^2}{4k^2\kappa^2}\sinh^2\kappa a\right]^{-1} \quad (4.133)$$

For example, for an electron with energy $E = 5$ eV incident on a barrier of height $V_0 = 10$ eV and width $a = 0.53 \times 10^{-10}$ m, $T = 0.70$. Thus tunneling can be a very common occurrence on the microscopic scale. On the other hand, for a macroscopic particle the transmission coefficient is miniscule. See Example 4.4.

For a "thick" barrier, one for which $\kappa a \gg 1$,

$$\sinh \kappa a = \frac{e^{\kappa a} - e^{-\kappa a}}{2} \cong \frac{1}{2}e^{\kappa a} \gg 1 \quad (4.134)$$

and therefore

$$T \cong \left(\frac{4\kappa k}{k^2+\kappa^2}\right)^2 e^{-2\kappa a} \quad (4.135)$$

Notice that for $\kappa a \gg 1$ the tunneling probability is sensitively dependent on the thickness a of the barrier and the value of κ (which depends on the magnitude of $V_0 - E$). For example, suppose $e^{-2\kappa a} = 10^{-10}$ for the tunneling of an electron through an oxide layer of thickness $a = 1$ nm separating two wires. Since there are on the order of 10^{22} conduction electrons per cubic centimeter, many electrons would tunnel between the wires. If the oxide layer were just five times thicker, $a = 5$ nm instead of 1 nm, then $e^{-2\kappa a} = 10^{-50}$ in this case and tunneling would effectively be prohibited. It is this extreme sensitivity to changes in distance that have led in the past twenty years to the development of scanning tunneling microscopy (STM). Electrons in the surface of the medium being surveyed tunnel across a gap in vacuum to a sharp conducting tip of the microscope. See Fig. 4.24. A change in the distance between the surface and the tip by as little as 0.001 nm can lead to a 2% change in the tunneling current. An image of the surface can be obtained by "dragging" the tip over the surface with a feedback loop designed to keep the current constant. The tip moves up and down, tracing out the contour of the surface being probed.

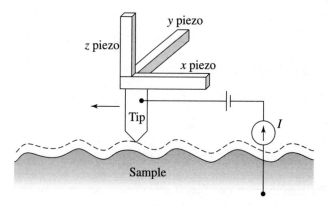

Figure 4.24 A schematic diagram of a scanning tunneling microscope operating in a constant-current mode. Piezoelectric crystals whose dimensions change slightly when an electric field is applied are used to control the motion of the tip.

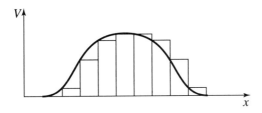

Figure 4.25 A barrier that varies slowly with position can be approximated as a series of "thick" square barriers.

Field emission of electrons: Tunneling through a non-square barrier

For $\kappa a \gg 1$,

$$\ln T \cong \ln\left(\frac{4\kappa k}{k^2 + \kappa^2}\right)^2 - 2\kappa a \approx \text{constant} - 2\kappa a \tag{4.136}$$

where the last step follows since the logarithmic term will have at best a modest dependence on energy [this is another way of saying what is happening in the exponent of the exponential in (4.135) plays the dominant role in transmission through a barrier]. If the barrier through which the particle is tunneling varies slowly with position in comparison with $1/\kappa$, then the barrier can be approximated as a series of square barriers that are thick enough that we can make the approximations that resulted in (4.135). See Fig. 4.25. In this case, it is not a bad overall approximation to write

$$\ln T \approx \text{constant} - 2\int \sqrt{\frac{2m[V(x) - E]}{\hbar^2}}\, dx \tag{4.137}$$

for the whole barrier, where the limits of integration include the region for which $V(x) > E$. This approximation is clearly going to work least well in the vicinity of the classical turning points where $V(x) \approx E$ and hence $\kappa \approx 0$.

An interesting application of the result (4.137) is the field emission of electrons. We noted in the discussion of the photoelectric effect in Section 1.3 that the energy required to eject the most energetic electrons from the metal is the work function W. Another way to liberate these electrons is to change the shape of the potential energy well and allow the electrons to tunnel out. If some negative charge is put on the metal, then near the surface of the metal there will be a constant electric field of magnitude $\mathcal{E}$ (which is proportional to the surface charge density). Since the force $e\mathcal{E}$ is presumed to be constant, the potential energy in the vicinity of the surface is simply $W - e\mathcal{E}x$, where x is the perpendicular distance from the surface and W is potential energy at the surface, as illustrated in Fig. 4.26. Thus the transmission probability is roughly given by

$$T \approx Ce^{-2\sqrt{\frac{2m}{\hbar^2}} \int_0^L \sqrt{W - e\mathcal{E}x}\, dx} \tag{4.138}$$

where C is a constant and $L = W/e\mathcal{E}$ is the distance that the most energetic electrons must tunnel before reaching a region outside the metal where $E > V$. Carrying out the integral in the exponent, we find

$$T \approx Ce^{-\frac{4\sqrt{2m}}{3\hbar e\mathcal{E}} W^{3/2}} \tag{4.139}$$

144   Chapter 4: One-Dimensional Potentials

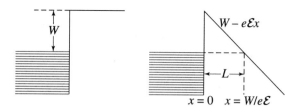

Figure 4.26 The potential energy of an electron in a metal that carries a negative charge. The most energetic electrons in the metal tunnel through a triangular barrier of length $L = W/e\mathcal{E}$, where W is the work function of the metal.

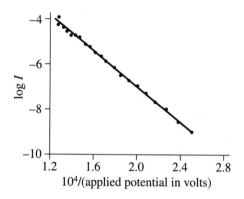

Figure 4.27 The current emitted by a solid which carries a negative charge. The data are from R. A. Millikan and C. C. Lauritsen, *Proc. Nat. Acad. Sci.* **14**, 45 (1928).

This formula is called the Fowler–Nordheim relation. It was one of the early predictions of quantum mechanics. Figure 4.27 shows the data for the logarithm of the current generated by this tunneling as a function of the applied voltage. The straight line drawn through the data, which cover five orders of magnitude variation in the tunneling current, is consistent with the linear behavior of the exponent with respect to the reciprocal of the voltage applied to the metal. We will see another interesting application of tunneling in Chapter 9 when we discuss alpha particle emission from nuclei.

EXAMPLE 4.4 This is an Enrico Fermi problem. Suppose a car of mass $m = 10^3$ kg very slowly (kinetic energy essentially equal to zero) approaches the speed bump shown in Fig. 4.28. Take the length of the bump to be 1 m and the height to be 10 cm. Estimate the probability that the car tunnels through the bump?

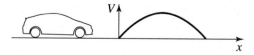

Figure 4.28 Tunneling through a speed bump.

SOLUTION The transmission probability is given approximately by

$$T \approx e^{-(2/\hbar) \int \sqrt{2m(V-E)}\,dx}$$

where the integral is understood to be over the region in which $V > E$. To determine the tunneling probability for the car, we must first evaluate the integral

$$\int \sqrt{V-E}\,dx$$

We set $E = 0$ since the kinetic energy of the car can be taken to be zero. To make the integral easy, we choose the spatial dependence to be $V(x) = mgb \sin^2(\pi x/L)$, where b is the height of the potential energy barrier and L is its length. Then

$$\int_0^L \sqrt{V}\,dx = \int_0^L \sqrt{mgb}\,\sin(\pi x/L)\,dx = L\sqrt{mgb}/\pi$$

and therefore

$$T \approx e^{-2mL\sqrt{2gb}/\hbar\pi}$$

Plugging in the numerical values

$$T \approx e^{-8.5 \times 10^{36}} = 10^{-3.7 \times 10^{36}}$$

Physical quantities don't come much smaller than this! On the other hand, if we replace the 1-tonne car with an electron, the exponent is reduced by a factor of 10^{34}.

4.8 Summary

This chapter is a chapter of examples, and therefore not an easy chapter to summarize. In order to gain experience with how nature behaves in the microscopic world, we have solved the time-independent Schrödinger equation for bound-state solutions for the finite square well, the simple harmonic oscillator, and the Dirac delta function potential. We have also investigated the bound-state solutions of a double-well system, a primitive model for a molecule. We have then looked at scattering in one-dimensional quantum mechanics. Our results are consistent with the general features of the energy eigenfunctions that we outlined in Section 4.1. The energy eigenfunctions are oscillatory in regions in which $E > V$ and exponential-like when $V > E$. The wave functions must be continuous and, unless the potential energy is unusually singular, have a continuous first derivative as well. Physically admissible wave functions are normalizable and hence approach zero as $|x| \to \infty$. For a particle bound in a potential well, the allowed energies take on discrete values; they are quantized. As the energy increases, the wave function becomes increasingly oscillatory in the region in which $E > V$, with an increasing number of nodes. The amplitude of the wave function in the $E > V$ region grows as the kinetic energy of the particle decreases. And for a particle that is not bound, the allowed energies form a continuum. While it is possible to generate normalizable wave functions from the continuum of energy eigenfunctions by superposing eigenfunctions with different energies to form a wave packet, we can also analyze scattering as a steady-state phenomenon with the aid of the probability current.

Problems

4.1. Figure 4.29 shows four wave functions in the region $x > 0$. Indicate for each wave function whether the wave function is an acceptable or unacceptable wave function for an actual physical system. If the wave function is not acceptable, explain why.

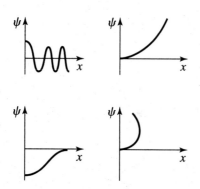

Figure 4.29 Four possible wave functions.

4.2. Are the allowed values of the energy in the ranges $E < 0, 0 < E < V_1, V_1 < E < V_2$, and $V_2 < E < V_3$ for the potential energy in Fig. 4.30 discrete or continuous?

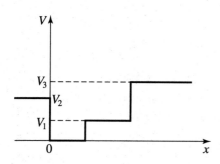

Figure 4.30 The potential energy $V(x)$ for Problem 4.2.

4.3. Sketch the energy eigenfunctions ψ_1, ψ_2, and ψ_3 corresponding to the three lowest energy eigenvalues E_1, E_2, and E_3 for the potential energy shown in Fig. 4.31.

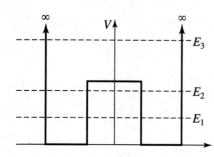

Figure 4.31 A square potential energy bump in the infinite square well.

4.4. The curve in Fig. 4.32 is alleged to be the wave function for the fifth energy level for a particle confined in the one-dimensional potential energy well shown. Explain which aspects of the wave function are qualitatively correct. Indicate the ways in which the wave function is not qualitatively correct. Explain your reasoning fully for each error that you find.

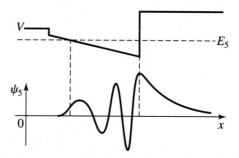

Figure 4.32 A sloped potential energy well and an alleged energy eigenfunction.

4.5. For which of the potential energies in Fig. 4.33 could $\psi(x)$ be an energy eigenfunction? Sketch any lower energy eigenfunctions that are bound to exist.

4.6. Verify that

$$\psi(x) = Ax^n e^{-m\omega x^2/2\hbar}$$

where n is an integer, is an approximate solution to the differential equation

$$-\frac{\hbar^2}{2m}\frac{d^2\psi}{dx^2} + \frac{1}{2}m\omega^2 x^2 \psi = E\psi$$

for sufficiently large $|x|$.

4.7. Normalize the wave function (4.21).

4.8. The energy eigenfunctions ψ_1, ψ_2, ψ_3, and ψ_4 corresponding to the four lowest energy states for a particle confined in the finite potential well

$$V(x) = \begin{cases} -V_0 & |x| < a/2 \\ 0 & |x| > a/2 \end{cases}$$

are sketched in Fig. 4.4. For which of these energy eigenfunctions would the probability of finding the particle outside the well, that is, in the region $|x| > a/2$, be greatest? Explain. Justify your reasoning using the solution to the Schrödinger equation in the region $x > a/2$.

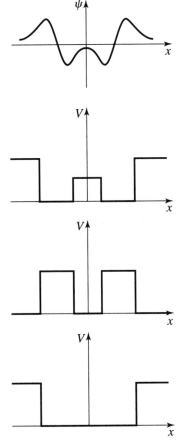

Figure 4.33 A wave function and three possible potential energy functions.

4.9. A particle of mass m is bound in a one-dimensional well with one impenetrable wall. The potential energy is given by

$$V(x) = \begin{cases} \infty & x < 0 \\ -V_0 & 0 < x < a \\ 0 & x > a \end{cases}$$

(a) Solve the Schrödinger equation for $E < 0$ inside and outside the well (See Fig. 4.34). Apply the boundary conditions at $x = 0$ and $x = a$ to obtain an equation that determines the allowed values of E. (b) Show that there will be no bound state unless $2mV_0a^2/\hbar^2 \geq \pi^2/4$. *Suggestion*: Note the similarity of this problem to the solution of the finite square well for the odd wave functions. Follow the procedure outlined in Section 4.1. (c) This potential energy well is used in first attempts to describe the deuteron as a bound state of a proton and a neutron. The problem is, of course, really three dimensional, but the Schrödinger equation for states with zero angular momentum is the same as that given in (a) with the radius r replacing x, and m replaced with $m_p m_n/(m_p + m_n)$, the reduced mass of the proton–neutron system. This system has just one bound state, the deuteron. Take the width of the well to be $a = 1.4 \times 10^{-15}$ m and assume the deuteron is just barely bound. Obtain a numerical value for the depth of the well. The observed binding energy of the deuteron is $E = -2.2$ MeV. Is your assumption that $V_0 \gg |E|$ consistent? Sketch the ground-state wave function.

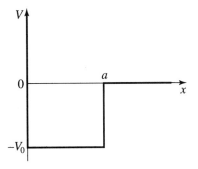

Figure 4.34 A square well bounded by an infinite potential energy barrier at the origin.

4.10. The energy eigenvalues and eigenfunctions of the simple harmonic oscillator are given in Section 4.3. What are the energy eigenvalues for the "half" harmonic oscillator potential energy

$$V(x) = \begin{cases} \infty & x < 0 \\ \frac{1}{2}m\omega^2 x^2 & x > 0 \end{cases}$$

shown in Fig. 4.35. Sketch the corresponding energy eigenfunctions for the three lowest energy states.

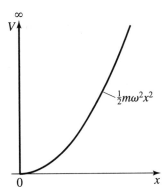

Figure 4.35 A half harmonic oscillator potential energy function.

4.11. In Section 4.3 we saw that

$$\int_{-\infty}^{\infty} e^{-bx^2} dx = \sqrt{\frac{\pi}{b}}$$

(a) By taking the derivative of this equation with respect to b, show that

$$\int_{-\infty}^{\infty} x^2 e^{-bx^2} dx = \frac{1}{2}\sqrt{\frac{\pi}{b^3}}$$

(b) Evaluate

$$\int_{-\infty}^{\infty} x^4 e^{-bx^2} dx$$

4.12. A particle of mass m moves in the harmonic oscillator potential energy

$$V(x) = \frac{1}{2} m\omega^2 x^2$$

The normalized energy eigenfunctions are denoted by $\psi_n(x)$ and the corresponding energies are $E_n = (n + \frac{1}{2})\hbar\omega$, $n = 0, 1, 2, \ldots$ Suppose that at time $t = 0$ the wave function of the particle is

$$\Psi(x) = \frac{\sqrt{3}}{2}\psi_0(x) + \frac{1-i}{2\sqrt{2}}\psi_1(x)$$

(a) Determine the time dependence of the wave function. That is, what is $\Psi(x, t)$? (b) What is the probability of obtaining $\hbar\omega/2$ if a measurement of the energy of the particle is made? Of obtaining $3\hbar\omega/2$? Of obtaining $5\hbar\omega/2$? Do these probabilities vary with time? Justify your answer. What are $\langle E \rangle$ and ΔE at time t?

4.13. Use the result of Problem 4.11a to determine Δx for the ground state of the simple harmonic oscillator.

4.14. (a) Without using exact mathematics—using only arguments of curvature, symmetry, and semiquantitative estimates of wavelength—sketch the energy eigenfunctions for the ground state and first excited and second excited states of a particle in the potential energy well

$$V(x) = a|x|$$

See Fig. 4.36. This potential energy has been suggested as arising from the force exerted by one quark on another quark. (b) Estimate the ground-state energy of a particle of mass m for the potential energy

$$V(x) = a|x|$$

by using the uncertainty principle.

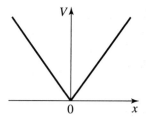

Figure 4.36 A linear potential energy function.

4.15. The one-dimensional time-independent Schrödinger equation for the potential energy discussed in Problem 4.14 is

$$\frac{d^2\psi}{dx^2} + \frac{2m}{\hbar^2}(E - a|x|)\psi = 0$$

Define $E = \varepsilon(\hbar^2 a^2/m)^{1/3}$ and $x = z(\hbar^2/ma)^{1/3}$.
(a) Show that ε and z are dimensionless. (b) Show that the Schrödinger equation can be expressed in the form

$$\frac{d^2\psi}{dz^2} + 2(\varepsilon - |z|)\psi = 0$$

(c) Numerically integrate this equation for various values of ε, beginning with $d\psi/dz = 0$ at $z = 0$, to find the value of ε corresponding to the ground-state eigenfunction. *Note*: To solve the Schrödinger equation numerically, we convert the differential equation into a finite difference equation through the transposition $z \to j\delta$, where δ is the step size of the iteration and j is an integer, that is, $\psi(z) \to \psi(z_j) = \psi_j$ and $V(z) \to V(z_j) = V_j$. We can determine ψ_{j+1} from ψ_j. Using the Taylor series expansions

$$\psi(z + \delta) = \psi(z) + \left(\frac{d\psi}{dz}\right)_z \delta + \frac{1}{2}\left(\frac{d^2\psi}{dz^2}\right)_z \delta^2 + \cdots$$

and

$$\psi(z - \delta) = \psi(z) - \left(\frac{d\psi}{dz}\right)_z \delta + \frac{1}{2}\left(\frac{d^2\psi}{dz^2}\right)_z \delta^2 + \cdots$$

and adding these two series together, we obtain

$$\psi(z + \delta) = -\psi(z - \delta) + 2\psi(z) + \left(\frac{d^2\psi}{dz^2}\right)_z \delta^2 + \cdots$$

which is good to order δ^4 since the δ^3 terms cancel just as did the terms of order δ. Replacing the second derivative from the Schrödinger equation, we find

$$\psi_{j+1} = -\psi_{j-1} + 2\psi_j - 2\delta^2\left(\epsilon - |z_j|\right)\psi_j$$

To get started, that is, to calculate ψ_2, we need to know ψ_0 and ψ_1. For the ground state, $(d\psi/dz)_{z=0} = 0$. Thus $\psi_0 = \psi_1$ through first order. Or you can use the Schrödinger equation itself to find ψ_1 to second order from the Taylor series:

$$\psi_1 = \psi_0 - \delta^2 \epsilon \psi_0$$

Suggestion: The value for ϵ from part (*b*) of Problem 4.14 is a good starting point for your calculation. Move up or down until you find a region for which a small change in ϵ causes the wave function to change from diverging "up" to diverging "down," as suggested by Fig. 4.9.

4.16. A particle of mass m rests upon an impenetrable horizontal floor located at $z = 0$ in a uniform gravitational field. (*a*) What is the one-dimensional Hamiltonian (energy operator) for this system? Sketch the potential energy. (*b*) Sketch the wave function for the ground state of the Hamiltonian as a function of z. Indicate on your sketch the approximate location of $\langle z \rangle$, the average height of the particle above the floor. (*c*) Assume you can approximate $\langle z \rangle \cong \Delta z$, where Δz is the uncertainty in the position of the particle. Use the Heisenberg uncertainty principle

$$\Delta z \Delta p_z \geq \hbar/2$$

to estimate $\langle z \rangle$ for the lowest energy state, the ground state. Obtain numerical values for $\langle z \rangle$ for $m = 1$ kg (a macroscopic particle) and $m = 10^{-30}$ kg (an electron).

4.17. Suppose a molecule is in the superposition of states

$$\Psi(x) = \frac{1}{\sqrt{2}} \psi_0(x) + \frac{1}{\sqrt{2}} \psi_1(x)$$

where ψ_0 is the ground-state wave function with energy E_0 and ψ_1 is the first-excited-state wave function with energy E_1. What is $\Psi(x, t)$? Show that the motion is periodic and determine the period and hence frequency of oscillation. Show that this frequency is the same as that of a photon that would be emitted in the transition between these two energy states.

4.18. (*a*) Sketch the wave function for the first excited state of a double delta function potential energy well. (*b*) Show that the transcendental equation that results from satisfying the boundary conditions for the potential energy (4.89) may or may not have a solution depending on the value of α.

4.19. The eigenfunctions of the position operator satisfy

$$x_{op} \psi_{x_0}(x) = x_0 \psi_{x_0}(x)$$

Argue that the position eigenfunction corresponding to the particular position x_0 is proportional to the Dirac delta function

$$\psi_{x_0}(x) = \delta(x - x_0)$$

4.20. (*a*) At time $t = 0$ a position measurement locates a particle in the potential energy box

$$V(x) = \begin{cases} 0 & 0 < x < L \\ \infty & \text{elsewhere} \end{cases}$$

to be in the vicinity of the center of the box, $x = L/2$. Assume that we approximate the particle's (unnormalized) wave function to be $\delta(x - L/2)$, a Dirac delta function. That is, we take $\Psi(x) = \delta(x - L/2)$. Find the relative probabilities P_n that a measurement of the particles energy will yield E_n, for all n. (*b*) Determine $\Psi(x, t)$, the (unnormalized) wave function of the particle at time t. Do the probabilities that you determined in (*a*) vary with time? Explain why or why not.

4.21. Show that

$$\frac{d\Theta(x)}{dx} = \delta(x)$$

where

$$\Theta(x) = \begin{cases} 1 & x > 0 \\ 0 & x < 0 \end{cases}$$

is a step function and $\delta(x)$ is a Dirac delta function. *Suggestion*: Start with

$$\int_{-\infty}^{x} \delta(y)\, dy = \begin{cases} 1 & x > 0 \\ 0 & x < 0 \end{cases}$$

4.22. Discuss qualitatively how the energy difference between the ground state and the first excited state varies with separation of the potential energy wells shown in Fig. 4.16.

4.23. In Chapter 1 we saw in the reflection of light from an interface with a medium in which the light slows down that there is a phase change of π upon reflection. Show in the analogous situation for the Schrödinger equation that there is no phase change. When does a phase change occur? *Suggestion*: See Section 4.6.

4.24. (*a*) Verify that the wave function $\Psi = Ce^{-\kappa x - iEt/\hbar}$ (see Example 4.3) in the region $x > 0$ for the step potential of Section 4.6 leads to zero probability current in this region. (*b*) Use the conservation of probability equation

$$\frac{\partial \Psi^* \Psi}{\partial t} = -\frac{\partial j_x}{\partial x}$$

to argue that the probability current vanishes in the region $x < 0$ as well for this energy eigenfunction. What can you therefore conclude about the magnitude of the reflection coefficient?

4.25. Solve the time-independent Schrödinger equation for a particle of mass m and energy $E > V_0$ incident from the left on the step potential

$$V(x) = \begin{cases} V_0 & x < 0 \\ 0 & x > 0 \end{cases}$$

See Fig. 4.37. Determine the reflection coefficient R and the transmission coefficient T. Verify that probability is conserved.

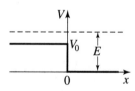

Figure 4.37 A particle with energy $E > V_0$ is incident from the left on the step potential.

4.26. Solve equations (4.128) through (4.131) for the ratio A/C and verify that the transmission coefficient T for tunneling through a square barrier is given by (4.133), namely

$$T = \left[1 + \frac{(k^2 + \kappa^2)^2}{4k^2\kappa^2} \sinh^2 \kappa a\right]^{-1}$$

where

$$k = \frac{\sqrt{2mE}}{\hbar} \quad \text{and} \quad \kappa = \frac{\sqrt{2m(V_0 - E)}}{\hbar}$$

4.27. There is a fair amount of algebra involved in doing Problem 4.26. But if you have done this algebra, it is not so difficult to determine the probability of transmission for a particle with energy $E > V_0$ incident on a square barrier of width a and height V_0. Use the "substitution strategy" outlined in Example 4.3 to show that

$$T = \left[1 + \frac{(k^2 - k_0^2)^2}{4k^2 k_0^2} \sin^2 k_0 a\right]^{-1}$$

where

$$k = \frac{\sqrt{2mE}}{\hbar} \quad \text{and} \quad k_0 = \frac{\sqrt{2m(E - V_0)}}{\hbar}$$

in this case. *Caution*: Since the substitution strategy involves the change from real to complex exponentials in the solution in the region $0 < x < a$, it is safest to make the substitution in the ratio A/C.

4.28. (*a*) Explain why it is okay to make the substitution $V_0 \to -V_0$ in the results of Problem 4.27 to determine the probability of transmission for a particle with energy E incident on the potential energy well

$$V(x) = \begin{cases} 0 & x < 0 \\ -V_0 & 0 < x < a \\ 0 & x > a \end{cases}$$

(*b*) Show that for certain incident energies there is 100% transmission at just those values of the kinetic energy $E + V_0$ of the particle in the region $0 < x < a$ that would be energy eigenvalues for the particle were it trapped in a potential well of width a extending upward to infinity from $-V_0$. (*c*) The diameter of the krypton atom is about 4.1 Å. Suppose that an electron with 0.7 eV kinetic energy encounters a one-dimensional square well with a width of 4.1 Å, representing a crude model of the krypton atom. What must be the depth of the well for 100% transmission? This unusual transparency to electron scattering of certain noble gases is known as the Ramsauer–Townsend effect.

4.29. (*a*) Solve the time-independent Schrödinger equation

$$-\frac{\hbar^2}{2m}\frac{d^2\psi}{dx^2} + V\psi = E\psi$$

for a particle of mass m with energy $0 < E < V_1$ for the potential energy

$$V(x) = \begin{cases} \infty & x < 0 \\ -V_0 & 0 < x < a \\ V_1 & a < x < b \\ 0 & x > b \end{cases}$$

as indicated in Fig. 4.38. Write your solution in a form that is appropriate for particles incident from the right.

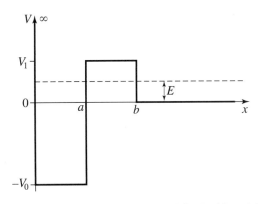

Figure 4.38 The potential energy $V(x)$ for Problem 4.29.

Define clearly any parameters you introduce in solving the equation. (*b*) Apply the appropriate boundary conditions to obtain a set of equations involving any arbitrary constants that appear in your solution. Do *not* solve the resulting set of equations. (*c*) A beam of particles with energy $E < V_1$ is incident from the right. What combination of constants appearing in your solution would you evaluate to determine the reflection coefficient? Without explicit evaluation, can you say what the reflection coefficient would be? Explain your reasoning.

4.30. Calculate the reflection and transmission coefficients for scattering from the potential energy barrier

$$\frac{2mV(x)}{\hbar^2} = \frac{\alpha}{a}\delta(x)$$

where α/a is a constant and $\delta(x)$ is a Dirac delta function. Assume the particle with mass m and energy E is incident on the barrier from the left. Recall from Section 4.4 that the derivative of the wave function is not continuous at the origin but rather satisfies

$$\left(\frac{d\psi}{dx}\right)_{0^+} - \left(\frac{d\psi}{dx}\right)_{0^-} = \frac{\alpha}{a}\psi(0)$$

4.31. A particle of mass m is projected with energy $E = V_0$ at the potential energy barrier

$$V(x) = \begin{cases} 0 & x < 0 \\ V_0 & 0 < x < a \\ 0 & x > a \end{cases}$$

See Fig. 4.39. Determine the most general solution to the time-independent Schrödinger equation in the regions $x < 0$, $0 < x < a$, and $x > a$. Apply the appropriate boundary conditions and determine the transmission coefficient for a particle projected at the barrier from the left. Check that your result behaves appropriately in the limit $a \to 0$. *Reminder*: The most general solution to a second-order differential equation has two arbitrary constants. This is true in each of the regions $x < 0$, $0 < x < a$, and $x > a$.

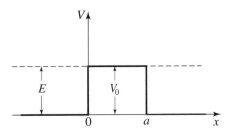

Figure 4.39 A potential energy barrier of height V_0 for scattering of a particle with $E = V_0$.

CHAPTER 5

Principles of Quantum Mechanics

Operators play a fundamental role in quantum mechanics. So far we have introduced the position, momentum, and energy operators. After adding the parity operator to our quiver of operators, we go on to examine some of the fundamental properties of these so-called Hermitian operators. We will also examine the fundamental role played by commutation relations of operators in quantum mechanics.

5.1 The Parity Operator

We have noted that the time-independent Schrödinger equation

$$-\frac{\hbar^2}{2m}\frac{\partial^2 \psi}{\partial x^2} + V(x)\psi = E\psi \tag{5.1}$$

is also the energy eigenvalue equation

$$H\psi_E = E\psi_E \tag{5.2}$$

where

$$H = -\frac{\hbar^2}{2m}\frac{\partial^2}{\partial x^2} + V(x) \tag{5.3}$$

is the energy operator and we have added a subscript E on the wave function in (5.2) to emphasize that this particular eigenfunction corresponds to the eigenvalue E. As we have seen in Chapter 4, the energy eigenfunctions and eigenvalues can be determined once we know the potential energy $V(x)$. We have also noted in the previous chapter that when the potential energy $V(x)$ is an even function of x [$V(x) = V(-x)$], then the energy eigenfunctions are either even or odd functions. This behavior of the wave functions under $x \to -x$ is closely connected with the parity operator.

The parity operator Π is defined by

$$\Pi\psi(x) = \psi(-x) \tag{5.4}$$

that is, when the parity operator acts on a wave function it inverts the coordinates. The eigenvalue equation for the parity operator

$$\Pi\psi_\lambda(x) = \lambda\psi_\lambda(x) \tag{5.5}$$

where λ is a constant, can be easily solved by noting that inverting the coordinates twice must get us right back where we started, that is, Π^2 is the identity operator.[1] Since

$$\Pi^2\psi_\lambda(x) = \Pi\lambda\psi_\lambda(x) = \lambda\Pi\psi_\lambda(x) = \lambda^2\psi_\lambda(x) \tag{5.6}$$

and $\Pi^2 = 1$, we see that

$$\lambda^2 = 1 \quad\text{or}\quad \lambda = \pm 1 \tag{5.7}$$

The eigenfunction corresponding to the eigenvalue $\lambda = +1$ satisfies

$$\Pi\psi_{\lambda=1}(x) = \psi_{\lambda=1}(x) \tag{5.8}$$

but given (5.4), this means

$$\psi_{\lambda=1}(-x) = \psi_{\lambda=1}(x) \tag{5.9}$$

Thus this eigenfunction is an even function. Similarly, for $\lambda = -1$

$$\Pi\psi_{\lambda=-1}(x) = -\psi_{\lambda=-1}(x) \tag{5.10}$$

and hence

$$\psi_{\lambda=-1}(-x) = -\psi_{\lambda=-1}(x) \tag{5.11}$$

Thus the eigenfunction corresponding to $\lambda = -1$ is an odd function. The eigenfunctions of the parity operator are simply even and odd functions.

It is easy to show that any function can be written as a superposition of even and odd functions. Note that

$$\psi(x) = \frac{1}{2}[\psi(x) + \psi(-x)] + \frac{1}{2}[\psi(x) - \psi(-x)] \tag{5.12}$$

and the first term in the brackets is an even function while the second term is an odd one under the transformation $x \to -x$. Thus the eigenfunctions of the parity operator form a complete set.

EXAMPLE 5.1 Argue that the eigenfunctions of the parity operator corresponding to distinct eigenvalues satisfy the orthogonality condition

$$\int_{-\infty}^{\infty} \psi^*_{\lambda=-1}(x)\psi_{\lambda=1}(x)\,dx = 0$$

---

[1] A couple of words about our notation are probably in order. We use the symbol Π instead of P for the parity operator to avoid confusion with momentum. Since the eigenvalues are dimensionless, there is no obvious symbol to assign to them. In linear algebra, the symbol λ (not to be confused with wavelength) is often used in the general eigenvalue problem, and we have followed suit here.

SOLUTION Since $\psi_{\lambda=1}(x)$ is an even function of x and $\psi_{\lambda=-1}(x)$ is an odd function of x, the product $\psi^*_{\lambda=-1}(x)\psi_{\lambda=1}(x)$ is an odd function. The integral in the orthogonality condition is the area under the curve of $\psi^*_{\lambda=-1}(x)\psi_{\lambda=1}(x)$ from $-\infty$ to ∞. Since the integrand is an odd function and the integral is over all x, there is exactly as much negative area as positive area under the curve.

Alternatively, make the change of variables $x' = -x$:

$$\int_{-\infty}^{\infty} \psi^*_{\lambda=-1}(x)\psi_{\lambda=1}(x)\,dx = -\int_{\infty}^{-\infty} \psi^*_{\lambda=-1}(-x')\psi_{\lambda=1}(-x')\,dx'$$

$$= -\int_{-\infty}^{\infty} \psi^*_{\lambda=-1}(x')\psi_{\lambda=1}(x')\,dx'$$

In going from the first line to the second line, we have switched the limits of integration (which absorbs one minus sign) and taken advantage of the fact that $\psi_{\lambda=1}(-x') = \psi_{\lambda=1}(x')$ and $\psi_{\lambda=-1}(-x') = -\psi_{\lambda=-1}(x')$, which yields a second minus sign. Thus, the expression on the right is equal to the negative of itself (you can now set dummy variable x' to x on the right-hand side to make this more evident) and consequently must be zero.

5.2 Observables and Hermitian Operators

The parity operator has real eigenvalues. The eigenfunctions corresponding to distinct eigenvalues are orthogonal. And the eigenfunctions form a complete set. These properties are true for a general class of operators called **Hermitian operators**. We postulate that the operators in quantum mechanics that correspond to observables, to the things we can measure (such as energy), are Hermitian operators. A linear operator A_{op} corresponding to the observable A is Hermitian provided it satisfies

$$\int_{-\infty}^{\infty} \phi^* \left(A_{op}\psi\right) dx = \int_{-\infty}^{\infty} \left(A_{op}\phi\right)^* \psi \, dx \tag{5.13}$$

where ϕ and ψ are physical wave functions. We insert the parentheses in the right-hand side of (5.13) to indicate that A_{op} acts on the function ϕ but not on ψ and that we are taking the complex conjugate of the function $A_{op}\phi$. We could rewrite the left-hand side as $\int_{-\infty}^{\infty} \phi^* A_{op}\psi\, dx$, without the parentheses shown on the left-hand side of (5.13). The parentheses aren't strictly necessary since the only function to the right of A_{op} on which the operator can act is ψ.

The defining relationship (5.13) for a Hermitian operator looks pretty abstract. But if we take the special case of $\phi = \psi$, then (5.13) reduces to

$$\int_{-\infty}^{\infty} \psi^* \left(A_{op}\psi\right) dx = \int_{-\infty}^{\infty} \left(A_{op}\psi\right)^* \psi \, dx \tag{5.14}$$

Recall that the left-hand side of (5.14) is the expectation value of A, while the right-hand side is just the complex conjugate of the left-hand side since the complex conjugate of ψ^* is ψ and the complex conjugate of the function $A_{op}\psi$ is $\left(A_{op}\psi\right)^*$. Therefore (5.14) states that

$$\langle A \rangle = \langle A \rangle^* \tag{5.15}$$

Thus a Hermitian operator is one that yields real expectation values, which is certainly a necessary condition for an observable.[2]

Next consider the eigenvalue equation

$$A_{op}\psi_a = a\psi_a \qquad (5.16)$$

where A_{op} is a Hermitian operator and ψ_a is an eigenfunction of A_{op} with eigenvalue a. If we calculate the expectation value of A assuming the wave function is an eigenfunction, we obtain

$$\langle A \rangle = \int_{-\infty}^{\infty} \psi_a^* A_{op} \psi_a \, dx = \int_{-\infty}^{\infty} \psi_a^* a \psi_a \, dx = a \qquad (5.17)$$

presuming that the eigenfunction is normalized. Thus the requirement that $\langle A \rangle = \langle A \rangle^*$ means that the eigenvalues are real: $a = a^*$.

It is also straightforward to show that the eigenfunctions of a Hermitian operator corresponding to distinct eigenvalues are orthogonal. If we assume that ψ_1 and ψ_2 are eigenfunctions corresponding to distinct eigenvalues a_1 and a_2, respectively

$$A_{op}\psi_1 = a_1 \psi_1 \qquad \text{and} \qquad A_{op}\psi_2 = a_2 \psi_2 \qquad (5.18)$$

then

$$\int_{-\infty}^{\infty} \psi_2^* A_{op} \psi_1 \, dx = \int_{-\infty}^{\infty} (A_{op}\psi_2)^* \psi_1 \, dx \qquad (5.19)$$

since A_{op} is presumed to be Hermitian. But taking advantage of (5.18), this equation becomes

$$\int_{-\infty}^{\infty} \psi_2^* a_1 \psi_1 \, dx = \int_{-\infty}^{\infty} (a_2 \psi_2)^* \psi_1 \, dx \qquad (5.20)$$

or

$$(a_1 - a_2) \int_{-\infty}^{\infty} \psi_2^* \psi_1 \, dx = 0 \qquad (5.21)$$

Therefore, since the eigenvalues are distinct, we must have

$$\int_{-\infty}^{\infty} \psi_2^* \psi_1 \, dx = 0 \qquad a_1 \neq a_2 \qquad (5.22)$$

This is the same definition of orthogonality that we used in our discussion of the properties of the energy eigenfunctions for the particle in a box in Chapter 3 [see (3.45)]. But now we see that it applies to the eigenfunctions of any Hermitian operator corresponding to an observable.

The eigenfunctions $\{\psi_n\}$ form an orthonormal basis, at least if the eigenvalues are discrete. Orthonormality is a mix of orthogonality and normalization and can be conveniently expressed as

$$\int_{-\infty}^{\infty} \psi_m^* \psi_n \, dx = \delta_{mn} \qquad (5.23)$$

---

[2]While (5.14) looks less general than (5.13), we can start from the requirement that the expectation value be real for an arbitrary square-integrable wave function and derive (5.13). See Problem 5.6.

where again

$$\delta_{mn} = \begin{cases} 1 & m = n \\ 0 & m \neq n \end{cases} \quad (5.24)$$

is the Kronecker delta. The fact that the eigenfunctions forms a basis means that any wave function Ψ can be expressed as a superposition of the eigenfunctions

$$\Psi = c_1\psi_1 + c_2\psi_2 + c_3\psi_3 + \cdots = \sum_n c_n \psi_n \quad (5.25)$$

that is, the eigenfunctions are complete. In Section 5.1 we demonstrated that the eigenfunctions of the parity operator are complete.

In general, proving completeness is not as easy as proving orthogonality, but completeness is an essential part of quantum mechanics. For if Ψ is normalized, then

$$\begin{aligned} 1 &= \int_{-\infty}^{\infty} \Psi^* \Psi \, dx = \int_{-\infty}^{\infty} \left(\sum_m c_m^* \psi_m^* \right) \left(\sum_n c_n \psi_n \right) dx \\ &= \sum_m c_m^* \sum_n c_n \int_{-\infty}^{\infty} \psi_m^* \psi_n \, dx \\ &= \sum_{m,n} c_m^* c_n \delta_{mn} = \sum_n c_n^* c_n = \sum_n |c_n|^2 \end{aligned} \quad (5.26)$$

Thus it is natural to identify $|c_n|^2$ as the probability of obtaining the eigenvalue a_n if a measurement of the observable A is carried out. While we made a similar statement earlier, in Section 3.3 for the particle in a box, here we are extending this ansatz to the eigenfunctions of any operator corresponding to an observable. If the eigenfunctions were not a complete set, if we could not presume that we could write any wave function as a superposition of the eigenfunctions as shown in (5.25), then we would have no way of calculating these probabilities for an arbitrary wave function. Because of the orthonormality of the eigenfunctions, it is straightforward to determine c_n given a particular wave function Ψ:

$$\begin{aligned} \int_{-\infty}^{\infty} \psi_n^* \Psi \, dx &= \int_{-\infty}^{\infty} \psi_n^* \left(\sum_m c_m \psi_m \right) dx \\ &= \sum_m c_m \int_{-\infty}^{\infty} \psi_n^* \psi_m \, dx = \sum_m c_m \delta_{nm} = c_n \end{aligned} \quad (5.27)$$

Of course, since $|c_n|^2$ is the probability of obtaining the eigenvalue a_n, then the expectation value of the observable A is given by

$$\langle A \rangle = \sum_n |c_n|^2 a_n \quad (5.28)$$

However,

$$\begin{aligned} \int_{-\infty}^{\infty} \Psi^* A_{\text{op}} \Psi \, dx &= \int_{-\infty}^{\infty} \left(\sum_m c_m^* \psi_m^* \right) A_{\text{op}} \left(\sum_n c_n \psi_n \right) dx \\ &= \int_{-\infty}^{\infty} \left(\sum_m c_m^* \psi_m^* \right) \left(\sum_n c_n a_n \psi_n \right) dx \\ &= \sum_m c_m^* \sum_n a_n c_n \int_{-\infty}^{\infty} \psi_m^* \psi_n \, dx \\ &= \sum_{m,n} c_m^* c_n a_n \delta_{mn} = \sum_n c_n^* c_n a_n = \sum_n |c_n|^2 a_n \end{aligned} \quad (5.29)$$

and therefore we see quite generally that

$$\langle A \rangle = \int_{-\infty}^{\infty} \Psi^* A_{\text{op}} \Psi \, dx \qquad (5.30)$$

Similarly

$$\langle A^2 \rangle = \int_{-\infty}^{\infty} \Psi^* A_{\text{op}}^2 \Psi \, dx \qquad (5.31)$$

and therefore

$$(\Delta A)^2 = \langle A^2 \rangle - \langle A \rangle^2 = \int_{-\infty}^{\infty} \Psi^* A_{\text{op}}^2 \Psi \, dx - \left(\int_{-\infty}^{\infty} \Psi^* A_{\text{op}} \Psi \, dx \right)^2 \qquad (5.32)$$

EXAMPLE 5.2 Show that the linear momentum operator

$$p_{x_{\text{op}}} = \frac{\hbar}{i} \frac{\partial}{\partial x}$$

is a Hermitan operator.

SOLUTION Our goal is to show that

$$\int_{-\infty}^{\infty} \phi^* p_{x_{\text{op}}} \psi \, dx = \int_{-\infty}^{\infty} (p_{x_{\text{op}}} \phi)^* \psi \, dx$$

Start with the expression on the left and integrate by parts (see footnote 12 in Section 2.9)

$$\int_{-\infty}^{\infty} \phi^* \frac{\hbar}{i} \frac{\partial \psi}{\partial x} \, dx = -\frac{\hbar}{i} \int_{-\infty}^{\infty} \left(\frac{\partial \phi}{\partial x} \right)^* \psi \, dx + \frac{\hbar}{i} \phi^* \psi \Big|_{-\infty}^{\infty}$$

$$= \int_{-\infty}^{\infty} \left(\frac{\hbar}{i} \frac{\partial \phi}{\partial x} \right)^* \psi \, dx$$

where we have taken advantage of the fact that ϕ and ψ vanish as $|x| \to \infty$ to eliminate the "surface" term. Notice that the fact that the linear momentum operator is Hermitian depends on the nature of the operator itself (consider whether this operator would be Hermitian without the "i", for example), but it also depends on the behavior of the wave functions on which this operator acts, namely that these wave functions vanish at infinity.

EXAMPLE 5.3 Assume the operator A_{op} corresponding to an observable of a particle has just two eigenfunctions $\psi_1(x)$ and $\psi_2(x)$ with distinct eigenvalues a_1 and a_2, respectively. Thus a wave function corresponding to an arbitrary state of the particle can be written as

$$\psi(x) = c_1 \psi_1(x) + c_2 \psi_2(x)$$

An operator B_{op} is defined according to

$$B_{\text{op}} \psi(x) = c_2 \psi_1(x) + c_1 \psi_2(x)$$

What are the eigenvalues and normalized eigenfunctions of B_{op}?

SOLUTION Start with the eigenvalue equation $B_{op}\psi = b\psi$, that is

$$B_{op}\psi = c_2\psi_1(x) + c_1\psi_2(x) = b\left[c_1\psi_1(x) + c_2\psi_2(x)\right]$$

Since ψ_1 and ψ_2 are orthogonal basis functions, the coefficients multiplying ψ_1 on both sides of this equation must be equal (alternatively, multiply both sides by ψ_1^* and integrate over all x, taking advantage of the orthogonality of ψ_1 and ψ_2). Similarly, the coefficients multiplying ψ_2 must be equal. Thus we really have two equations

$$c_2 = bc_1 \quad \text{and} \quad c_1 = bc_2$$

If we substitute the second of these equations into the first, for example, we obtain $c_2 = b^2 c_2$. Thus $b^2 = 1$, or $b = \pm 1$. For $b = 1$, $c_2 = c_1$, while for $b = -1$, $c_2 = -c_1$. Thus the two normalized eigenfunctions are

$$\psi_{b=1} = \frac{1}{\sqrt{2}}\psi_1 + \frac{1}{\sqrt{2}}\psi_2 \quad \text{and} \quad \psi_{b=-1} = \frac{1}{\sqrt{2}}\psi_1 - \frac{1}{\sqrt{2}}\psi_2$$

There is an alternative strategy for solving this problem, a strategy that introduces some of the key ideas of matrix mechanics. Although this example may seem to be an abstract one, it will turn out to be very useful to us when we turn our attention to intrinsic spin angular momentum in the next chapter. To set the stage, note that we have determined the eigenfunctions and eigenvalues of B_{op} without actually knowing the explicit form of the eigenfunctions ψ_1 and ψ_2. In the same way that we can write an ordinary two-dimensional vector $\mathbf{V} = V_x\mathbf{i} + V_y\mathbf{j}$ as $\mathbf{V} = (V_x, V_y)$, we use a more compact notation that expresses $\psi = c_1\psi_1 + c_2\psi_2$ as a two-dimensional column vector

$$\psi = \begin{pmatrix} c_1 \\ c_2 \end{pmatrix}$$

Similarly, we can write the operator B_{op} as a 2×2 matrix, namely

$$B_{op} = \begin{pmatrix} 0 & 1 \\ 1 & 0 \end{pmatrix}$$

since

$$B_{op}\psi = \begin{pmatrix} 0 & 1 \\ 1 & 0 \end{pmatrix}\begin{pmatrix} c_1 \\ c_2 \end{pmatrix} = \begin{pmatrix} c_2 \\ c_1 \end{pmatrix}$$

consistent with $B_{op}\psi(x) = c_2\psi_1(x) + c_1\psi_2(x)$

In this matrix mechanics, the eigenvalue problem $B_{op}\psi = b\psi$ takes the familiar form

$$\begin{pmatrix} 0 & 1 \\ 1 & 0 \end{pmatrix}\begin{pmatrix} c_1 \\ c_2 \end{pmatrix} = b\begin{pmatrix} c_1 \\ c_2 \end{pmatrix}$$

which can be rewritten as

$$\begin{pmatrix} 0 & 1 \\ 1 & 0 \end{pmatrix}\begin{pmatrix} c_1 \\ c_2 \end{pmatrix} = \begin{pmatrix} b & 0 \\ 0 & b \end{pmatrix}\begin{pmatrix} c_1 \\ c_2 \end{pmatrix} \quad \text{or} \quad \begin{pmatrix} -b & 1 \\ 1 & -b \end{pmatrix}\begin{pmatrix} c_1 \\ c_2 \end{pmatrix} = 0$$

For a nontrivial solution the determinant of the coefficients of this last 2×2 matrix must vanish. Hence

$$\begin{vmatrix} -b & 1 \\ 1 & -b \end{vmatrix} = 0$$

Thus $b^2 = 1$. As before $c_2 = c_1$ when $b = 1$ and $c_2 = -c_1$ when $b = -1$. Thus the normalized eigenvectors are given by the column vectors

$$\psi_{b=1} = \begin{pmatrix} 1/\sqrt{2} \\ 1/\sqrt{2} \end{pmatrix} \quad \text{and} \quad \psi_{b=-1} = \begin{pmatrix} 1/\sqrt{2} \\ -1/\sqrt{2} \end{pmatrix}$$

consistent with our earlier results.

5.3 Commuting Operators

When one multiplies *numbers* such as a and b together, the order in which we carry out the multiplication does not matter, namely $ab = ba$. We say that the numbers commute under multiplication. But in quantum mechanics observables are associated with Hermitian operators and the ordering of these operators matters a great deal. As we will see in this section and the next, whether the operators corresponding to observables commute or not is of utmost importance. In fact, one can make the case that the commutation relations satisfied by these operators are at the very heart of quantum mechanics. The **commutator** of two operators A_{op} and B_{op} is defined by

$$[A_{\text{op}}, B_{\text{op}}] \equiv A_{\text{op}} B_{\text{op}} - B_{\text{op}} A_{\text{op}} \tag{5.33}$$

Let's start with an example in which the commutator is zero, that is $A_{\text{op}} B_{\text{op}} = B_{\text{op}} A_{\text{op}}$. Consider the commutator of the parity operator and the Hamiltonian for the harmonic oscillator. In order to evaluate the commutator, we apply the commutator, which is after all an operator, to an arbitrary wave function $\psi(x)$:

$$[H_{\text{SHO}}, \Pi] \psi(x)$$

$$= \left(-\frac{\hbar^2}{2m} \frac{\partial^2}{\partial x^2} + \frac{1}{2} m\omega x^2 \right) \Pi \psi(x) - \Pi \left(-\frac{\hbar^2}{2m} \frac{\partial^2}{\partial x^2} + \frac{1}{2} m\omega x^2 \right) \psi(x)$$

$$= \left(-\frac{\hbar^2}{2m} \frac{\partial^2}{\partial x^2} + \frac{1}{2} m\omega x^2 \right) \psi(-x) - \left(-\frac{\hbar^2}{2m} \frac{\partial^2}{\partial (-x)^2} + \frac{1}{2} m\omega (-x)^2 \right) \psi(-x)$$

$$= \left(-\frac{\hbar^2}{2m} \frac{\partial^2}{\partial x^2} + \frac{1}{2} m\omega x^2 \right) \psi(-x) - \left(-\frac{\hbar^2}{2m} \frac{\partial^2}{\partial x^2} + \frac{1}{2} m\omega x^2 \right) \psi(-x)$$

$$= 0 \tag{5.34}$$

where the key step is noting that the transformation $x \to -x$ leaves the Hamiltonian unchanged. Thus the commutator vanishes. This will be true for any Hamiltonian for which $V(x) = V(-x)$.

What are the implications of two Hermitian operators A_{op} and B_{op} commuting? If we apply the operator B_{op} to the eigenvalue equation

$$A_{\text{op}} \psi_a = a \psi_a \tag{5.35}$$

we obtain

$$B_{op}A_{op}\psi_a = aB_{op}\psi_a \qquad (5.36)$$

But since A_{op} and B_{op} commute (i.e., $A_{op}B_{op} = B_{op}A_{op}$), we can flip the order of operators on the left-hand side to obtain

$$A_{op}B_{op}\psi_a = aB_{op}\psi_a \qquad (5.37)$$

To make the meaning of this equation clearer, let's add some parentheses to guide the eye:

$$A_{op}\left(B_{op}\psi_a\right) = a\left(B_{op}\psi_a\right) \qquad (5.38)$$

This equation tells us that $B_{op}\psi_a$ is also an eigenfunction of the operator A_{op} with eigenvalue a.

Now there are two cases to consider. First, let's assume there exists a single eigenfunction ψ_a with eigenvalue a. In this case we say the eigenfunction is **nondegenerate**. Then since $B_{op}\psi_a$ is an eigenfunction of A_{op} with eigenvalue a as well, $B_{op}\psi_a$ differs from ψ_a by at most a multiplicative constant. If we call this constant b, then

$$B_{op}\psi_a = b\psi_a \qquad (5.39)$$

But (5.39) simply states that ψ_a is an eigenfunction of B_{op} with eigenvalue b. Hence ψ_a is simultaneously an eigenfunction of A_{op} with eigenvalue a and of B_{op} with eigenvalue b. We can therefore relabel the eigenfunction as $\psi_{a,b}$, indicating the eigenvalues for each of the operators. The simple harmonic oscillator provides a nice illustration. Since the Hamiltonian for the harmonic oscillator commutes with the parity operator and there exists a single energy eigenfunction for each energy eigenvalue, each energy eigenfunction must also be an eigenfunction of the parity operator. That is, each eigenfunction must be either an even or an odd function, as we saw in Section 4.3.

If there is more than one eigenfunction with the eigenvalue a, we say there is **degeneracy**. What then can we conclude? Let's take an example involving the Hamiltonian for a free particle to illustrate. As Example 5.4 shows, this Hamiltonian commutes with the linear momentum operator. But there are two eigenfunctions of the Hamiltonian with a particular eigenvalue E, namely,

$$\psi(x) = A\sin kx \qquad \text{and} \qquad \psi(x) = B\cos kx \qquad (5.40)$$

where

$$E = \frac{\hbar^2 k^2}{2m} \qquad (5.41)$$

We say, in this case, there is two-fold degeneracy. You might at first be tempted to think that these energy eigenfunctions should also be eigenfunctions of the momentum operator, but this is clearly not the case since

$$p_{x_{op}}A\sin kx = \frac{\hbar}{i}kA\cos kx \qquad \text{and} \qquad p_{x_{op}}A\cos kx = -\frac{\hbar}{i}kA\sin kx \qquad (5.42)$$

A single derivative does not return the same function when applied to a sine or a cosine. Nonetheless, there are linear combinations of these energy eigenfunctions that are simultaneously eigenfunctions of the momentum operator, namely the complex exponentials

$$\psi(x) = Ae^{ikx} \quad \text{and} \quad \psi(x) = Be^{-ikx} \tag{5.43}$$

that we used in our analysis of scattering. These two functions are not only eigenfunctions of the Hamiltonian with the eigenvalue $\hbar^2 k^2/2m$ but they are also eigenfunctions of the momentum operator with eigenvalues $\hbar k$ and $-\hbar k$, respectively. This illustrates the general result: when there exists degeneracy of the eigenfunctions of A_{op}, it is always possible to choose a linear combination of the degenerate eigenfunctions that are simultaneous eigenfunctions of B_{op} when A_{op} and B_{op} commute. We will not give a general proof. The interested reader is referred to a textbook on linear algebra.

In summary, when two Hermitian operators commute, they have a complete set of eigenfunctions in common. We can then know both of these dynamical variables simultaneously, without uncertainty.

EXAMPLE 5.4 Show that the Hamiltonian for a free particle and the linear momentum operator commute.

SOLUTION For a free particle, we can set the potential energy V to zero. Then

$$H = \frac{(p_{x_{\text{op}}})^2}{2m}$$

To evaluate the commutator of this Hamiltonian and the linear momentum operator, let the commutator act on an arbitrary wave function $\psi(x)$:

$$[H, p_{x_{\text{op}}}]\psi(x) = \left[-\frac{\hbar^2}{2m}\frac{\partial^2}{\partial x^2}, \frac{\hbar}{i}\frac{\partial}{\partial x}\right]\psi(x)$$

$$= -\frac{\hbar^3}{2mi}\left(\frac{\partial^2}{\partial x^2}\frac{\partial}{\partial x} - \frac{\partial}{\partial x}\frac{\partial^2}{\partial x^2}\right)\psi(x)$$

$$= -\frac{\hbar^3}{2mi}\left(\frac{\partial^3 \psi(x)}{\partial x^3} - \frac{\partial^3 \psi(x)}{\partial x^3}\right) = 0$$

5.4 Noncommuting Operators and Uncertainty Relations

What if the commutator of two operators is not zero? What are the implications? A classic example of a nonvanishing commutator is the commutator of the position and the momentum operators:

$$[x_{\text{op}}, p_{x_{\text{op}}}] = x_{\text{op}} p_{x_{\text{op}}} - p_{x_{\text{op}}} x_{\text{op}} \tag{5.44}$$

To evaluate this commutator, we again let the commutator act on an arbitrary function $\psi(x)$.

$$[x_{\text{op}}, p_{x_{\text{op}}}]\psi = \left[x, \frac{\hbar}{i}\frac{\partial}{\partial x}\right]\psi \tag{5.45}$$

Therefore

$$\left[x, \frac{\hbar}{i}\frac{\partial}{\partial x}\right]\psi = \left(x\frac{\hbar}{i}\frac{\partial}{\partial x} - \frac{\hbar}{i}\frac{\partial}{\partial x}x\right)\psi$$

$$= x\frac{\hbar}{i}\frac{\partial}{\partial x}\psi - \frac{\hbar}{i}\frac{\partial}{\partial x}(x\psi)$$

$$= x\frac{\hbar}{i}\frac{\partial\psi}{\partial x} - \frac{\hbar}{i}\psi - x\frac{\hbar}{i}\frac{\partial\psi}{\partial x}$$

$$= i\hbar\psi \qquad (5.46)$$

Since ψ is arbitrary, we have shown that

$$[x_{op}, p_{x_{op}}] = i\hbar \qquad (5.47)$$

Notice that the key element in this derivation was that the momentum operator differentiates everything to its right. In the first term in the commutator, this corresponds to the wave function ψ while for the second term of the commutator the function to the right of the momentum operator is $x\psi$.

The product of two operators, say $A_{op}B_{op}$, is itself an operator. Consequently, the commutator of two operators is also an operator. We will now show that if two Hermitian operators do not commute and their commutator is expressed as

$$[A_{op}, B_{op}] = iC_{op} \qquad (5.48)$$

where C_{op} is a Hermitian operator, then an uncertainty relation of the form

$$\Delta A \Delta B \geq \frac{|\langle C \rangle|}{2} \qquad (5.49)$$

must hold.[3] Comparing (5.47) with (5.48), we see for the position–momentum commutator that $C_{op} = \hbar$ and the uncertainty relation (5.49) becomes

$$\Delta x \Delta p_x \geq \frac{\hbar}{2} \qquad (5.50)$$

the famous Heisenberg uncertainty relation.

We now turn to the proof of (5.49). This general uncertainty relation is a very important one. The proof is presented so that you will have confidence in the result, but in the end it is more worthwhile to focus on the result itself rather than on the details of the derivation. Recall from (5.32) that

$$(\Delta A)^2 = \langle (A - \langle A \rangle)^2 \rangle = \langle A^2 \rangle - \langle A \rangle^2 \quad \text{and} \quad (\Delta B)^2 = \langle (B - \langle B \rangle)^2 \rangle = \langle B^2 \rangle - \langle B \rangle^2 \qquad (5.51)$$

We start by defining two operators

$$U_{op} \equiv A_{op} - \langle A \rangle \qquad \text{and} \qquad V_{op} \equiv B_{op} - \langle B \rangle \qquad (5.52)$$

Then

$$\int_{-\infty}^{\infty} \psi^* U_{op}^2 \psi \, dx = (\Delta A)^2 \qquad \text{and} \qquad \int_{-\infty}^{\infty} \psi^* V_{op}^2 \psi \, dx = (\Delta B)^2 \qquad (5.53)$$

---

[3] It is not hard to show that $C_{op} = -i[A_{op}, B_{op}]$ is Hermitian provided A_{op} and B_{op} are Hermitian. See Problem 5.11.

and since
$$[U_{op}, V_{op}] = iC_{op} \tag{5.54}$$

we also have
$$\int_{-\infty}^{\infty} \psi^* [U_{op}, V_{op}] \psi \, dx = \int_{-\infty}^{\infty} \psi^* i C_{op} \psi \, dx = i \langle C \rangle \tag{5.55}$$

Now we know for any complex function ϕ that
$$\int_{-\infty}^{\infty} \phi^* \phi \, dx \geq 0 \tag{5.56}$$

Let's choose as our ϕ
$$\phi = U_{op} \psi + i\lambda V_{op} \psi \tag{5.57}$$

where the constant λ is presumed to be real, and define
$$I(\lambda) = \int_{-\infty}^{\infty} \phi^* \phi \, dx \tag{5.58}$$

Since A_{op} and B_{op} are Hermitian, so are U_{op} and V_{op}. Thus
$$\begin{aligned} I(\lambda) &= \int_{-\infty}^{\infty} (U_{op}\psi + i\lambda V_{op}\psi)^* (U_{op}\psi + i\lambda V_{op}\psi) \, dx \\ &= \int_{-\infty}^{\infty} \psi^* (U_{op}^2 + \lambda^2 V_{op}^2 + i\lambda [U_{op}, V_{op}]) \psi \, dx \\ &= (\Delta A)^2 + \lambda^2 (\Delta B)^2 - \lambda \langle C \rangle \geq 0 \end{aligned} \tag{5.59}$$

The minimum occurs when
$$\frac{dI}{d\lambda} = 2\lambda (\Delta B)^2 - \langle C \rangle = 0 \quad \Rightarrow \quad \lambda = \frac{\langle C \rangle}{2(\Delta B)^2} \tag{5.60}$$

Substituting this value for λ into the last line of (5.59), we obtain
$$(\Delta A)^2 + \frac{\langle C \rangle^2}{4(\Delta B)^2} - \frac{\langle C \rangle^2}{2(\Delta B)^2} \geq 0 \tag{5.61}$$

or
$$(\Delta A)^2 (\Delta B)^2 \geq \frac{\langle C \rangle^2}{4} \tag{5.62}$$

Finally, taking the square root of this equation, we obtain the general uncertainty relation
$$\Delta A \Delta B \geq \frac{|\langle C \rangle|}{2} \tag{5.63}$$

As we have noted, not only does this result lead to the Heisenberg uncertainty principle, but it also can be used to derive many other important uncertainty relations. In the next section, we examine another such relation, the Heisenberg energy–time uncertainty relation.

EXAMPLE 5.5 Determine the matrix that represents the operator A_{op} in Example 5.3. Evaluate the commutator of A_{op} and B_{op} in matrix mechanics. What do you conclude about the eigenstates of A_{op} and B_{op}?

SOLUTION Since ψ_1 and ψ_2 are eigenstates of A_{op} with eigenvalues a_1 and a_2, respectively, the matrix representation of A_{op} is given by

$$A_{op} = \begin{pmatrix} a_1 & 0 \\ 0 & a_2 \end{pmatrix}$$

as can be verified by applying the matrix to the column vectors

$$\begin{pmatrix} 1 \\ 0 \end{pmatrix} \quad \text{and} \quad \begin{pmatrix} 0 \\ 1 \end{pmatrix}$$

From Example 5.3

$$B_{op} = \begin{pmatrix} 0 & 1 \\ 1 & 0 \end{pmatrix}$$

Therefore the commutator of A_{op} and B_{op} is given by

$$A_{op}B_{op} - B_{op}A_{op} = \begin{pmatrix} a_1 & 0 \\ 0 & a_2 \end{pmatrix}\begin{pmatrix} 0 & 1 \\ 1 & 0 \end{pmatrix} - \begin{pmatrix} 0 & 1 \\ 1 & 0 \end{pmatrix}\begin{pmatrix} a_1 & 0 \\ 0 & a_2 \end{pmatrix}$$

$$= \begin{pmatrix} 0 & a_1 \\ a_2 & 0 \end{pmatrix} - \begin{pmatrix} 0 & a_2 \\ a_1 & 0 \end{pmatrix}$$

$$= \begin{pmatrix} 0 & a_1 - a_2 \\ a_2 - a_1 & 0 \end{pmatrix} \neq 0$$

Since the operators do not commute (presuming $a_1 \neq a_2$), they do not have eigenstates in common, as we have seen in working out Example 5.3.

5.5 Time Development

The time-dependent Schrödinger equation

$$-\frac{\hbar^2}{2m}\frac{\partial^2 \Psi(x,t)}{\partial x^2} + V(x)\Psi(x,t) = i\hbar\frac{\partial \Psi(x,t)}{\partial t} \tag{5.64}$$

is the equation of motion for one-dimensional wave mechanics. It can also be written simply as

$$H\Psi(x,t) = i\hbar\frac{\partial \Psi(x,t)}{\partial t} \tag{5.65}$$

where H is the Hamiltonian. It is instructive to ask how expectation values vary with time. Let's start with

$$\frac{d\langle A \rangle}{dt} = \frac{d}{dt}\int_{-\infty}^{\infty} \Psi^* A_{op} \Psi \, dx \tag{5.66}$$

Since the limits of integration are independent of time, we can move the time derivative under the integral sign, being careful to replace the ordinary time derivative with a partial time derivative since the wave function depends on both x and t and we are only differentiating with respect to t. Thus, using the chain rule,

$$\frac{d\langle A\rangle}{dt} = \int_{-\infty}^{\infty} \frac{\partial \Psi^*}{\partial t} A_{\text{op}} \Psi \, dx + \int_{-\infty}^{\infty} \Psi^* \frac{\partial A_{\text{op}}}{\partial t} \Psi \, dx + \int_{-\infty}^{\infty} \Psi^* A_{\text{op}} \frac{\partial \Psi}{\partial t} \, dx \quad (5.67)$$

From (5.65)

$$\frac{\partial \Psi}{\partial t} = \frac{1}{i\hbar} H \Psi \quad (5.68)$$

and therefore

$$\frac{\partial \Psi^*}{\partial t} = -\frac{1}{i\hbar} (H\Psi)^* \quad (5.69)$$

Substituting (5.68) and (5.69) into (5.67), we obtain

$$\frac{d\langle A\rangle}{dt} = \frac{i}{\hbar} \int_{-\infty}^{\infty} (H\Psi)^* A_{\text{op}} \Psi \, dx + \int_{-\infty}^{\infty} \Psi^* \frac{\partial A_{\text{op}}}{\partial t} \Psi \, dx - \frac{i}{\hbar} \int_{-\infty}^{\infty} \Psi^* A_{\text{op}} H \Psi \, dx \quad (5.70)$$

Since the operator H is Hermitian, we can rewrite this result as

$$\frac{d\langle A\rangle}{dt} = \frac{i}{\hbar} \int_{-\infty}^{\infty} \Psi^* H A_{\text{op}} \Psi \, dx - \frac{i}{\hbar} \int_{-\infty}^{\infty} \Psi^* A_{\text{op}} H \Psi \, dx + \int_{-\infty}^{\infty} \Psi^* \frac{\partial A_{\text{op}}}{\partial t} \Psi \, dx \quad (5.71)$$

Finally, note that the first two terms can be rewritten in terms of the commutator $[H, A_{\text{op}}] = H A_{\text{op}} - A_{\text{op}} H$, leading to the final result

$$\frac{d\langle A\rangle}{dt} = \frac{i}{\hbar} \int_{-\infty}^{\infty} \Psi^* [H, A_{\text{op}}] \Psi \, dx + \int_{-\infty}^{\infty} \Psi^* \frac{\partial A_{\text{op}}}{\partial t} \Psi \, dx \quad (5.72)$$

In most cases, the operator corresponding to an observable, such as position, momentum, or parity, does not depend on time explicitly and the term involving $\partial A_{\text{op}}/\partial t$ vanishes, in which case

$$\frac{d\langle A\rangle}{dt} = \frac{i}{\hbar} \int_{-\infty}^{\infty} \Psi^* [H, A_{\text{op}}] \Psi \, dx \quad (5.73)$$

Expression (5.73) shows us that if the operator corresponding to an observable A commutes with the Hamiltonian, then that variable is a constant of the motion—its expectation value does not vary with time. In Section 5.3 we saw that the Hamiltonian for the simple harmonic oscillator and the parity operator commute. Thus parity is a constant of the motion; we say parity is conserved for this Hamiltonian.

On the other hand, the Hamiltonian and the momentum operator do not generally commute. As before, the commutator can be evaluated by letting it act on an arbitrary wave function ψ:

$$[H, p_{x_{\text{op}}}] \psi = \left[-\frac{\hbar^2}{2m} \frac{\partial^2}{\partial x^2} + V(x), \frac{\hbar}{i} \frac{\partial}{\partial x} \right] \psi$$

$$= \left[V(x), \frac{\hbar}{i} \frac{\partial}{\partial x} \right] \psi$$

$$= V(x) \frac{\hbar}{i} \frac{\partial \psi}{\partial x} - \frac{\hbar}{i} \frac{\partial}{\partial x} V(x) \psi$$

$$= -\frac{\hbar}{i} \left(\frac{\partial V}{\partial x} \right) \psi \quad (5.74)$$

where in going from the first line to the second line we have taken advantage of the fact that the linear momentum operator commutes with the kinetic energy portion of the Hamiltonian, as we saw in Example 5.4. In going from the third to the fourth line we have used the fact that the momentum operator acts on everything to the right of it, and thus differentiates both the potential energy and the wave function in the last term in the second line. If we now go back to (5.73), we see that

$$\frac{d\langle p_x \rangle}{dt} = \frac{i}{\hbar} \int_{-\infty}^{\infty} \Psi^* [H, p_{x_{\text{op}}}] \Psi \, dx$$

$$= -\int_{-\infty}^{\infty} \Psi^* \frac{\partial V}{\partial x} \Psi \, dx = \left\langle -\frac{\partial V}{\partial x} \right\rangle \quad (5.75)$$

Thus we see that the momentum is not a constant of the motion in general. We have also found a convenient way to derive the second of Ehrenfest's equations (see Section 2.9), which demonstrates how the time variation of the expectation value of the linear momentum obeys Newton's second law, $dp_x/dt = F_x = -\partial V/\partial x$.

The Energy–Time Uncertainty Relation

Perhaps the most misunderstood uncertainty relation is the Heisenberg uncertainty relation $\Delta E \Delta t \geq \hbar/2$. After all, we know what ΔE means; it is the uncertainty in the energy of the particle, which we calculate from $(\Delta E)^2 = \langle E^2 \rangle - \langle E \rangle^2$. But what is Δt? In nonrelativistic quantum mechanics, time is not an observable like position or energy. We don't have a time operator. Time is a parameter that always takes on a definite value. Nonetheless, we can use the general uncertainty relation (5.49) to give a definition of Δt. We choose the operator B_{op} in the commutator (5.48) to be the Hamiltonian H. In this case, (5.49) becomes

$$\Delta A \Delta E \geq \frac{|\langle C \rangle|}{2} \quad (5.76)$$

where

$$C_{\text{op}} = -i [A_{\text{op}}, H] \quad (5.77)$$

and therefore

$$|\langle C \rangle| = \left| i \int_{-\infty}^{\infty} \Psi^* [H, A_{\text{op}}] \Psi \, dx \right| = \left| \hbar \frac{d\langle A \rangle}{dt} \right| \quad (5.78)$$

where the last step follows from (5.73). Thus

$$\Delta A \Delta E \geq \frac{\hbar}{2} \left| \frac{d\langle A \rangle}{dt} \right| \quad (5.79)$$

Dividing through by $|d\langle A \rangle/dt|$, we find

$$\Delta E \frac{\Delta A}{\left| \frac{d\langle A \rangle}{dt} \right|} \geq \frac{\hbar}{2} \quad (5.80)$$

We define

$$\Delta t \equiv \frac{\Delta A}{\left| \frac{d\langle A \rangle}{dt} \right|} \quad (5.81)$$

in which case we obtain

$$\Delta E \Delta t \geq \frac{\hbar}{2} \quad (5.82)$$

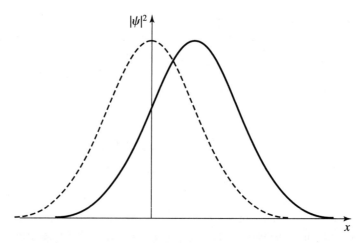

Figure 5.1 A wave packet with an uncertainty Δx in the position of the particle can be clearly said to have changed its location when the packet has shifted from its original position by an amount equal to Δx.

Let's take an example to illustrate the meaning of Δt. Consider a wave packet with a position uncertainty $\Delta x = 10^{-3}$ m, as depicted in Fig. 5.1. Suppose that $d\langle x\rangle/dt = 1$ m/s, that is the wave packet is moving along the x direction with a speed of 1 m/s. The time $\Delta t = \Delta x/|d\langle x\rangle/dt| = 10^{-3}$ s. This is the time it takes the wave packet to shift its position by an amount equal to the uncertainty in the position. We can say this is the time it is necessary to wait before we can be confident that the position of the particle has changed. We can call the time Δt an evolutionary time, the time required for the particle's position to change. Clearly if the wave function of the particle had been an energy eigenfunction instead of a superposition of energy eigenfunctions, then $\Delta E = 0$. But such a wave function corresponds to a stationary state since the time we would need to wait for the system to evolve is infinite.

Example 3.3 provides a good illustration of a wave function that is a superposition of two energy eigenfunctions. In this example, we examined the time dependence of the wave function for a particle in a box that initially is in the state

$$\Psi(x) = \frac{1}{\sqrt{2}}\psi_1(x) + \frac{1}{\sqrt{2}}\psi_2(x) \tag{5.83}$$

with a 50% probability that a measurement of the energy yields E_1 and a 50% chance that it yields E_2. The wave function at time t is given by

$$\Psi(x,t) = \frac{e^{-iE_1 t/\hbar}}{\sqrt{2}}\psi_1(x) + \frac{e^{-iE_2 t/\hbar}}{\sqrt{2}}\psi_2(x) \tag{5.84}$$

which can be rewritten as

$$\Psi(x,t) = e^{-iE_1 t/\hbar}\left[\frac{1}{\sqrt{2}}\psi_1(x) + \frac{e^{-i(E_2-E_1)t/\hbar}}{\sqrt{2}}\psi_2(x)\right] \tag{5.85}$$

In Example 3.3, we saw that $\Delta E = (E_2 - E_1)/2$. Thus we can also write the wave function as

$$\Psi(x,t) = e^{-iE_1 t/\hbar}\left[\frac{1}{\sqrt{2}}\psi_1(x) + \frac{e^{-i2\Delta E t/\hbar}}{\sqrt{2}}\psi_2(x)\right] \tag{5.86}$$

We can say the time we must wait before the wave function has clearly changed from its initial form (5.83) to a different wave function is the time Δt necessary for the relative phase to become of order unity, that is, $2\Delta E \Delta t/\hbar \approx 1$, or $\Delta E \Delta t \approx \hbar/2$, consistent with our earlier discussion.

This example illustrates how a nonzero ΔE leads to a finite evolutionary time Δt. We can, in fact, turn this argument around and, at the same time, make our discussion of excited states of the systems we have treated so far more realistic. Consider, for example, an electron in an excited state of an atom. Since the electron will eventually make a radiative transition to a lower energy state with a lifetime τ, there is a finite evolutionary time for the system in this excited state. For the first excited state of hydrogen, for example, $\tau = 1.6$ ns. Consequently, according to the energy–time uncertainty relation, the excited state must have a nonzero uncertainty in its energy, as indicated in Fig. 5.2. Therefore, the photon that is emitted in the transition between this excited state and the ground state does not have a definite frequency. This spread in frequencies, or wavelengths, is referred to as the **natural linewidth** for the state.

Figure 5.2 An energy-level diagram for the ground state and the first excited state requires modification when the effect of the finite lifetime of the excited state is taken into account. The spread in energy for the excited state is not drawn to scale.

EXAMPLE 5.6 Calculate the natural linewidth for the first excited state of hydrogen. Compare this spread in wavelengths with the primary wavelength of the transition. Note: $E_2 - E_1 = 10.2$ eV.

SOLUTION

$$E_2 - E_1 = h\nu = \frac{hc}{\lambda}$$

The principal wavelength for the transition is 121.5 nm. But since the excited state has a lifetime $\tau = 1.6 \times 10^{-9}$ s, the uncertainty in energy for this excited state is

$$\Delta E \approx \frac{\hbar}{\tau}$$

The uncertainty in the energy corresponds to a spread in wavelength

$$\Delta E \approx \frac{hc}{\lambda^2} \Delta \lambda$$

Given the uncertainty in ΔE, we see that $\Delta \lambda = 5 \times 10^{-15}$ m, or roughly 4 parts in 10^8 of the wavelength of the transition. Thus the natural linewidth is tough to observe in this case. Of course, if the lifetime were shorter, the effect would be easier to observe. We will see a striking example in Chapter 10.

5.6 EPR, Schrödinger's Cat, and All That

One of the arresting features of quantum mechanics is that particles or systems of particles do not in general have definite attributes. When we express the wave function as

$$\Psi = c_1 \psi_1 + c_2 \psi_2 \tag{5.87}$$

where ψ_1 and ψ_2 are eigenfunctions of an operator A_{op} corresponding to an observable A, we are saying that before we make a measurement of A the particle does not have a definite value of that attribute. The probability that a measurement yields a_1 is, of course, $|c_1|^2$. After a measurement yielding a_1, the wave function collapses to ψ_1, since a measurement of A immediately thereafter again yields the value a_1. How this collapse happens is a mystery. It is referred to as the measurement problem. We will return to this issue at the end of this section.[4]

Not everyone has been happy with the idea that particles do not necessarily have definite attributes before a measurement is made. The most famous malcontent was almost certainly Albert Einstein. In his view, writing a wave function in the form (5.87) was really expressing our lack of knowledge of the state of the particle. Some particles have attribute a_1 and others have attribute a_2, but we are not able to distinguish one particle from the other. It is as if the attribute, or variable, that would allow us to make this distinction is hidden from us, hence a hidden variable theory of quantum mechanics. Such a view, while no doubt appealing to some, seems to raise troubling issues as well. After all, if the helium atom in the double-slit experiment described in Section 2.1 really had a definite position before its position was measured in the detection plane, then one could logically presume it also had a definite position just moments before this, and so forth. Thus we would be led to presume that the atom followed a definite trajectory between the source and the detector, passing through one slit or the other, which makes it hard to understand how interference can occur. Nonetheless, Einstein persisted in his view that the description of nature provided by quantum mechanics was incomplete, that there was more to nature than quantum mechanics presumed.

To sharpen his criticism of quantum mechanics, Einstein, together with Boris Podolsky and Nathan Rosen, proposed a thought experiment that, in his view, showed how crazy quantum mechanics really is.[5] The key aspect of quantum mechanics that EPR focused on was something that we now call an **entangled state**. Consider the two-particle wave function

$$\Psi = \frac{1}{\sqrt{2}} \left[\psi_{a_1}(1)\psi_{a_2}(2) - \psi_{a_2}(1)\psi_{a_1}(2) \right] \tag{5.88}$$

as an example. In this wave function $\psi_{a_1}(1)$ means that one of the particles, particle 1, is in state with value a_1 of the observable A. The wave function (5.88) shows there is a 50% probability that a measurement of A on particle 1 will yield this value. But then we will subsequently know that particle 2 is in a state with value a_2 for the observable A. What was really distressing, at least to Einstein, was that not only do particles 1 and 2 not have definite attributes, but the value of the observable that particle 2 takes on is determined by a measurement on particle 1. After all, if the measurement on particle 1 yielded the value a_2 instead of a_1, then we would know that particle 2 is in the state with value a_1. And particles 1 and 2 do not have to be situated together in space for this to occur.

A particularly interesting example is the two-photon "wave function"

$$\Psi = \frac{1}{\sqrt{2}} \left[\psi_R(1)\psi_R(2) + \psi_L(1)\psi_L(2) \right] \tag{5.89}$$

---

[4]If the operator A_{op} corresponding to the observable A commutes with the Hamiltonian, then A is a constant of the motion. Thus once the system collapses to a particular eigenfunction, it will remain in this state.

[5]A. Einstein, B. Podolsky, and N. Rosen, *Phys. Rev.* **47**, 777 (1935).

where the subscripts R and L refer to right and left circularly polarized photons. Such a two-photon state is generated in the cascade decay of the excited state of the calcium atom that we discussed in Section 1.5. Thus if a measurement on one of the photons finds it to be right-circularly polarized, then the other photon must be right-circularly polarized as well. But if we express the circularly polarized states in terms of the appropriate linearly polarized states:

$$\psi_R(1) = \frac{1}{\sqrt{2}}\left[\psi_x(1) + i\psi_y(1)\right] \quad (5.90)$$

and

$$\psi_L(1) = \frac{1}{\sqrt{2}}\left[\psi_x(1) - i\psi_y(1)\right] \quad (5.91)$$

assuming that photon 1 is traveling along the z axis, we find

$$\Psi = \frac{1}{\sqrt{2}}\left[\psi_x(1)\psi_x(2) + \psi_y(1)\psi_y(2)\right] \quad (5.92)$$

as can be readily verified by substituting the expressions (5.90) and (5.91) for the circularly polarized wave functions for photon 1 and the corresponding expressions for photon 2 into (5.89). See Problem 5.12. In this form the wave function indicates that if a measurement of the linear polarization shows photon 1 to be x polarized (and there is a 50% chance of obtaining this result), then photon 2 will necessarily be x polarized if a measurement of its linear polarization is carried out. Similarly, if a measurement of linear polarization on photon 1 finds it to be y polarized, then photon 2 is necessarily y polarized. The state (5.92) exhibits the same sort of entanglement that we saw in our discussion of circular polarization for the state (5.89). But what is striking here is that photon 1 cannot be both right circularly polarized and at the same time linearly polarized along the x-axis. Thus it is impossible to think of the photons as having a definite state of polarization before a measurement is carried out. In the original EPR argument, the two incompatible variables were position and momentum, not circular and linear polarization. As we have seen, the operators corresponding to position and momentum do not commute with each other. Thus it is impossible to imagine a particle as having both a definite position and a definite momentum, just like we cannot think of a photon as having a definite circular and linear polarization. Nonetheless, states of the form (5.89) or (5.92) seem to be saying that not only do particles not have definite attributes but these attributes can be determined by measurements on a different particle altogether, one that may be spatially quite separated from the other. To Einstein, this "spooky action at a distance" was unacceptable, something that "no reasonable definition of reality" should permit.

In the 1960s John S. Bell realized that two-particle states such as (5.88) could provide an experimental test as to whether a particle really has definite attributes before a measurement is carried out. It was just necessary to measure the correlations between measurements carried out on the two particles. These were challenging experiments, at least initially. But as the technology has improved, the results have become more and more striking, to the point that the most recent results are in clear agreement with the predictions of quantum mechanics and disagree, at the level of 250 standard deviations, with local hidden variable theory.

A classic example that seems to illustrate the problem with taking this line of reasoning that particles do not necessarily have definite attributes too far is Schrödinger's cat. As

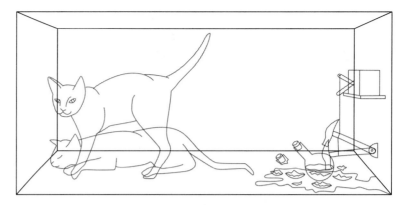

Figure 5.3 Schrödinger's cat.

a thought experiment, Schrödinger suggested placing a cat in a box with a radioactive isotope that had a half-life of one hour. As indicated in Fig. 5.3, the decay of a single nucleus, with the emission of, say, an alpha particle, would trigger a hammer to break a flask of prussic acid, a highly volatile and toxic substance. Thus, in the language of quantum mechanics, one might write the wave function for the cat at the one-hour mark as

$$\Psi(1 \text{ hour}) = \frac{1}{\sqrt{2}}\psi_{\text{alive}}^{\text{cat}} + \frac{1}{\sqrt{2}}\psi_{\text{dead}}^{\text{cat}} \tag{5.93}$$

Quantum mechanics seemed to say that only when the box was opened at the one-hour mark and the cat was observed to be alive or dead was the wave function collapsed to one of these two states. This indeed seemed silly, since there was little doubt that the cat was truly alive or dead independent of the measurement. Thus thinking of a macroscopic object such as a cat as genuinely not having a definite attribute, such as being alive or dead, seems not to be correct. But on the other hand, we might think of the cat as a macroscopic measuring device that has become entangled with the corresponding state of the nucleus, say in the form

$$\Psi(1 \text{ hour}) = \frac{1}{\sqrt{2}}\psi_{\text{nodecay}}^{\text{nucleus}}\psi_{\text{alive}}^{\text{cat}} + \frac{1}{\sqrt{2}}\psi_{\text{decayed}}^{\text{nucleus}}\psi_{\text{dead}}^{\text{cat}} \tag{5.94}$$

Such a wave function would still suggest that the cat was genuinely in a superposition of two states. But how would we know if this were indeed the case?

Let's go back to the double-slit experiment with helium atoms (Section 2.1). Since there are two paths each atom can take between the source and the detector, we might write the wave function as

$$\Psi = \psi_1 + \psi_2 = \psi_1 + e^{i\phi}\psi_1 \tag{5.95}$$

where ϕ is the phase difference between the amplitude for the atom to reach the detector by passing through slit 2 relative to reaching it by passing through slit 1. If we then calculate the probability density for detecting a helium atom, we find

$$\Psi^*\Psi = \left(\psi_1^* + e^{-i\phi}\psi_1^*\right)\left(\psi_1 + e^{i\phi}\psi_1\right) = 4\psi_1^*\psi_1 \cos^2\frac{\phi}{2} \tag{5.96}$$

which shows the interference fringes that we saw in the data. If there are really two amplitudes contributing to a process, we should expect to see interference effects if we conduct the appropriate experiment. The largest objects for which we have seen interference fringes in a double-slit experiment are C_{60} molecules, buckyballs. Physicists are trying to push the envelope here, to move toward macroscopic objects that would truly exhibit interference effects. Such states, if they exist, are called **Schrödinger cat states**. But so far, no one has been able to observe interference effects with even Schrödinger kittens. Why not? Our best guess is that interactions with the environment, which are inevitable for a macroscopic object such as a cat, cause the wave function to lose this relative phase information. We say the state decoheres. And perhaps this **decoherence** in some as yet unexplained way leads to the collapse of the system to one or the other of the states making up the superposition. But this seems inconsistent with the fundamental principles of quantum mechanics, since the equation of time development, the Schrödinger equation, is a linear differential equation. If the initial state is a superposition of states, so too should be the state at a later time. This is the crux of the measurement problem.

5.7 Summary

Let us summarize the principles of quantum mechanics that we have discussed in this chapter:

1. In one-dimensional wave mechanics, the state of a particle is given by a wave function $\Psi(x, t)$ that contains all that can be known about the particle.[6]

    If the question is "where is the particle," then $\Psi^*\Psi\, dx$ is the probability of finding the particle between x and $x + dx$ if a measurement of the position of the particle is carried out at time t, provided the wave function is normalized, namely

    $$\int_{-\infty}^{\infty} \Psi^*\Psi\, dx = 1 \tag{5.97}$$

    Note that the particle does not have a definite location before a measurement is carried out. But what if we want to know something else about the particle such as its energy? We start with a second principle:

2. Each dynamical variable, or observable, A is associated with a linear, Hermitian operator A_{op}, an operator for which

    $$\int_{-\infty}^{\infty} \phi^* A_{\text{op}} \psi\, dx = \int_{-\infty}^{\infty} (A_{\text{op}}\phi)^* \psi\, dx \tag{5.98}$$

    The eigenvalue equation is of the form

    $$A_{\text{op}} \psi_a = a \psi_a \tag{5.99}$$

    where the constant a is the eigenvalue and ψ_a is the eigenfunction. The only possible result of a measurement for the observable A is one of the eigenvalues.

---

[6]In Chapter 6 we extend our discussion of quantum mechanics to include three-dimensional systems, at which point we discuss intrinsic spin, a "degree of freedom" that is not specified by a wave function.

So far, we have considered observables such as the momentum and the energy. For the energy the corresponding operator is referred to as the Hamiltonian H and the energy eigenvalue equation is the time-independent Schrödinger equation. In the next chapter, where we will venture into three dimensions, we will introduce additional operators such as the orbital angular momentum operators.

3. The eigenfunctions ψ_a form an orthonormal basis. Therefore any wave function Ψ can be written as

$$\Psi = \sum_a c_a \psi_a \tag{5.100}$$

The probability of obtaining eigenvalue a is given by

$$|c_a|^2 = \left| \int_{-\infty}^{\infty} \psi_a^* \Psi \, dx \right|^2 \tag{5.101}$$

where ψ_a is the eigenfunction of A_{op} with eigenvalue a.[7] The average value, or expectation value, of the observable in this state is then given by

$$\langle A \rangle = \sum_a |c_a|^2 a = \int_{-\infty}^{\infty} \Psi^* A_{\text{op}} \Psi \, dx \tag{5.102}$$

The results that follow from the commutation relations of two operators can be derived from the general properties of Hermitian operators. Nonetheless, these results are of such importance that it is worth singling them out.

4. The commutator of two operators A_{op} and B_{op} is defined by

$$[A_{\text{op}}, B_{\text{op}}] = A_{\text{op}} B_{\text{op}} - B_{\text{op}} A_{\text{op}} \tag{5.103}$$

If the operators commute, that is the commutator vanishes, then it is possible to label the basis states as $\psi_{a,b}$, namely as simultaneous eigenfunctions of the two operators, which we are assuming to be Hermitian. If the operators do not commute, but

$$[A_{\text{op}}, B_{\text{op}}] = i C_{\text{op}} \tag{5.104}$$

then

$$\Delta A \Delta B \geq \frac{|\langle C \rangle|}{2} \tag{5.105}$$

The most famous example is the Heisenberg uncertainty principle

$$\Delta x \Delta p_x \geq \frac{\hbar}{2} \tag{5.106}$$

which follows from the commutator

$$[x_{\text{op}}, p_{x_{\text{op}}}] = i\hbar \tag{5.107}$$

---

[7] If the eigenvalue spectrum is continuous rather than discrete, then the probability of obtaining a result between a and $a+da$ is given by $|c_a|^2 da$ provided the eigenfunctions satisfy $\int_{-\infty}^{\infty} \psi_a^* \psi_{a'} \, dx = \delta(a-a')$ where $\delta(a-a')$ is a Dirac delta function.

5. Time dependence is determined by the Schrödinger equation

$$H\Psi(x,t) = i\hbar \frac{\partial \Psi(x,t)}{\partial t} \quad (5.108)$$

where the Hamiltonian H is the energy operator. In one-dimensional wave mechanics

$$H = -\frac{\hbar^2}{2m}\frac{\partial^2}{\partial x^2} + V(x) \quad (5.109)$$

We can use the Schrödinger equation to show that expectation values vary with time according to

$$\frac{d\langle A\rangle}{dt} = \frac{i}{\hbar}\int_{-\infty}^{\infty}\Psi^*\left[H, A_{\text{op}}\right]\Psi\,dx + \int_{-\infty}^{\infty}\Psi^*\frac{\partial A_{\text{op}}}{\partial t}\Psi\,dx \quad (5.110)$$

Thus if the Hamiltonian commutes with the operator corresponding to the observable A and $\partial A_{\text{op}}/\partial t = 0$, then $\langle A \rangle$ is independent of time and A is referred to as a constant of the motion. One consequence of (5.110) that is obtained by choosing the operator B_{op} in (5.104) to be the Hamiltonian is

$$\Delta E \frac{\Delta A}{\left|\frac{d\langle A\rangle}{dt}\right|} \geq \frac{\hbar}{2} \quad (5.111)$$

which can be written as

$$\Delta E \Delta t \geq \frac{\hbar}{2} \quad (5.112)$$

the Heisenberg energy–time uncertainty relation. In this relation, Δt is called an evolutionary time for the system, the time that is necessary for the system to change in a significant way.

In general, the number of eigenfunctions is infinite and the space spanned by these basis functions is an infinite dimensional vector space called a Hilbert space.[8]

Problems

5.1. (a) Prove that the parity operator is Hermitian.
(b) Show that the eigenfunctions of the parity operator corresponding to different eigenvalues are orthogonal.

5.2. Show that the operator

$$\frac{\partial}{\partial x}$$

is *not* Hermitian.

5.3. We have argued that a Hermitian operator corresponds to each observable. Physically, why is it essential that the eigenvalues be real?

5.4. A particle of mass m moves in the potential energy $V(x) = \frac{1}{2}m\omega^2 x^2$. The ground-state wave function is

$$\psi_0(x) = \left(\frac{a}{\pi}\right)^{1/4} e^{-ax^2/2}$$

and the first excited-state wave function is

$$\psi_1(x) = \left(\frac{4a^3}{\pi}\right)^{1/4} x e^{-ax^2/2}$$

where $a = m\omega/\hbar$. What is the average value of the parity for the state

$$\Psi(x) = \frac{\sqrt{3}}{2}\psi_0(x) + \frac{1-i}{2\sqrt{2}}\psi_1(x)$$

---

[8]Note this is not ordinary space. It is sometimes said that in Hilbert space no one can hear you scream.

5.5. For a particle in a harmonic oscillator potential, it is known that there is a one-third chance of obtaining the ground-state energy E_0, a one-third chance of obtaining the first-excited-state energy E_1, and a one-third chance of obtaining the second-excited-state energy E_2 if a measurement of the energy is carried out. If a measurement of the parity is carried out and the value -1 is obtained, what value will a subsequent measurement of the energy yield? If a measurement of the parity yields the value $+1$, what values can a subsequent measurement of the energy yield? What are the probabilities of obtaining those energies?

5.6. By writing the wave function Ψ in

$$\int_{-\infty}^{\infty} \Psi^* A_{\text{op}} \Psi \, dx = \int_{-\infty}^{\infty} (A_{\text{op}} \Psi)^* \Psi \, dx$$

as $\psi + \lambda \phi$ where λ is an arbitrary complex number, show that

$$\int_{-\infty}^{\infty} \phi^* A_{\text{op}} \psi \, dx = \int_{-\infty}^{\infty} (A_{\text{op}} \phi)^* \psi \, dx$$

Thus the requirement that an operator corresponding to an observable has real expectation values is equivalent to the definition of a Hermitian operator given in (5.98). *Suggestion*: Take advantage of the fact that λ and λ^* are linearly independent.

5.7. Show that the two wave functions

$$\psi_{b=1} = \frac{1}{\sqrt{2}} \psi_1 + \frac{1}{\sqrt{2}} \psi_2$$

and

$$\psi_{b=-1} = \frac{1}{\sqrt{2}} \psi_1 - \frac{1}{\sqrt{2}} \psi_2$$

from Example 5.3 can be expressed in the form

$$\psi_{b=1} = \cos\theta \, \psi_1 + \sin\theta \, \psi_2$$

and

$$\psi_{b=-1} = \sin\theta \, \psi_1 - \cos\theta \, \psi_2$$

if the appropriate choice for the angle θ is made. What is the value of θ for these wave functions?

5.8. Let the operator A_{op} correspond to an observable of a particle. It is assumed to have just two eigenfunctions $\psi_1(x)$ and $\psi_2(x)$ with distinct eigenvalues. The function corresponding to an arbitrary state of the particle can be written as

$$\psi(x) = c_1 \psi_1(x) + c_2 \psi_2(x)$$

An operator B_{op} is defined according to

$$B_{\text{op}} \psi(x) = c_2 \psi_1(x) + c_1 \psi_2(x)$$

Prove that B_{op} is Hermitian.

5.9. Show that if the operator A_{op} corresponding to the observable A is Hermitian then

$$\langle A^2 \rangle \geq 0$$

5.10. If A_{op} and B_{op} are Hermitian operators, prove that $A_{\text{op}} B_{\text{op}}$ is Hermitian only if A_{op} and B_{op} commute.

5.11. Suppose that A_{op} and B_{op} are Hermitian operators that do not commute:

$$[A_{\text{op}}, B_{\text{op}}] = iC_{\text{op}}$$

Prove that C_{op} is Hermitian.

5.12. Use the definitions (5.90) and (5.91) of the right and left circular polarized states to show that the two-photon state (5.89) becomes (5.92) when expressed in terms of the linearly polarized states. *Caution*: Since the photons are traveling back to back and photon 1 is traveling in the positive z direction, photon 2 is traveling in the negative z direction. Consequently, for photon 2

$$\psi_R(2) = \frac{1}{\sqrt{2}} [\psi_x(2) - i\psi_y(2)]$$

and

$$\psi_L(2) = \frac{1}{\sqrt{2}} [\psi_x(2) + i\psi_y(2)]$$

CHAPTER 6

Quantum Mechanics in Three Dimensions

As you have no doubt noticed, we do not live in a one-dimensional world. Nonetheless, through the technique of separation of variables, the Schrödinger equation for some important three-dimensional systems can be reduced to one-dimensional ordinary differential equations. As a simple but useful example, we will start by analyzing a particle in a cubic box. We will then go on to examine central potentials, which will lead us naturally to a discussion of angular momentum in quantum mechanics, which is certainly central to our understanding of how the world works. One central potential in particular, namely the Coulomb potential, plays a vital role in our understanding of atomic physics, not just for the special case of the hydrogen atom but also for our understanding of multielectron atoms. We will also take our first look in this chapter at the important topic of intrinsic spin angular momentum.

6.1 The Three-Dimensional Box

As we did in one-dimensional quantum mechanics in Chapter 3, we focus our attention in three dimensions first on a particle in a box. In Chapter 7 we will see how this potential plays a key role in determining the behavior of electrons in a metal. Since the energy of a particle is the sum of the kinetic energy and the potential energy:

$$E = \frac{\mathbf{p}^2}{2m} + V = \frac{p_x^2 + p_y^2 + p_z^2}{2m} + V(x, y, z) \quad (6.1)$$

the corresponding Hamiltonian (the energy operator) is given by

$$H = -\frac{\hbar^2}{2m}\left(\frac{\partial^2}{\partial x^2} + \frac{\partial^2}{\partial y^2} + \frac{\partial^2}{\partial z^2}\right) + V(x, y, z) \quad (6.2)$$

where we have introduced partial derivatives in x, y, and z for the linear momentum operators in the x, y, and z directions, respectively. The energy eigenvalue equation, the time-independent Schrödinger equation, is given by

$$-\frac{\hbar^2}{2m}\left(\frac{\partial^2}{\partial x^2} + \frac{\partial^2}{\partial y^2} + \frac{\partial^2}{\partial z^2}\right)\psi(x, y, z) + V(x, y, z)\psi(x, y, z) = E\psi(x, y, z) \quad (6.3)$$

a partial differential equation in three variables. For the particle in a cubic box having rigid walls, we are assuming that the potential energy is zero within the box and infinite outside it, namely

$$V(x, y, z) = \begin{cases} 0 & 0 < x, y, z < L \\ \infty & \text{elsewhere} \end{cases} \tag{6.4}$$

Thus the wave function vanishes outside the box.

Given the introductory comments in this chapter about using the technique of separation of variables, it is natural to try a solution of the form

$$\psi(x, y, z) = X(x)Y(y)Z(z) \tag{6.5}$$

that is, as a product of a function of x, a function of y, and a function of z. The goal is to separate the partial differential equation (6.3) into three ordinary differential equations. While separation of variables is not guaranteed to work, it is simple to try. Moreover, this approach does work in a number of important cases that we will examine in this chapter, including orbital angular momentum and the hydrogen atom, in addition to the particle in a three-dimensional cubic box. Substituting (6.5) into (6.3) inside the box, where $V = 0$, we obtain

$$-\frac{\hbar^2}{2m}\left(YZ\frac{d^2 X}{dx^2} + XZ\frac{d^2 Y}{dy^2} + XY\frac{d^2 Z}{dz^2}\right) = E\,XYZ \tag{6.6}$$

Dividing this equation by the wave function $\psi = XYZ$, we obtain

$$-\frac{\hbar^2}{2m}\left(\frac{1}{X}\frac{d^2 X}{dx^2} + \frac{1}{Y}\frac{d^2 Y}{dy^2} + \frac{1}{Z}\frac{d^2 Z}{dz^2}\right) = E \tag{6.7}$$

Thus the equation has indeed separated on the left-hand side into the *sum* of three functions, one a function of x, one a function of y, and the third a function of z. Since the equation must hold for all x, y, and z independently, each of the terms on the left-hand side must itself be a constant. Given that the right-hand side is also a constant, namely the energy E, we choose to call these constants E_x, E_y, and E_z such that

$$E_x + E_y + E_z = E \tag{6.8}$$

The resulting three differential equations are then

$$-\frac{\hbar^2}{2m}\frac{d^2 X}{dx^2} = E_x X$$

$$-\frac{\hbar^2}{2m}\frac{d^2 Y}{dy^2} = E_y Y$$

$$-\frac{\hbar^2}{2m}\frac{d^2 Z}{dz^2} = E_z Z \tag{6.9}$$

each of which is the energy eigenvalue equation for a particle in a one-dimensional box. Thus we can use the results of Section 3.2 to write the energy eigenvalues as

$$E_{n_x, n_y, n_z} = E_{n_x} + E_{n_y} + E_{n_z}$$
$$= \frac{(n_x^2 + n_y^2 + n_z^2)\hbar^2 \pi^2}{2mL^2} \qquad n_x, n_y, n_z = 1, 2, 3, \ldots \tag{6.10}$$

with corresponding energy eigenfunctions

$$\psi_{n_x,n_y,n_z} = \left(\frac{2}{L}\right)^{3/2} \sin\frac{n_x\pi x}{L} \sin\frac{n_y\pi y}{L} \sin\frac{n_z\pi z}{L} \qquad 0 < x, y, z < L \qquad (6.11)$$

namely the product of the three wave functions for the particle in a one-dimensional box. As we noted, the energy eigenfunctions vanish outside the well, since we are taking the potential energy there to be infinite. The ground state corresponds to the state for which $n_x = n_y = n_z = 1$. Thus there is a single state with energy $3\hbar^2\pi^2/2mL^2$. For the first excited state, which has energy $6\hbar^2\pi^2/2mL^2$, there are three possibilities: $n_x = 2$ and $n_y = n_z = 1$, $n_y = 2$ and $n_x = n_z = 1$, or $n_z = 2$ and $n_x = n_z = 1$. Thus the first excited state is three-fold degenerate. As we will see, such degeneracy is quite common in three-dimensional systems that have certain symmetries. If we were to consider a rectangular box, with sides of different lengths, the degeneracy that is present for the cubic box would disappear. See Problem 6.1.

6.2 Orbital Angular Momentum

The example of the previous section nicely illustrates how a partial differential equation in three variables can be reduced to three one-dimensional ordinary differential equations by separation of variables. In the next section, we will examine the hydrogen atom. For the hydrogen atom the potential energy is given by $-e^2/4\pi\epsilon_0 r$, where $r = \sqrt{x^2 + y^2 + z^2}$ is the distance of the electron from the proton. Consequently, the time-independent Schrödinger equation is not separable in the Cartesian coordinates x, y, and z in this case. But the energy eigenvalue equation is separable in spherical coordinates as long as $V = V(r)$. Thus the results of this section have a broad degree of applicability since they apply to any central potential. Moreover, we are led naturally to the eigenfunctions and eigenvalues of orbital angular momentum, a critically important dynamical variable.

To see how to proceed, notice that in three dimensions the momentum operator is, of course, a vector operator with components

$$\mathbf{p}_{op} = \frac{\hbar}{i}\left(\frac{\partial}{\partial x}\mathbf{i} + \frac{\partial}{\partial y}\mathbf{j} + \frac{\partial}{\partial z}\mathbf{k}\right) = \frac{\hbar}{i}\nabla \qquad (6.12)$$

where ∇ is the usual gradient operator. Consequently

$$\mathbf{p}_{op}^2 = -\hbar^2\nabla^2 \qquad (6.13)$$

where the operator

$$\nabla^2 = \frac{\partial^2}{\partial x^2} + \frac{\partial^2}{\partial y^2} + \frac{\partial^2}{\partial z^2} \qquad (6.14)$$

is called the Laplacian. Expressed in the spherical coordinates r, θ, and ϕ shown in Fig. 6.1, the Laplacian becomes

$$\nabla^2 = \frac{1}{r^2}\frac{\partial}{\partial r}\left(r^2\frac{\partial}{\partial r}\right) + \frac{1}{r^2\sin\theta}\frac{\partial}{\partial \theta}\left(\sin\theta\frac{\partial}{\partial \theta}\right) + \frac{1}{r^2\sin^2\theta}\frac{\partial^2}{\partial \phi^2} \qquad (6.15)$$

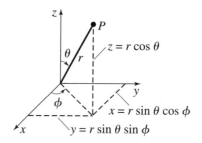

Figure 6.1 Spherical coordinates.

Thus when the potential energy $V = V(r)$, the Schrödinger equation is given by

$$-\frac{\hbar^2}{2m}\left[\frac{1}{r^2}\frac{\partial}{\partial r}\left(r^2\frac{\partial}{\partial r}\right) + \frac{1}{r^2\sin\theta}\frac{\partial}{\partial \theta}\left(\sin\theta\frac{\partial}{\partial \theta}\right) + \frac{1}{r^2\sin^2\theta}\frac{\partial^2}{\partial \phi^2}\right]\Psi + V(r)\Psi = E\Psi \tag{6.16}$$

Although this partial differential equation looks quite complicated, it too can be reduced to three ordinary differential equations by using the separation of variables technique and writing

$$\Psi(r, \theta, \phi) = R(r)\Theta(\theta)\Phi(\phi) \tag{6.17}$$

As we will see in this section, the operator $\mathbf{L}_{\text{op}}^2$ corresponding to the magnitude squared of the orbital angular momentum is given by

$$\mathbf{L}_{\text{op}}^2 = -\hbar^2\left[\frac{1}{\sin\theta}\frac{\partial}{\partial \theta}\left(\sin\theta\frac{\partial}{\partial \theta}\right) + \frac{1}{\sin^2\theta}\frac{\partial^2}{\partial \phi^2}\right] \tag{6.18}$$

and therefore the angular term in the Hamiltonian

$$-\frac{\hbar^2}{2m}\left[\frac{1}{r^2\sin\theta}\frac{\partial}{\partial \theta}\left(\sin\theta\frac{\partial}{\partial \theta}\right) + \frac{1}{r^2\sin^2\theta}\frac{\partial^2}{\partial \phi^2}\right] = \frac{\mathbf{L}_{\text{op}}^2}{2mr^2} \tag{6.19}$$

is simply the rotational kinetic energy.[1] If we take the angular part of the wave function (6.17) to be an eigenfunction of $\mathbf{L}_{\text{op}}^2$, then we will be able to replace $\mathbf{L}_{\text{op}}^2$ in (6.16) with its eigenvalue, thus reducing the equation to an ordinary differential equation in the variable r.

In classical physics, the angular momentum $\mathbf{L}$ of a particle is defined by the relation

$$\mathbf{L} = \mathbf{r} \times \mathbf{p} \tag{6.20}$$

or written out in terms of its Cartesian components

$$L_x = yp_z - zp_y$$
$$L_y = zp_x - xp_z$$
$$L_z = xp_y - yp_x \tag{6.21}$$

---

[1] Recall from classical mechanics that $L = I\omega$ and therefore the rotational kinetic energy $I\omega^2/2 = L^2/2I$. In this case $I = mr^2$ for a mass m moving at a distance r from the center of force.

In quantum mechanics, these components of the angular momentum become the operators

$$L_{x\,\text{op}} = y\frac{\hbar}{i}\frac{\partial}{\partial z} - z\frac{\hbar}{i}\frac{\partial}{\partial y}$$
$$L_{y\,\text{op}} = z\frac{\hbar}{i}\frac{\partial}{\partial x} - x\frac{\hbar}{i}\frac{\partial}{\partial z}$$
$$L_{z\,\text{op}} = x\frac{\hbar}{i}\frac{\partial}{\partial y} - y\frac{\hbar}{i}\frac{\partial}{\partial x} \tag{6.22}$$

Perhaps not surprisingly, since we are discussing *angular* momentum, it is best to express these operators in terms of angles. In spherical coordinates

$$x = r\sin\theta\cos\phi$$
$$y = r\sin\theta\sin\phi$$
$$z = r\cos\theta \tag{6.23}$$

Consequently

$$\frac{\partial x}{\partial \phi} = -y$$
$$\frac{\partial y}{\partial \phi} = x$$
$$\frac{\partial z}{\partial \phi} = 0 \tag{6.24}$$

and therefore

$$\frac{\partial}{\partial \phi} = \frac{\partial x}{\partial \phi}\frac{\partial}{\partial x} + \frac{\partial y}{\partial \phi}\frac{\partial}{\partial y} + \frac{\partial z}{\partial \phi}\frac{\partial}{\partial z}$$
$$= -y\frac{\partial}{\partial x} + x\frac{\partial}{\partial y} \tag{6.25}$$

Comparing the last line of (6.25) with the expression for $L_{z\,\text{op}}$ in (6.22), we see that

$$L_{z\,\text{op}} = \frac{\hbar}{i}\frac{\partial}{\partial \phi} \tag{6.26}$$

This form for $L_{z\,\text{op}}$ is relatively simple because rotations about the z axis involve changes solely in the azimuthal angle ϕ, whereas rotations about the x and y axes require changes in both θ and ϕ. Consequently, the x and y components of the orbital angular momentum are not as simple as the z component. As shown in Problem 6.8, they are given by

$$L_{x\,\text{op}} = \frac{\hbar}{i}\left(-\sin\phi\frac{\partial}{\partial \theta} - \cot\theta\cos\phi\frac{\partial}{\partial \phi}\right) \tag{6.27}$$

and

$$L_{y\,\text{op}} = \frac{\hbar}{i}\left(\cos\phi\frac{\partial}{\partial \theta} - \cot\theta\sin\phi\frac{\partial}{\partial \phi}\right) \tag{6.28}$$

If we evaluate the operator corresponding to the magnitude squared of the angular momentum

$$\mathbf{L}^2_{\text{op}} = L^2_{x\,\text{op}} + L^2_{y\,\text{op}} + L^2_{z\,\text{op}} \tag{6.29}$$

in terms of the angular derivatives using (6.26), (6.27), and (6.28), we find

$$\mathbf{L}_{op}^2 = -\hbar^2 \left[\frac{1}{\sin\theta} \frac{\partial}{\partial\theta} \left(\sin\theta \frac{\partial}{\partial\theta} \right) + \frac{1}{\sin^2\theta} \frac{\partial^2}{\partial\phi^2} \right] \quad (6.30)$$

Our goal is to find the eigenvalues and eigenfunctions of $\mathbf{L}_{op}^2$. Let us start by noting that the angle ϕ enters into (6.30) only through the second derivative with respect to ϕ. There are no $\sin\phi$ or $\cos\phi$ terms as there are in (6.27) and (6.28). Thus since $L_{z\,op}$ involves only the derivative with respect to ϕ, we see that

$$\left[\mathbf{L}_{op}^2, L_{z\,op}\right] = 0 \quad (6.31)$$

that is, these operators commute.[2] We can therefore find simultaneous eigenfunctions of these two operators. Using the technique of separation of variables again, we can write the eigenfunction as $Y(\theta, \phi) = \Theta(\theta)\Phi(\phi)$. Then the eigenvalue equation

$$L_{z\,op} Y(\theta, \phi) = L_z Y(\theta, \phi) \quad (6.32)$$

becomes

$$\frac{\hbar}{i} \frac{\partial}{\partial\phi} \Theta(\theta)\Phi(\phi) = L_z \Theta(\theta)\Phi(\phi) \quad (6.33)$$

where L_z is the corresponding eigenvalue. Thus the ϕ dependence of the eigenfunction is determined by solving the ordinary differential equation

$$\frac{\hbar}{i} \frac{d\Phi}{d\phi} = L_z \Phi \quad (6.34)$$

which yields

$$\Phi(\phi) = N e^{iL_z\phi/\hbar} \quad (6.35)$$

where N is a constant that can be fixed by normalization (see Example 6.1). This angular momentum eigenfunction is reminiscent of the linear momentum eigenfunction $Ae^{ip_xx/\hbar}$. But there is a significant difference. Here when ϕ changes by 2π, we must return to the same point in space. If the wave function is not single valued, for example if $\Phi(2\pi) \neq \Phi(0)$, then the derivative of the wave function is not well defined, say at $\phi = 0$, just as the spatial derivative for the linear momentum operator would be infinite if the wave function were not continuous. But if we require

$$N e^{iL_z(\phi+2\pi)/\hbar} = N e^{iL_z\phi/\hbar} \quad (6.36)$$

we find that

$$e^{iL_z 2\pi/\hbar} = 1 \quad (6.37)$$

Hence,

$$L_z = m_l \hbar \qquad m_l = 0, \pm 1, \pm 2, \ldots \quad (6.38)$$

Thus we see that the z component of the *orbital* angular momentum is quantized, always! The allowed values are integer multiples of $\hbar$.

---

[2]While it is far from self-evident from the form of the operators for $L_{x\,op}$ and $L_{y\,op}$ given in (6.27) and (6.28), you can verify that $\mathbf{L}_{op}^2$ commutes with these operators as well as with $L_{z\,op}$.

How do we determine the θ dependence of the eigenfunction $Y(\theta, \phi)$? The eigenvalue equation

$$\mathbf{L}_{\text{op}}^2 Y(\theta, \phi) = \lambda \hbar^2 Y(\theta, \phi) \tag{6.39}$$

yields the differential equation

$$-\hbar^2 \left[\frac{1}{\sin\theta} \frac{d}{d\theta} \left(\sin\theta \frac{d\Theta}{d\theta} \right) - \frac{m_l^2}{\sin^2\theta} \Theta \right] = \lambda \hbar^2 \Theta \tag{6.40}$$

where we have used the explicit ϕ dependence $Y(\theta, \phi) = \Theta(\theta) e^{im_l\phi}$ and written the eigenvalue in the form $\lambda \hbar^2$, where λ is a dimensionless constant. Equation (6.40) is not solved as easily as (6.34)! The solution is discussed in Section B.2, where it is shown that the Θ that solves (6.40) will be a satisfactory wave function, namely not blowing up at $\theta = 0$ or $\theta = \pi$, only if the eigenvalue satisfies

$$\lambda = l(l+1) \qquad l = 0, 1, 2, \ldots \tag{6.41}$$

Thus the magnitude squared as well as the z component of the orbital angular momentum is quantized. The eigenfunctions are referred to as the **spherical harmonics** and are denoted by $Y_{l,m_l}(\theta, \phi)$. They satisfy the two equations

$$L_{z\,\text{op}} Y_{l,m_l}(\theta, \phi) = m_l \hbar Y_{l,m_l}(\theta, \phi) \qquad m_l = -l, -l+1, \ldots, l-1, l \tag{6.42}$$

and

$$\mathbf{L}_{\text{op}}^2 Y_{l,m_l}(\theta, \phi) = l(l+1)\hbar^2 Y_{l,m_l}(\theta, \phi) \qquad l = 0, 1, 2, \ldots \tag{6.43}$$

As examples, the spherical harmonics with $l = 0$, $l = 1$, and $l = 2$ are

$$Y_{0,0}(\theta, \phi) = \sqrt{\frac{1}{4\pi}} \tag{6.44}$$

$$Y_{1,\pm 1}(\theta, \phi) = \mp \sqrt{\frac{3}{8\pi}} \sin\theta \, e^{\pm i\phi} \tag{6.45}$$

$$Y_{1,0}(\theta, \phi) = \sqrt{\frac{3}{4\pi}} \cos\theta \tag{6.46}$$

$$Y_{2,\pm 2}(\theta, \phi) = \sqrt{\frac{15}{32\pi}} \sin^2\theta \, e^{\pm 2i\phi} \tag{6.47}$$

$$Y_{2,\pm 1}(\theta, \phi) = \mp \sqrt{\frac{15}{8\pi}} \sin\theta \cos\theta \, e^{\pm i\phi} \tag{6.48}$$

$$Y_{2,0}(\theta, \phi) = \sqrt{\frac{5}{16\pi}} \left(3\cos^2\theta - 1\right) \tag{6.49}$$

The spherical harmonics satisfy the orthonormality condition

$$\int_0^{2\pi} \int_0^{\pi} Y_{l',m_{l'}}^* Y_{l,m_l} \sin\theta \, d\theta \, d\phi = \delta_{l'l} \delta_{m_{l'} m_l} \tag{6.50}$$

The factor of $\sin\theta \, d\theta \, d\phi$ arises from the fact that the differential volume element in spherical coordinates is given by $r^2 \sin\theta \, dr \, d\theta \, d\phi = r^2 dr d\Omega$, as shown in Fig. 6.2. As is customary, we have chosen to normalize the angular part of the wave function separately from the radial part, which will then satisfy the condition

$$\int_0^\infty |R(r)|^2 r^2 dr = 1 \tag{6.51}$$

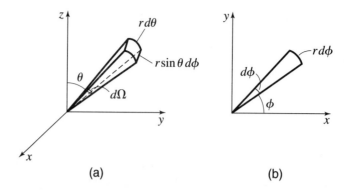

Figure 6.2 (a) The solid angle $d\Omega$ in three dimensions is defined as the surface area dS subtended at radius r divided by r^2: $d\Omega = dS/r^2 = (rd\theta)(r\sin\theta d\phi)/r^2$. (b) The ordinary angle $d\phi$ in two dimensions is defined as the arc length ds subtended at radius r divided by the radius: $ds/r = d\phi$.

Example 6.2 verifies that $Y_{1,1}(\theta,\phi)$ satisfies the eigenvalue equation (6.43) and is appropriately normalized. Figure 6.3 shows plots of $|Y_{l,m_l}|^2$ as functions of θ and ϕ.

One of the most striking aspects of these angular momentum eigenfunctions is that the magnitude of the orbital angular momentum

$$|\mathbf{L}| = \sqrt{l(l+1)}\hbar \tag{6.52}$$

is always bigger than the maximum projection of the angular momentum on the z axis

$$L_{z\,\text{max}} = l\hbar \tag{6.53}$$

since m_l is always less than or equal to l. Why is this? As we will now show, the commutation relations of the orbital angular momentum operators prohibit the angular momentum, which is of course a vector, from pointing in any particular direction, since that would mean that we know all three components of the angular momentum simultaneously.

To see the profound difference between linear momentum, which is also a vector, and angular momentum, consider the following commutator:

$$[p_{x\,\text{op}}, p_{y\,\text{op}}] = \left[\frac{\hbar}{i}\frac{\partial}{\partial x}, \frac{\hbar}{i}\frac{\partial}{\partial y}\right] = 0 \tag{6.54}$$

This commutator vanishes because

$$\frac{\partial}{\partial x}\frac{\partial}{\partial y} = \frac{\partial}{\partial y}\frac{\partial}{\partial x} \tag{6.55}$$

Thus it is possible to specify both the x and y components of the linear momentum simultaneously. On the other hand,

$$\begin{aligned}
[L_{x\,\text{op}}, L_{y\,\text{op}}] &= \left[y\frac{\hbar}{i}\frac{\partial}{\partial z} - z\frac{\hbar}{i}\frac{\partial}{\partial y}, z\frac{\hbar}{i}\frac{\partial}{\partial x} - x\frac{\hbar}{i}\frac{\partial}{\partial z}\right] \\
&= \left[y\frac{\hbar}{i}\frac{\partial}{\partial z}, z\frac{\hbar}{i}\frac{\partial}{\partial x}\right] + \left[z\frac{\hbar}{i}\frac{\partial}{\partial y}, x\frac{\hbar}{i}\frac{\partial}{\partial z}\right] \\
&= y\left[\frac{\hbar}{i}\frac{\partial}{\partial z}, z\right]\frac{\hbar}{i}\frac{\partial}{\partial x} + \frac{\hbar}{i}\frac{\partial}{\partial y}\left[z, \frac{\hbar}{i}\frac{\partial}{\partial z}\right]x \\
&= i\hbar\left(-y\frac{\hbar}{i}\frac{\partial}{\partial x} + x\frac{\hbar}{i}\frac{\partial}{\partial y}\right) = i\hbar L_{z\,\text{op}}
\end{aligned} \tag{6.56}$$

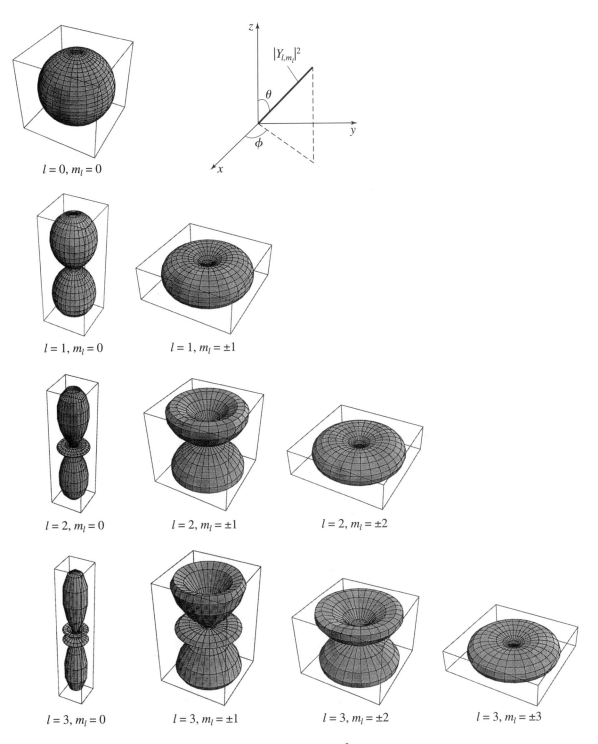

Figure 6.3 Plots of $\left|Y_{l,m_l}(\theta,\phi)\right|^2$ for $l = 0, 1, 2, 3$.

where we have used the fact [see the argument leading up to (5.47)] that

$$[z_{\text{op}}, p_{z\,\text{op}}] = \left[z, \frac{\hbar}{i}\frac{\partial}{\partial z}\right] = i\hbar \qquad (6.57)$$

As we saw in Section 5.4, when the commutator does not vanish as in (6.56), we end up with an uncertainty relation, which in the case of (6.56) takes the form

$$\Delta L_x \Delta L_y \geq \frac{\hbar}{2}|\langle L_z \rangle| \qquad (6.58)$$

Suppose the state of a particle is an eigenstate of $L_{z\,\text{op}}$ with eigenvalue $m_l \hbar \neq 0$. Then the z component of the particle's orbital angular momentum is known with certainty, namely it is $m_l \hbar$. But since $\langle L_z \rangle \neq 0$, then (6.58) shows that ΔL_x and ΔL_y must both be nonzero. Thus there is an inherent uncertainty in the values for the x and y components of the orbital angular momentum when the z component is known. Consequently, we must give up on the classical notion that the angular momentum points in a particular direction, since that would imply that we know L_x, L_y, and L_z simultaneously.

EXAMPLE 6.1 Show that the azimuthal wave function

$$\psi(\phi) = \frac{1}{\sqrt{\pi}}\cos\phi$$

is properly normalized. What are the possible results of a measurement of the z component of the orbital angular momentum for a particle in this state? What are the probabilities of obtaining each result? What is $\langle L_z \rangle$ for this state?

SOLUTION

$$\int_0^{2\pi} |\psi(\phi)|^2\,d\phi = \frac{1}{\pi}\int_0^{2\pi} \cos^2\phi\,d\phi = \frac{1}{\pi}\left(\frac{\phi}{2}+\frac{\sin 2\phi}{4}\right)\Big|_0^{2\pi} = 1$$

so ψ is properly normalized. Next note that

$$L_{z\,\text{op}}\psi(\phi) = -\frac{\hbar}{i\sqrt{\pi}}\sin\phi$$

Thus $\psi(\phi)$ is not an eigenfunction of $L_{z\,\text{op}}$. Rather it must be a superposition of the normalized eigenfunctions of $L_{z\,\text{op}}$, namely

$$\psi_{m_l}(\phi) = \frac{1}{\sqrt{2\pi}} e^{im_l\phi} \qquad m_l = 0, \pm 1, \pm 2, \ldots$$

Since

$$\psi(\phi) = \frac{1}{\sqrt{\pi}}\cos\phi = \frac{1}{\sqrt{\pi}}\left(\frac{e^{i\phi}+e^{-i\phi}}{2}\right) = \frac{1}{\sqrt{2}}\psi_1 + \frac{1}{\sqrt{2}}\psi_{-1}$$

a measurement of L_z can yield either $\hbar$ or $-\hbar$ each with a probability $(1/\sqrt{2})^2 = 1/2$. Therefore

$$\langle L_z \rangle = \sum_{m_l} |c_{m_l}|^2 m_l \hbar = \frac{1}{2}\hbar + \frac{1}{2}(-\hbar) = 0$$

Alternatively

$$\langle L_z \rangle = \int_0^{2\pi} \psi^*(\phi)\frac{\hbar}{i}\frac{\partial \psi(\phi)}{\partial \phi}\,d\phi = \frac{i\hbar}{\pi}\int_0^{2\pi}\cos\phi\sin\phi\,d\phi = \frac{\hbar}{4\pi i}\cos 2\phi\Big|_0^{2\pi} = 0$$

EXAMPLE 6.2 Verify that $Y_{1,-1}(\theta, \phi) = \sqrt{\frac{3}{8\pi}} \sin\theta \, e^{-i\phi}$ is a normalized eigenfunction of $\mathbf{L}_{op}^2$ with the appropriate eigenvalue.

SOLUTION

$$\mathbf{L}_{op}^2 Y_{1,-1} = -\hbar^2 \left[\frac{1}{\sin\theta} \frac{\partial}{\partial \theta} \left(\sin\theta \frac{\partial}{\partial \theta} \right) + \frac{1}{\sin^2\theta} \frac{\partial^2}{\partial \phi^2} \right] \left(\sqrt{\frac{3}{8\pi}} \sin\theta \, e^{-i\phi} \right)$$

$$= -\hbar^2 \sqrt{\frac{3}{8\pi}} \left[\frac{1}{\sin\theta} \frac{\partial}{\partial \theta} \left(\sin\theta \cos\theta \, e^{-i\phi} \right) - \frac{1}{\sin^2\theta} \sin\theta \, e^{-i\phi} \right]$$

$$= -\hbar^2 \sqrt{\frac{3}{8\pi}} \left[\frac{1}{\sin\theta} \left(\cos^2\theta - \sin^2\theta \right) e^{-i\phi} - \frac{1}{\sin\theta} e^{-i\phi} \right]$$

$$= -\hbar^2 \sqrt{\frac{3}{8\pi}} \left(\frac{\cos^2\theta - \sin^2\theta - 1}{\sin\theta} \right) e^{-i\phi}$$

$$= 2\hbar^2 \sqrt{\frac{3}{8\pi}} \sin\theta \, e^{-i\phi} = 2\hbar^2 \, Y_{1,-1}$$

The eigenvalue is $2\hbar^2 = l(l+1)\hbar^2$ with $l = 1$. And the normalization checks, too:

$$\int_0^{2\pi} \int_0^{\pi} |Y_{1,-1}(\theta, \phi)|^2 \sin\theta \, d\theta \, d\phi = \frac{3}{8\pi} \int_0^{2\pi} \int_0^{\pi} \sin^2\theta \sin\theta \, d\theta \, d\phi$$

$$= \frac{3}{4} \int_0^{\pi} \left(1 - \cos^2\theta \right) \sin\theta \, d\theta$$

$$= \frac{3}{4} \left(-\cos\theta + \frac{\cos^3\theta}{3} \right) \Big|_0^{\pi} = 1$$

6.3 The Hydrogen Atom

If you had to pick one problem to solve in three-dimensional quantum mechanics, it would probably be the hydrogen atom, the simplest atomic system. Not only does the Schrödinger equation for the Coulomb potential have an exact solution, but the energy levels and the wave functions for the hydrogen atom have much to tell us about multielectron atoms as well. You have probably heard of the Bohr model—Niels Bohr's attempt to "explain" the spectrum of hydrogen with its discrete energy levels. The Bohr model is of much historical interest (see Problem 6.18), but apart from getting the right answer, at least in terms of the allowed energies, almost everything else about the model is incorrect. In particular, the electron does not follow a definite trajectory, as is often suggested by the planetary-like models of the atom. Rather, there are certain allowed wave functions, the energy eigenfunctions, which result from solving the Schrödinger equation

$$-\frac{\hbar^2}{2m} \left[\frac{1}{r^2} \frac{\partial}{\partial r} \left(r^2 \frac{\partial}{\partial r} \right) + \frac{1}{r^2 \sin\theta} \frac{\partial}{\partial \theta} \left(\sin\theta \frac{\partial}{\partial \theta} \right) + \frac{1}{r^2 \sin^2\theta} \frac{\partial^2}{\partial \phi^2} \right] \Psi - \frac{Ze^2}{4\pi \epsilon_0 r} \Psi = E\Psi \quad (6.59)$$

where we have taken the charge on the nucleus to be Ze. This permits us to consider not only hydrogen, for which $Z = 1$, but also singly ionized helium ($Z = 2$), et cetera. Because the mass of the proton in the hydrogen atom is so much bigger than that of the electron, we can think of the electron as "moving" about a fixed center of force. Strictly, this is a two-body problem and the mass m in the Schrödinger equation should really be the reduced mass $\mu = m_1 m_2/(m_1 + m_2)$, where m_1 and m_2 are the masses of the two constituents.

If we write the wave function

$$\Psi(r, \theta, \phi) = R(r) Y_{l,m_l}(\theta, \phi) \qquad (6.60)$$

then (6.59) with the aid of (6.43) reduces to the radial differential equation

$$-\frac{\hbar^2}{2m}\left[\frac{1}{r^2}\frac{d}{dr}\left(r^2\frac{dR}{dr}\right)\right] + \frac{l(l+1)\hbar^2}{2mr^2}R - \frac{Ze^2}{4\pi\epsilon_0 r}R = ER \qquad (6.61)$$

This radial equation simplifies if we make the change of variables

$$u(r) = rR(r) \qquad (6.62)$$

We then end up with the differential equation

$$-\frac{\hbar^2}{2m}\frac{d^2 u}{dr^2} + \left[\frac{l(l+1)\hbar^2}{2mr^2} - \frac{Ze^2}{4\pi\epsilon_0 r}\right]u = Eu \qquad (6.63)$$

This is an interesting way to write the radial equation because it looks just like the one-dimensional Schrödinger equation with an effective potential energy

$$V_{\text{eff}} = \frac{l(l+1)\hbar^2}{2mr^2} - \frac{Ze^2}{4\pi\epsilon_0 r} \qquad (6.64)$$

Figure 6.4 shows a plot of the effective potential energy for various values of the orbital angular momentum quantum number l. The term $l(l+1)\hbar^2/2mr^2$ is referred to as the **centrifugal barrier**, since for nonzero l it tends to keep the electron away from the origin, where the proton is located. For $E < 0$ the electron is confined in a potential well and thus we expect to find discrete energies.

The differential equation (6.63) is certainly not as simple as the one-dimensional differential equations that we solved in Chapter 4. We can learn a lot, nonetheless, by examining the behavior of the solutions for large and small r. For large r,

$$\frac{1}{r} \to 0 \qquad \text{and} \qquad \frac{1}{r^2} \to 0 \qquad (6.65)$$

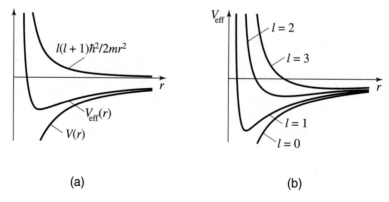

Figure 6.4 (a) The Coulomb potential energy $V(r) = -Ze^2/4\pi\epsilon_0 r$ and the centrifugal barrier $l(l+1)\hbar^2/2mr^2$ add together to produce the effective potential energy V_{eff} given in (6.64). (b) The effective potential energy for several values of l.

and thus the differential equation reduces to

$$-\frac{\hbar^2}{2m}\frac{d^2u}{dr^2} \cong Eu \qquad r \to \infty \qquad (6.66)$$

which for $E < 0$ has the solutions

$$u = e^{\pm\sqrt{-2mE/\hbar^2}\,r} \qquad r \to \infty \qquad (6.67)$$

For small r, on the other hand, the differential equation is dominated by the centrifugal barrier

$$\frac{d^2u}{dr^2} \cong \frac{l(l+1)}{r^2}u \qquad r \to 0 \qquad (6.68)$$

which has as solutions

$$u = r^{l+1} \quad \text{and} \quad u = r^{-l} \qquad r \to 0 \qquad (6.69)$$

In solving (6.66) we will ignore the solution with a radially increasing exponential, while in solving (6.68) we will ignore the solution that behaves as r^{-l} since it too leads to problems with normalization.[3]

Taking advantage of what we know about the small-r and large-r behavior of the wave function suggests searching for an exact solution of the form

$$u(r) = r^{l+1} F(r) e^{-\sqrt{-2mE/\hbar^2}\,r} \qquad (6.70)$$

and hence

$$R(r) = r^l F(r) e^{-\sqrt{-2mE/\hbar^2}\,r} \qquad (6.71)$$

The corresponding differential equation for $F(r)$ can be solved as a power series in r. In Section B.3 it is shown that the only wave functions that are physically acceptable occur when $F(r)$ consists of polynomials (called associated Laguerre polynomials) in r. The corresponding energy eigenvalues are given by

$$E_n = -\frac{mZ^2e^4}{(4\pi\epsilon_0)^2 2\hbar^2 n^2} = -\frac{(13.6 \text{ eV})Z^2}{n^2} \qquad n = 1, 2, 3, \ldots \qquad (6.72)$$

as indicated in Fig. 6.5.[4]

The integer n in (6.72) is called the principal quantum number. When the hydrogen atom makes a transition from a state with principal quantum number n_i to one with principal quantum number n_f, a photon of frequency ν is emitted, where

$$h\nu = E_{n_i} - E_{n_f} \qquad (6.73)$$

---

[3] An exception is $l = 0$, in which case u approaches a constant for small r, meaning that R behaves like $1/r$ for small r, which is not a solution to (6.61) at the origin.

[4] A nice way to write (and hence remember) the expression for the allowed energies is $E_n = -mc^2 Z^2 \alpha^2 / 2n^2$ (or, more accurately, $E_n = -\mu c^2 Z^2 \alpha^2 / 2n^2$, where μ is the reduced mass of the electron–proton system). The constant $\alpha = e^2/4\pi\epsilon_0\hbar c$ is called the fine-structure constant. The numerical value of α is approximately 1/137, which is the reason the energies of the hydrogen atom are measured in eV rather than MeV, since $mc^2 = 0.511$ MeV for an electron.

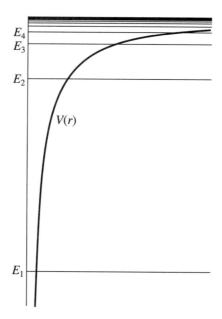

Figure 6.5 The energy levels of the hydrogen atom superposed on a graph of the Coulomb potential energy.

Figure 6.6 The visible spectrum of hydrogen, showing the Balmer series. Adapted from W. Finkelnburg, *Structure of Matter*, Springer-Verlag, 1964.

Figure 6.6 shows the spectrum of hydrogen when transitions take place from $n_i > 2$ to $n_f = 2$. These transitions, the lowest four of which are in the visible portion of the electromagnetic spectrum, are referred to as the Balmer series. Transitions that end on $n_f = 1$ produce more energetic photons, in the ultraviolet portion of the electromagnetic spectrum (the Lyman series), while transitions that end on $n_f = 3$ are in the infrared portion of the spectrum (the Paschen series).

One of the striking aspects of (6.72) is that the allowed energies depend on the integer n, the principal quantum number, but seemingly not on the values of l and m_l. Of course, the lack of dependence on m_l is not unexpected, since m_l did not enter into the differential equation (6.61). This degeneracy—the fact that different values of m_l and hence different wave functions correspond to the same energy—arises from the fact that the potential energy is spherically symmetric. Thus it shouldn't matter which axis is called the z axis, and therefore the energy should not depend on the projection of the angular momentum on this axis. On the other hand, the lack of dependence on l is surprising. The differential equation (6.61) clearly depends on l, with different values for l generating different effective potentials. The place where l enters is actually hidden in the result (6.72). For a particular value of n, the allowed values of l are given by

$$l = 0, 1, 2, \ldots, n - 1 \tag{6.74}$$

Figure 6.7 The $n = 1$ through $n = 4$ energy levels of the hydrogen atom, showing the degeneracy.

as shown in Fig. 6.7. Thus for the ground state, for which $n = 1, l = 0$ only, while for $n = 2, l = 0$ and $l = 1$, and so forth. The radial functions $R_{n,l}(r)$ thus require two subscripts to label the corresponding values of n and l. As an illustration, the normalized radial function for the ground state is given by

$$R_{1,0}(r) = 2\left(\frac{Z}{a_0}\right)^{3/2} e^{-Zr/a_0} \tag{6.75}$$

As Example 6.3 shows, the parameter a_0 is given by

$$a_0 = \frac{4\pi\epsilon_0\hbar^2}{me^2} \tag{6.76}$$

For hydrogen, $a_0 = 0.53$ Å, which is precisely the value of the radius for the circular orbit that Bohr presumed in the Bohr model. Hence a_0 is known as the Bohr radius. Example 6.3 illustrates how this value is determined in quantum mechanics. Of course, in quantum mechanics the electron does not follow a circular orbit. In fact, since $l = 0$ for the ground state, the electron in the ground state does not possess any orbital angular momentum at all. It does have a radial wave function that is nonzero for all r, one for which the probability of finding the electron between r and $r + dr$ has a maximum at a_0 (see Example 6.4).

The normalized radial wave functions for the first excited states ($n = 2$) are

$$R_{2,0}(r) = 2\left(\frac{Z}{2a_0}\right)^{3/2}\left(1 - \frac{Zr}{2a_0}\right) e^{-Zr/2a_0} \tag{6.77}$$

$$R_{2,1}(r) = \frac{1}{\sqrt{3}}\left(\frac{Z}{2a_0}\right)^{3/2} \frac{Zr}{a_0} e^{-Zr/2a_0} \tag{6.78}$$

and for the second excited states ($n = 3$) are

$$R_{3,0}(r) = 2\left(\frac{Z}{3a_0}\right)^{3/2}\left[1 - \frac{2Zr}{3a_0} + \frac{2(Zr)^2}{27a_0^2}\right] e^{-Zr/3a_0} \tag{6.79}$$

$$R_{3,1}(r) = \frac{4\sqrt{2}}{9}\left(\frac{Z}{3a_0}\right)^{3/2} \frac{Zr}{a_0}\left(1 - \frac{Zr}{6a_0}\right) e^{-Zr/3a_0} \tag{6.80}$$

$$R_{3,2}(r) = \frac{2\sqrt{2}}{27\sqrt{5}}\left(\frac{Z}{3a_0}\right)^{3/2}\left(\frac{Zr}{a_0}\right)^2 e^{-Zr/3a_0} \tag{6.81}$$

where

$$\psi_{n,l,m_l}(r,\theta,\phi) = R_{n,l}(r) Y_{l,m_l}(\theta,\phi) \tag{6.82}$$

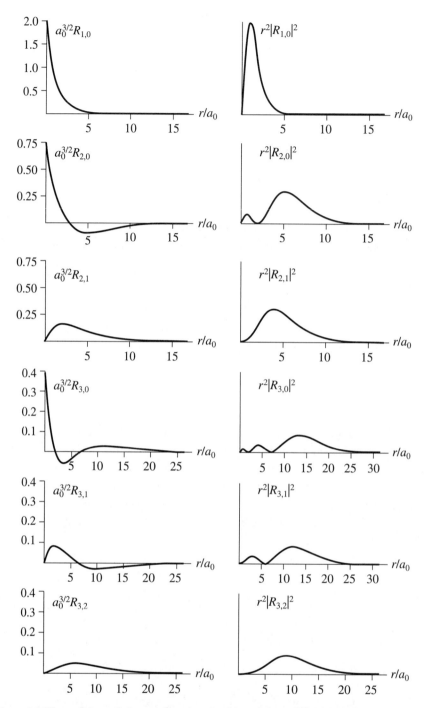

Figure 6.8 Plots of the radial wave function $R_{n,l}(r)$ and the radial probability densities $r^2 \left| R_{n,l}(r) \right|^2$ for the wave functions with $n = 1, 2, 3$.

These radial functions as well as the radial probability densities $r^2 \left| R_{n,l}(r) \right|^2$ are plotted in Fig. 6.8. Notice how the probability of finding the electron at increasing r varies with the quantum numbers n and l. This is why it is often said that the electron resides in different shells depending on the principal quantum number n. Some properties of the radial functions that you may see in Fig. 6.8 include the fact that the number of nodes is given by $n_r = n - l - 1$, the exponential dependence depends on n as e^{-Zr/na_0}, and the behavior for small r varies as r^l. In order to get a sense of how the probability of finding

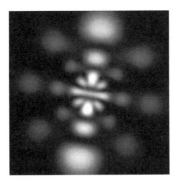

Figure 6.9 A plot of the probability density for the $n = 7$, $l = 4$, $m_l = 0$ hydrogen atom wave function. To "see" the three-dimensional nature of the eigenfunctions up through $n = 7$, including this figure, check out the award-winning Atom in a Box program created by Dean Dauger, one of my former students, at http://daugerresearch.com/orbitals/.

the electron varies in three dimensions, we need to combine the radial wave function and the orbital wave function. See Fig. 6.9 for an example.

EXAMPLE 6.3 Verify that (6.75) is the ground-state energy eigenfunction for the hydrogen atom with the appropriate ground-state energy.

SOLUTION Since the ground-state has $l = 0$, the radial Schrödinger equation (6.61) is given by

$$\left[-\frac{\hbar^2}{2m}\left(\frac{d^2}{dr^2} + \frac{2}{r}\frac{d}{dr}\right) - \frac{e^2}{4\pi\epsilon_0 r} \right] R_{1,0} = E R_{1,0}$$

or

$$\left[-\frac{\hbar^2}{2m}\left(\frac{d^2}{dr^2} + \frac{2}{r}\frac{d}{dr}\right) - \frac{e^2}{4\pi\epsilon_0 r} \right] e^{-r/a_0} = E e^{-r/a_0}$$

where we have divided out the normalization constant from both sides of the equation and have set $Z = 1$. Taking the derivatives, we obtain

$$\left[-\frac{\hbar^2}{2m}\left(\frac{1}{a_0^2} - \frac{2}{a_0 r}\right) - \frac{e^2}{4\pi\epsilon_0 r} \right] e^{-r/a_0} = E e^{-r/a_0}$$

Notice that the $1/r$ terms multiplying the wave function on the left-hand side must cancel so we end up with a constant. Therefore,

$$\frac{\hbar^2}{m a_0} = \frac{e^2}{4\pi\epsilon_0}$$

consistent with (6.76). Consequently, the remaining constant on the left-hand side must equal the energy E:

$$E = -\frac{\hbar^2}{2m a_0^2} = -\frac{m e^4}{2(4\pi\epsilon_0)^2 \hbar^2}$$

consistent with (6.72). Notice how solving the Schrödinger equation determines both the energy and the size (through a_0) of the atom.

EXAMPLE 6.4 At what radius is the probability of finding the particle a maximum in the ground state of the hydrogen atom?

SOLUTION Given the normalization condition (6.51), the probability of finding the particle between r and $r + dr$ is given by $R_{1,0}^2 r^2 \, dr$. To find where the probability is a maximum, we differentiate the radial probability density with respect to r and set the result to zero:

$$\frac{d}{dr} r^2 e^{-2r/a_0} = 2r e^{-2r/a_0} - 2\frac{r^2}{a_0} e^{-2r/a_0} = 0$$

which is satisfied if $r = a_0$.

EXAMPLE 6.5 What is the ionization energy of positronium?

SOLUTION Positronium is an "atom" in which the nucleus is the positron, the antiparticle of the electron. Since the positron has the same mass as the electron, the reduced mass of the electron–positron system

$$\frac{m_1 m_2}{m_1 + m_2} = \frac{m^2}{2m} = \frac{m}{2}$$

where m is the mass of the electron. Thus the allowed energies of positronium are given by (6.72) with the replacement $m \to m/2$. Consequently the ionization energy of positronium is $(13.6/2)$ eV $= 6.8$ eV.

It may strike you as strange to be talking about an atom for which the nucleus is a positron, the antiparticle of the electron, since the eventual fate of a positron is to annihilate with the electron. But we have seen that the radial wave functions $R(r)$ for the hydrogen atom vanish at $r = 0$ except for $l = 0$. Thus the electron and positron, which are point-like particles, do not overlap with each other in space and do not have a chance to annihilate unless the electron is in a state such as the ground state that has orbital angular momentum $l = 0$.

6.4 The Zeeman Effect

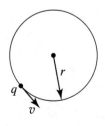

Figure 6.10 A particle with charge q moving with speed v in a circular orbit of radius r behaves like a loop of current with magnetic moment μ.

Consider a particle with charge q and mass m moving in a circular orbit of radius r with speed v, as illustrated in Fig. 6.10. The motion, which is periodic with a period $T = (2\pi r)/v$, effectively generates a current loop with area $A = \pi r^2$ and current $I = q/T = qv/(2\pi r)$. From classical electromagnetic theory, we know that such a current loop possesses a magnetic dipole moment μ of magnitude

$$\mu = IA = \frac{qrv}{2} \tag{6.83}$$

If we insert a factor of m in the numerator and denominator of this expression, we then have rmv in the numerator, which is just the magnitude of $\mathbf{L} = \mathbf{r} \times \mathbf{p}$, the orbital angular momentum of the particle. We can even write the resulting equation as the vector equation

$$\boldsymbol{\mu} = \frac{q}{2m}\mathbf{L} \tag{6.84}$$

since the sign of q determines whether the magnetic moment and the angular momentum are parallel or antiparallel. If this current loop is immersed in an external magnetic field **B**, there will be an energy of interaction

$$-\boldsymbol{\mu} \cdot \mathbf{B} = -\frac{q}{2m}\mathbf{L} \cdot \mathbf{B} \tag{6.85}$$

This expression suggests that if we were to immerse a hydrogen atom in an external magnetic field in the z direction, there will be a term added to the Hamiltonian of the form

$$\frac{eB}{2m}L_{z_{\text{op}}} \tag{6.86}$$

where we have substituted $q = -e$ as the charge of the electron.

The argument in the preceding paragraph, which starts with the principles of classical physics, makes a leap at the end in suggesting a specific form for the modification to the Hamiltonian for an atom in an external magnetic field.[5] What is the effect of a term such as (6.86) in the Hamiltonian? Notice that the energy eigenfunctions $\psi_{n,l,m_l}(r,\theta,\phi) = R_{n,l}(r)Y_{l,m_l}(\theta,\phi)$ of the hydrogen atom are energy eigenfunctions of this additional term as well, since the $Y_{l,m_l}(\theta,\phi)$'s are eigenfunctions of $L_{z_{\text{op}}}$ with eigenvalues $m_l \hbar$. Hence the energies of the hydrogen atom are modified according to

$$E_n \to E_n + \frac{eBm_l\hbar}{2m} = E_n + \mu_B B m_l \tag{6.87}$$

where we have introduced the **Bohr magneton**

$$\mu_B = \frac{e\hbar}{2m} = 5.79 \times 10^{-5} \text{ eV/T} \tag{6.88}$$

Thus the net effect of this additional term is to break the degeneracy of the energy levels with respect to m_l, as expected since the magnetic field destroys the rotational symmetry by picking out a direction in space. Transitions between a state with $l = 1$ and one with $l = 0$, say between the first excited state and the ground state in hydrogen, result in the emission of photons with three different frequencies, as illustrated in Fig. 6.11. Given the size of the Bohr magneton, you can see that magnetic fields in the tesla range produce quite small energy shifts in comparison, for example, with the 10.2 eV difference in energy between these states in the absence of the magnetic field. This splitting of the spectral lines in the presence of an external magnetic field is called the Zeeman effect, after Pieter Zeeman who first observed it in 1896. Zeeman, who was inspired to make these measurements by H. A. Lorentz, took advantage of the high resolution of the diffraction grating, as discussed in Section 1.7. Lorentz and Zeeman shared the Nobel Prize for this work in 1902.[6]

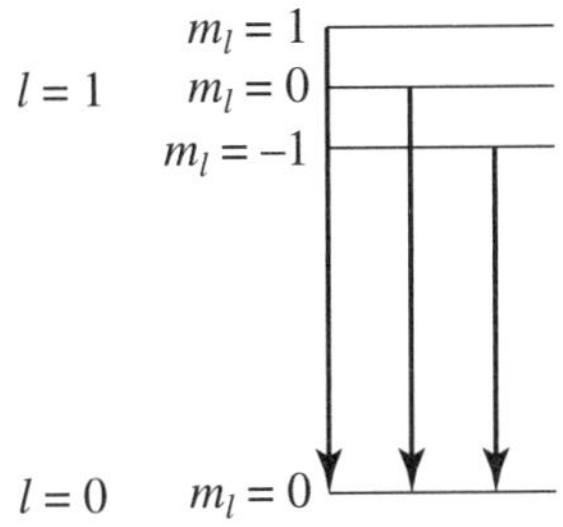

Figure 6.11 An external magnetic field breaks the degeneracy with respect to m_l of the energy levels of the hydrogen atom. Thus a transition from $l = 1$ to $l = 0$ results in three different frequencies for the emitted light instead of one frequency.

---

[5]It is probably worth re-emphasizing a point made earlier, namely that quantum mechanics cannot be derived from classical physics. The argument leading to (6.86) is really at best a heuristic one. We have used a toy classical model, since we know that electrons in the atom do not move in circles and do not occupy a particular point in space. Unfortunately, I do not know of a better way to motivate (6.84) and (6.86). As we will see in the next section, this result is incomplete, since it doesn't include the effect of the intrinsic spin (and corresponding magnetic moment) of the constituents of the atom.

[6]It should be noted that in these transitions there is a **selection rule** in that transitions occur that satisfy the condition $\Delta m_l = 0, \pm 1$, where Δm_l is the change in the value of m_l. Thus a transition from a state with $m_l = 2$, say for $l = 2$, to $m_l = 0$, say for $l = 1$, would not be observed. For a discussion of another selection rule ($\Delta l = \pm 1$) and its role in Bohr's **correspondence principle**, see Problem 6.14.

> **EXAMPLE 6.6** An electron with the azimuthal wave function
>
> $$\psi(\phi) = \frac{1}{\sqrt{\pi}} \cos\phi$$
>
> is placed in an external magnetic field. Take the Hamiltonian to be
>
> $$H = \frac{eB}{2m} L_{z_{\text{op}}} = \omega_0 L_{z_{\text{op}}}$$
>
> where $\omega_0 = eB/2m$. Determine $\psi(\phi, t)$.
>
> SOLUTION In Example 6.1 we saw that
>
> $$\psi(\phi) = \frac{1}{\sqrt{\pi}} \cos\phi = \frac{1}{\sqrt{\pi}} \left(\frac{e^{i\phi} + e^{-i\phi}}{2} \right)$$
>
> $$= \frac{1}{\sqrt{2}} \psi_1 + \frac{1}{\sqrt{2}} \psi_{-1}$$
>
> namely the wave function is a superposition of eigenfunctions of $L_{z_{\text{op}}}$ with eigenvalues $\pm\hbar$. Since the eigenfunctions of $L_{z_{\text{op}}}$ are also eigenfunctions of the Hamiltonian (H and $L_{z_{\text{op}}}$ differ by only the multiplicative constant ω_0), the second line of this equation expresses the wave function as a superposition of two energy eigenfunctions with the distinct eigenvalues $\pm\hbar\omega_0$. Tacking on the time-dependent factor $e^{-iEt/\hbar}$ for each energy state, we find
>
> $$\psi(\phi, t) = \frac{e^{-i\omega_0 t}}{\sqrt{2}} \psi_1 + \frac{e^{i\omega_0 t}}{\sqrt{2}} \psi_{-1}$$
>
> $$= \frac{1}{\sqrt{\pi}} \left(\frac{e^{i(\phi - \omega_0 t)} + e^{-i(\phi - \omega_0 t)}}{2} \right)$$
>
> $$= \frac{1}{\sqrt{\pi}} \cos(\phi - \omega_0 t)$$
>
> We see that the argument of the cosine is rotating about the z axis with angular speed ω_0. This is in accord with the behavior that we would expect in classical physics for a magnetic moment in an external magnetic field in the z direction, namely, the magnetic moment would precess about the z axis with angular frequency ω_0.

6.5 Intrinsic Spin

Atomic spectra are more complex than our discussion so far suggests. For hydrogen, for example, what should be a single wavelength (a single line in the spectrum) corresponding to the $n = 2$ to $n = 1$ transition in the absence of an external magnetic field turns out to be a doublet: two closely spaced lines that can be resolved with a good spectrograph.[7] And when an external magnetic field is applied, the response is more complex than was

---

[7] A similar doublet with a larger spacing between the lines is formed by the famous sodium D-lines.

suggested in the previous section, leading to what was historically referred to as the anomalous Zeeman effect.[8] In 1925, in an effort to explain atomic spectra, two Dutch graduate students, Samuel Goudsmit (an experimentalist who had worked in Zeeman's laboratory) and George Uhlenbeck (a theorist), suggested that the electron had its own intrinsic spin angular momentum **S** with a corresponding magnetic moment

$$\boldsymbol{\mu} = g\left(\frac{-e}{2m}\right)\mathbf{S} \tag{6.89}$$

Thus if an atom such as hydrogen is placed in an external magnetic field **B**, the interaction energy (6.85) of the electron with this magnetic field becomes

$$-\boldsymbol{\mu}\cdot\mathbf{B} = \left(\frac{e}{2m}\mathbf{L} + g\frac{e}{2m}\mathbf{S}\right)\cdot\mathbf{B} \tag{6.90}$$

where we have included the contributions of the orbital and the spin magnetic moments. The factor of g in (6.89) is known, perhaps not too imaginatively, as the g factor. In nonrelativistic quantum mechanics it is a "fudge" factor (another technical term) that must be inserted in order to give good agreement with experiment. Relativistic quantum mechanics (via the Dirac equation) predicts $g = 2$ for the electron, exactly.[9]

What are we to make of this intrinsic spin? You are probably thinking that the electron is a ball of charge spinning about its axis very much as the earth spins about its axis as it revolves around the sun. That is basically what Goudsmit and Uhlenbeck thought, at least initially.[10] But this model of the electron's spin *cannot* be correct. After all, we can calculate spin angular momentum of a rotating ball by integrating the orbital angular momentum of each small piece of mass dm as it moves about its rotation axis ($I\boldsymbol{\omega} = \int \mathbf{r} \times dm\mathbf{v}$). Thus this sort of spin angular momentum is just orbital angular momentum, too; we simply give it a different name. But we have seen that the L_z eigenvalues of orbital angular momentum are restricted to be $m_l\hbar$ with m_l an integer (so that the wave function $\psi(r,\theta,\phi)$ is single valued). Since the observed values of S_z for an electron are $\pm\hbar/2$ (or $m_s\hbar$ with $m_s = \pm 1/2$, clearly not an integer), we cannot think of the electron's spin as arising in some way from the motion of the particle ($\mathbf{S} \neq \mathbf{r} \times \mathbf{p}$). We must give up on the notion of a wave function that tells us, for example, something about the orientation of the electron. In fact, as far as we can tell, the electron itself is a point particle. This spin angular momentum is an intrinsic attribute of the particle, like its charge. Any attempt to search for a deeper physical model that generates this spin is probably not appropriate. As an antidote to such attempts, you should note that particles

---

[8] It turns out that there were classical arguments that seemed to "explain" the normal Zeeman effect, but the origin of the anomalous Zeeman effect was a mystery before the advent of quantum mechanics and the discovery of the role of intrinsic spin.

[9] The observed value is 2.00232. This apparent discrepancy between the prediction of the Dirac equation and the observed value is beautifully reconciled through quantum electrodynamics, as we will discuss in Section 10.1.

[10] When Goudsmit and Uhlenbeck proposed the idea of electron spin to P. Ehrenfest, he encouraged them to write up their results and to talk with H. Lorentz. After some analysis, Lorentz pointed out that a classical model of a spinning electron required that the speed at the surface be approximately ten times the speed of light in order to obtain the observed magnetic moment. When Goudsmit and Uhlenbeck went to tell Ehrenfest of their foolishness, he informed them that he had already submitted their paper for publication. He assured them they shouldn't worry since they were "both young enough to be able to afford a stupidity." *Physics Today*, June 1976, p. 40.

besides the electron have their own intrinsic spin. Photons, for example, have values of S_z in the direction of propagation of the photon equal to $\pm\hbar$. And you are probably much less inclined to think of a photon as a spinning ball.

How does quantum mechanics handle intrinsic spin? It simply posits that there are spin operators $S_{x\,\text{op}}$, $S_{y\,\text{op}}$, and $S_{z\,\text{op}}$ that obey the same commutation relations as do the orbital angular momentum operators, namely

$$\begin{aligned}[S_{x\,\text{op}}, S_{y\,\text{op}}] &= i\hbar S_{z\,\text{op}} \\ [S_{y\,\text{op}}, S_{z\,\text{op}}] &= i\hbar S_{x\,\text{op}} \\ [S_{z\,\text{op}}, S_{x\,\text{op}}] &= i\hbar S_{y\,\text{op}}\end{aligned} \quad (6.91)$$

For the electron, there are two basis states, one with $S_z = \hbar/2$ (called spin up) and one with $S_z = -\hbar/2$ (called spin down). Look back at Example 5.3, where there were also two basis states that we simply specified as being eigenstates of some unspecified A_{op} with distinct eigenvalues. In the latter part of Example 5.3 we showed how we could represent these states as column vectors. Here too we can write the spin-up state as a column vector

$$\chi_+ = \begin{pmatrix} 1 \\ 0 \end{pmatrix} \quad (6.92)$$

the spin-down state as a column vector

$$\chi_- = \begin{pmatrix} 0 \\ 1 \end{pmatrix} \quad (6.93)$$

and the operator for the z component of the spin angular momentum as a 2×2 matrix

$$S_{z\,\text{op}} = \frac{\hbar}{2} \begin{pmatrix} 1 & 0 \\ 0 & -1 \end{pmatrix} \quad (6.94)$$

with the S_z eigenvalues on the diagonal so that

$$\begin{aligned}S_{z\,\text{op}}\chi_+ &= \frac{\hbar}{2} \begin{pmatrix} 1 & 0 \\ 0 & -1 \end{pmatrix} \begin{pmatrix} 1 \\ 0 \end{pmatrix} \\ &= \frac{\hbar}{2} \begin{pmatrix} 1 \\ 0 \end{pmatrix} = \frac{\hbar}{2}\chi_+\end{aligned} \quad (6.95)$$

and

$$\begin{aligned}S_{z\,\text{op}}\chi_- &= \frac{\hbar}{2} \begin{pmatrix} 1 & 0 \\ 0 & -1 \end{pmatrix} \begin{pmatrix} 0 \\ 1 \end{pmatrix} \\ &= -\frac{\hbar}{2} \begin{pmatrix} 0 \\ 1 \end{pmatrix} = -\frac{\hbar}{2}\chi_-\end{aligned} \quad (6.96)$$

Namely, χ_+ and χ_- are eigenstates (eigenvectors) of $S_{z\,\text{op}}$ with eigenvalues $\hbar/2$ and $-\hbar/2$, respectively, as desired.

What about $S_{x\,\text{op}}$ and $S_{y\,\text{op}}$? You can check (see Example 6.7) that the 2×2 matrices

$$S_{x\,\text{op}} = \frac{\hbar}{2} \begin{pmatrix} 0 & 1 \\ 1 & 0 \end{pmatrix} \quad (6.97)$$

and

$$S_{y\,\mathrm{op}} = \frac{\hbar}{2}\begin{pmatrix} 0 & -i \\ i & 0 \end{pmatrix} \tag{6.98}$$

do the job in that they satisfy the commutation relations (6.91). Moreover, given the explicit form of these matrices, it is not hard to verify that

$$\mathbf{S}_{\mathrm{op}}^2 = S_{x\,\mathrm{op}}^2 + S_{y\,\mathrm{op}}^2 + S_{z\,\mathrm{op}}^2$$

$$= \frac{3\hbar^2}{4}\begin{pmatrix} 1 & 0 \\ 0 & 1 \end{pmatrix} \tag{6.99}$$

that is, just $3\hbar^2/4$ times the identity matrix. Since the identity matrix commutes with all 2×2 matrices, we see that

$$\left[S_{z\,\mathrm{op}}, \mathbf{S}_{\mathrm{op}}^2\right] = 0 \tag{6.100}$$

just as the orbital angular momentum operators $L_{z\,\mathrm{op}}$ and $\mathbf{L}_{\mathrm{op}}^2$ commute.[11] Thus the eigenvectors (6.92) and (6.93) are simultaneous eigenstates of $S_{z\,\mathrm{op}}$ and $\mathbf{S}_{\mathrm{op}}^2$. The eigenvalue $3\hbar^2/4$ of $\mathbf{S}_{\mathrm{op}}^2$ can be written as $s(s+1)\hbar^2$ with $s=1/2$. For this reason we say the electron is a spin-$1/2$ particle even though the magnitude of its spin is $(\sqrt{3}/2)\hbar$.

Since we did not include the intrinsic spin of the electron in our discussion of the hydrogen atom in Section 6.3, the energy eigenfunctions given in (6.82) are therefore incomplete. We can label the energy eigenstates by the total energy, the magnitude of the orbital angular momentum, the z component of the orbital angular momentum, and the z component of the intrinsic spin angular momentum, since the operators corresponding to each of these observables commute with each other. Consequently, we can express the wave functions in the form $R_{n,l}(r)Y_{l,m_l}(\theta,\phi)\chi_\pm$. A shorthand notation that takes the intrinsic spin into account is typically used to label the different energy levels. The orbital angular momentum states are labeled by the letters $s(l=0)$, $p(l=1)$, $d(l=2)$, $f(l=3)$, $g(l=4)$, $h(l=5)$, etc. As you can see, from $l=3$ upward, the labeling is alphabetical.[12] The sum of the intrinsic spin angular momentum and the orbital angular momentum of the electron generates a total angular momentum labeled by the quantum number j, meaning that the magnitude squared of the total angular momentum is equal to $j(j+1)\hbar^2$. When the intrinsic spin $s=1/2$ of the electron is combined with the electron's orbital angular momentum l, the resulting j takes on the two values $j=l\pm 1/2$ (with the exception of $l=0$ where the total angular momentum is solely $j=1/2$).[13] The energy levels of hydrogen are then labeled by $1s_{1/2}, 2s_{1/2}, 2p_{1/2}, 2p_{3/2}, \ldots$, where the letters $s,p,d,\ldots$ label the orbital angular momentum, the number in front of the letter labels the value n of the principal quantum number, and the subscript labels the value of j.

---

[11] The spin operators $S_{x\,\mathrm{op}}$ and $S_{y\,\mathrm{op}}$ also commute with $\mathbf{S}_{\mathrm{op}}^2$ just as $L_{x\,\mathrm{op}}$ and $L_{y\,\mathrm{op}}$ commute with $\mathbf{L}_{\mathrm{op}}^2$, although it is easier to see this explicitly from the matrices (6.97), (6.98), and (6.99) for the intrinsic spin operators than from the forms (6.27), (6.28), and (6.30) for the orbital angular momentum operators.

[12] This s, p, d, and f terminology arose from characteristics of the spectrum, namely *sharp*, *principal*, *diffuse*, and *fundamental*.

[13] In the discussion of the helium atom in Section 7.3, we will see an example that illustrates how this addition of angular momenta works in quantum mechanics.

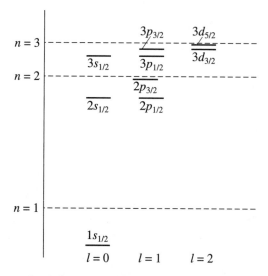

Figure 6.12 An energy-level diagram for the $n = 1$, $n = 2$, and $n = 3$ levels of hydrogen, including fine structure, which is exaggerated in scale by roughly a factor of 10^4. States with different values for l and the same j and n are degenerate.

If an external magnetic field is applied to the atom, then the Hamiltonian will have the additional term (6.90) and the different intrinsic spin states will have different energies. Even without an external magnetic field, there are internal magnetic fields in the atom that arise from the relative (orbital) motion of the electron and the nucleus. This relativistic interaction is called **spin–orbit coupling** and it contributes together with two other relativistic effects to what is termed a **fine structure** of the energy levels, as illustrated in Fig. 6.12. Moreover, the proton in the hydrogen atom has its own intrinsic spin and magnetic moment, which interacts with the magnetic moment of the electron generating a hyperfine structure on the energy levels. Finally, when quantum field theory is taken into account, there is an additional even smaller shift (the Lamb shift) in the positioning of the energy levels that, for example, breaks the degeneracy of the $2s_{1/2}$ and $2p_{1/2}$ energy levels. The success of quantum mechanics in providing such a complete and accurate description of the hydrogen atom is without doubt one of its most notable triumphs and has contributed significantly to our confidence in the theory.

EXAMPLE 6.7 Verify that the matrices (6.94), (6.97), and (6.98) satisfy the commutation relation

$$\left[S_{x\,\text{op}}, S_{y\,\text{op}}\right] = i\hbar S_{z\,\text{op}}$$

SOLUTION

$$\begin{aligned}
\left[S_{x\,\text{op}}, S_{y\,\text{op}}\right] &= S_{x\,\text{op}} S_{y\,\text{op}} - S_{y\,\text{op}} S_{x\,\text{op}} \\
&= \frac{\hbar}{2}\begin{pmatrix} 0 & 1 \\ 1 & 0 \end{pmatrix} \frac{\hbar}{2}\begin{pmatrix} 0 & -i \\ i & 0 \end{pmatrix} - \frac{\hbar}{2}\begin{pmatrix} 0 & -i \\ i & 0 \end{pmatrix} \frac{\hbar}{2}\begin{pmatrix} 0 & 1 \\ 1 & 0 \end{pmatrix} \\
&= \frac{\hbar^2}{2}\begin{pmatrix} i & 0 \\ 0 & -i \end{pmatrix} = i\hbar S_{z\,\text{op}}
\end{aligned}$$

The Stern–Gerlach Experiment

This is a good point to describe an atomic beam experiment carried out in 1922 by Otto Stern and Walther Gerlach with silver atoms. This classic experiment gives clear evidence of the intrinsic spin of the electron. As we will see in the next chapter, intrinsic spin is an attribute that plays a very important role in the structure of multielectron atoms. For now, we simply note that although a silver atom is a pretty complex entity with 47 electrons bound to a nucleus, 46 of these electrons fill up shells in such a way that their spin and orbital angular momentum sum to zero. Thus the magnetic properties of a neutral silver atom are determined essentially by the properties of a single (valence) electron, which it turns out is bound in an orbital angular momentum $l = 0$ state. Therefore the interaction energy (6.90) of the atom in a magnetic field reduces to

$$-\boldsymbol{\mu} \cdot \mathbf{B} = \frac{ge}{2m} \mathbf{S} \cdot \mathbf{B} \tag{6.101}$$

In the Stern–Gerlach experiment silver is vaporized by heating it in an oven. The atoms exit the oven through a small opening, are collimated into a beam by traversing a narrow slit, and travel down the center of a long magnet, as indicated in Fig. 6.13. Notice how one of the pole pieces of the magnet is pointed, so that the magnetic field between the poles is not uniform. Therefore there will be a force on the atom given by the negative gradient of the interaction energy (6.101). If we orient our z axis so that the magnetic field gradient points in this direction, then the z component of the force is given by

$$\begin{aligned} F_z &= -\frac{\partial}{\partial z} \frac{ge}{2m} \mathbf{S} \cdot \mathbf{B} \\ &\cong -\frac{ge}{2m} S_z \frac{\partial B_z}{\partial z} \end{aligned} \tag{6.102}$$

where we have assumed that $\partial B_x / \partial z$ and $\partial B_y / \partial z$ are much smaller in magnitude than $\partial B_z / \partial z$. If, as illustrated in the figure, $\partial B_z / \partial z < 0$, then if $S_z = \hbar/2$, the atom will be deflected in the positive z direction. On the other hand, if $S_z = -\hbar/2$, the atom is deflected in the negative z direction. After traversing the magnetic field, the silver atoms are collected on a glass plate. Figure 6.14 shows the data obtained by Stern and Gerlach.

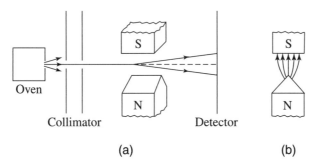

Figure 6.13 (a) A schematic diagram of the Stern–Gerlach experiment. (b) A cross sectional view of the pole pieces of the magnet depicting the inhomogeneous magnetic field they produce.

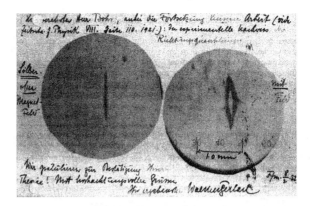

Figure 6.14 A postcard from Walther Gerlach to Niels Bohr, dated February 8, 1922. Note that the image on the postcard has been rotated by 90° relative to Fig. 6.13, where the collimating slit is horizontal. The left-hand image, with the magnetic field turned off, shows the finite width of the collimating slit. The right-hand image shows the beam profile with the magnetic field turned on. Only in the center of the apparatus is the magnetic field gradient sufficiently strong to cause splitting. Photograph reproduced with permission from the Niels Bohr Archive.

Atoms are indeed deflected up or down, consistent with the two possible values of S_z for the spin-1/2 electron.[14]

In our discussion of measurement in quantum mechanics (see Section 5.6 in particular), we have been quite vague about what constitutes a measurement. The Stern–Gerlach experiment provides us with a concrete example as to how one might measure at least one observable, namely, S_z. From our discussion, we know that silver atoms that are deflected upward have $S_z = \hbar/2$ and silver atoms that are deflected downward have $S_z = -\hbar/2$. We can confirm these results by repeating the experiment. Namely, if instead of allowing the atoms to collide with a glass plate, we take the ones that are deflected upward and send them through another Stern–Gerlach device with its inhomogeneous magnetic field also oriented in the z direction, we find that all the atoms that were deflected upward in traversing the first device are deflected upward by the second device as well. They do indeed still have $S_z = \hbar/2$. See Fig. 6.15a. Similarly, the ones that were deflected downward in traversing the first device are all deflected downward in traversing the second Stern–Gerlach device as well. Does this mean that the atoms all have either $S_z = \hbar/2$ or $S_z = -\hbar/2$ before they enter the first device? The answer is that we can't be certain if we aren't sure how the initial state was prepared. But the Stern–Gerlach experiment gives us a good way to prepare the state of the atoms. If, for example, we orient the inhomogeneous magnetic field in the first device along the x axis instead of the z axis, we find the atoms are deflected either to the left or the right and hence have $S_x = \hbar/2$ or $S_x = -\hbar/2$ upon exiting this device. But if we take the atoms that are deflected to the right (the ones with $S_x = \hbar/2$) and send them into the second device with its inhomogeneous field in the z direction, we find 50% of the atoms are deflected upward and 50% of the atoms

---

[14] Stern and Gerlach thought their results confirmed the quantization of orbital angular momentum that had been suggested by Bohr in 1913. But given what we now know about the quantum mechanics of orbital angular momentum, if the deflection had been due to the magnetic moment arising from the orbital angular momentum, there should have been an odd number of traces since for a particular value of l there are $2l + 1$ m_l values.

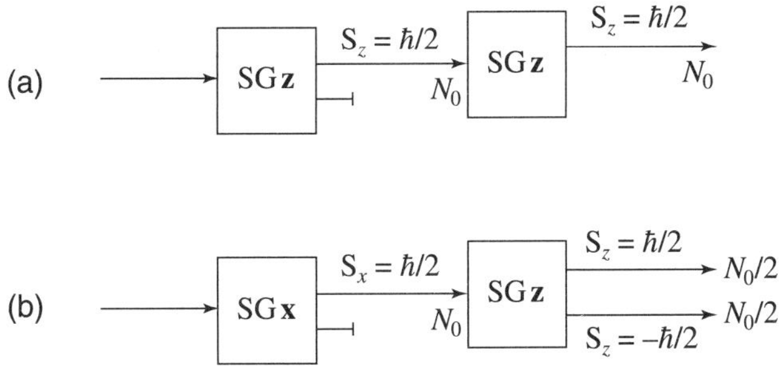

Figure 6.15 (a) A block diagram of an experiment in which spin-1/2 atoms pass through two Stern–Gerlach devices, each of which has its inhomogeneous magnetic field oriented in the z direction. (b) Atoms that exit a Stern–Gerlach device with its inhomogeneous magnetic field in the x direction and are deflected, say, to the right have a 50% probability of being deflected up upon exiting a second Stern–Gerlach device with its inhomogeneous magnetic field oriented in the z direction.

are deflected downward. See Fig. 6.15b. It isn't hard to show (see Example 6.8) that the state with $S_x = \hbar/2$ is a particular superposition of the S_z eigenstates, namely

$$\frac{1}{\sqrt{2}}\chi_+ + \frac{1}{\sqrt{2}}\chi_- \tag{6.103}$$

Thus the measurement by this second Stern–Gerlach device with its inhomogeneous magnetic field in the z direction has collapsed the state from this superposition to one or the other of the eigenstates of S_z. How this collapse occurs, why the superposition disappears and we end up with an either–or situation, is an open question that is known as the measurement question. In the Copenhagen interpretation of quantum mechanics this collapse occurs in the interaction with a macroscopic measuring device, such as the magnet/detection plate in the Stern–Gerlach experiment. Such macroscopic measuring devices are sufficiently complex that we are not able to treat them quantum mechanically from first principles.[15]

EXAMPLE 6.8 Verify that

$$\frac{1}{\sqrt{2}}\chi_+ + \frac{1}{\sqrt{2}}\chi_-$$

is an eigenstate of $S_{x\,\text{op}}$.

SOLUTION

$$\frac{\hbar}{2}\begin{pmatrix} 0 & 1 \\ 1 & 0 \end{pmatrix} \frac{1}{\sqrt{2}} \begin{pmatrix} 1 \\ 1 \end{pmatrix} = \frac{\hbar}{2} \frac{1}{\sqrt{2}} \begin{pmatrix} 1 \\ 1 \end{pmatrix}$$

---

[15] It is time to get on to the applications of quantum mechanics. If you would like to work through more of the details of intrinsic spin as well as see an elegant operator approach to the angular momentum eigenvalue problem, check out John S. Townsend, *A Modern Approach to Quantum Mechanics*, University Science Books, Sausalito, CA, 2000. This book is highly recommended for your next course in quantum mechanics!

6.6 Summary

In many ways, we have just scratched the surface of three-dimensional quantum mechanics. Nonetheless, we have covered the main ideas that will be needed as we move on to discuss applications of quantum mechanics to solid-state, nuclear, and particle physics. We have analyzed a particle of mass bound in two three-dimensional potential energy wells: a cubic box of length L on a side for which the energy eigenvalues are given by

$$E_{n_x,n_y,n_z} = \frac{(n_x^2 + n_y^2 + n_z^2)\hbar^2\pi^2}{2mL^2} \qquad n_x, n_y, n_z = 1, 2, 3, \ldots \qquad (6.104)$$

and the hydrogenic atom for which the allowed energies are given by

$$E_n = -\frac{mZ^2e^4}{(4\pi\epsilon_0)^2 2\hbar^2 n^2} = -\frac{(13.6 \text{ eV})Z^2}{n^2} \qquad n = 1, 2, 3, \ldots \qquad (6.105)$$

In the case of the hydrogenic atom, for each value of the principal quantum number n the allowed values of l are given by $l = 0, 1, 2, \ldots, n-1$. These l values are connected to the allowed eigenvalues of the orbital angular momentum operator $\mathbf{L}_{\text{op}}^2 = L_{x\,\text{op}}^2 + L_{y\,\text{op}}^2 + L_{z\,\text{op}}^2$ via

$$\mathbf{L}_{\text{op}}^2 Y_{l,m_l}(\theta, \phi) = l(l+1)\hbar^2 Y_{l,m_l}(\theta, \phi) \qquad l = 0, 1, 2, \ldots \qquad (6.106)$$

The eigenfunctions $Y_{l,m_l}(\theta, \phi)$, the spherical harmonics, also satisfy

$$L_{z\,\text{op}} Y_{l,m_l}(\theta, \phi) = m_l \hbar Y_{l,m_l}(\theta, \phi) \qquad m_l = -l, -l+1, \ldots, l-1, l \qquad (6.107)$$

The energies (6.105) are independent of m_l because of the rotational symmetry of the potential energy $-Ze^2/4\pi\epsilon_0 r$.

One of the striking features of orbital angular momentum is that $l\hbar$, the maximum value for L_z, is always less than the magnitude of the orbital angular momentum itself, namely $\sqrt{l(l+1)}\hbar$. The angular momentum cannot completely align itself along one axis (in which case all three components would be known simultaneously) because of the commutation relations

$$\left[L_{x\,\text{op}}, L_{y\,\text{op}}\right] = i\hbar L_{z\,\text{op}}$$
$$\left[L_{y\,\text{op}}, L_{z\,\text{op}}\right] = i\hbar L_{x\,\text{op}}$$
$$\left[L_{z\,\text{op}}, L_{x\,\text{op}}\right] = i\hbar L_{y\,\text{op}} \qquad (6.108)$$

The first of these equations, for example, leads to the uncertainty relation

$$\Delta L_x \Delta L_y \geq \frac{\hbar}{2}|\langle L_z \rangle| \qquad (6.109)$$

which shows that if $\langle L_z \rangle \neq 0$ then the values of L_x and L_y are both uncertain.

While the commutation relations (6.108) can be derived from the explicit form of the orbital angular momentum $\mathbf{L} = \mathbf{r} \times \mathbf{p}$ expressed in terms of position and momentum operators, we can take these commutation relations as the defining relationship for angular momentum in quantum mechanics. In addition to orbital angular momentum, there is another form of angular momentum called intrinsic spin angular momentum. Particles have their own intrinsic spin angular momentum $\mathbf{S}$, which is present even when the particle is at rest. We say the electron is a spin-1/2 particle since the analogues of (6.106) and (6.107) are

$$\mathbf{S}^2_{\text{op}}\chi_\pm = s(s+1)\hbar^2\chi_\pm \qquad s = 1/2 \qquad (6.110)$$

and

$$S_{z\,\text{op}}\chi_\pm = \pm\frac{\hbar}{2}\chi_\pm \qquad (6.111)$$

where $\chi_\pm$ are two-dimensional column vectors, in contrast to the $Y_{l,m}(\theta,\phi)$ wave functions. Note the similarity between the form of the eigenvalue for $\mathbf{L}^2_{\text{op}}$ and for $\mathbf{S}^2_{\text{op}}$ except that for intrinsic spin the value of s need not be integral. Associated with this intrinsic spin angular momentum is a magnetic moment

$$\boldsymbol{\mu} = g\left(\frac{-e}{2m}\right)\mathbf{S} \qquad (6.112)$$

where $g = 2.00232$ for an electron. Thus internal magnetic fields in an atom or external magnetic fields imposed on the atom couple to this magnetic moment and hence to the spin angular momentum of the particle.

Problems

6.1. What are the energy eigenvalues for a particle of mass m confined in a rectangular box with sides of length a, b, and c? If $a < b < c$, what is the energy of the first excited state? What is the degeneracy of this energy level?

6.2. Solve the three-dimensional harmonic oscillator for which

$$V(r) = \frac{1}{2}m\omega^2\left(x^2+y^2+z^2\right)$$

by separation of variables in Cartesian coordinates. Assume that the one-dimensional oscillator has eigenfunctions $\psi_n(x)$ with corresponding energy eigenvalues $E_n = (n+1/2)\hbar\omega$. What is the degeneracy of the first excited state of the oscillator?

6.3. What is the normalized ground-state energy eigenfunction for the three-dimensional harmonic oscillator

$$V(r) = \frac{1}{2}m\omega^2 r^2$$

Suggestion: Use the separation of variables strategy outlined in Problem 6.2. Express the wave function in spherical coordinates. What is the orbital angular momentum of the ground state? Explain your reasoning.

6.4. The normalized energy eigenfunction for the first excited state of the one-dimensional harmonic oscillator is given by

$$\psi_1(x) = \left(\frac{4m^3\omega^3}{\pi\hbar^3}\right)^{1/4} x e^{-m\omega x^2/2\hbar}$$

with corresponding energy $E_1 = 3\hbar\omega/2$. What is the energy of the first excited state of the three-dimensional isotropic harmonic oscillator? What is the degeneracy for this energy eigenvalue? What is the orbital angular momentum of the particle in this excited state? Explain your reasoning.

6.5. Prove that the operator

$$L_{z\,\text{op}} = \frac{\hbar}{i}\frac{\partial}{\partial\phi}$$

is Hermitian. *Suggestion*: Follow the procedure outlined in Example 5.2. Keep in mind that the wave function must be single valued.

6.6. Suppose that

$$\Psi(\phi) = \sqrt{\frac{1}{3\pi}} + \sqrt{\frac{1}{6\pi}}e^{i\phi}$$

Show that $\Psi(\phi)$ is properly normalized. What are the possible results of a measurement of L_z for a particle whose wave function is $\Psi(\phi)$ and what are the probabilities of obtaining those results? *Suggestion*: See Example 6.1.

6.7. Verify that
$$Y_{2,2}(\theta, \phi) = \sqrt{\frac{15}{32\pi}} \sin^2\theta\, e^{2i\phi}$$
is an eigenfunction of $\mathbf{L}_{\text{op}}^2$ and $L_{z\,\text{op}}$ with the appropriate eigenvalues.

6.8. Show that the orbital angular momentum operators
$$L_{x\,\text{op}} = y\frac{\hbar}{i}\frac{\partial}{\partial z} - z\frac{\hbar}{i}\frac{\partial}{\partial y}$$
and
$$L_{y\,\text{op}} = z\frac{\hbar}{i}\frac{\partial}{\partial x} - x\frac{\hbar}{i}\frac{\partial}{\partial z}$$
can be written in spherical coordinates as
$$L_{x\,\text{op}} = \frac{\hbar}{i}\left(-\sin\phi\frac{\partial}{\partial\theta} - \cot\theta\cos\phi\frac{\partial}{\partial\phi}\right)$$
and
$$L_{y\,\text{op}} = \frac{\hbar}{i}\left(\cos\phi\frac{\partial}{\partial\theta} - \cot\theta\sin\phi\frac{\partial}{\partial\phi}\right)$$
Suggestion: Since $\mathbf{L} = \mathbf{r} \times \mathbf{p}$, expressed in terms of spherical coordinates
$$\mathbf{L}_{\text{op}} = \mathbf{r} \times \frac{\hbar}{i}\nabla$$
$$= \frac{\hbar}{i}\left[r\mathbf{u}_r \times \left(\mathbf{u}_r\frac{\partial}{\partial r} + \mathbf{u}_\theta\frac{1}{r}\frac{\partial}{\partial\theta} + \mathbf{u}_\phi\frac{1}{r\sin\theta}\frac{\partial}{\partial\phi}\right)\right]$$
where $\mathbf{u}_r$, $\mathbf{u}_\theta$, and $\mathbf{u}_\phi$ are unit vectors in the r, θ, and ϕ directions, respectively. Express these unit vectors in terms of $\mathbf{i}$, $\mathbf{j}$, and $\mathbf{k}$ to obtain the x and y components of $\mathbf{L}_{\text{op}}$.

6.9. The spherical harmonic $Y_{0,0}$ is an eigenfunction of $L_{z\,\text{op}}$ with eigenvalue 0. Show it is also an eigenfunction of $L_{x\,\text{op}}$ and $L_{y\,\text{op}}$ with eigenvalue zero as well. Thus for $l = 0$, all three components of the orbital angular momentum are zero simultaneously. Is this consistent with the uncertainty relation
$$\Delta L_x \Delta L_y \geq \frac{\hbar}{2}|\langle L_z \rangle|$$

6.10. Show that in addition to (6.56), the following commutation relations must hold for the orbital angular momentum operators
$$[L_{y\,\text{op}}, L_{z\,\text{op}}] = i\hbar L_{x\,\text{op}} \quad \text{and} \quad [L_{z\,\text{op}}, L_{x\,\text{op}}] = i\hbar L_{y\,\text{op}}$$
Verify that these commutation relations generate the additional uncertainty relations
$$\Delta L_y \Delta L_z \geq \frac{\hbar}{2}|\langle L_x \rangle| \quad \text{and} \quad \Delta L_z \Delta L_x \geq \frac{\hbar}{2}|\langle L_y \rangle|$$

6.11. The normalized angular wave function for a three-dimensional rigid rotator is given by
$$\psi(\theta, \phi) = \sqrt{\frac{3}{8\pi}}(-\sin\theta\cos\phi + i\cos\theta)$$
Show that this wave function is an eigenfunction of
$$L_{y\,\text{op}} = \frac{\hbar}{i}\left(\cos\phi\frac{\partial}{\partial\theta} - \cot\theta\sin\phi\frac{\partial}{\partial\phi}\right)$$
the operator corresponding to the y component of the orbital angular momentum. What is the corresponding eigenvalue for this wave function?

6.12. Show that the uncertainty relation
$$\Delta L_x \Delta L_y \geq \frac{\hbar}{2}|\langle L_z \rangle|$$
is satisfied for the wave function $\psi(\theta, \phi)$ given in Problem 6.11. *Suggestion*: In evaluating the left-hand side of the uncertainty relation, first calculate ΔL_y. When evaluating the right-hand side of the uncertainty relation, you may find it useful to take advantage of the fact that the wave function $\psi(\theta, \phi)$ can be expressed in terms of the spherical harmonics as
$$\psi(\theta, \phi) = \frac{1}{2}Y_{1,1} + \frac{i}{\sqrt{2}}Y_{1,0} - \frac{1}{2}Y_{1,-1}$$

6.13. If the rotator in Problem 6.11 is immersed in an external magnetic field B_0 in the z direction, the Hamiltonian becomes
$$H = \frac{\mathbf{L}_{\text{op}}^2}{2I} + \omega_0 L_{z\,\text{op}}$$
where I and ω_0 are constants. If at $t = 0$
$$\psi(\theta, \phi) = \frac{1}{2}Y_{1,1} + \frac{i}{\sqrt{2}}Y_{1,0} - \frac{1}{2}Y_{1,-1}$$
what is $\psi(\theta, \phi, t)$?

6.14. In a diatomic molecule the atoms can rotate about each other. This rotation can be shown to be equivalent to a reduced mass $\mu = m_1 m_2/(m_1 + m_2)$ rotating in three dimensions about a fixed point. According to classical physics, the energy of a three-dimensional rigid rotator is given by $E = \mathbf{L}^2/2I$, where I is the moment of inertia and $\mathbf{L}^2 = L_x^2 + L_y^2 + L_z^2$ is the magnitude squared of the orbital angular momentum. (a) What is the energy operator for this three-dimensional rotator? What are the energy eigenfunctions and corresponding energy eigenvalues? Take the moment of inertia I to be a constant. (A complication is that molecules "stretch" as they rotate faster, so the moment of inertia is not a constant.) Show in the limit of large l ($l \gg 1$) that

$$\frac{E_l - E_{l-1}}{E_l} \to 0$$

which means that the discrete nature of the energy levels becomes less apparent as l increases. (b) Determine the frequency of the photon that would be emitted if the rotator makes a transition from one energy level labeled by the angular momentum quantum number l to one labeled by the quantum number $l - 1$. (c) Show in the limit in which the orbital angular momentum quantum number l is large ($l \gg 1$) that the frequency of the photon from part (b) coincides with the classical frequency of rotation of the rotator (recall $L = I\omega$). This result illustrates the correspondence principle, namely how the results of classical physics and the results of quantum mechanics can coincide in the appropriate limit. It also shows how we might use the correspondence principle to deduce the existence of a selection rule, in this case $\Delta l = \pm 1$. *Note:* Here Δl means the change in l, not the uncertainty in l. (d) Sketch the spectrum (in terms of the allowed frequencies) assuming the selection rule $\Delta l = \pm 1$ holds.

6.15. The ammonia molecule, NH_3, can be treated as a symmetric rigid rotator, with the three hydrogen atoms residing in the x-y plane and the nitrogen atom above the plane on the z axis, as shown in Fig. 6.16. If we call the moment of inertia about the z axis I_3, and the moments about the pair of axes perpendicular to the z axis I_1, the rotational energy of the molecule can be written as

$$E = \frac{L_x^2}{2I_1} + \frac{L_y^2}{2I_1} + \frac{L_z^2}{2I_3} = \frac{\mathbf{L}^2 - L_z^2}{2I_1} + \frac{L_z^2}{2I_3}$$

Suppose that at time $t = 0$ the wave function of the ammonia molecule is

$$\psi(\theta, \phi) = \frac{1}{\sqrt{2}} Y_{0,0} + \frac{1}{\sqrt{2}} Y_{1,1}$$

where the $Y_{l,m_l}(\theta, \phi)$ are the spherical harmonics. (a) What are the energy eigenvalues for this symmetric rigid rotator? (b) Determine $\psi(\theta, \phi, t)$, the wave function at time t. (c) Determine the probability that a measure of the rotational energy of the ammonia molecule yields the value 0 for the state ψ. What is $\langle E \rangle$ for this state?

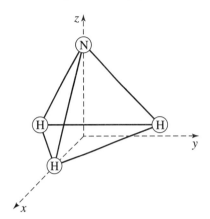

Figure 6.16 The ammonia molecule.

6.16. Estimate the moment of inertia of the HCl molecule. Use your value for I to estimate the frequency of the photon that would be emitted in the transition $l = 1$ to $l = 0$. In what region of the electromagnetic spectrum does this photon reside? See Problem 6.14.

6.17. The energy for a rigid rotator constrained to rotate in the x-y plane is given by

$$E = \frac{L_z^2}{2I}$$

where the moment of inertia I is a constant. (a) What is the Hamiltonian? Show that the Hamiltonian and $L_{z_{\text{op}}}$ commute. What are the allowed energies and normalized energy eigenfunctions of the rigid rotator? (b) At time $t = 0$ the normalized wave function for the rotator is

$$\psi(\phi) = \sqrt{\frac{4}{3\pi}} \sin^2 \frac{\phi}{2} = \sqrt{\frac{4}{3\pi}} \left(\frac{1 - \cos\phi}{2} \right)$$

What values would be obtained if a measurement of L_z were carried out? Determine the probabilities of obtaining these results. (c) What is $\psi(\phi, t)$? (d) What is the probability that a measurement of the energy at time t will yield $\hbar^2/2I$? (e) What is the uncertainty ΔE in the energy of the rotator in the state $\psi(\phi, t)$?

6.18. The Bohr model played an important historical role in the transition from classical mechanics to quantum mechanics. Although most of the physical assumptions underlying the Bohr model turned out to be incorrect or incomplete, the model did succeed in giving the "correct" quantized energies of the hydrogen atom. (a) Use the idea that the allowed orbits of an electron in the hydrogen atom are determined by the condition that an integral number of de Broglie wavelengths fit in each orbit to show that the orbital angular momentum $L = rmv = n\hbar$, where $n = 1, 2, 3, \ldots$, and is therefore quantized. (b) Show that the allowed orbits have radii given by

$$r_n = \frac{4\pi\epsilon_0 n^2 \hbar^2}{me^2}$$

and energies given by

$$E_n = -\frac{me^4}{(4\pi\epsilon_0)^2 2\hbar^2 n^2}$$

6.19. Consider the Earth–Sun system, which is bound by the gravitational force $F = Gm_1m_2/r^2$, to be a large "gravity atom," as compared with the hydrogen atom, which is bound by the electrostatic force $F = e^2/(4\pi\epsilon_0 r^2)$. Using the results of the Bohr model from Problem 6.18, determine the principal quantum number n of the Earth–Sun system, given that the distance between Earth and the Sun is 1.5×10^{11} m. *Note:* $m_{\text{Earth}} = 6 \times 10^{24}$ kg, $m_{\text{Sun}} = 2 \times 10^{30}$ kg, and $G = 6.67 \times 10^{-11}$ m^3/(kg·s^2).

6.20. Verify that

$$R_{2,1}(r) = \frac{1}{\sqrt{3}} \left(\frac{Z}{2a_0}\right)^{3/2} \frac{Zr}{a_0} e^{-Zr/2a_0}$$

is an energy eigenfunction for the hydrogen atom with the appropriate eigenvalue. *Suggestion:* See Example 6.3.

6.21. An electron outside a dielectric is attracted to the surface by a force $F = -A/x^2$, where x is the perpendicular distance from the electron to the surface, and A is a positive constant. Electrons are prevented from crossing the surface, since there are no quantum states in the dielectric for them to occupy. Assume that the surface is infinite in extent, so that the problem is effectively one-dimensional. Write the Schrödinger equation for an electron outside the surface, that is for $x > 0$. What is the appropriate boundary condition at $x = 0$? Obtain a formula for the allowed energy levels of the system. *Suggestion:* Compare the equation for the wave function $\psi(x)$ with that satisfied by the wave function $u(r) = rR(r)$ for a hydrogenic atom.

6.22. A particle of mass m is trapped in an infinite spherical potential energy well with

$$V(r) = \begin{cases} 0 & r < a \\ \infty & r > a \end{cases}$$

The ground-state wave function is a spherically symmetric solution, that is, $\psi = \psi(r)$. (a) Make a change of variables to $u(r) = r\psi(r)$ and solve for $u(r)$ and the value of the ground-state energy. The boundary condition on the wave function at $r = 0$ is $u(0) = 0$. (b) Compare the ground-state energy for a particle in a spherical well of radius a with the ground-state energy for a particle in a cubic box of side a. Which well has the lower ground-state energy? Is your result consistent with the Heisenberg uncertainty principle? *Suggestion:* Compare the volumes of the spherical and cubic wells.

6.23. The unnormalized wave function for a negatively charged pion bound to a proton in an energy eigenstate is given by

$$\psi = (x + y + z)e^{-\sqrt{x^2+y^2+z^2}/2b_0}$$

where b_0 is a constant for this "pionic" atom that has the dimensions of length. Show that the pion is in a p orbital. What is the magnitude of the orbital angular momentum of the pion? What is the probability that a measurement of L_z will yield the value 0? *Suggestion:* Express the wave function in spherical coordinates. *Note:* It is not necessary to normalize the wave function to assess the relative probabilities of obtaining the possible values of L_z.

6.24. Use the time-independent Schrödinger equation to determine the energy of the eigenfunction given in Problem 6.23 and the constant b_0 in terms of fundamental constants. One of these constants may be taken to be μ, the reduced mass of the pion–proton system.

6.25. If the pionic atom described in Problem 6.23 is immersed in an external magnetic field B_0 in the z direction, a term of the form $\omega_0 L_{z_{\text{op}}}$ is added to the Hamiltonian. Explain the origin of this additional term and show how ω_0 is related to B_0. Given the wave function ψ in Problem 6.23, what is $\psi(t)$?

6.26. Show that the matrix operators

$$S_{x\,\text{op}} = \frac{\hbar}{2}\begin{pmatrix} 0 & 1 \\ 1 & 0 \end{pmatrix} \quad S_{y\,\text{op}} = \frac{\hbar}{2}\begin{pmatrix} 0 & -i \\ i & 0 \end{pmatrix} \quad S_{z\,\text{op}} = \frac{\hbar}{2}\begin{pmatrix} 1 & 0 \\ 0 & -1 \end{pmatrix}$$

satisfy the commutation relation

$$\left[S_{y\,\text{op}},\, S_{z\,\text{op}}\right] = i\hbar S_{x\,\text{op}}$$

Note: The three 2×2 matrices

$$\sigma_x = \begin{pmatrix} 0 & 1 \\ 1 & 0 \end{pmatrix} \quad \sigma_y = \begin{pmatrix} 0 & -i \\ i & 0 \end{pmatrix} \quad \sigma_z = \begin{pmatrix} 1 & 0 \\ 0 & -1 \end{pmatrix}$$

are often referred to as the **Pauli spin matrices**.

6.27. Determine the eigenvalues and eigenvectors of the operator

$$S_{x\,\text{op}} = \frac{\hbar}{2}\begin{pmatrix} 0 & 1 \\ 1 & 0 \end{pmatrix}$$

Suggestion: If you haven't solved problems such as this in linear algebra, follow the procedure outlined in Example 5.3.

6.28. Determine the eigenvalues and eigenvectors of the operator

$$S_{y\,\text{op}} = \frac{\hbar}{2}\begin{pmatrix} 0 & -i \\ i & 0 \end{pmatrix}$$

What is the probability of obtaining $S_z = \hbar/2$ if a measurement is carried out on a silver atom that is known to be in a state with $S_y = -\hbar/2$?

CHAPTER 7

Identical Particles

It is difficult to overemphasize the importance of identical particles in quantum mechanics. In classical physics, so-called identical particles are distinguishable, since it is possible in theory to keep track of the particles (without influencing their motion). In quantum mechanics, identical particles are truly indistinguishable. The consequences of this indistinguishability are, as we will see, profound, accounting for, for example, the stability and the chemical properties of the elements, the electrical and mechanical properties of crystalline solids, the end phase in the evolution of stars, the frequency distribution of the cosmic background radiation that permeates the universe, the phenomenon of superfluidity, and the coherence properties of lasers.

7.1 Multiparticle Systems

In order to see how to handle systems containing two or more particles, we consider two noninteracting particles confined in a one-dimensional box of length L. The Hamiltonian is given by

$$H = -\frac{\hbar^2}{2m_1}\frac{\partial^2}{\partial x_1^2} + V(x_1) - \frac{\hbar^2}{2m_2}\frac{\partial^2}{\partial x_2^2} + V(x_2) \tag{7.1}$$

where $V(x)$ is the potential energy of the box. We are using the coordinate x_1 for particle 1 and the coordinate x_2 for particle 2. Here $\Psi(x_1, x_2)^*\Psi(x_1, x_2)dx_1 dx_2$ is the probability of finding particle 1 between x_1 and $x_1 + dx_1$ *and* particle 2 between x_2 and $x_2 + dx_2$, assuming that $\Psi(x_1, x_2)$ satisfies the normalization condition

$$\int_{-\infty}^{\infty}\int_{-\infty}^{\infty} |\Psi(x_1, x_2)|^2 dx_1\, dx_2 = 1 \tag{7.2}$$

We can solve the time-independent Schrödinger equation

$$H\Psi = E\Psi \tag{7.3}$$

by the technique of separation of variables. If we write

$$\Psi(x_1, x_2) = \psi(x_1)\phi(x_2) \tag{7.4}$$

and divide (7.3) by the wave function itself, we obtain

$$\frac{1}{\psi(x_1)}\left(-\frac{\hbar^2}{2m_1}\frac{d^2\psi}{dx_1^2}\right) + V(x_1) + \frac{1}{\phi(x_2)}\left(-\frac{\hbar^2}{2m_2}\frac{d^2\phi}{dx_2^2}\right) + V(x_2) = E \quad (7.5)$$

As we have discussed earlier, since the first two terms on the left-hand side of this equation contain all the x_1 dependence and the second two terms contain all the x_2 dependence, the only way this equation can hold for all x_1 and x_2 is for each of these pairs of terms to be constant, namely

$$-\frac{\hbar^2}{2m_1}\frac{d^2\psi}{dx_1^2} + V(x_1)\psi = E_1\psi \quad (7.6)$$

and

$$-\frac{\hbar^2}{2m_2}\frac{d^2\phi}{dx_2^2} + V(x_2)\phi = E_2\phi \quad (7.7)$$

We have called the constants E_1 and E_2 since

$$E = E_1 + E_2 \quad (7.8)$$

You can see from (7.6) and (7.7) that the wave functions ψ and ϕ are the energy eigenfunctions for a single particle confined in a one-dimensional box, which is reasonable since we have assumed the particles do not interact with each other. Given (7.4), the two-particle wave function Ψ is therefore the *product* of the energy eigenfunctions for a single particle

$$\Psi(x_1, x_2) = \psi_{n_1}(x_1)\psi_{n_2}(x_2) \quad (7.9)$$

and the total energy is the *sum* of the energies for a single particle confined in the box:

$$E = \frac{\hbar^2\pi^2 n_1^2}{2m_1 L^2} + \frac{\hbar^2\pi^2 n_2^2}{2m_2 L^2} \quad (7.10)$$

If we are solely interested in determining the probability of finding particle 1 in some region in the box, we would integrate the probability density $|\Psi(x_1, x_2)|^2$ over all possible values of the coordinate x_2, leaving us with $|\psi_{n_1}(x_1)|^2 dx_1$ as the probability of finding particle 1 between x_1 and x_1+dx_1. Similarly, the probability of finding particle 2 between x_2 and $x_2 + dx_2$ independent of the position of particle 1 is $|\psi_{n_2}(x_2)|^2 dx_2$.

7.2 Identical Particles in Quantum Mechanics

But what if the particles are identical? Not only should we set $m_1 = m_2$ in the Hamiltonian [and in the energy eigenvalues (7.10)], but we should expect to get the same result for the probability of finding one of the particles in a certain region of the box, whether it be particle 1 or particle 2. After all, if the particles are identical, we cannot distinguish one of the particles from the other. Hence the wave functions should not permit such a distinction either.

To see how to implement this requirement, we introduce the **exchange operator** P_{12}, which is defined by its action on a two-particle state

$$P_{12}\Psi(1, 2) = \Psi(2, 1) \quad (7.11)$$

where by this we mean exchange all the coordinates (or labels) for particle 1 with those for particle 2. For example,

$$P_{12}\psi_{n_1}(x_1)\psi_{n_2}(x_2) = \psi_{n_1}(x_2)\psi_{n_2}(x_1) \qquad (7.12)$$

When the particles have intrinsic spin, the exchange operator exchanges the spin labels as well as the position coordinates of the particles. You might be tempted to assume that if the particles are identical (and hence indistinguishable) that we should require $P_{12}\Psi(1,2) = \Psi(1,2)$. But we have seen in quantum mechanics that two states that differ only by an overall phase factor are really the same state. Thus for identical particles we will insist that

$$P_{12}\Psi(1,2) = \lambda\Psi(1,2) \qquad (7.13)$$

where $\lambda = e^{i\delta}$. Note that

$$P_{12}^2 \Psi(1,2) = \lambda^2 \Psi(1,2)$$
$$= \Psi(1,2) \quad \Rightarrow \quad \lambda^2 = 1 \qquad (7.14)$$

since exchanging the particles twice returns us to the initial state. Thus the allowed eigenvalues of the exchange operator are

$$\lambda = \pm 1 \qquad (7.15)$$

This is all reminiscent of the parity operator (see Section 5.1) in that performing the operation twice (in the case of parity, inverting the coordinates twice) is equivalent to the identity operator. Thus the eigenvalues for the parity operator are also ± 1, corresponding to even and odd wave functions under an inversion of coordinates. Here the corresponding states are symmetric under particle exchange:

$$P_{12}\Psi_S(1,2) = \Psi_S(1,2) \qquad \text{when} \qquad \lambda = 1 \qquad (7.16)$$

and antisymmetric under particle exchange:

$$P_{12}\Psi_A(1,2) = -\Psi_A(1,2) \qquad \text{when} \qquad \lambda = -1 \qquad (7.17)$$

One straightforward example of a two-particle state that is symmetric under exchange is the state

$$\Psi_S(1,2) = \psi_\alpha(1)\psi_\alpha(2) \qquad (7.18)$$

where particle 1 and particle 2 are in the same quantum state label by the quantum number(s) α. Another example where the two particles are in different states with quantum numbers α and β is given by

$$\Psi_S(1,2) = \frac{1}{\sqrt{2}} \left[\psi_\alpha(1)\psi_\beta(2) + \psi_\beta(1)\psi_\alpha(2) \right] \qquad (7.19)$$

since switching the labels 1 and 2 returns exactly the same state. We have put a factor of $1/\sqrt{2}$ in front so that the state is properly normalized. For a two-particle state that is antisymmetric under exchange, the only possibility is for the two particles to be in different states, namely

$$\Psi_A(1,2) = \frac{1}{\sqrt{2}} \left[\psi_\alpha(1)\psi_\beta(2) - \psi_\beta(1)\psi_\alpha(2) \right] \qquad (7.20)$$

Here switching the labels on the particles returns the state multiplied by a minus sign.

For three identical particles, the corresponding symmetric and antisymmetric states are given by

$$\Psi_S(1,2,3) = \frac{1}{\sqrt{6}} \big[\psi_\alpha(1)\psi_\beta(2)\psi_\gamma(3) + \psi_\beta(1)\psi_\gamma(2)\psi_\alpha(3)$$
$$+ \psi_\gamma(1)\psi_\alpha(2)\psi_\beta(3) + \psi_\gamma(1)\psi_\beta(2)\psi_\alpha(3)$$
$$+ \psi_\beta(1)\psi_\alpha(2)\psi_\gamma(3) + \psi_\alpha(1)\psi_\gamma(2)\psi_\beta(3)\big] \quad (7.21)$$

and

$$\Psi_A(1,2,3) = \frac{1}{\sqrt{6}} \big[\psi_\alpha(1)\psi_\beta(2)\psi_\gamma(3) + \psi_\beta(1)\psi_\gamma(2)\psi_\alpha(3)$$
$$+ \psi_\gamma(1)\psi_\alpha(2)\psi_\beta(3) - \psi_\gamma(1)\psi_\beta(2)\psi_\alpha(3)$$
$$- \psi_\beta(1)\psi_\alpha(2)\psi_\gamma(3) - \psi_\alpha(1)\psi_\gamma(2)\psi_\beta(3)\big] \quad (7.22)$$

These states are symmetric or antisymmetric under the exchange of any two of the particles in the state. The antisymmetric state can be neatly expressed in terms of a determinant, known as the Slater determinant:

$$\Psi_A(1,2,3) = \frac{1}{\sqrt{6}} \begin{vmatrix} \psi_\alpha(1) & \psi_\alpha(2) & \psi_\alpha(3) \\ \psi_\beta(1) & \psi_\beta(2) & \psi_\beta(3) \\ \psi_\gamma(1) & \psi_\gamma(2) & \psi_\gamma(3) \end{vmatrix} \quad (7.23)$$

The determinant has the property that it changes sign when any two columns (or rows, for that matter) are interchanged. Interchanging two columns corresponds to switching the labels corresponding to the particles for those two columns and therefore (7.23) is manifestly antisymmetric under exchange.

What makes these results so important is that there is a connection between the intrinsic spin of a particle and the sort of states that are permitted for multi-particle systems:

(1) Particles with integral intrinsic spin ($s = 0, 1, 2, \ldots$) are called **bosons**. The total "wave function" for a system of identical bosons must be symmetric under exchange of any two of the particles.

(2) Particles with half-integral intrinsic spin ($s = 1/2, 3/2, 5/2, \ldots$) are called **fermions**. The total "wave function" for a system of identical fermions must be antisymmetric under exchange of any two of the particles.

Your first reaction to statements (1) and (2) may be to say, "Where do they come from?" We could take these results to be "postulates," postulates that one might call a generalized **Pauli Principle** since one consequence is that for half-integral spin particles no two identical particles can be in the same state. After all, if you try to put the two identical fermions in the same quantum state by setting $\alpha = \beta$ in (7.20), the state vanishes identically. But postulates (1) and (2) actually are a natural outgrowth of combining special relativity and quantum mechanics together in the form of quantum field theory. Unfortunately, no one has figured out an easy way to explain this spin-symmetry connection, which may mean we don't understand it as fully as we would like.[1]

---

[1] In Section 7.5 we will see that fermions and bosons obey different types of statistics. In 2002 the editor of the *American Journal of Physics* noted an earlier challenge by Richard Feynman seeking an elementary proof of the spin–statistics connection. Despite numerous efforts in the intervening years, he concluded that no one had yet met the challenge.

7.3 Multielectron Atoms

There are a lot of different atoms. In this section we will focus primarily on low-Z elements to see the important role that the identical nature of electrons plays in atomic structure. It is not difficult to generalize our results to higher Z atoms. After hydrogen, helium is the next simplest atom. If we include only the Coulomb interactions between the nucleus and the two electrons in helium, the Hamiltonian is given by

$$H = -\frac{\hbar^2}{2m}\nabla_1^2 - \frac{Ze^2}{4\pi\epsilon_0 r_1} - \frac{\hbar^2}{2m}\nabla_2^2 - \frac{Ze^2}{4\pi\epsilon_0 r_2} \quad (7.24)$$

where we have ignored the contribution of the kinetic energy of the nucleus and, more importantly, the Coulomb repulsion $e^2/4\pi\epsilon_0|\mathbf{r}_1 - \mathbf{r}_2|$ between the two electrons. This has the effect of making the problem mathematically tractable (the three-body problem is a tough one to solve, in either classical or quantum mechanics) and it does not obscure the fundamental underlying physics. Notice that we have introduced subscripts on the derivative operators to indicate whether we are taking the derivative with respect to the coordinates $\mathbf{r}_1$ or $\mathbf{r}_2$ just as we did in the one-dimensional Hamiltonian in Section 7.1. And of course for helium we must set $Z = 2$.

Without the Coulomb repulsion of the electrons, the Schrödinger equation with the Hamiltonian (7.24) can be solved by separation of variables. The energy eigenfunctions can be written as a product of the single-particle hydrogenic energy eigenfunctions. But since we are dealing with electrons, we must include the spin state for each electron in the "wave function," making sure that the overall state is antisymmetric under exchange of the two particles. It is easiest to see what is going on with some explicit examples. Let us start with the ground state, in which each of the electrons is in the ground state of the corresponding single-particle Hamiltonian:

$$\Psi = \psi_{1s}(\mathbf{r}_1)\psi_{1s}(\mathbf{r}_2)\frac{1}{\sqrt{2}}[\chi_+(1)\chi_-(2) - \chi_-(1)\chi_+(2)] \quad (7.25)$$

where we have used the abbreviated notation $\psi_{1s} = \psi_{1,0,0}$, that is, $n = 1, l = 0, m_l = 0$. We can think of the exchange operator as consisting of the product of two operators, one that exchanges the positions of the two electrons and one that exchanges their spins. Note that the spatial part of the wave function (7.25) is symmetric under the exchange $\mathbf{r}_1 \leftrightarrow \mathbf{r}_2$, while the spin state picks up an overall minus sign when the 1 and the 2 labels are exchanged. The product of a symmetric state and an antisymmetric state is antisymmetric, just as the product of an even function and an odd function is itself an odd function.

As is shown in Example 7.3, this particular spin state is an eigenstate of the total spin angular momentum $\mathbf{S}^2 = (\mathbf{S}_1 + \mathbf{S}_2)^2$ with eigenvalue 0. Thus we say this **singlet** state is a total-spin-0 state.[2] The total orbital angular momentum of the ground state is zero as well, since each electron has zero orbital angular momentum. Consequently, the total angular momentum, which is the sum of the total spin angular momentum and the total

---

[2] The spin state in (7.25) is a beautiful example of the sort of two-particle entangled state that we discussed in Section 5.6. Notice that it is not possible to say which spin state is occupied by particle 1 or particle 2, but a measurement of the spin state of one of the particles, say particle 1, is sufficient to determine the spin state of the other particle, particle 2.

orbital angular momentum, is also zero. It is common to label states by the spectroscopic notation $^{2S+1}L_J$, where S is the quantum number that specifies the total spin angular momentum squared in the form $\mathbf{S}^2 = S(S+1)\hbar^2$ (and therefore the $2S+1$ superscript indicates the multiplicity of the total spin states), L labels the total orbital angular momentum in the form $\mathbf{L}^2 = L(L+1)\hbar^2$, and J is the quantum number specifying the resultant of adding the total spin and total orbital angular momenta together in the form $\mathbf{J}^2 = J(J+1)\hbar^2$. As we discussed in Section 6.5, the orbital angular momentum states are labeled by the letters such as s, p, and d, short for $l=0, l=1$, and $l=2$ respectively. Consequently, in spectroscopic notation, the ground state of helium is referred to as the $^1S_0$ state. Here capital letters are used for the total orbital angular momentum, so the S in $^1S_0$ refers to the fact that $L=0$ and not to the total spin.

Let's next look at the first excited states. Instead of putting both electrons in the ground states of the single-particle Hamiltonians, we will put one of them in the first excited state. Of course, since the electrons are identical, we can't specify which of the electrons is in the ground state and which is in the excited state. There are a number of possibilities. For example, one possibility is

$$\Psi = \frac{1}{\sqrt{2}}[\psi_{1s}(\mathbf{r}_1)\psi_{2s}(\mathbf{r}_2) + \psi_{2s}(\mathbf{r}_1)\psi_{1s}(\mathbf{r}_2)]\frac{1}{\sqrt{2}}[\chi_+(1)\chi_-(2) - \chi_-(1)\chi_+(2)] \quad (7.26)$$

where $\psi_{2s} = \psi_{2,0,0}$. Here the spatial state is symmetric under the exchange of positions of the electrons ($\mathbf{r}_1 \leftrightarrow \mathbf{r}_2$) and the spin state, the singlet spin state, is antisymmetric under exchange of the spin states of the electrons (switch the 1 and the 2 labels on the spin states). In spectroscopic notation this is a $^1S_0$ state, like the ground state.

Another possibility is to make the spatial state antisymmetric and the spin state symmetric, namely

$$\Psi = \frac{1}{\sqrt{2}}[\psi_{1s}(\mathbf{r}_1)\psi_{2s}(\mathbf{r}_2) - \psi_{2s}(\mathbf{r}_1)\psi_{1s}(\mathbf{r}_2)]\frac{1}{\sqrt{2}}[\chi_+(1)\chi_-(2) + \chi_-(1)\chi_+(2)] \quad (7.27)$$

But once we allow for a symmetric spin state, there are two additional possible spin states in which each of the electrons is in the same spin state:

$$\Psi = \frac{1}{\sqrt{2}}[\psi_{1s}(\mathbf{r}_1)\psi_{2s}(\mathbf{r}_2) - \psi_{2s}(\mathbf{r}_1)\psi_{1s}(\mathbf{r}_2)]\chi_+(1)\chi_+(2) \quad (7.28)$$

and

$$\Psi = \frac{1}{\sqrt{2}}[\psi_{1s}(\mathbf{r}_1)\psi_{2s}(\mathbf{r}_2) - \psi_{2s}(\mathbf{r}_1)\psi_{1s}(\mathbf{r}_2)]\chi_-(1)\chi_-(2) \quad (7.29)$$

These three symmetric spin states turn out to be eigenstates of the total spin angular momentum $(\mathbf{S}_1 + \mathbf{S}_2)^2$ with eigenvalue $2\hbar^2$, namely $S(S+1)\hbar^2$ with $S=1$. The $2S+1=3$ spin states are said to form a **triplet** of spin-1 states, as is shown in Problem 7.1 and Problem 7.2. The addition of total spin angular momentum $S=1$ and total orbital angular momentum $L=0$ generates a state with total angular momentum $J=1$. Thus the spectroscopic notation for these three states is $^3S_1$.

Even though the Hamiltonian does not have any direct dependence on intrinsic spin, these total-spin-0 and total-spin-1 states can end up with quite different energies once we take into account the Coulomb repulsion of the electrons. Notice that the spatial part of the wave function for the total-spin-1 states, which is antisymmetric under the exchange $\mathbf{r}_1 \leftrightarrow \mathbf{r}_2$, vanishes when $\mathbf{r}_1 = \mathbf{r}_2$. On the other hand, the two pieces of the spatial part

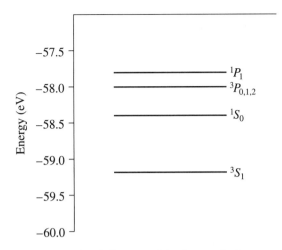

Figure 7.1 An energy-level diagram of the first excited states of helium.

of the wave function for the total-spin-0 state (7.26) add together constructively when $\mathbf{r}_1 = \mathbf{r}_2$. Thus the electrons are much more likely to overlap each other in space for the total-spin-0 state than for the total-spin-1 states.[3] Therefore, the Coulomb repulsion between the electrons will tend to raise the energy of the total spin-0 state relative to that of the total spin-1 states, as indicated in Fig. 7.1. Consequently, the lower energy state—the one nature tends to favor—is one in which the spins are "aligned." This is the sort of mechanism that is presumably responsible for ferromagnetism, in which the spins (and therefore the magnetic moments) of certain valence electrons in atoms such as iron align.

What happens if we move on from helium with its two electrons to lithium with its three electrons? What is the ground state? If we try to put all three electrons in the ground state (the $1s$ state) of the single-particle Hamiltonian, we end up with a spatial state that is symmetric under exchange of any two of the electrons. But since there are only two possibilities for the intrinsic spin state of each electron, namely the spin-up state χ_+ or spin-down state χ_-, at least two of the electrons must be in the same spin state, which means the state cannot be antisymmetric under exchange. Thus in the ground state of lithium one of the electrons must be in the $2s$ state. Similarly, in beryllium ($Z = 4$), we can put two electrons in the $1s$ state and two electrons in the $2s$ state, but if we try to move up to boron ($Z = 5$), then the fifth electron must go into another energy state, the $2p$ state. In this way, we see how the exchange symmetry dictates the structure of the periodic table, and thus chemistry!

You may be wondering why the $2s$ energy level fills before the $2p$ level, since in the hydrogen atom the $2s$ and $2p$ states have the same energy. In multielectron atoms, the potential energy that any one electron experiences is not as simple as that in the hydrogen atom. For small r the potential energy approaches $-Ze^2/4\pi\epsilon_0 r$ as the electron "sees" the full charge of the nucleus, while for large r the potential energy should approach $-e^2/4\pi\epsilon_0 r$ since the electron is screened from the nucleus by the other $Z - 1$ electrons in the atom. For nonzero values of l, the centrifugal barrier $l(l+1)\hbar^2/2mr^2$ tends to keep the electron away from the origin, where the potential energy is the most negative.

---

[3] The profound effect of the symmetry of the exchange symmetry of the spatial wave functions on the likelihood of finding the particles in the same region of space is illustrated in Problem 7.3.

I	II											III	IV	V	VI	VII	VIII
$_1$H $1s^1$ $^2S_{1/2}$																	$_2$He $1s^2$ $^1S_0$
$_3$Li $2s^1$ $^2S_{1/2}$	$_4$Be $2s^2$ $^1S_0$											$_5$B $2p^1$ $^2P_{1/2}$	$_6$C $2p^2$ $^3P_0$	$_7$N $2p^3$ $^4S_{3/2}$	$_8$O $2p^4$ $^3P_2$	$_9$F $2p^5$ $^2P_{3/2}$	$_{10}$Ne $2p^6$ $^1S_0$
$_{11}$Na $3s^1$ $^2S_{1/2}$	$_{12}$Mg $3s^2$ $^1S_0$											$_{13}$Al $3p^1$ $^2P_{1/2}$	$_{14}$Si $3p^2$ $^3P_0$	$_{15}$P $3p^3$ $^4S_{3/2}$	$_{16}$S $3p^4$ $^3P_2$	$_{17}$Cl $3p^5$ $^2P_{3/2}$	$_{18}$Ar $3p^6$ $^1S_0$
$_{19}$K $4s^1$ $^2S_{1/2}$	$_{20}$Ca $4s^2$ $^1S_0$	$_{21}$Sc $3d^1$ $^2D_{3/2}$	$_{22}$Ti $3d^2$ $^3F_2$	$_{23}$V $3d^3$ $^4F_{3/2}$	$_{24}$Cr $4s^13d^5$ $^7S_3$	$_{25}$Mn $3d^5$ $^6S_{5/2}$	$_{26}$Fe $3d^6$ $^5D_4$	$_{27}$Co $3d^7$ $^4F_{9/2}$	$_{28}$Ni $3d^8$ $^3F_4$	$_{29}$Cu $4s^13d^{10}$ $^2S_{1/2}$	$_{30}$Zn $3d^{10}$ $^1S_0$	$_{31}$Ga $4p^1$ $^2P_{1/2}$	$_{32}$Ge $4p^2$ $^3P_0$	$_{33}$As $4p^3$ $^4S_{3/2}$	$_{34}$Se $4p^4$ $^3P_2$	$_{35}$Br $4p^5$ $^2P_{3/2}$	$_{36}$Kr $4p^6$ $^1S_0$
$_{37}$Rb $5s^1$ $^2S_{1/2}$	$_{38}$Sr $5s^2$ $^1S_0$	$_{39}$Y $4d^1$ $^2D_{3/2}$	$_{40}$Zr $4d^2$ $^3F_2$	$_{41}$Nb $5s^14d^4$ $^6D_{1/2}$	$_{42}$Mo $5s^14d^5$ $^7S_3$	$_{43}$Tc $5s^14d^6$ $^6D_{9/2}$	$_{44}$Ru $5s^14d^7$ $^5F_5$	$_{45}$Rh $5s^14d^8$ $^4F_{9/2}$	$_{46}$Pd $5s^04d^{10}$ $^1S_0$	$_{47}$Ag $5s^14d^{10}$ $^2S_{1/2}$	$_{48}$Cd $4d^{10}$ $^1S_0$	$_{49}$In $5p^1$ $^2P_{1/2}$	$_{50}$Sn $5p^2$ $^3P_0$	$_{51}$Sb $5p^3$ $^4S_{3/2}$	$_{52}$Te $5p^4$ $^3P_2$	$_{53}$I $5p^5$ $^2P_{3/2}$	$_{54}$Xe $5p^6$ $^1S_0$
$_{55}$Cs $6s^1$ $^2S_{1/2}$	$_{56}$Ba $6s^2$ $^1S_0$	57–71 Rare earths	$_{72}$Hf $5d^2$ $^3F_2$	$_{73}$Ta $5d^3$ $^4F_{3/2}$	$_{74}$W $5d^4$ $^5D_0$	$_{75}$Re $5d^5$ $^6S_{5/2}$	$_{76}$Os $5d^6$ $^5D_4$	$_{77}$Ir $5d^7$ $^4F_{9/2}$	$_{78}$Pt $6s^15d^9$ $^3D_3$	$_{79}$Au $6s^15d^{10}$ $^2S_{1/2}$	$_{80}$Hg $5d^{10}$ $^1S_0$	$_{81}$Tl $6p^1$ $^2P_{1/2}$	$_{82}$Pb $6p^2$ $^3P_0$	$_{83}$Bi $6p^3$ $^4S_{3/2}$	$_{84}$Po $6p^4$ $^3P_2$	$_{85}$At $6p^5$ $^2P_{3/2}$	$_{86}$Rn $6p^6$ $^1S_0$
$_{87}$Fr $7s^1$ $^2S_{1/2}$	$_{88}$Ra $7s^2$ $^1S_0$	89–103 Actinides	$_{104}$Rf $5f^{14}6d^2$	$_{105}$Db $6d^3$	$_{106}$Sg $6d^4$	$_{107}$Bh $6d^5$	$_{108}$Hs $6d^6$	$_{109}$Mt $6d^7$	$_{110}$Ds $7s^16d^9$	$_{111}$Rg $7s^26d^9$							

Rare earths (Lanthanides)	$_{57}$La $5d^1$ $^2D_{3/2}$	$_{58}$Ce $4f^2$ $^3H_4$	$_{59}$Pr $4f^3$ $^4I_{9/2}$	$_{60}$Nd $4f^4$ $^5I_4$	$_{61}$Pm $4f^5$ $^6H_{5/2}$	$_{62}$Sm $4f^6$ $^7F_0$	$_{63}$Eu $4f^7$ $^8S_{7/2}$	$_{64}$Gd $5d^14f^7$ $^9D_2$	$_{65}$Tb $6s^14f^9$ $^6H_{15/2}$	$_{66}$Dy $4f^{10}$ $^5I_8$	$_{67}$Ho $4f^{11}$ $^4I_{15/2}$	$_{68}$Er $4f^{12}$ $^3H_6$	$_{69}$Tm $4f^{13}$ $^2F_{7/2}$	$_{70}$Yb $4f^{14}$ $^1S_0$	$_{71}$Lu $5d^14f^{14}$ $^2D_{3/2}$
Actinides	$_{89}$Ac $6d^1$ $^2D_{3/2}$	$_{90}$Th $6d^2$ $^3F_2$	$_{91}$Pa $6d^15f^2$ $^4K_{11/2}$	$_{92}$U $6d^15f^3$ $^5L_6$	$_{93}$Np $6d^15f^4$ $^6L_{11/2}$	$_{94}$Pu $5f^6$ $^7F_0$	$_{95}$Am $5f^7$ $^8S_{7/2}$	$_{96}$Cm $6d^15f^7$ $^9D_2$	$_{97}$Bk $6d^15f^8$ $^6G_{15/2}$	$_{98}$Cf $5f^{10}$ $^5I_8$	$_{99}$Es $5f^{11}$ $^4I_{15/2}$	$_{100}$Fm $5f^{12}$ $^3H_6$	$_{101}$Md $5f^{13}$ $^2F_{7/2}$	$_{102}$No $5f^{14}$ $^1S_0$	$_{103}$Lr $5f^{14}6d^1$ $^2D_{3/2}$

Figure 7.2 The periodic table, including the electron configuration and angular momentum $^{2S+1}L_J$ of the ground state.

Thus smaller values of l correspond to lower values of the energy and the degeneracy of the energy levels with different values of l is broken.

If we continue along the second row of the periodic table (see Fig. 7.2), after boron come carbon, nitrogen, oxygen, fluorine, and neon, filling up the $2p$ states. The $2p$ states can accommodate six electrons, since for $l = 1$ there are three different possible values for m_l ($m_l = 0, \pm 1$) and there are two possible spin states for each of these orbital states. Like helium, neon is relatively inert, since it has a closed (or filled) shell. Helium and neon are both referred to as noble gases. Moving from $Z = 10$ to $Z = 11$, from neon to sodium, leads to a pronounced change in the chemical properties of these elements. For sodium, the eleven electrons not only fill up the $n = 1$ and $n = 2$ levels, but one electron must occupy the $3s$ level. This last electron, the valence electron, is electrically shielded from the nucleus by the other ten electrons. Thus this valence electron is not so tightly bound to the atom, which makes sodium very reactive chemically. As we noted in our discussion of the photoelectric effect in Section 1.3, sodium, like the other alkali metals, oxidizes readily. Or, as another illustration, sodium can donate an electron to an element such as fluorine, which needs one electron to complete the $n = 2$ shell, or to chlorine, which needs one electron to complete the $3p$ subshell. The result is a crystalline solid, which in the case of sodium chloride is ordinary table salt.

EXAMPLE 7.1 The negatively charged pion π^- is a spin-0 particle roughly 280 times more massive than the electron. Suppose the two electrons in a helium atom are replaced with two negatively charged pions, generating a pionic atom. Which of the following wave functions are valid wave functions for pionic helium?

(a)
$$\Psi = \psi_{1s}(\mathbf{r}_1)\psi_{1s}(\mathbf{r}_2)$$

(b)
$$\Psi = \psi_{1s}(\mathbf{r}_1)\psi_{2s}(\mathbf{r}_2)$$

(c)
$$\Psi = \frac{1}{\sqrt{2}}[\psi_{1s}(\mathbf{r}_1)\psi_{2s}(\mathbf{r}_2) + \psi_{2s}(\mathbf{r}_1)\psi_{1s}(\mathbf{r}_2)]$$

(d)
$$\Psi = \frac{1}{\sqrt{2}}[\psi_{1s}(\mathbf{r}_1)\psi_{2s}(\mathbf{r}_2) - \psi_{2s}(\mathbf{r}_1)\psi_{1s}(\mathbf{r}_2)]$$

SOLUTION Since the pion is a spin-0 particle, it is a boson. Consequently, the two-pion wave function must be symmetric under particle exchange. Since the pion doesn't have intrinsic spin, the only attribute to exchange is the position of the pion, namely $\mathbf{r}_1 \leftrightarrow \mathbf{r}_2$. Wave functions (a) and (c) are symmetric under $\mathbf{r}_1 \leftrightarrow \mathbf{r}_2$, while wave function (d) is antisymmetric. Wave function (b) does not have any definite symmetry under exchange. Thus only (a) and (c) are allowed wave functions for pionic helium.

EXAMPLE 7.2 It is often stated that the Pauli Principle asserts that no two electrons can be in the same state in an atom. What then is wrong with the two-electron "wave function"

$$\Psi = \psi_{1s}(\mathbf{r}_1)\psi_{1s}(\mathbf{r}_2)\chi_+(1)\chi_-(2)$$

in which one of the electrons is in the ground state with its spin up and the other is in the ground state with its spin down?

SOLUTION Because electrons are spin-1/2 particles, a two-electron wave function must be antisymmetric under exchange. The wave function Ψ is symmetric under the exchange of the positions of the two particles ($\mathbf{r}_1 \leftrightarrow \mathbf{r}_2$), but if the spin states of the two particles are exchanged $\chi_+(1)\chi_-(2) \leftrightarrow \chi_-(1)\chi_+(2)$. Thus the spin state is not antisymmetric under exchange of the spins, which is required to make the overall state antisymmetric under the combined operation of exchanging the positions and spins of the two particles. The correct Ψ is

$$\Psi = \psi_{1s}(\mathbf{r}_1)\psi_{1s}(\mathbf{r}_2)\frac{1}{\sqrt{2}}[\chi_+(1)\chi_-(2) - \chi_-(1)\chi_+(2)]$$

EXAMPLE 7.3 Show that the two-particle spin state

$$\frac{1}{\sqrt{2}} [\chi_+(1)\chi_-(2) - \chi_-(1)\chi_+(2)]$$

is a state with total spin angular momentum equal to zero.

SOLUTION For a single spin-1/2 particle, we know that

$$S_{x\,op}\chi_+ = \frac{\hbar}{2} \begin{pmatrix} 0 & 1 \\ 1 & 0 \end{pmatrix} \begin{pmatrix} 1 \\ 0 \end{pmatrix} = \frac{\hbar}{2} \begin{pmatrix} 0 \\ 1 \end{pmatrix} = \frac{\hbar}{2}\chi_-$$

and

$$S_{x\,op}\chi_- = \frac{\hbar}{2} \begin{pmatrix} 0 & 1 \\ 1 & 0 \end{pmatrix} \begin{pmatrix} 0 \\ 1 \end{pmatrix} = \frac{\hbar}{2} \begin{pmatrix} 1 \\ 0 \end{pmatrix} = \frac{\hbar}{2}\chi_+$$

Also

$$S_{y\,op}\chi_+ = \frac{\hbar}{2} \begin{pmatrix} 0 & -i \\ i & 0 \end{pmatrix} \begin{pmatrix} 1 \\ 0 \end{pmatrix} = i\frac{\hbar}{2} \begin{pmatrix} 0 \\ 1 \end{pmatrix} = i\frac{\hbar}{2}\chi_-$$

and

$$S_{y\,op}\chi_- = \frac{\hbar}{2} \begin{pmatrix} 0 & -i \\ i & 0 \end{pmatrix} \begin{pmatrix} 0 \\ 1 \end{pmatrix} = -i\frac{\hbar}{2} \begin{pmatrix} 1 \\ 0 \end{pmatrix} = -i\frac{\hbar}{2}\chi_+$$

And of course χ_+ and χ_- are eigenfunctions of $S_{z\,op}$ with eigenvalues $\hbar/2$ and $-\hbar/2$, respectively. With these results in mind, we are ready to consider the action of the operator

$$\mathbf{S}^2_{op} = \mathbf{S}^2_{1_{op}} + \mathbf{S}^2_{2_{op}} + 2\,\mathbf{S}_{1_{op}} \cdot \mathbf{S}_{2_{op}}$$
$$= \mathbf{S}^2_{1_{op}} + \mathbf{S}^2_{2_{op}} + 2\left(S_{1x_{op}}S_{2x_{op}} + S_{1y_{op}}S_{2y_{op}} + S_{1z_{op}}S_{2z_{op}}\right)$$

on the two-particle spin state

$$\chi_{0,0} = \frac{1}{\sqrt{2}} [\chi_+(1)\chi_-(2) - \chi_-(1)\chi_+(2)]$$

where we have labeled the superposition with the subscript 0, 0 (a special case of the more general result χ_{S,M_S}) since we going to establish that this state has total spin equal to zero (just like the spherical harmonic $Y_{0,0}$ has orbital angular momentum equal to zero). Since

$$S_{1z_{op}}S_{2z_{op}}\chi_+(1)\chi_-(2) = -\frac{\hbar^2}{4}\chi_+(1)\chi_-(2)$$

and

$$S_{1z_{op}}S_{2z_{op}}\chi_-(1)\chi_+(2) = -\frac{\hbar^2}{4}\chi_-(1)\chi_+(2)$$

the state $\chi_{0,0}$ is an eigenstate of $S_{1z_{op}}S_{2z_{op}}$ with eigenvalue $-\hbar^2/4$. However, since

$$S_{1x_{op}}S_{2x_{op}}\chi_+(1)\chi_-(2) = \frac{\hbar^2}{4}\chi_-(1)\chi_+(2) \quad \text{and} \quad S_{1x_{op}}S_{2x_{op}}\chi_-(1)\chi_+(2) = \frac{\hbar^2}{4}\chi_+(1)\chi_-(2)$$

as well as

$$S_{1y_{op}}S_{2y_{op}}\chi_+(1)\chi_-(2) = \frac{\hbar^2}{4}\chi_-(1)\chi_+(2) \quad \text{and} \quad S_{1y_{op}}S_{2y_{op}}\chi_-(1)\chi_+(2) = \frac{\hbar^2}{4}\chi_+(1)\chi_-(2)$$

the individual terms in the superposition $\chi_{0,0}$ are *not* eigenstates of $S_{1x_{op}}S_{2x_{op}}$ and $S_{1y_{op}}S_{2y_{op}}$ but the overall superpositon $\chi_{0,0}$ is an eigenstate, namely

$$S_{1x_{op}}S_{2x_{op}}\chi_{0,0} = -\frac{\hbar^2}{4}\chi_{0,0}$$

and

$$S_{1y_{op}}S_{2y_{op}}\chi_{0,0} = -\frac{\hbar^2}{4}\chi_{0,0}$$

Therefore

$$\begin{aligned}\mathbf{S}^2_{op}\chi_{0,0} &= \left[\mathbf{S}^2_{1_{op}} + \mathbf{S}^2_{2_{op}} + 2\left(S_{1x_{op}}S_{2x_{op}} + S_{1y_{op}}S_{2y_{op}} + S_{1z_{op}}S_{2z_{op}}\right)\right]\chi_{0,0}\\ &= \left[\frac{1}{2}\left(\frac{1}{2}+1\right) + \frac{1}{2}\left(\frac{1}{2}+1\right) - 2\left(\frac{1}{4} + \frac{1}{4} + \frac{1}{4}\right)\right]\hbar^2\chi_{0,0}\\ &= \left(\frac{3}{4} + \frac{3}{4} - \frac{3}{2}\right)\hbar^2\chi_{0,0}\\ &= 0\end{aligned}$$

In addition, $\chi_{0,0}$ is an eigenstate of $S_{1z_{op}} + S_{2z_{op}}$ with eigenvalue 0 as well, which is why we have two subscripts on $\chi_{0,0}$.

This result is a striking one. You should check that if we simply change the two-particle spin state to

$$\chi_{1,0} = \frac{1}{\sqrt{2}}\left[\chi_+(1)\chi_-(2) + \chi_-(1)\chi_+(2)\right]$$

then the eigenvalue equation yields

$$\begin{aligned}\mathbf{S}^2_{op}\chi_{1,0} &= \left[\mathbf{S}^2_{1_{op}} + \mathbf{S}^2_{2_{op}} + 2\left(S_{1x_{op}}S_{2x_{op}} + S_{1y_{op}}S_{2y_{op}} + S_{1z_{op}}S_{2z_{op}}\right)\right]\chi_{1,0}\\ &= \left[\frac{1}{2}\left(\frac{1}{2}+1\right) + \frac{1}{2}\left(\frac{1}{2}+1\right) + 2\left(\frac{1}{4} + \frac{1}{4} - \frac{1}{4}\right)\right]\hbar^2\chi_{1,0}\\ &= \left(\frac{3}{4} + \frac{3}{4} + \frac{1}{2}\right)\hbar^2\chi_{1,0}\\ &= 2\hbar^2\chi_{1,0}\end{aligned}$$

The eigenvalue is $S(S+1)\hbar^2$ with $S=1$. Thus even though the two states $\chi_{0,0}$ and $\chi_{1,0}$ are both superpositions of spin states for two spin-1/2 particles with opposite values for the projection of the spin along the z axis for the two particles, the relative phase in the superposition matters a great deal. A simple change of sign from minus to plus changes the state from one with $S=0$ to $S=1$ for the total spin.

7.4 The Fermi Energy

In the next chapter, we will examine the quantum mechanics of crystalline solids. There we will see that for a solid such as metallic sodium the valence electrons are essentially free to travel throughout the solid and are not tied to any particular atom. A useful approximation, referred to as the **free-electron model**, is to treat the electrons as if they are simply confined in the three-dimensional potential energy box that we analyzed in Section 6.1. In that example, we took the height of the potential energy barrier confining the particle to be infinite, which may seem like a rather bad approximation for electrons confined in a solid. But if you look at the detailed solutions to the one-dimensional finite square well in Section 4.1, you will notice that the results of the finite square well reduce to the results of the infinite square well if *either* the height of the barrier *or* the width of the well approaches infinity. Since a typical linear dimension of a solid is in the centimeter or longer range (and therefore is at least a factor of 10^8 larger than the size of a typical atom), it is not a bad approximation to use the results of the infinite well when treating a solid.

Typically, for a solid like sodium there are on the order of 10^{23} valence electrons per cubic centimeter. This is a large number! If we assume that each electron is in the lowest energy state available, the electrons will fill up the states up to a maximum energy called the **Fermi energy**. Recall that the energy eigenvalues for the cubic box of side L are given by

$$E_{n_x,n_y,n_z} = \frac{\left(n_x^2 + n_y^2 + n_z^2\right)\hbar^2\pi^2}{2mL^2} \qquad n_x, n_y, n_z = 1, 2, 3, \ldots \tag{7.30}$$

We can put two electrons in the energy state $E_{1,1,1}$ in a fashion similar to the way we constructed the ground state wave function for helium. Similarly, a third electron can go into the state corresponding to either $E_{2,1,1}$, $E_{1,2,1}$, or $E_{1,1,2}$, since all these energies are the same. Obviously figuring out the ground state for, say, 10^{23} electrons is not as easy as figuring out the ground state configuration of lithium. A good strategy for dealing with this problem is to calculate the number of states between energy E and $E + dE$. Since each allowed energy state is labeled by the integers n_x, n_y, and n_z, we can show each allowed energy state as a point in the lattice pictured in Fig. 7.3.

It is convenient to introduce a dimensionless radius

$$r = \sqrt{n_x^2 + n_y^2 + n_z^2} \tag{7.31}$$

in this lattice. Figure 7.4 shows the projection in the $n_x - n_y$ plane of such a lattice. Notice how the number of states between r and $r + \Delta r$ grows with the magnitude of r. In the full three-dimensional lattice, rather than circular arcs of radii r and $r + \Delta r$, we choose

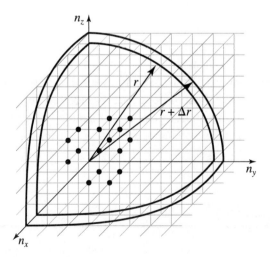

Figure 7.3 Each allowed state is labeled by point in the lattice with axes labeled by n_x, n_y, and n_z. For the sake of clarity, only a few points are shown. The number of points between r and $r + \Delta r$ is given by the volume of the spherical shell between r and $r + \Delta r$.

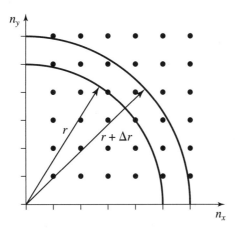

Figure 7.4 The allowed states in the n_x-n_y plane. The number of states between r and $r + \Delta r$ grows with increasing r. This effect is more pronounced in three dimensions.

a spherical shell of inner radius r and outer radius $r + \Delta r$. The volume of this spherical shell is given by $4\pi r^2 \Delta r$. Since the number of particles filling these individual states is very large, we can, to a very good approximation, replace Δr with the differential dr. From (7.30), we see that

$$r = \frac{\sqrt{2mE}\,L}{\hbar\pi} \tag{7.32}$$

and hence

$$dr = \frac{\sqrt{2m}\,L}{2\hbar\pi}\frac{dE}{\sqrt{E}} \tag{7.33}$$

Since by construction there is one allowed energy state per unit volume in the lattice, the number of states between E and $E + dE$ is given by

$$\begin{aligned} D(E)\,dE &= 2\left(\frac{1}{8}\right) 4\pi r^2 dr \\ &= \frac{V(2m)^{3/2}}{2\hbar^3 \pi^2}\sqrt{E}\,dE \end{aligned} \tag{7.34}$$

where $V = L^3$ is the volume of the box. The factor of $1/8$ on the first line of this equation arises from the fact that the integers n_x, n_y, and n_z take on only positive values, and thus the number of states is the volume of the spherical shell restricted to the first octant of the lattice. The factor of 2 on the right-hand side of the first line of (7.34) arises from the fact that there are two spin states, spin up and spin down, for each allowed energy for a spin-1/2 particle like the electron. The quantity $D(E)$ is called the **density of states**. It is the number of states per unit energy. The important thing to notice in the result (7.34) is that the density of states grows as the square root of the energy.

Let us assume there are N electrons in the box. In the ground state, we put one electron in each state until we run out of electrons, which occurs at an energy E_F, which we call the Fermi energy. Therefore

$$\begin{aligned} N &= \int_0^{E_F} D(E)\,dE \\ &= \int_0^{E_F} \frac{V(2m)^{3/2}}{2\hbar^3 \pi^2}\sqrt{E}\,dE \end{aligned} \tag{7.35}$$

Evaluating the integral, we obtain

$$N = \frac{V(2mE_F)^{3/2}}{3\hbar^3 \pi^2} \qquad (7.36)$$

and therefore

$$E_F = \frac{\hbar^2}{2m}\left(\frac{3\pi^2 N}{V}\right)^{2/3} \qquad (7.37)$$

What happens if we take this box and squeeze it, attempting to decrease the volume V of the solid? As you would expect, the solid resists such a compression, since according to (7.37) reducing the volume of the box increases the energy of the electrons confined in the box. From conservation of energy, $dE_{\text{total}} = -PdV$ (the analogue of $dW = Fdx$), where E_{total} is the total energy of the electrons in the solid and the pressure P is the force/area with which the solid pushes back when it is compressed. This pressure is generally referred to as electron **degeneracy pressure**.[4] In the Fermi gas model of the solid (see Problem 7.4),

$$E_{\text{total}} = \frac{3}{5}NE_F \qquad (7.38)$$

where E_{total} is the sum of the energy of the N electrons contained in the box. Therefore

$$P = -\frac{dE_{\text{total}}}{dV} = \frac{2}{5}(3\pi^2)^{2/3}\frac{\hbar^2}{2m}\left(\frac{N}{V}\right)^{5/3} \qquad (7.39)$$

Using this result, we can compute the compressibility K of a solid

$$K = \left(-V\frac{dP}{dV}\right)^{-1} \qquad (7.40)$$

As noted in Example 7.4, for alkali metals such as sodium the electron degeneracy pressure can account for a significant fraction of the resistance of the metal to compression.

EXAMPLE 7.4 Determine the Fermi energy E_F and the Fermi velocity v_F, the speed of an electron at the Fermi energy, for sodium. The molecular weight for sodium is 23 and the density of metallic sodium is 0.97 g/cm^3. There is one valence electron per atom for sodium.

SOLUTION

$$\frac{N}{V} = \frac{0.97 \text{ g/cm}^3}{23 \text{ g/mole}}(6.02 \times 10^{23} \text{ atoms/mole}) = 2.5 \times 10^{22} \text{ atoms/cm}^3$$

Substituting this result into (7.37), we find that $E_F = 3.1$ eV for sodium. Since this is the kinetic energy of the most energetic electron in the box, the Fermi velocity is determined from

$$E_F = \frac{1}{2}mv_F^2$$

For sodium, $v_F = 1.0 \times 10^6$ m/s. Thus the most energetic electrons are nonrelativistic, which is necessary for self-consistency since we have used the nonrelativistic

---

[4] Here the term degeneracy is used to indicate that the electrons are in their lowest energy state.

Schrödinger equation to determine the allowed energies. But the speed of these most energetic electrons is appreciable, namely 0.3% of the speed of light.

Given the value for N/V it is straightforward to calculate the compressibility K as well as the Fermi energy for sodium. We obtain a value of 1.2×10^{-10} m²/N, which should be compared with the measured value of 1.56×10^{-10} m²/N.

White Dwarf and Neutron Stars

This electron degeneracy pressure also plays a big role in determining the end state of stellar evolution for stars that are not too massive. As we will see in Chapter 9, stars like the Sun are "burning" hydrogen in a process of nuclear fusion that converts hydrogen into helium. The energy released in these fusion reactions produces a pressure that balances the gravitational pull that would otherwise cause the star to collapse. But what happens after the nuclear fuel is exhausted?

Relatively nearby, 8.6 light-years away, is the star Sirius, the brightest star in the night sky. Sirius has a companion, the star Sirius B, which orbits around it with a period of 50 years. The mass of Sirius B is very close to the mass of our Sun, but its radius is 6,000 km, just slightly less than the radius of the Earth (see Example 7.7). Consequently, the density of Sirius B is 2×10^6 g/cm³. This is to be contrasted with the Sun, which has a radius of 7×10^5 km and a density of 1.4 g/cm³, which is near the density of water. Sirius B is an example of a white dwarf star, the end phase of evolution of stars like the Sun. How can such an enormous density be supported? Certainly not by the sort of repulsion between neighboring atoms that leads to stability of ordinary matter on Earth, for which the average density is 5.5 g/cm³. The answer is electron degeneracy pressure.

In 1930, Subramanyan Chandrasekhar, then 19 years old, was on a sea voyage from India to Cambridge, England, where he planned to begin graduate work. Chandrasekhar was interested in exploring the consequences of quantum mechanics for astrophysics. During his trip, he analyzed how the density, pressure, and gravity in a white dwarf star vary with radius. For a star like Sirius B Chandrasekhar found that the Fermi velocity of inner electrons approaches the speed of light. Consequently, he found it necessary to redo the calculation of the Fermi energy taking relativistic effects into account. In the limiting case in which the electrons in the star are ultrarelativistic, he found that the electron degeneracy pressure varies as $(N/V)^{4/3}$ instead of as $(N/V)^{5/3}$, which is what we obtained in (7.39) for the nonrelativistic case. (See Problem 7.7.) Thus, for example, a 1% increase in the density produces a 4/3% increase in the pressure in contrast to the 5/3% increase that would be expected if the electrons were nonrelativistic. Therefore the degeneracy pressure of relativistic electrons does not offer as much resistance to collapse as does the degeneracy pressure due to nonrelativistic electrons. Chandrasekhar deduced that a high-density, high-mass star cannot support itself against gravitational collapse unless the mass of the star is less than 1.4 solar masses. This finding was quite controversial within the astronomical community and it was 54 years before Chandrasekhar was awarded the Nobel Prize for this work.[5]

---

[5]For an interesting discussion of the history associated with Chrandrasekhar's discovery, see Kip S. Thorne, *Black Holes and Time Warps*, W. W. Norton, New York, 1994, pp. 140–163.

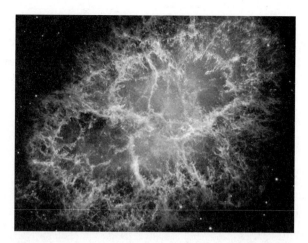

Figure 7.5 The Crab Nebula as seen by the Hubble Space Telescope. At the center of the nebula is a pulsar, a neutron star rotating at a rate of 30 revolutions per second. (Courtesy NASA)

The natural question to raise is what happens if the mass of the star exceeds this 1.4 solar mass limit. As the star collapses, the size of the box confining the electrons decreases and, consequently, the energy of the electrons confined in the box increases. When the energy of the electrons reaches the point that is sufficient to initiate the reaction $e^- + p \rightarrow \nu_e + n$, that is, an electron combines with a proton to produce the more massive neutron (a spin-1/2 particle) and an associated neutrino, the inner core of the star collapses to a neutron star. Calculations similar in spirit to the ones Chandrasekhar did for white dwarf stars show that for neutron stars a typical radius is on the order of 10 km and the density is on the order of 10^{14} g/cm^3 (the density of nuclear matter). Moreover, there is an upper limit on the mass of a neutron star of roughly 1.5 to 3 solar masses.[6] Beyond that limit, the star either collapses to a black hole or ejects mass in a catastrophic explosion known as a supernova, often leaving a neutron star surrounded by ejected gas as a remnant, as is the case for the Crab nebula, shown in Fig. 7.5.

7.5 Quantum Statistics

In our discussion of the Fermi energy we have focused on systems such as the valence electrons in a solid or the neutrons in a neutron star in which the number of particles that make up the system is very large, ranging from, say, 10^{23} to 10^{55}. It is not practical or, for that matter, even useful to attempt to write a wave function for such a large number of particles. The most we can reasonably do is to obtain probabilistic information about how the particles are distributed on average in terms of energy. In other words, we will treat the particles statistically.

The fundamental postulate of statistical mechanics states that an isolated system in thermal equilibrium is found with equal probability in each of its accessible **microstates**. To illustrate, let's examine the microstates for a system of 6 *distinguishable* particles with total energy $8E_1$ in a system for which the possible energy states for each particle are

---

[6]The uncertainty in the upper limit for the mass of neutron stars arises from our lack of detailed knowledge of the nature of the interactions between neutrons.

	E_1	$E_2 = 2E_1$	$E_3 = 3E_1$	
(a)	ABCDE		F	$E = 8E_1$
(b)	ABCD	EF		$E = 8E_1$

Figure 7.6 (a) A microstate of six particles in which five particles reside in the ground state and one particle is in the second excited state, satisfying the constraint that the total energy of the six particles is $8E_1$. Since any of the six particles (labeled with the letters A through F) can be in the state with energy $3E_1$, there are six such microstates if the particles are presumed to be distinguishable. (b) A microstate of the six particles with total energy $8E_1$ in which four of the particles are in the ground state and two of the particles are in the first excited state. There are fifteen such microstates if the particles are presumed to be distinguishable.

E_1, $E_2 = 2E_1$, $E_3 = 3E_1$, and so forth. One possible configuration of the particles with the requisite energy is to have five particles in the lowest energy state and the sixth particle in the state with energy $E_3 = 3E_1$, as illustrated in Fig. 7.6a. Since any of the six particles might be the one with energy E_3, there are six different microstates ($6 = 6!/5!$) corresponding to this configuration of the particles. Another way to distribute the six particles so that the total energy is $8E_1$ is to have 4 particles each with energy E_1 and two particles each with energy $E_2 = 2E_1$, as illustrated in Fig. 7.6b. In this case, there are $6!/4!2! = 15$ different microstates.[7] Given the constraint that the total energy be equal to $8E_1$, there are no additional microstates possible. Thus there are a total of $6 + 15 = 21$ microstates. Notice that in 6 of these 21 microstates there are 5 particles in the state with energy E_1, and in 15 of these 21 microstates there are 4 particles in the state with energy E_1. Thus the average number of particles in this energy state is $(6/21)5 + (15/21)4 = 4.3$. Similarly, the average number of particles in the state with energy E_2 is $(15/21)2 = 1.4$. And finally, the average number of particles in the state with energy E_3 is $(6/21)1 = 0.3$. Clearly, there would be no particles in any higher energy states if the total energy is constrained to be $8E_1$. These average numbers are shown in Fig. 7.7 together with a curve of $Ae^{-\beta E}$ with appropriate values of A and β. Even though the number of particles in our example is not large, the beginnings of an exponential fall off seem to be already in place.

If we were to extend our discussion to a very large number of distinguishable particles in thermal equilibrium at a particular temperature T (which is equivalent to specifying the average energy of a system in thermal equilibrium with a reservoir), we would find the average number of particles in each energy state to be equal to

$$n(E) = Ae^{-E/k_B T} \tag{7.41}$$

where the constant A is chosen so that sum of (7.41) over all possible energy states is equal to the total number of particles N in the system. The factor $e^{-E/k_B T}$ is called the **Boltzmann factor**, named after Ludwig Boltzmann, who along with Maxwell was one of

---

[7]These results for the number of microstates are special cases of "N choose N_1," the number of different ways of choosing N_1 particles from a group of N particles, namely

$$\binom{N}{N_1} = \frac{N!}{N_1!(N-N_1)!}$$

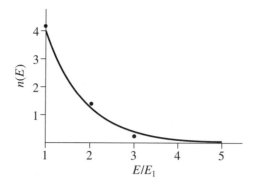

Figure 7.7 The average number of particles in each of the energy states for the microstates illustrated in Fig. 7.6. The solid curve is a graph of $Ae^{-\beta E}$ with A and β chosen to give a reasonable fit.

the founders of statistical mechanics. The constant k_B in the exponent is the **Boltzmann constant**. It has the value

$$k_B = 1.38 \times 10^{-23} \text{ J/K} \tag{7.42}$$

If you are not familiar with the Boltzmann constant and the Boltzmann factor, check out Example 7.5 at the end of this section. There you will see an illustration of how the Boltzmann factor arises in the case of an ideal gas in a uniform gravitational field.

If you examine the counting arguments that we used in our discussion of the allowed microstates, you see that our treating the particles as distinguishable played a crucial role. For example, we pointed out that there were six microstates corresponding to the configuration with five particles in the state with energy E_1 and one particle in the state with energy E_3 since we were treating each of the six possible ways we could put one of the particles in the state with energy E_3 as distinguishable. Similarly, there are fifteen microstates in which four particles are in the state with energy E_1 and two particles are in the state with energy E_2. Since all microstates are presumed to occur with equal probability, the configuration with four particles in the ground state and two particles in the first excited state is $15/6 = 2.5$ times more likely to occur than the configuration with five particles in the ground state and one particle in the second excited state. But if the particles are indistinguishable, then there is only one microstate corresponding to each of these configurations, as shown in Fig. 7.8. Hence instead of a 2.5 ratio of allowed microstates for distinguishable particles for these two configurations, there is a 1 to 1 ratio if the particles are indistinguishable. This difference in the counting of the ratio of the number of allowed microstates (the statistical weight, if you will) for distinguishable versus indistinguishable particles becomes increasingly pronounced as

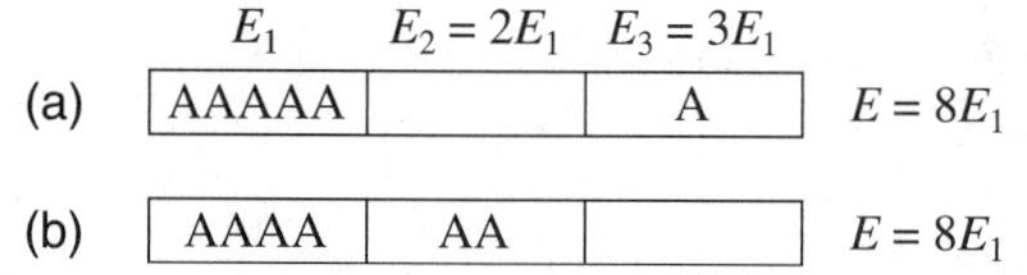

	E_1	$E_2 = 2E_1$	$E_3 = 3E_1$	
(a)	AAAAA		A	$E = 8E_1$
(b)	AAAA	AA		$E = 8E_1$

Figure 7.8 (a) There is a single microstate for six indistinguishable particles (each labeled by the letter A) in which five particles reside in the ground state and one particle resides in the second excited state. (b) Similarly, there is a single microstate of six indistinguishable particles in which four of the particles are in the ground state and two of the particles are in the first excited state.

the number of particles increases. See Problem 7.12. Since our counting of the number of microstates for distinguishable particles suggested the Boltzmann factor, you can foresee that we should end up with quite different results for indistinguishable particles. This is the rationale for calling this subject quantum statistics.

In our discussion of indistinguishable particles in Section 7.2, we saw that the quantum states of identical particles are either symmetric or antisymmetric under particle exchange. For states that are antisymmetric under particle exchange, there can be no more than one particle in any quantum state, while for states that are symmetric under particle exchange there is no limit on the number of particles that can be in any quantum state. Thus the indistinguishable particles in the example shown in Fig. 7.8 must be bosons since more than one particle resides in a number of the energy states. To illustrate the important role that intrinsic spin plays consider a simple example in which we distribute two particles in three possible energy states E_1, E_2, and E_3. There are three alternatives.

1. **Maxwell–Boltzmann statistics**: If we call the particles A and B, treating them as distinguishable, then there are nine possible microstates, since each of the two particles can be placed in any one of the three states, as shown in Fig. 7.9a. Note that in three of these nine microstates the two particles are in the same energy state.

2. **Bose–Einstein statistics**: If the particles are indistinguishable and there are no limits on the number of particles that can be put in any one of the energy states, then there are six possible microstates, with three possible ways of putting the particles in the same energy state and three possible ways of putting them in different energy states, as illustrated in Fig. 7.9b.

3. **Fermi–Dirac statistics**: If the particles are indistinguishable and no more than one particle can be put in a particular energy state, then there are just three possible microstates, as shown in Fig. 7.9c.

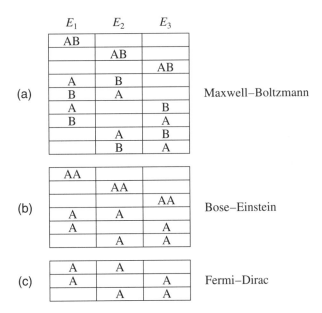

Figure 7.9 The microstates for two particles distributed in three energy states. In (a) the particles are distinguishable (denoted by A and B). In (b) the particles are indistinguishable with no limitation on the number of particles that can be in the same state. And in (c) the particles are indistinguishable with the limitation that no more than one particle can be in any particular state.

Notice in particular that the probability that the the two particles are found in the same state varies from $3/9 = 1/3$ for the Maxwell–Boltzmann statistics to $3/6 = 1/2$ for Bose–Einstein statistics to 0 for Fermi–Dirac statistics.

The examples that we have examined so far have involved a small number of particles. As was noted earlier, statistical mechanics is typically applied to systems with a very large number of particles. The guiding principle of statistical mechanics is that each accessible microstate occurs with equal probability. Thus the most probable configuration of the particles is the one corresponding to the maximum number of microstates. While counting the number of microstates in the general case is beyond the scope of this text, once the number of microstates is determined the most probable configuration is obtained by maximizing this number subject to the constraints that the total number of particles in the system is N and the total energy of the system is E. As with any maximization problem in which there are constraints, the constraints can be handled by the method of Lagrange multipliers. Since there are two constraints, there are two Lagrange multipliers, typically called α and β, that are associated with the constraints on the number of particles and the total energy, respectively. The good news is that as the number of particles in the system grows this most probable configuration becomes overwhelmingly more likely than its competitors.[8]

For distinguishable particles, we obtain the **Maxwell–Boltzmann distribution function**

$$n(E) = \frac{1}{e^\alpha e^{E/k_B T}} \tag{7.43}$$

namely the exponential distribution (7.41) suggested by our example with six distinguishable particles (with the identification $A = e^{-\alpha}$). Example 7.5 shows how the energy is related to the temperature for the ideal gas provided we choose the Lagrange multiplier $\beta = 1/k_B T$. For indistinguishable particles with integer intrinsic spin angular momentum, namely bosons, the corresponding **Bose–Einstein distribution function** is given by

$$n(E) = \frac{1}{e^\alpha e^{E/k_B T} - 1} \tag{7.44}$$

For indistinguishable particles with half-integer intrinsic spin, namely fermions, the corresponding **Fermi–Dirac distribution function** is given by

$$n(E) = \frac{1}{e^\alpha e^{E/k_B T} + 1} \tag{7.45}$$

Figure 7.10 shows each of these distribution functions plotted as a function of $E/k_B T$. At high energies ($E \gg k_B T$), both the Bose–Einstein and the Fermi–Dirac distribution functions reduce to the Maxwell–Boltzmann distribution function. In this case, the probability of a particle being in any one energy state is small so the differences arising from what happens when multiple particles are in that state are negligible. At low energies ($E \ll k_B T$), where the number in a particular state is comparable to or larger than one, the differences between the distributions are striking, with the quantum distributions falling on opposite sides of the Maxwell–Boltzmann distribution.

---

[8]A concise derivation of the most probable configuration for identical particles is given by D. J. Griffiths, *Introduction to Quantum Mechanics*, Prentice-Hall, Englewood Cliffs, NJ (1995), pp. 204–218.

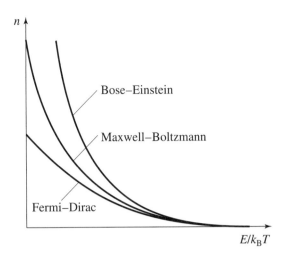

Figure 7.10 The Maxwell–Boltzmann, Fermi–Dirac, and Bose–Einstein distribution functions plotted as a function of $E/k_B T$ for the value $\alpha = 0$.

We conclude this section with a comment on the Fermi–Dirac distribution function. It is convenient to write the Fermi–Dirac distribution function in a form that brings in the Fermi energy. We define $E_F = -\alpha k_B T$ in (7.45), in which case

$$n(E) = \frac{1}{e^{(E-E_F)/k_B T} + 1} \tag{7.46}$$

Given this form for $n(E)$, you can see that $n(E_F) = 0.5$. The number of fermions with energy between E and $E + dE$ for identical fermions in a box is given by the product of $n(E)$ and the number of states with energy between E and $E + dE$, namely

$$n(E)D(E)\,dE = \frac{1}{e^{(E-E_F)/k_B T} + 1} \frac{V(2m)^{3/2}}{2\hbar^3 \pi^2} \sqrt{E}\,dE \tag{7.47}$$

The value of the Fermi energy is then determined from the requirement that the sum over all possible energy states equals N, the total number of fermions:

$$N = \int_0^\infty \frac{1}{e^{(E-E_F)/k_B T} + 1} \frac{V(2m)^{3/2}}{2\hbar^3 \pi^2} \sqrt{E}\,dE \tag{7.48}$$

As $T \to 0$, $n(E) = 1$ for $E < E_F$ and $n(E) = 0$ for $E > E_F$. In this case, the integral (7.48) is straightforward to evaluate, as we saw in Section 7.4. As indicated in Fig. 7.11, the value of the Fermi energy decreases with increasing temperature as fermions from

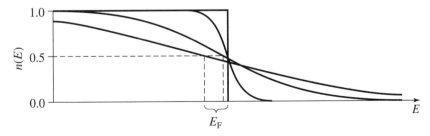

Figure 7.11 The Fermi–Dirac distribution function for four different temperatures. Note how the location of the Fermi energy [the energy for which $n(E) = 0.5$] changes with increasing temperature.

the lower energy states are shifted into higher energy states. Thus the energy E at which $n(E) = 0.5$ shifts to lower energies as these lower energy states become depopulated.

It is convenient to define a **Fermi temperature** through

$$E_F = k_B T_F \tag{7.49}$$

For sodium, we saw in the previous section that $E_F = 3.1$ eV and therefore $T_F = 36 \times 10^3$ K. Now you can see why it is an excellent approximation to treat the solid as if it were effectively at absolute zero. At room temperature, $k_B T = \frac{1}{40}$ eV $\ll k_B T_F$, so that only a small fraction of the electrons, those roughly within $k_B T$ of the Fermi energy, are excited. The vast majority of the electrons are "frozen" in the lower energy states.

EXAMPLE 7.5 A well-known example of a macroscopic equation of state involving pressure and temperature is the ideal gas law

$$PV = \frac{N}{N_A} RT$$

where P is the pressure of the gas, V is the volume, T is the temperature, and N is the total number of molecules of gas. The constant N_A is Avogadro's number and the constant R is the ideal gas constant ($R = 8.315$ J/mole·K). The fraction N/N_A is simply the number of moles of the gas. Physicists like to express the ideal gas law in the form

$$PV = N k_B T$$

where $k_B = R/N_A$ is the Boltzmann constant. Suppose this ideal gas at temperature T is in a uniform gravitational field. Show that the molecules in the gas are distributed as a function of the height z according to the Boltzmann factor.

SOLUTION In static equilibrium a pressure gradient arises in the gas, since the pressure pushing upward from below on a small slab of gas of thickness dz must be greater than the pressure pushing downward from above to balance the weight of the gas contained in the slab, as illustrated in Fig. 7.12. Taking the mass of each molecule in the gas to be m and calling g the acceleration due to gravity, we find that setting the sum of the forces on this slab of gas to zero leads to $PA - (P + dP)A - mg\tilde{n} A dz = 0$, where A is the cross-sectional area of the slab and $\tilde{n}$ is the number of molecules per unit volume of the gas. Since the ideal gas law can be written in terms of the density $\tilde{n} = N/V$ of molecules as $P = \tilde{n} k_B T$, the force equation becomes

$$d\tilde{n} \, k_B T = -mg\tilde{n} \, dz$$

or

$$\frac{d\tilde{n}}{\tilde{n}} = -\frac{mg \, dz}{k_B T}$$

This equation can be easily integrated to yield

$$\tilde{n} = \tilde{n}(0) e^{-mgz/k_B T}$$

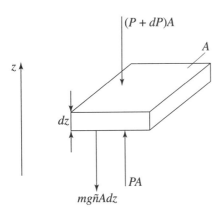

Figure 7.12 The external forces on a small slab of gas due to pressure and gravity must balance in equilibrium.

where $\tilde{n}(0)$ is the number of molecules per unit volume at $z = 0$. Note that mgz is simply the gravitational potential energy of a molecule at height z. This equation of state is often referred to as the **law of atmospheres**. At a height z such that $mgz = k_B T$ the density is reduced to $1/e$ of its value at $z = 0$. The larger the temperature, the more slowly the density decreases with height.

It is natural to ask at this point what role the kinetic energy of the molecules plays. To see the connection between kinetic energy and temperature for an ideal gas, consider the gas at temperature T confined in the volume V shown in Fig. 7.13. The pressure P exerted by the gas on, say, the right-hand wall of area A arises from collisions of the molecules with the wall. We calculate an average pressure due to the collision of a single molecule and then sum the results for the N molecules contained in the volume. Assuming a molecule approaches the wall with speed v_x in the x direction and collides elastically with the wall, the change in the momentum of the molecule is $2mv_x$. Thus the force per unit area exerted by this molecule on the wall is given by

$$\frac{F_x}{A} = \frac{1}{A}\frac{\Delta p_x}{\Delta t} = \frac{2mv_x}{A\Delta t}$$

where $\Delta p_x/\Delta t$ is the change in the x component of the momentum in the time interval Δt. Since the molecules are distributed randomly in the volume, whether the molecule strikes the wall in time Δt depends on whether it is within a distance

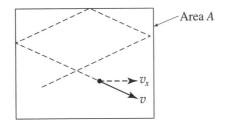

Figure 7.13 A molecule with horizontal speed v_x exerts pressure on the right-hand wall as a consequence of collisions with the wall.

$v_x \Delta t$ of the wall. The probability that it strikes the wall is then

$$\frac{v_x \Delta t A}{2V}$$

namely the volume $v_x \Delta t A$ divided by $2V$, where the factor of 2 arises from the fact that the molecule is equally likely to be headed away from the wall as toward the wall. Multiplying F_x/A by the probability of making a collision, we obtain

$$\frac{F_x}{A} = \frac{mv_x^2}{V}$$

Thus the pressure arising from the collisions of the N molecules in the gas is

$$P = \frac{m}{V}\left(v_{x_1}^2 + v_{x_2}^2 + \cdots + v_{x_N}^2\right)$$

$$= \frac{N}{V} m \overline{v_x^2}$$

where

$$\overline{v_x^2} = \frac{1}{N}\left(v_{x_1}^2 + v_{x_2}^2 + \cdots + v_{x_N}^2\right)$$

is the average square of the x component of the velocity. Since the ideal gas law states that $P = Nk_\text{B}T/V$, we see that

$$m\overline{v_x^2} = k_\text{B}T$$

Finally, since $\overline{v_x^2} = \overline{v_y^2} = \overline{v_z^2}$, the average kinetic energy of the molecules in the gas is

$$\frac{1}{2}m\overline{v^2} = \frac{1}{2}m\overline{v_x^2} + \frac{1}{2}m\overline{v_y^2} + \frac{1}{2}m\overline{v_z^2} = \frac{3}{2}k_\text{B}T$$

Thus in an isothermal ideal gas the average kinetic energy of the molecules is simply proportional to the temperature.[9]

Returning to the derivation of the law of atmospheres in which we took into account the influence of gravity on the molecules in the gas, the average kinetic energy of the molecules does not vary with the height z in the gas since the gas is presumed to be isothermal. Therefore the change in energy of the molecules in moving to higher elevations in the gas results solely from the change in gravitational potential energy. The law of atmospheres can thus be seen as an interesting manifestation of the Boltzmann factor $e^{-E/k_\text{B}T}$.

EXAMPLE 7.6 Calculate the de Broglie wavelength λ_F of an electron in a solid at the Fermi energy. Compare this length with the spacing between electrons in the solid.

SOLUTION The de Broglie wavelength of an electron at the Fermi energy is

$$\lambda_\text{F} = \frac{h}{p_\text{F}}$$

---

[9]This result is an illustration of the equipartition theorem, namely that in thermal equilibrium the average kinetic energy in each (quadratic) degree of freedom is $k_\text{B}T/2$.

where p_F is the momentum of an electron at the Fermi energy. From (7.37) we know that

$$E_F = \frac{p_F^2}{2m} = \frac{\hbar^2}{2m}\left(\frac{3\pi^2 N}{V}\right)^{2/3}$$

Therefore

$$\frac{h^2}{2m\lambda_F^2} = \frac{\hbar^2}{2m}\left(\frac{3\pi^2 N}{V}\right)^{2/3}$$

where N/V is the number of electrons per unit volume. The spacing between electrons is roughly $(V/N)^{1/3}$ and is thus given by

$$\left(\frac{V}{N}\right)^{1/3} = \frac{1}{2}\left(\frac{3}{\pi}\right)^{1/3}\lambda_F$$

Therefore the spacing between atoms in the crystal is roughly half the de Broglie wavelength of the most energetic conduction electrons. Lower energy electrons have de Broglie wavelengths that are of course even larger. A good rule of thumb is that quantum effects become pronounced in multiparticle systems when the spacing between the particles is comparable to the de Broglie wavelength of these particles. We will see some striking examples involving bosons in Section 7.7.

7.6 Cavity Radiation

In the remaining sections of this chapter we will examine some interesting applications of the Bose–Einstein distribution function. One of the most common and important applications is the phenomenon of **cavity radiation**. If you crack open the door to a kiln (a thermally insulated oven) that is at room temperature and look in, the contents are black. If the kiln is heated, the contents glow red. At higher temperatures the contents appear yellow and then white. The increasing thermal energy causes increasing excitation of the atoms and molecules that constitute the walls of the kiln. The atoms and molecules emit and absorb electromagnetic radiation. Thus we can think of the kiln as a box containing photons in thermal equilibrium at a particular temperature T. Figure 7.14 shows a plot of the spectral radiancy emitted as a function of frequency for three different temperatures. The radiancy, the total energy per unit time per unit area emitted, is the area under the curve.

Cavity radiation is often referred to as **blackbody radiation**. To see why, consider a cavity with a very small hole leading to the outside. If light falls on the hole, it will be almost entirely absorbed, since the odds that the light is reflected back through the hole are very small, as illustrated in Fig. 7.15. Thus the hole looks extremely black. As a demonstration, take a couple of pieces of the blackest paper you can find. Cut a small hole in one of the pieces and hold that piece in front of the other piece with some spacing between the two pieces. You will observe how much blacker the hole is than the surrounding paper. The reason that cavity radiation, or blackbody radiation, attracted the attention of physicists like Max Planck is that the radiation emitted from a blackbody is universal in character. In particular, the spectrum is entirely independent of the size and

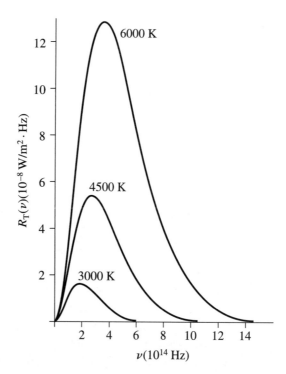

Figure 7.14 The spectral radiancy $R_T(\nu)$ for a blackbody radiator for three temperatures. $R_T(\nu)\,d\nu$ is the energy per unit area per unit time in the frequency interval between ν and $\nu + d\nu$ emitted by the blackbody.

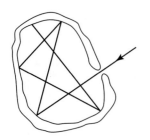

Figure 7.15 A cavity connected by a small hole to a region outside. Light incident on the hole has a very small probability of being reflected back out and thus the hole is very black.

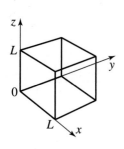

Figure 7.16 A cubic cavity.

the shape of the cavity as well as the composition of the material in which the cavity is contained. Thus there is something truly fundamental about this cavity radiation.

We start by calculating the energy density (the energy per unit volume) of a cavity filled with electromagnetic radiation. We can take advantage of the universal character of the spectrum to choose an especially straightforward cavity to analyze, namely a cubic cavity (shown in Fig. 7.16) in a metal with a very high conductivity. Of course, since we are concerned with electromagnetic radiation contained within the cavity, the equations that we are solving are Maxwell's equations or, more precisely, the wave equation for the electric field $\mathcal{E}$:

$$\frac{\partial^2 \mathcal{E}}{\partial x^2} + \frac{\partial^2 \mathcal{E}}{\partial y^2} + \frac{\partial^2 \mathcal{E}}{\partial z^2} - \frac{1}{c^2}\frac{\partial^2 \mathcal{E}}{\partial t^2} = 0 \qquad (7.50)$$

The solutions to this wave equation satisfying the boundary condition that the tangential component of the electric field vanishes at the boundary are given by

$$\mathcal{E}_x = \mathcal{E}_{x_0} \cos\frac{n_x \pi x}{L} \sin\frac{n_y \pi y}{L} \sin\frac{n_z \pi z}{L} \sin\omega t$$

$$\mathcal{E}_y = \mathcal{E}_{y_0} \sin\frac{n_x \pi x}{L} \cos\frac{n_y \pi y}{L} \sin\frac{n_z \pi z}{L} \sin\omega t$$

$$\mathcal{E}_z = \mathcal{E}_{z_0} \sin\frac{n_x \pi x}{L} \sin\frac{n_y \pi y}{L} \cos\frac{n_z \pi z}{L} \sin\omega t \qquad (7.51)$$

which are reminiscent of the solutions (6.11) to the Schrödinger equation for a nonrelativistic particle in a cubic box that we determined in Section 6.1. Unlike the wave function, the electric field is a vector quantity, so we need to specify all three of its components. See Problem 7.14. The important thing to notice is that if any one of the components of $\mathcal{E}$ is substituted into (7.50), we obtain the condition

$$\frac{\omega^2}{c^2} = \frac{\pi^2}{L^2}\left(n_x^2 + n_y^2 + n_z^2\right) \tag{7.52}$$

or in terms of the ordinary frequency ν

$$\nu = \frac{c}{2L}\sqrt{n_x^2 + n_y^2 + n_z^2} \tag{7.53}$$

Thus the allowed frequencies are specified by the positive integers n_x, n_y, and n_z, just as were the allowed energies in our derivation of the density of states for fermions in a box in Section 7.4.

To determine the density of states, we similarly define a radius in the lattice of allowed states by

$$r = \sqrt{n_x^2 + n_y^2 + n_z^2} \tag{7.54}$$

Hence from (7.53)

$$r = \frac{2L}{c}\nu \tag{7.55}$$

and the number of states between ν and $\nu + d\nu$ is given by

$$\begin{aligned} D(\nu)\,d\nu &= 2\frac{1}{8}4\pi r^2\,dr \\ &= \frac{8\pi V}{c^3}\nu^2\,d\nu \end{aligned} \tag{7.56}$$

where $D(\nu)$ is the number of states per unit frequency and $V = L^3$ is the volume of the cavity. Note that here the factor of $1/8$ in the first line of this equation arises from the fact that we want to include only the volume of the spherical shell between r and $r + dr$ in the first octant of the lattice pictured in Fig. 7.3. Here too, as in (7.34), we have inserted an extra factor of two in counting the number of allowed states. In this case, the factor of two arises from the fact that there are two polarization states for each mode.[10]

We now have almost everything we need to determine the energy density of a photon gas in thermal equilibrium at temperature T. The energy of each photon with frequency ν is of course $h\nu$. Since photons have integral intrinsic spin ($s = 1$), we must use the Bose–Einstein distribution function (7.44). However, since photons are emitted and absorbed by the atoms in the walls of the cavity and consequently their number is not conserved, we do not have any constraint on the number of photons in the cavity. This lack of constraint means we can set the Lagrange multiplier α to zero in (7.44), in which case the Bose–Einstein distribution function becomes

$$n(E) = \frac{1}{e^{E/k_\mathrm{B}T} - 1} \tag{7.57}$$

Since $E = h\nu$ for photons, the distribution function becomes

$$n(h\nu) = \frac{1}{e^{h\nu/k_\mathrm{B}T} - 1} \tag{7.58}$$

---

[10] Since $E = h\nu$ for photons, the density of states is proportional to E^2 in contrast to the $\sqrt{E}$ dependence for nonrelativistic particles that we saw in (7.34).

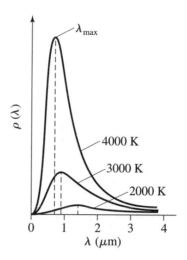

Figure 7.17 The energy density $\rho(\lambda)$.

Consequently, the energy in the cavity between ν and $\nu + d\nu$ is the product of three factors

$$h\nu n(h\nu)D(\nu)\,d\nu \tag{7.59}$$

namely, the energy of each photon, the average number of photons in each state, and the number of allowed states between ν and $\nu + d\nu$. Finally, dividing by the volume V of the cavity, we obtain the energy density (energy per unit volume) between ν and $\nu + d\nu$:

$$\rho(\nu)\,d\nu = \frac{8\pi h\nu^3}{c^3\left(e^{h\nu/k_B T} - 1\right)}\,d\nu \tag{7.60}$$

This energy density can also be expressed in terms of the wavelength λ of the radiation as[11]

$$\rho(\lambda)\,d\lambda = \frac{8\pi hc}{\lambda^5\left(e^{hc/\lambda k_B T} - 1\right)}\,d\lambda \tag{7.61}$$

Figure 7.17 shows a plot of $\rho(\lambda)$.

The expression for $\rho(\lambda)$ in (7.61) is often referred to as the **Planck function**. Max Planck first obtained this distribution function in 1900 in an attempt to fit the blackbody spectrum of cavity radiation, a fit that yielded a value for the constant h that was close to the currently accepted value. In an attempt to derive the Planck function, Planck was forced to assume that the atoms in the walls of the cavity had discrete, evenly spaced energy levels (which we now recognize as the energy levels of the harmonic oscillator, apart from the zero-point energy). It was Einstein who, in his 1905 paper on the photoelectric effect, first recognized that a consequence of Planck's assumption was that electromagnetic energy in the cavity "consists of a finite number of energy quanta localized at points in space, moving without dividing and capable of being absorbed or generated only as entities."

---

[11] Since $\nu = c/\lambda$, $d\nu = -c\,d\lambda/\lambda^2$, which is why the energy density per unit wavelength is proportional to $1/\lambda^5$ while the energy density per unit frequency is proportional to ν^3. Strictly speaking, we should use a symbol other than $\rho(\lambda)$ for the energy density expressed in terms of wavelength since $\rho(\nu)$ and $\rho(\lambda)$ are different functions.

Given these results, it is straightforward to calculate the total energy density in the cavity:

$$\int_0^\infty \rho(\nu)\,d\nu = \int_0^\infty \frac{8\pi h \nu^3}{c^3\left(e^{h\nu/k_B T}-1\right)}\,d\nu$$

$$= \frac{8\pi k_B^4 T^4}{c^3 h^3}\int_0^\infty \frac{x^3}{(e^x-1)}\,dx \qquad (7.62)$$

where in the last step we introduced the dimensionless variable $x = h\nu/k_B T$. Notice how the factor of T^4 pops out in front of the integral (because the integrand depends on ν^3 with a factor of ν^2 coming from the density of states and a factor of ν from the fact that the energy of a photon is $h\nu$). Thus the energy density varies as the fourth power of the temperature. Doubling the temperature increases the energy density by a factor of 16. Since

$$\int_0^\infty \frac{x^3}{(e^x-1)}\,dx = \frac{\pi^4}{15} \qquad (7.63)$$

we obtain

$$\int_0^\infty \rho(\nu)\,d\nu = aT^4 \qquad (7.64)$$

where

$$a = \frac{8\pi^5 k_B^4}{15 c^3 h^3} \qquad (7.65)$$

As shown in Problem 7.13, the energy density of cavity radiation and the radiancy of a blackbody are related by

$$R_T(\nu)\,d\nu = \frac{c}{4}\rho(\nu)\,d\nu \qquad (7.66)$$

where $R_T(\nu)d\nu$ is the (energy/area)/time with frequency between ν and $\nu + d\nu$. Thus the total energy per unit area per unit time radiated by a blackbody is

$$\int_0^\infty R_T(\nu)\,d\nu = \sigma T^4 \qquad (7.67)$$

where

$$\sigma = \frac{ac}{4} = 5.67 \times 10^{-8} \text{ W/m}^2\cdot\text{K}^4 \qquad (7.68)$$

is called the Stefan–Boltzmann constant. The T^4 dependence for the total radiancy is commonly called the **Stefan–Boltzmann law**.

Finally, it is not difficult to derive an additional "law," **Wien's law**, that follows from the form (7.61) of the energy density. Namely, the wavelength $\lambda_{\max}$ at which (7.61) has its maximum (see Fig. 7.17) satisfies the relation

$$\lambda_{\max} T = 2.9 \times 10^{-3} \text{ m}\cdot\text{K} \qquad (7.69)$$

Thus as the temperature of the blackbody increases, $\lambda_{\max}$ decreases. The derivation of (7.69) is left as an exercise (see Problem 7.15).

Astrophysics/Cosmology Implications

Apart from emission or absorption lines at some specific wavelengths or frequencies, the spectrum of a star is to a good approximation that of a blackbody. Although stars are not black, they behave as blackbodies to the extent that they absorb all radiation incident upon them. The Sun is a pale yellow because its surface temperature is 5800 K, which corresponds to $\lambda_{\max} = 500$ nm. The North Star, on the other hand, must be a substantially hotter star ($T = 8300$ K), since for this star $\lambda_{\max} = 350$ nm, in the ultraviolet part of the spectrum. The North Star is significantly bluer than the Sun (and fortunately, given its temperature, much farther away, too).

The most important astrophysical example of blackbody radiation is the **cosmic background radiation**. The universe emerged 13.7 billon years ago from a very hot dense state of matter and energy. At 4 s after this so-called Big Bang, the temperature had cooled to $T = 6 \times 10^9$ K, a temperature such that $k_B T = m_e c^2$, where m_e is the mass of an electron. Roughly 400,000 years later the temperature had cooled to 3,000 K, a temperature for which $k_B T = 0.2$ eV. At this temperature the photons do not have sufficient energy to ionize atoms, so that atoms (and therefore matter) are said to decouple from the radiation. It is this radiation that forms the cosmic background radiation. Since then the universe has expanded in scale by a factor of 1000. It is this expansion of space itself that has led to a shift in the wavelengths of the cavity radiation, very much as the distance between two fixed points on the surface of a balloon increases as the balloon is being blown up. From Wien's law, for example, we can see that these longer wavelengths correspond to a temperature for the cosmic background radiation that is consequently a factor of 1000 lower than when the radiation decoupled from matter ($T = 3000$ K $\rightarrow$ 3 K).

The existence of this cosmic background radiation was inadvertently discovered in 1962 by Arno Penzias and Robert Wilson. Penzias and Wilson were trying to understand the origin of background noise at $\lambda = 7.35$ cm in a radio telescope, a noise that was interfering with the first communication satellites that had been launched somewhat earlier. The noise, or hiss, was observed as they pointed the telescope (see Fig. 7.18) in all directions, including in regions of apparently empty sky. Their measurement was consistent with a background radiation of 3 K, but it was, after all, only one data point. It was not possible to map out the entire blackbody distribution function, which peaks at a wavelength of 1 mm, since the earth's atmosphere is strongly absorbing for wavelengths of order 1 mm. In 1989 the situation changed with the launching of the COBE satellite. The first data from COBE are shown in Fig. 7.19. Later data and error bars from COBE reveal the most precisely measured blackbody spectrum in nature, one for which the data

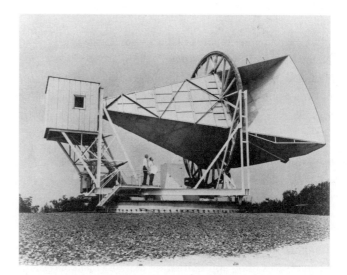

Figure 7.18 The Horn reflector antenna at Bell Telephone Laboratories in Holmdel, New Jersey. This antenna was built in 1959 for satellite communication. (Courtesy NASA)

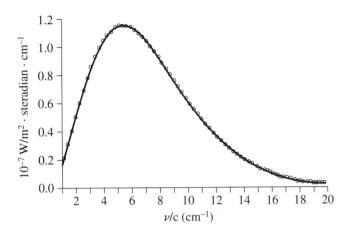

Figure 7.19 Data from NASA's Cosmic Background Explorer satellite fits a blackbody spectrum with a temperature of 2.735 ± 0.06 K. Adapted from J. C. Mather et al., *Astro. Jour.* **354**, L37 (1990).

points and error bars are obscured by the theoretical blackbody curve at a temperature of 2.725 K. More recently the Wilkinson Microwave Anisotropy Probe (WMAP) has revealed inhomogeneities in the temperature at the level of 1 part in 10^5. These hot and cold spots are the result of small density fluctuations in the early universe. Regions that were slightly compressed at that time had a higher density and a higher temperature. Over time these regions of higher density attracted nearby matter and grew even denser, eventually giving rise to clouds of gas, stars, and galaxies.

EXAMPLE 7.7 (*a*) The white dwarf star Sirius B has a surface temperature of 25,000 K. At what wavelength does the star emit the most radiation? (*b*) Sirius B is 8.6 light years from Earth. The flux of radiation from this star at Earth is 1.4×10^{-10} W/m². Determine the radius of Sirius B.

SOLUTION (*a*) Using $\lambda_{\max} T = 2.9 \times 10^{-3}$ m·K,

$$\lambda_{\max} = \frac{2.9 \times 10^{-3} \text{ m·K}}{25000 \text{ K}} = 1.2 \times 10^{-7} \text{ m} = 120 \text{ nm}$$

which is in the ultraviolet part of the spectrum. (*b*) The flux at Earth from a star of temperature T and radius r a distance D away is

$$\sigma T^4 \frac{4\pi r^2}{4\pi D^2}$$

Therefore

$$\frac{[5.67 \times 10^{-8} \text{ W/(m}^2 \cdot \text{K}^4)](2.5 \times 10^4 \text{ K})^4 r^2}{[8.6(9.46 \times 10^{15} \text{ m})]^2} = 1.4 \times 10^{-10} \text{ W/m}^2$$

Solving for r, we obtain

$$r = 6000 \text{ km}$$

just slightly less than the radius of the earth. However, as we noted in Section 7.4, the mass of Sirius B is very close to the mass of the Sun.

7.7 Bose–Einstein Condensation

In 1924, the Indian physicist S. N. Bose sent Albert Einstein a manuscript in which he derived Planck's blackbody radiation formula solely from the fact that the radiation consisted of a gas of indistinguishable massless quanta: photons. Einstein translated the article into German for publication, telling Bose in a postcard that his paper "signifies an important step forward and pleases me very much." In a note appended to his translation and published with the paper Einstein indicated that Bose's work could be extended to a gas of atoms. Einstein subsequently published three papers on this subject, the third of which predicted the possibility of a new state of matter that arises from the indistinguishability of atoms in a gas. It is this subject to which we now turn.

In our discussion of identical fermions in Section 7.5, we expressed the Fermi–Dirac distribution function

$$n(E) = \frac{1}{e^\alpha e^{E/k_B T} + 1} \quad (7.70)$$

in the form

$$n(E) = \frac{1}{e^{(E-E_F)/k_B T} + 1} \quad (7.71)$$

where E_F is the Fermi energy. For identical bosons, it is common to write the Bose–Einstein distribution function

$$n(E) = \frac{1}{e^\alpha e^{E/k_B T} - 1} \quad (7.72)$$

in the form

$$n(E) = \frac{1}{e^{(E-\mu)/k_B T} - 1} \quad (7.73)$$

where μ is called the **chemical potential**. Since $n(E)$ is the number of bosons in a state with energy E, $n(E)$ must be a positive quantity. Hence $(E - \mu)/k_B T$ must be greater than zero. If the energy of the ground state is taken to be zero, then μ must be negative. This is in sharp contrast to the Fermi energy, which is an inherently positive quantity.

We wish to apply this Bose–Einstein distribution function to a gas of identical atoms confined to a box. The requirement that

$$\sum_i n(E_i) = N \quad (7.74)$$

where N is the total number of atoms in the box, leads to the equation

$$N = \int_0^\infty \frac{1}{e^{(E-\mu)/k_B T} - 1} \frac{V(2m)^{3/2}}{4\hbar^3 \pi^2} \sqrt{E}\, dE \quad (7.75)$$

where m is the mass of the atom and V is the volume of the box. This equation is the analogue of (7.48) for N identical fermions in a box with the replacement of (7.71) by (7.73) and the suppression of the factor of two in the density of states that we inserted to account for the two different spin states for a spin-1/2 particle. For simplicity, we are assuming here that the atoms in the box have intrinsic spin 0. Otherwise, we would need to insert a factor of $2s + 1$, where s is the spin of the atom, in the density of states. Equation (7.75) shows how the number of atoms in the box determines the value of the chemical potential.

If we change the variable of integration to $x = E/k_B T$, (7.75) becomes

$$N = \frac{V(2mk_B T)^{3/2}}{4\hbar^3 \pi^2} \int_0^\infty \frac{\sqrt{x}}{e^{x-\mu/k_B T} - 1} dx \qquad (7.76)$$

Notice that the left-hand side of this equation is the number of atoms in the box, which is independent of T. But the right-hand side has a factor of $T^{3/2}$ in front of the integral. This factor can be made arbitrarily small as $T \to 0$. The only way to compensate for a decrease in temperature is to decrease the value of $-\mu/k_B T$ in the integrand. Since μ is a negative quantity if the ground-state energy is taken to be zero, increasing μ, say from -2 eV to -1 eV, causes $-\mu/k_B T$ to decrease. But μ is limited in how much it can grow. It can be infinitesimally close to zero, but it cannot be positive since $(E - \mu)/k_B T$ must be greater than zero. So setting μ to zero is the best we can do in terms of making the integral in (7.76) as large as possible. But the temperature-dependent prefactor in (7.76) can be made arbitrarily small by making T small. Thus (7.76) cannot be satisfied at sufficiently low temperature.

Einstein recognized that the solution to this conundrum was to separate out the ground state, which for the moment we will take to have energy E_0. Assuming that N_0 of the atoms are in the ground state, then the first term in the sum in (7.74) yields

$$\frac{1}{e^{(E_0-\mu)/k_B T} - 1} = N_0 \qquad (7.77)$$

Therefore,

$$\frac{E_0 - \mu}{k_B T} = \ln\left(1 + \frac{1}{N_0}\right) \simeq \frac{1}{N_0} \qquad (7.78)$$

where in the last step we have assumed that the number of atoms in the ground state is large. For a gas with a large number of atoms, say $N \sim 10^{23}$ (as originally considered by Einstein), $1/N_0$ is a very small number even if a relatively small fraction of the atoms have condensed to the ground state.[12] What we see in (7.78) is that μ differs from E_0 by a small amount in comparison with $k_B T$. For $E_0 = 0$ (7.78) becomes

$$\mu \simeq -\frac{k_B T}{N_0} \qquad (7.79)$$

Thus there is a simple relationship between the chemical potential μ and the number of atoms N_0 in the ground state.

For states with energy E other than the ground state

$$E - \mu \simeq E + \frac{k_B T}{N_0} \qquad (7.80)$$

For sufficiently large N_0, we can safely make the approximation

$$E - \mu \simeq E \qquad (7.81)$$

and hence

$$n(E) \simeq \frac{1}{e^{E/k_B T} - 1} \qquad (7.82)$$

---

[12] This approximation works well even for collections of trapped atoms with N as small as 10^6, which has been achieved experimentally.

in the integral over these excited states in (7.74) now that the ground state has been taken care of separately. Notice how (7.82) is reminiscent of the Bose–Einstein distribution for photons [see (7.57)]. In our discussion of cavity radiation, we set α (and hence μ) to zero in the Bose–Einstein distribution function on the grounds that the number of photons in the cavity is not conserved since photons can be emitted and absorbed by the material comprising the walls of the cavity. The same distribution function applies for atoms in the excited energy states of the box because the number of atoms in these excited states is not conserved, since atoms can be exchanged between the excited states and the ground state.

With the ground state separated out and μ set to zero in the integral, (7.76) becomes

$$N = N_0 + \frac{V(2mk_BT)^{3/2}}{4\hbar^3\pi^2} \int_0^\infty \frac{\sqrt{x}}{e^x - 1} dx \tag{7.83}$$

The integral

$$\int_0^\infty \frac{\sqrt{x}}{e^x - 1} dx \tag{7.84}$$

is proportional to the Riemann zeta function $\zeta(3/2)$. Its value is

$$\int_0^\infty \frac{\sqrt{x}}{e^x - 1} dx = 1.3\sqrt{\pi} \tag{7.85}$$

Thus we end up with the result

$$N_0 = N - \frac{1.3V}{4\hbar^3}\left(\frac{2mk_BT}{\pi}\right)^{3/2} \tag{7.86}$$

We define a critical temperature T_C through the relationship

$$\frac{N}{V} = \frac{1.3}{4\hbar^3}\left(\frac{2mk_BT_C}{\pi}\right)^{3/2} \tag{7.87}$$

or

$$T_C = \frac{2\pi\hbar^2}{mk_B}\left(\frac{N}{2.6V}\right)^{2/3} \tag{7.88}$$

Expressed in terms of this critical temperature, (7.86) becomes

$$\frac{N_0}{N} = 1 - \left(\frac{T}{T_C}\right)^{3/2} \tag{7.89}$$

as shown in Fig. 7.20. Since this result was derived under the supposition that the number of atoms in the ground state is large, we need to exercise some caution in applying this result in the close vicinity of T_C. Nonetheless, provided the number of atoms in the box is large, the temperature doesn't need to deviate much from T_C to make N_0 sufficiently

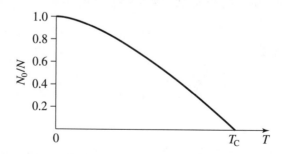

Figure 7.20 The number of atoms in a Bose–Einstein condensate as a function of temperature.

large for our approximation to hold. For example, if T/T_C is 0.99, then (7.89) shows that 1.5% of the atoms would be in the ground state. This jump in the number of atoms in the ground state from essentially zero to a significant fraction of the total number N as the temperature decreases below T_C is a phase transition, reminiscent of the transition that takes place when a gas is cooled below a certain critical temperature and condenses to a liquid. Consequently, the atoms in the ground state are referred to as a **Bose–Einstein condensate**. Unlike the gas-to-liquid phase transition in which the attractive interactions between the atoms or molecules in the gas play a sizable role, Bose–Einstein condensation in a gas is purely a result of the quantum statistics that describes these atoms, since the strength of interaction of the atoms does not appear in our derivation (although some interaction is implicit to the extent that the atoms in the gas are presumed to be in thermal equilibrium).

Experimental confirmation of Einstein's 1925 prediction of Bose–Einstein condensation did not take place until 1995. E. Cornell and C. Wieman used laser and evaporative cooling to lower the temperature of a dilute gas of rubidium atoms to 20 nK, a temperature roughly 300 times lower than any that had been reached previously. They succeeded in trapping several thousand rubidium atoms in a potential energy well roughly 10 μm on a side. The critical temperature for Bose–Einstein condensation was observed to be $T_C = 190 \times 10^{-9}$ K. See Fig. 7.21. A few months later W. Ketterle succeeded in obtaining Bose–Einstein condensation in a gas of 500,000 sodium atoms at densities in excess of 10^{14} cm^{-3}. In this case the critical temperature was ~ 2 μK.

Each of the atoms in the condensate is, of course, in the same quantum state, the ground state of the potential energy well confining these atoms. There is a single wave function for this state. Hence each atom in the condensate occupies the same region of space. The condensate is a macroscopic quantum state, meaning that its quantum properties are observable on a macroscopic scale. Ketterle provided a beautiful demonstration by creating a double-well arrangement in which two separate Bose–Einstein condensates

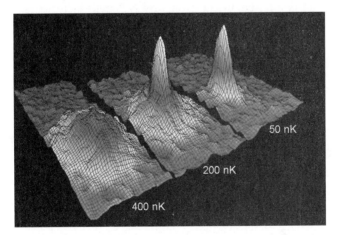

Figure 7.21 Bose–Einstein condensation in a gas of rubidium atoms. The density of atoms in the vapor is plotted vertically as a function of horizontal position. The field of view is 200×270 μm^2. The pictures were taken by illuminating the condensate with a laser and imaging the scattered light. The act of measuring the atom density with the laser destroys the condensate, so a new condensate must be generated for each picture. Because there are thousands of atoms in the condensate with which the light can interact, a single laser flash can map out the probability density. The sharp peak in atom density indicates the formation of the condensate at 200 nK. See M. H. Anderson et al., *Science* **269**, 198 (1995). (Courtesy M. Matthews)

Figure 7.22 The interference fringes with a period of 15μm that are observed when two freely expanding Bose–Einstein condensates of atomic sodium overlap. Reproduced with permission from M. R. Andrews et al., *Science* **275**, 637 (1997). Copyright 1997 American Association for the Advancement of Science.

resided. When the potential confining the atoms was switched off, the two condensates expanded and overlapped, producing interference fringes, as shown in Fig. 7.22. The position of the fringes shifted from one repetition to the next as the relative phase between the wave functions for the two condensates varied.[13]

As the 70-year interval between the prediction of Bose–Einstein condensation and its experimental realization attests, it has not been easy to achieve Bose–Einstein condensation in a dilute gas. And it is a very strange phenomenon as well, which is probably one of the reasons it has attracted so much attention. At first, you may be tempted to think that getting all the atoms to pile into the ground state should be straightforward. Unlike the case with fermions in which it is not possible to have more than one fermion in a state, bosons have no such inhibitions. Thus you might expect that if the temperature were lowered sufficiently the atoms would naturally end up in the ground state, presuming, of course that they have integral intrinsic spin. But for a macroscopic container, say the one in the Cornell–Wieman experiment, the spacing between energy levels is extremely small, roughly 10^{-13} eV. Setting this energy difference equal to $k_B T$ yields a temperature that is less than a nanokelvin. Thus it would appear that in order to have a significant fraction of the atoms in the ground state would require temperatures that are smaller still. But Bose–Einstein condensation is taking place at temperatures that are as much as a thousand times hotter. At these "high" temperatures (in the microkelvin range), whether atoms are fermions or bosons should have little impact on the distribution because each energy state should be sparsely populated. Moreover, the spacing between atoms in these containers, or traps as they are typically called, is roughly 10,000 times the size of the atoms themselves, which is why it is a reasonable approximation to neglect the interaction of these atoms. But this raises the question as to how the atoms "know" to pile into the ground state once the critical temperature is reached.

To give us more insight into Bose–Einstein condensation, it is instructive to express the condition required for condensation in terms of the de Broglie wavelength $\lambda = h/p$. As the temperature is lowered, the de Broglie wavelength increases since the average kinetic energy of the atoms is given by

$$\frac{1}{2}mv^2 = \frac{3}{2}k_B T \tag{7.90}$$

Thus the average de Broglie wavelength for the atoms in the gas is given by

$$\lambda = \frac{h}{mv} = \frac{h}{\sqrt{3mk_B T}} \tag{7.91}$$

Expressed in terms of this wavelength, the critical temperature for Bose–Einstein condensation is reached when

$$\frac{N}{V} \sim \frac{1}{\lambda^3} \tag{7.92}$$

that is, when the spacing between atoms is comparable to their de Broglie wavelength.[14] This is a good figure of merit for assessing when quantum effects become pronounced.

---

[13] A nice discussion of atom trapping and laser cooling is given by C. Wieman, *Am. J. Phys.* **64**, 847 (1996). Wieman, Cornell, and Ketterle were awarded the Nobel Prize in physics in 2001 for their achievements in Bose–Einstein condensation. Ketterle's Nobel Prize lecture is well worth reading, too.

[14] The experts in the field define a thermal de Broglie wavelength by the relationship $\lambda = h/\sqrt{2\pi m k_B T}$, in which case condensation occurs when $N/V = 2.6/\lambda^3$.

As we have seen, achieving this condition in a dilute gas requires temperatures in the microkelvin range. In liquid helium, on the other hand, the separation between atoms in the liquid is comparable with the de Broglie wavelength of these atoms when the temperature is in the kelvin range (see Problem 7.30), a substantially higher temperature than that required for Bose–Einstein condensation in a gas. Helium was first liquified in 1908 by Kamerlingh Onnes. The transition from gas to liquid occurs at 4.2 K. At 2.2 K helium atoms begin to form a condensate, a macroscopic quantum state with truly unusual properties. Most strikingly, the condensate is a **superfluid**, a fluid with zero viscosity, that is, no resistance to fluid flow. An analogous phase transition occurs in solids that exhibit superconductivity. We will look briefly at this phenomenon at the end of the next chapter.

7.8 Lasers

We conclude this chapter with a discussion of the laser, another striking example of a macroscopic quantum state that results from the indistinguishability of identical bosons, in this case photons. In 1917, in an article titled "On the Quantum Theory of Radiation," Albert Einstein laid out the basic principles of absorption and of spontaneous and stimulated emission of electromagnetic radiation in its interaction with atoms. Einstein's work established the fundamental principles responsible for laser action, although the first operational laser was not invented until 1960.[15] As we will see, Einstein took good advantage of the Planck blackbody, or cavity, radiation formula.

For simplicity, consider atoms with two nondegenerate energy states with energies E_1 and E_2, as indicated in Fig. 7.23. In making a transition from the higher energy state to the lower one the atom emits a photon with energy $h\nu = E_2 - E_1$. Suppose that these atoms make up the walls of a cavity at temperature T. In thermal equilibrium, the ratio of the number of atoms in each of these energy states is given by the Boltzmann distribution (7.43):

$$\frac{N_2}{N_1} = e^{-(E_2 - E_1)/k_\text{B} T} = e^{-h\nu/k_\text{B} T} \tag{7.93}$$

These equilibrium populations are maintained by the absorption and the emission of radiation in the cavity. There are three processes to consider. First, the atoms in the state

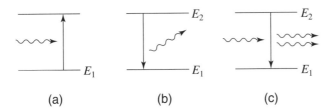

Figure 7.23 A schematic energy-level diagram illustrating (a) absorption, (b) emission, and (c) stimulated emission.

---

[15]Since then the laser has become such an invaluable and commonplace tool in science and technology that we may be tempted to take its unusual quantum properties for granted.

with energy E_1 can absorb radiation and make a transition to the excited state. The rate for this process is proportional to N_1, the number of atoms in the state with energy E_1, and to the energy density of the radiation in the cavity with the right frequency, namely $\rho(\nu, T)$. The constant of proportionality is referred to as the Einstein B coefficient, here called B_{12} to emphasize that we are talking about transitions from E_1 to E_2. This absorption of radiation and excitation of the atom are balanced by two processes. First there is the possibility of induced, or stimulated, emission of radiation. As was the case for absorption, the rate for this process is proportional to N_2, the number of atoms in the excited energy state, and to the energy density $\rho(\nu, T)$. The constant of proportionality is taken to be B_{21}, since the transition starts in the state with energy E_2 and ends in the state with energy E_1. In addition to stimulated emission, there is a rate for spontaneous emission, emission that would take place even if the atoms were isolated and not interacting with the radiation in the cavity. The rate for spontaneous emission is proportional to N_2. The coefficient of proportionality is A_{21}, the Einstein A coefficient. Putting everything together, we see that in thermal equilibrium

$$N_1 \rho(\nu, T) B_{12} = N_2 [B_{21} \rho(\nu, T) + A_{21}] \tag{7.94}$$

Taking advantage of (7.93), we find

$$\rho(\nu, T) = \frac{A_{21}}{B_{12} e^{h\nu/k_B T} - B_{21}} \tag{7.95}$$

If we compare this result with the cavity radiation formula

$$\rho(\nu) = \frac{8\pi h \nu^3}{c^3 \left(e^{h\nu/k_B T} - 1\right)} \tag{7.96}$$

we see that

$$B_{21} = B_{12} = B \tag{7.97}$$

and

$$\frac{A_{21}}{B} = \frac{8\pi h \nu^3}{c^3} \tag{7.98}$$

With these results, we find the total rate for emission of a photon to be

$$N_2 [B_{21} \rho(\nu, T) + A_{21}] = N_2 B \frac{8\pi h \nu^3}{c^3} \left(\frac{1}{e^{h\nu/k_B T} - 1} + 1 \right)$$
$$= N_2 B \frac{8\pi h \nu^3}{c^3} [n(h\nu) + 1] \tag{7.99}$$

The important conclusion that we can draw from this result is that the combined rate for spontaneous and stimulated emission of radiation with frequency ν is proportional to $n + 1$, where n is the number of photons in the gas with that frequency.

Although our derivation of this $n + 1$ factor depended on thermal equilibrium between the atoms and the cavity radiation, this result is more general than our derivation suggests. This enhancement factor really results from the fact that photons are bosons, that once there are n photons in a state the odds of putting an additional photon in that same state are larger by a factor of $n + 1$ than they would be without the presence of the n photons.

While a rigorous proof requires the use quantum field theory, we can get the basic idea of how the overall symmetry under exchange contributes to this enhancement by examining the position–space wave functions

$$\Psi_S(x_1, x_2) = \frac{1}{\sqrt{2}} \left[\psi_\alpha(x_1)\psi_\beta(x_2) + \psi_\beta(x_1)\psi_\alpha(x_2) \right] \quad (7.100)$$

and

$$\Psi_S(x_1, x_2, x_3) = \frac{1}{\sqrt{6}} \left[\psi_\alpha(x_1)\psi_\beta(x_2)\psi_\gamma(x_3) + \psi_\beta(x_1)\psi_\gamma(x_2)\psi_\alpha(x_3) \right.$$
$$+ \psi_\gamma(x_1)\psi_\alpha(x_2)\psi_\beta(x_3) + \psi_\gamma(x_1)\psi_\beta(x_2)\psi_\alpha(x_3)$$
$$\left. + \psi_\beta(x_1)\psi_\alpha(x_2)\psi_\gamma(x_3) + \psi_\alpha(x_1)\psi_\gamma(x_2)\psi_\beta(x_3) \right] \quad (7.101)$$

These multiparticle wave functions could apply to identical spin-0 particles, since the wave functions for such bosons must be symmetric under exchange. Notice that if we ask for the probability density for two bosons to have the same position ($x_1 = x_2$) in the two-particle wave function (7.100), we obtain

$$|\Psi_S(x_1, x_1)|^2 = 2|\psi_\alpha(x_1)\psi_\beta(x_1)|^2 \quad (7.102)$$

and similarly for the three bosons to have the same position ($x_1 = x_2 = x_3$) in the three-particle wave function (7.101), we find

$$|\Psi_S(x_1, x_1, x_1)|^2 = 6|\psi_\alpha(x_1)\psi_\beta(x_1)\psi_\gamma(x_1)|^2 \quad (7.103)$$

Thus the probability density is $2 = 2 \times 1$ times and $6 = 3 \times 2 \times 1$ times greater, respectively, than it would be if the particles were distinguishable and the wave functions did not need to be symmetric under particle exchange. In general, the probability density for having n such bosons at the same position would be $n!$ larger than if the particles were distinguishable. Thus $P_{\text{boson}}^n = n! P_n$ and $P_{\text{boson}}^{n+1} = (n+1)! P_{n+1}$, where P_n and P_{n+1} are the probabilities for distinguishable particles. If we write $P_{n+1} = P_1 P_n$, where P_1 is the probability of adding an additional particle to a state with n particles if the particles are distinguishable, we see that $P_{\text{boson}}^{n+1} = (n+1)! P_{n+1} = (n+1) P_1 n! P_n = (n+1) P_1 P_{\text{boson}}^n$, indicating the probability of adding an additional boson to a collection of n bosons is enhanced by a factor of $n + 1$ beyond what we would expect if the particles were distinguishable.

It is this $n + 1$ enhancement factor that encourages atoms with integral intrinsic spin to avalanche into the ground state in Bose–Einstein condensation. It is also this enhancement factor that produces laser action. The term laser derives from the acronym Light Amplification by Stimulated Emission of Radiation. Typically, when an atom makes a transition from an excited state to one with a lower energy, it can emit a photon in many directions. If the atom is located in a resonant cavity that is formed by two mirrors, as shown in Fig. 7.24, then only those photons with a particular value of the wave vector **k** are trapped in the cavity. Photons emitted in other directions are eventually absorbed in the cavity walls. Thus the number of photons in the cavity with this particular value of **k** grows as more excited atoms emit photons. But this growth in the number of photons with this particular value of **k** eventually makes it overwhelmingly likely that the photons emitted in subsequent decays will be ones with the same value of **k** because of the $n + 1$ enhancement factor.

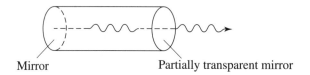

Figure 7.24 Two reflecting surfaces form a resonant cavity for specific modes of a laser.

Although the photons that are emitted by the atom arise from transitions between two states with different energies, say energies E_1 and E_2, it is not possible to construct a laser with a two-level atom. Recall from (7.94) that the rate for stimulated emission of radiation is proportional to the Einstein B coefficient B_{21} and the number of atoms N_2 in the excited state. The rate for absorption of radiation is proportional to the Einstein B coefficient B_{12} and the number of atoms N_1 in the lower energy state. Since $B_{21} = B_{12}$ and $N_1 > N_2$ in thermal equilibrium, as (7.93) shows, more photons will initially be absorbed than are emitted. This net absorption of energy will cause the populations N_1 and N_2 to equalize, at which point there will be as many photons emitted as absorbed. What is needed to obtain a net amplification of the radiation is a **population inversion** in which more of the atoms are in the higher energy state. As we will now show, it is possible to generate such an inversion with a three-level or a four-level system.

The first successful laser for visible light used a ruby rod containing chromium ions.[16] Figure 7.25a shows an energy-level diagram for a three-level laser such as chromium. In a process known as pumping, energy from a flash lamp is used to excited chromium ions to a short-lived excited state. Transitions then occur to a second state which is metastable, meaning this state lives for an unusually long time before it decays. In this way, a population inversion is obtained between the ground state and this excited metastable state. Transitions between these two states are used for lasing. When the ions have returned to the ground state, the process is repeated. Thus, this laser, like most solid-state lasers, is a pulsed laser.

A pulsed laser such as the chromium laser is inherently inefficient since the final state in the lasing transition is the ground state, in which most of the atoms tend to reside. It is therefore necessary to use a lot of energy to empty this state to the point where there are more atoms in the excited state than the ground state for laser action to occur. An alternative is the four-level system illustrated in Fig. 7.25b. In this case, it is much easier

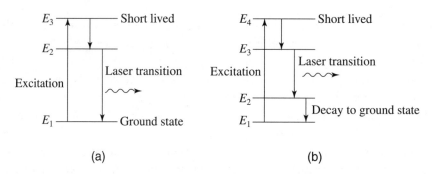

Figure 7.25 Energy-level diagrams for (a) a three-level laser and (b) a four-level laser.

---

[16] The laser was preceded by the maser, namely Microwave Amplification by Stimulated Emission of Radiation. The first maser was constructed by Charles Townes in 1954.

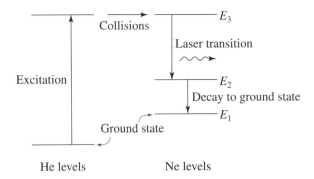

Figure 7.26 An energy-level diagram for the helium-neon laser.

to maintain a population inversion since the final state in the laser transition is an excited state which is not heavily populated.

The final state in the laser transition for the widely used He–Ne laser is also an excited state, in this case of the neon atom. The He–Ne laser is pumped by a steady electric discharge in the gas. Through collisions with electrons in the gas helium atoms are excited to a metastable excited state whose energy is very close to an excited state of neon. Collisions of helium and neon atoms have a high probability of transferring this energy to an excited state of neon (which itself is not readily excited by electron collisions). See Fig. 7.26. The laser transition is from this excited state to an energy level below it, which is mostly empty because of its fast transition to the ground state. Thus the condition $N_3 > N_2$ can be achieved with the modest power requirements needed to maintain the discharge. Gas lasers such as the He–Ne laser are called continuous wave (CW) lasers since their output is steady.

7.9 Summary

Identical particles are indistinguishable. This indistinguishability has profound consequences for the behavior of multiparticle systems in quantum mechanics.

Particles with integral intrinsic spin ($s = 0, 1, 2, \ldots$) are called bosons. The total "wave function" for a system of identical bosons must be symmetric under exchange of any two of the particles. Valid two-particle states include

$$\Psi_S(1, 2) = \psi_\alpha(1)\psi_\alpha(2) \tag{7.104}$$

and

$$\Psi_S(1, 2) = \frac{1}{\sqrt{2}}\left[\psi_\alpha(1)\psi_\beta(2) + \psi_\beta(1)\psi_\alpha(2)\right] \tag{7.105}$$

Bosons are more likely to be in the same state than are distinguishable particles. The average number of identical bosons in a state with energy E in thermal equilibrium at temperature T is given by

$$n(E) = \frac{1}{e^{(E-\mu)/k_B T} - 1} \tag{7.106}$$

where μ is the chemical potential. This distribution function is called the Bose–Einstein distribution function.

Particles with half-integral intrinsic spin ($s = 1/2, 3/2, 5/2, \ldots$) are called fermions. The total "wave function" for a system of identical fermions must be antisymmetric under exchange of any two of the particles. A valid two-particle state is given by

$$\Psi_A(1, 2) = \frac{1}{\sqrt{2}} \left[\psi_\alpha(1)\psi_\beta(2) - \psi_\beta(1)\psi_\alpha(2) \right] \quad (7.107)$$

Identical fermions cannot be in the same state, as antisymmetric states such as (7.107) vanish when the quantum numbers for two of the fermions are the same ($\alpha = \beta$). In this way we see how the Pauli principle arises. The average number of identical fermions in a state with energy E in thermal equilibrium is given by

$$n(E) = \frac{1}{e^{(E-E_F)/k_B T} + 1} \quad (7.108)$$

where E_F is the Fermi energy. This distribution function is called the Fermi–Dirac distribution function.

Consequences of the fact that electrons are fermions and are limited to one per state include the chemical properties of the elements, the degree of compressibility of metals, and the stability of white dwarf stars. In the latter example, the electron degeneracy pressure can stave off gravitational collapse of a star after the star has exhausted its nuclear fuel unless the mass of the star is too large.

Consequences of the fact that photons are bosons include the distribution of electromagnetic energy density in a cavity, the famous blackbody distribution function,

$$\rho(\nu)\, d\nu = \frac{8\pi h \nu^3}{c^3 \left(e^{h\nu/k_B T} - 1 \right)} d\nu \quad (7.109)$$

and the fact that stimulated emission can generate an intense beam of monochromatic (coherent) light in a laser. For atoms that are bosons, (7.106) coupled with conservation of particle number means that for temperatures less than a certain critical temperature the atoms pile into the ground state, forming a Bose–Einstein condensate. In the case of liquid helium, this condensate exhibits the property of superfluidity, that is fluid flow without viscosity.

Problems

7.1. Verify that the states

$$\chi_+(1)\chi_+(2)$$
$$\tfrac{1}{\sqrt{2}}(\chi_+(1)\chi_-(2) + \chi_-(1)\chi_+(2))$$
$$\chi_-(1)\chi_-(2)$$

are eigenstates of the z component of total spin $S_z = S_{1_z} + S_{2_z}$ with eigenvalues $\hbar$, 0, and $-\hbar$, respectively.

7.2. Following the procedure outlined in Example 7.3, show that $\chi_+(1)\chi_+(2)$ is an eigenstate of total spin with $S = 1$.

7.3. The spatial wave functions for two identical particles in the one-dimensional box (see Section 7.1) are given by

$$\Psi_S(x_1, x_2) = \frac{1}{\sqrt{2}} [\psi_1(x_1)\psi_2(x_2) + \psi_2(x_1)\psi_1(x_2)]$$

and

$$\Psi_A(x_1, x_2) = \frac{1}{\sqrt{2}} [\psi_1(x_1)\psi_2(x_2) - \psi_2(x_1)\psi_1(x_2)]$$

where one of the particles is in the ground state and one is in the first excited state. Calculate the probability that measurement of the positions of the two particles finds them both in the left-hand side of the box, that is measurement yields $0 < x_1 < L/2$ and $0 < x_2 < L/2$. Notice how the probability is significantly larger for the symmetric state Ψ_S than for the antisymmetric state Ψ_A. Thus the particles behave as if they attract each other in the symmetric state and repel each other in the antisymmetric state even though the Hamiltonian for the two particles does not include any interaction term. Heisenberg called these fictitious forces of attraction and repulsion **exchange forces**.

7.4. Show that the total ground-state energy of N fermions in a three-dimensional box is given by

$$E_{\text{total}} = \frac{3}{5} N E_F$$

Thus the average energy per fermion is $3E_F/5$.

7.5. Given that the mass of the Sun is $M = 2 \times 10^{30}$ kg, estimate the number of electrons in the Sun. Assume the Sun is composed primarily of hydrogen. In a typical white dwarf star, this number of electrons is contained in a sphere of radius 6000 km. Find the Fermi energy of the electrons in electron volts, assuming the electrons can be treated nonrelativistically. How does this energy compare with mc^2, the rest mass energy of the electron? Consequently, how reliable is your calculation of the Fermi energy?

7.6. (*a*) Calculate (i) the Fermi energy, (ii) the Fermi velocity, and (iii) the Fermi temperature for gold at 0 K. The density of gold is 19.32 g/cm^3 and the molar weight is 197 g/mole. Assume each gold atom contributes one "free" electron to the Fermi gas. (*b*) In a cube of gold 1 mm on an edge, calculate the approximate number of conduction electrons whose energies lie in the range from 4.000 to 4.025 eV.

7.7. (*a*) In the highly relativistic limit such that the total energy E of an electron is much greater than the electron's rest mass energy $(E \gg mc^2)$, $E \simeq pc = \hbar kc$, where $k = \sqrt{k_x^2 + k_y^2 + k_z^2}$. Determine the Fermi energy for a system for which essentially all the N electrons may be assumed to be highly relativistic. Show that (up to an overall multiplicative constant) the Fermi energy is roughly

$$E_F \sim \hbar c \left(\frac{N}{V}\right)^{1/3}$$

where N/V is the density of electrons. What is the multiplicative constant? *Note*: Take the allowed values of k_x, k_y, and k_z to be the same for the relativistic fermion gas, say in a cubic box, as for the nonrelativistic gas. (*b*) Calculate the zero-point pressure for the relativistic fermion gas. Compare the dependence on density for the nonrelativistic and highly relativistic approximations. Explain which gas is "stiffer," that is, more difficult to compress? Recall that

$$P = -\frac{dE_{\text{total}}}{dV}$$

7.8. (*a*) Solve the Schrödinger equation for an electron confined to a two-dimensional square box where the potential energy is given by

$$V(x, y) = \begin{cases} 0 & 0 < x < L, 0 < y < L \\ \infty & \text{elsewhere} \end{cases}$$

Determine the normalized energy eigenfunctions and eigenvalues. (*b*) Show that the Fermi energy for nonrelativistic electrons (treated as if they do not interact with each other) confined in the two-dimensional square box is given by

$$E_F = \frac{\pi \hbar^2}{m} \left(\frac{N}{L^2}\right)$$

where N is the number of electrons, L is the length of the side of the square, and m is the mass of an electron. Such confinement to a plane happens, for example, for electrons in the layered materials that are used to make high-temperature superconductors.

7.9. Determine the average energy of an electron in a two-dimensional square box containing N electrons.

7.10. (*a*) Determine the Fermi energy for a system of N electrons in a one-dimensional box of length L. (*b*) What is the average energy of an electron in this one-dimensional box?

7.11. Calculate the compressibility of lithium assuming that it arises completely from electron degeneracy pressure. Compare with the experimental value of 8.3×10^{-11} m^2/N. *Suggestion*: See Section 7.4. The density of lithium is 0.53 g/cm^3.

7.12. Consider the following two microstates for ten identical particles. In one of the microstates there are ten particles in the ground state and none in the excited state, while in the other microstate there are five particles in the ground state and five in the excited state. The "statistical weight" of these two microstates is 1 to 1. If the particles are distinguishable, there is still just one microstate corresponding to the ten particles being in the ground state. Calculate the number of microstates for the case of five particles in the ground state and five in the excited state if the particles are distinguishable. What is the statistical weight of these two configurations for distinguishable particles?

7.13. Show that the spectral radiancy of a blackbody $R_T(\nu)\,d\nu$ and the energy density $\rho(\nu)\,d\nu$ of cavity radiation are related by

$$R_T(\nu) = \frac{c}{4}\rho(\nu)$$

Suggestion: Place the origin of your coordinate system at the center of a hole of cross-sectional area ΔA. Consider the radiation in a hemisphere of radius $c\Delta t$ in the cavity. Note that the fraction of the radiation in a volume $r^2 dr \sin\theta\, d\theta\, d\phi$ that exits through the hole is

$$\frac{\Delta A \cos\theta}{4\pi r^2}$$

Integrate over the hemisphere and divide your result by $\Delta A \Delta t$ to obtain the energy per unit area per unit time radiated by the hole.

7.14. Verify that the components of the electric field in (7.51) satisfy the boundary condition that the tangential component of the electric field vanish at the boundaries of the box. Why should the normal components of the electric field not vanish at the boundaries?

7.15. Show that

$$\lambda_{\max} T = 0.2014 \frac{hc}{k_B}$$

Hint: The equation $x = 5 - 5e^{-x}$ can be solved iteratively.

7.16. Evaluate the blackbody (or cavity) radiation formula

$$\rho(\nu)\,d\nu = \frac{8\pi h \nu^3}{c^3 \left(e^{h\nu/k_B T} - 1\right)} d\nu$$

in the high-temperature limit. Show that the result reduces to the number of modes per unit volume between ν and $\nu + d\nu$ multiplied by $k_B T$. The resulting distribution, which could be derived from classical physics using the equipartition theorem, was referred to as the **ultraviolet catastrophe**. Can you explain why?

7.17. A tungsten sphere of 2.30 cm in diameter is heated to 2000°C. At this temperature tungsten radiates only about 30% of the energy radiated by a blackbody of the same size and temperature. (*a*) Calculate the temperature of a perfectly black spherical body of the same size that radiates at the same rate as the tungsten sphere. (*b*) Calculate the diameter of a perfectly black spherical body at the same temperature as the tungsten that radiates at the same rate.

7.18. Determine the radius of the star Procyon B from the following data: the flux of starlight reaching us from Procyon B is 1.7×10^{-12} W/m², the distance of the star from us is 11 light years, its surface temperature is 6600 K, and its mass is 65% of the mass of the Sun. Assume the star radiates like a blackbody.

7.19. The star ζ Pup in the constellation Puppus is a hot star, with a surface temperature that is 10 times hotter than that of the Sun. It has a radius that is a factor of 10 greater than that of the Sun. (*a*) Treat the star as a blackbody and estimate by comparison with the Sun a numerical value for the wavelength at which the emission of electromagnetic radiation from ζ Pup is a maximum. In what portion of the electromagnetic spectrum is this? (*b*) Compare the luminosity of ζ Pup—the total rate electromagnetic energy lost by the star—with the luminosity of the Sun, that is, evaluate the ratio of the luminosities.

7.20. (*a*) The star Stein 2051 B is 18 light-years from the earth. Its surface temperature is 7050 K. The flux of starlight reaching us on earth from Stein 2051 B is 2.8×10^{-13} W/m². From this information, determine the radius of the star assuming it radiates like a blackbody. (*b*) Compare the radius of Stein 2051 B with the radius of the earth (6.4×10^3 km). The mass of Stein 2051 B is $0.48 M_{\text{Sun}}$. This star is a white dwarf, the end phase of stars like the Sun after they have exhausted their nuclear fuel. Explain the mechanism responsible for stability of a white dwarf star. In particular, without nuclear fuel, how does the star withstand the gravitational pressure to collapse?

7.21. The average distance of the outermost "planet" Pluto from the Sun is one hundred times that of the innermost planet Mercury. Suppose that each planet absorbs and radiates energy as an ideal blackbody, that the Sun radiates as an ideal blackbody, and that the only significant source of energy on each of these planets comes from the absorbed sunlight. Find the ratio of the surface temperature (in kelvin) of Pluto to that of Mercury, assuming temperatures are uniform over the surface of each planet and that each planet is at its equilibrium temperature.

7.22. (*a*) Assuming that the Sun may be treated as a blackbody with a surface temperature of 5800 K, determine the mass lost per second to radiation by the Sun. Take the Sun's diameter to be 1.4×10^9 m. What fraction of the rest mass of the Sun is lost each year via electromagnetic radiation? Take the mass of the Sun to be 2×10^{30} kg. (*b*) The solar constant is the energy per unit area per unit time incident on the Earth from the Sun in a plane perpendicular to the incident radiation. Determine the value of the solar constant. The distance of the Earth from the Sun is 1.5×10^{11} m.

7.23. The star Hadar has a surface temperature $T = 24,000$ K and a luminosity that is roughly 39,000 times larger than the luminosity of the Sun. Determine the radius of Hadar (in terms of the radius of the Sun), assuming that Hadar radiates as a blackbody.

7.24. (*a*) Estimate the temperature of the Earth assuming the Sun and Earth are blackbodies. Take the surface temperature of the Sun to be $T_{\rm Sun} = 5800$ K, the radius of the Sun to be $R_{\rm Sun} = 7 \times 10^5$ km, and the Sun–Earth distance to be $D = 1.5 \times 10^8$ km. Does your answer depend on the size of the Earth? (*b*) Actually the Earth reflects 30% of the visible light incident upon it—that is, the Earth's albedo is 0.30. How does this fact alter your estimate of the Earth's temperature? (*c*) Finally, assume the Earth is surrounded by a thin spherical shell of gas that is transparent to the Sun's energy but opaque to infrared radiation. What then is the temperature of the Earth's surface? *Note:* The shell radiates energy both into space and back toward the surface of the Earth, as indicated in Fig. 7.27. This problem illustrates the origin of the greenhouse effect.

7.25. Determine the density of a gas of Rb atoms when it undergoes Bose–Einstein condensation at a critical temperature of 190×10^{-9} K.

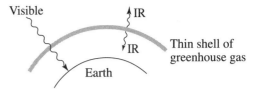

Figure 7.27 The greenhouse effect.

7.26. (*a*) The critical temperature at which Bose–Einstein condensation occurs for ^{87}Rb atoms confined in a cubic volume 10 μm on a side is 190×10^{-9} K. These are roughly the conditions that existed in the Cornell–Wieman experiment discussed in Section 7.7. What is the temperature when $N_0 = 2000$ atoms are in the condensate? (*b*) Compare the value of $\mu = -k_{\rm B}T/N_0$ with the energy of the first excited state of the box. Is it reasonable to neglect μ relative to E for the excited states?

7.27. Determine the fraction of atoms in the Bose–Einstein condensate when the temperature reaches one-half the critical temperature.

7.28. Determine the value of T such that $k_{\rm B}T$ is roughly equal to the spacing between the ground state and the first excited state for a cubic box 10 μm on a side. How does this value for T compare with the critical temperature for Bose–Einstein condensation for a gas of 5×10^5 sodium atoms confined in the box? These are roughly the conditions that existed in the Ketterle experiment discussed in Section 7.7.

7.29. By examining the behavior of the integrand in (7.83) for low energy, show that $x = 0$ (and hence $E = 0$) does not make a finite contribution to the integral. Thus the ground state is not being counted twice in the evaluation of (7.83).

7.30. The density of liquid helium 0.15 g/cm^3. Determine the temperature at which the de Broglie wavelength of the helium atoms satisfies the condition

$$\frac{N}{V} \sim \frac{1}{\lambda^3}$$

Note: A phase transition to a superfluid condensate is observed to occur in liquid helium at 2.2 K.

CHAPTER 8

Solid-State Physics

Given the large number of atoms in a solid, it may not seem feasible to apply basic quantum mechanics to such a complex system. However, for crystalline solids—solids for which the atoms exist in a periodic array, or lattice—the periodicity of the lattice dictates many of the key attributes of the solid. In particular, we will focus in this chapter on the capacity of these crystalline solids to conduct electricity. Crystalline solids that are semiconductors are especially important because they serve as the building blocks of the electronic/computer revolution in which we are all participating. In this chapter we will see how quantum mechanics provides the foundation for this revolution.

8.1 The Band Structure of Solids

We start our discussion of the quantum mechanics of crystalline solids with the Schrödinger equation

$$-\frac{\hbar^2}{2m}\frac{d^2\psi}{dx^2} + V\psi = E\psi \tag{8.1}$$

To keep things as simple as possible, we restrict ourselves to a one-dimensional solid. Figure 8.1 shows a sketch of the sort of potential energy $V(x)$ that might be appropriate. If we were dealing with solid hydrogen, a typical molecular solid, the regions where the potential energy drops sharply would be close to the nuclei (the protons for solid hydrogen), given the attractive force between the electrons in the solids and the positively charged nuclei. Solving the Schrödinger equation for this curvy potential energy is not easy. We might try replacing it with a series of square potential energy wells separated by square barriers, as shown in the middle of Fig. 8.1. This model, often referred to as the Kronig–Penney model, is reminiscent of our first pass at explaining molecular physics in Section 4.5. As we did in Section 4.5, we will go one step further and make the model even simpler by replacing the square barriers with Dirac delta functions. You may think we are far removed from a realistic model of even a one-dimensional solid, but the crucial feature that we have retained in our model is the periodic nature of the potential energy function. And this feature goes a surprisingly long way toward explaining the most important properties of a crystalline solid, as we will see.

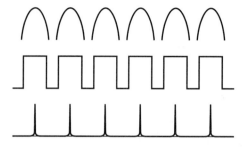

Figure 8.1 Periodic potentials. The top line is meant to convey the sort of realistic potential energy that an electron would experience in a one-dimensional solid. The bottom line is a rather extreme example of a periodic potential in which the square potential barriers shown in the middle are replaced with Dirac delta functions.

The natural extension of the potential energy of the delta function molecule in Section 4.5 to that for a solid is

$$\frac{2mV(x)}{\hbar^2} = \frac{\alpha}{a} \sum_{n=1}^{N} \delta(x - na) \tag{8.2}$$

where a is the separation between each of the delta functions, α is a dimensionless parameter that we are free to adjust, and N is the number of "atoms" in this one-dimensional "crystal." Of course, if we simply push ahead in our solution to the Schrödinger equation for even this potential energy, we face the daunting problem of solving the equation in a *large* number of regions between the delta function barriers and then applying the appropriate conditions on the continuity of the wave function and its derivative at each of the boundaries. Instead, we introduce the **Bloch ansatz**

$$\psi(x + a) = e^{i\theta}\psi(x) \tag{8.3}$$

As we would expect from the translational symmetry of the system as $x \to x + a$, this ansatz indicates that the probability of finding the particle in each of the regions between the barriers should be the same since

$$\psi^*(x + a)\psi(x + a) = e^{-i\theta}\psi^*(x)e^{i\theta}\psi(x) = \psi^*(x)\psi(x) \tag{8.4}$$

You may be troubled by starting our solution with an ansatz, a guess if you will, albeit a seemingly reasonable one. You can take comfort in the fact that if our ansatz were not a valid one, we would not be able to find a solution to the Schrödinger equation that satisfies the ansatz.[1]

We need to deal with one last issue before proceeding toward a solution. What happens when we get to the end of our one-dimensional solid? One nice strategy that retains the translational symmetry of the system and allows us to take full advantage of the Bloch ansatz is to presume that

$$\psi(x + Na) = \psi(x) \tag{8.5}$$

---

[1] The fact that this particular ansatz has a Nobel Prize winner's name associated with it should provide encouragement.

This would be equivalent to connecting our one-dimensional solid in a ring and requiring that doing one complete traversal of the ring should return us to the starting point. This sort of boundary condition is typically referred to as a periodic boundary condition and coupled with (8.3) leads to the requirement

$$e^{iN\theta} = 1 \tag{8.6}$$

which is satisfied if

$$N\theta = 2\pi n_x \qquad n_x = 0, \pm 1, \pm 2, \ldots \tag{8.7}$$

or

$$\theta = \frac{2\pi n_x}{N} \qquad n_x = 0, \pm 1, \pm 2, \ldots \tag{8.8}$$

Each allowed value of θ differs from the adjacent one by a factor of $2\pi/N$. Since the number N of atoms in a solid is very large, the allowed values of θ are discrete but very closely spaced. Also note that when $n_x = N$, $\theta = 2\pi$ and of course $e^{i2\pi} = 1$. Therefore for this value of θ the Bloch ansatz reduces to $\psi(x+a) = \psi(x)$, repeating the $\theta = 0$ condition. Thus there are N distinct values of θ ($n_x = 0, 1, \ldots, N-1$). We will now show that each of these N values of θ corresponds to a distinct value for the energy.

We write the Schrödinger equation in those regions in which $V = 0$ as

$$\frac{d^2\psi}{dx^2} = -k^2\psi \tag{8.9}$$

where we have defined

$$\frac{\sqrt{2mE}}{\hbar} = k \tag{8.10}$$

as usual. We could write the solution simply as

$$\psi(x) = A \sin kx + B \cos kx \tag{8.11}$$

but it is more convenient to write the solution in two adjacent regions (see Fig. 8.2) in the form

$$\psi(x) = A_n \sin k(x - na) + B_n \cos k(x - na) \qquad (n-1)a \leq x \leq na \tag{8.12}$$

and

$$\psi(x) = A_{n+1} \sin k[x - (n+1)a] + B_{n+1} \cos k[x - (n+1)a] \qquad na \leq x \leq (n+1)a \tag{8.13}$$

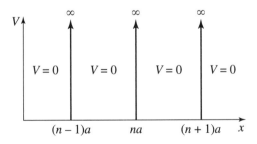

Figure 8.2 A view of the regions $(n-1)a < x < na$ and $na < x < (n+1)a$ in the periodic potential.

The wave functions (8.12) and (8.13) are equivalent to (8.11), as can be verified by using the identities $\sin(\theta - \phi) = \sin\theta\cos\phi - \cos\theta\sin\phi$ and $\cos(\theta - \phi) = \cos\theta\cos\phi + \sin\theta\sin\phi$. The rationale for writing the wave functions in this way is that it leads to the relatively straightforward result

$$A_{n+1} = e^{i\theta} A_n \qquad B_{n+1} = e^{i\theta} B_n \tag{8.14}$$

when the Bloch ansatz is imposed.

We are now set to impose the appropriate conditions on the wave function at the boundaries between the regions. Continuity of the wave function at $x = na$ dictates that

$$B_n = -A_{n+1} \sin ka + B_{n+1} \cos ka \tag{8.15}$$

As we saw in Section 4.4, the derivative of the wave function at $x = na$ is not continuous but rather satisfies the condition

$$\left(\frac{d\psi}{dx}\right)_{na^+} - \left(\frac{d\psi}{dx}\right)_{na^-} = \frac{\alpha}{a}\psi(na) \tag{8.16}$$

which leads to

$$k A_{n+1} \cos ka + k B_{n+1} \sin ka - k A_n = \frac{\alpha}{a} B_n \tag{8.17}$$

If we introduce the dimensionless parameter

$$\tilde{\alpha} = \frac{\alpha}{ka} \tag{8.18}$$

these equations can be written as

$$A_{n+1} = A_n \cos ka + (\tilde{\alpha}\cos ka - \sin ka) B_n \tag{8.19}$$

$$B_{n+1} = A_n \sin ka + (\tilde{\alpha}\sin ka + \cos ka) B_n \tag{8.20}$$

Finally, substituting in the Bloch ansatz constraints (8.14), we obtain two homogeneous equations in the two unknowns A_n and B_n:

$$\left(e^{i\theta} - \cos ka\right) A_n - (\tilde{\alpha}\cos ka - \sin ka) B_n = 0 \tag{8.21}$$

$$\sin ka\, A_n + \left(\tilde{\alpha}\sin ka + \cos ka - e^{i\theta}\right) B_n = 0 \tag{8.22}$$

As we noted in Example 5.3, a system of homogeneous equations in two unknowns will have a nontrivial solution only if the determinant of the coefficients of the corresponding matrix equation vanishes:

$$\begin{pmatrix} e^{i\theta} - \cos ka & -(\tilde{\alpha}\cos ka - \sin ka) \\ \sin ka & \tilde{\alpha}\sin ka + \cos ka - e^{i\theta} \end{pmatrix} \begin{pmatrix} A_n \\ B_n \end{pmatrix} = 0 \tag{8.23}$$

yielding the condition

$$\left(e^{i\theta} - \cos ka\right)\left(\tilde{\alpha}\sin ka + \cos ka - e^{i\theta}\right) + \sin ka(\tilde{\alpha}\cos ka - \sin ka) = 0 \tag{8.24}$$

Multiplying by $e^{-i\theta}$

$$\tilde{\alpha}\sin ka + \cos ka - e^{i\theta} - e^{-i\theta}(\cos^2 ka + \sin^2 ka) + \cos ka = 0 \tag{8.25}$$

or simply

$$\cos\theta = \cos ka + \frac{\alpha \sin ka}{2ka} \tag{8.26}$$

Note that as $ka \to 0$, the right-hand side of (8.26) approaches $1 + \alpha/2$, which is greater than 1 (presuming the parameter α is taken to be positive). Since the left-hand side of the equation is at most 1 in magnitude, there is no solution to this equation for $ka = 0$. But if ka increases, then the magnitude of $\sin ka/ka$ decreases to the point where the right-hand side of (8.26) is equal to one and a solution exists. Figure 8.3 shows a plot of the right-hand side of (8.26) as a function of k. Recall that there are N possible values of θ. Thus there are N allowed energy states in the region between k_1 and k_2 indicated in Fig. 8.3. From k_2 to k_3 the right-hand side of (8.26) is less than -1 and hence no allowed energy states exist. The next band of allowed energy states occurs from k_3 to k_4. You can see how the band of allowed energies increases in width as k increases, since the $\alpha \sin ka/2ka$ term grows smaller in magnitude as k increases. Figure 8.4 shows an energy-level diagram for this system. *The crucial point here is that we have bands of allowed energies separated by gaps in which there are no allowed energy states.* Moreover, the bands become broader (and consequently the gaps narrower) as the energy increases.

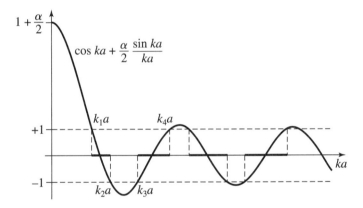

Figure 8.3 Determination of allowed energies.

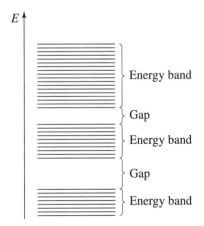

Figure 8.4 Bands of allowed energy states separated by gaps.

Qualitative Discussion

Our discussion of the band structure of crystalline solids has focused on the solution to the Schrödinger equation for a periodic potential, albeit a one-dimensional one. It is instructive to examine how this band structure arises more qualitatively. Let us go back to our discussion of the energy levels of a two-well system in Section 4.5. If the two wells are very far apart, there is two-fold degeneracy of the ground-state energy since the electron can be placed in either one of the two potential energy wells, as indicated in Fig. 8.5a. Consequently, the even and odd wave functions shown in Fig. 8.5b are degenerate too. However, as the wells are brought closer together, the difference in curvature between the even and odd wave functions increases and therefore so does the difference in energy between the two states. See Fig. 8.5c. If N wells are considered, there are N energy levels that are degenerate when the wells are widely separated (corresponding to putting the electron in any of these wells). As the wells are brought closer together, this N-fold degenerate energy level splits into N separate energies. As for the double well, the degree of splitting depends on the proximity of the wells, as indicated in Fig. 8.6.

Now take a look at Fig. 8.7, which shows a calculation of the energy levels for solid sodium as a function of the separation of the sodium atoms in the lattice. We have seen in our discussion of multielectron atoms that the electrons reside in shells at increasing distance from the nucleus. Since the electrons in the lower energy levels such as the $1s$ or $2s$ levels are bound quite closely to the nucleus, the degree of splitting of these levels is not large unless the atoms in the lattice are brought quite close together. On the other hand, the energy levels for the valence electrons—those in the $3s$ states—are split into

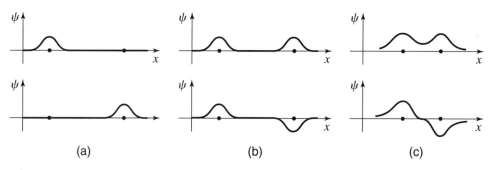

Figure 8.5 (a) The wave functions for an electron placed in one or the other of two widely separated potential energy wells. The even and odd linear combinations of these wave functions, shown in (b), are degenerate in energy. (c) As the separation between the wells decreases, the separation in energy between the even and odd wave functions increases because of the difference in curvature of the two wave functions.

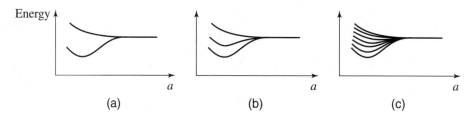

Figure 8.6 A schematic illustration of how the energy levels for (a) two, (b) three, and (c) N potential energy wells vary with separation of the wells.

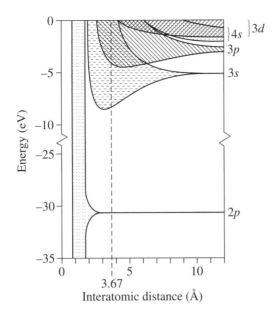

Figure 8.7 The energy levels for solid sodium. At the equilibrium separation $r_0 = 3.67$ Å of the sodium atoms the $3s$ and $3p$ bands overlap. Adapted from J. C. Slater, *Phys. Rev.* **45**, 794 (1934).

sizable bands when the sodium atoms are at their equilibrium locations in the solid. Each sodium atom has, of course, eleven electrons. Two of the electrons fill the $1s$ level, two fill the $2s$ level and six fill the $2p$ level. Consequently, the $3s$ level is only half filled, since there is just one remaining valence electron for each sodium atom. As we will explain in the next section, a partially filled valence band is referred to as the conduction band.

8.2 Electrical Properties of Solids

We can group crystalline solids into four categories: molecular, ionic, covalent, and metallic, corresponding to the type of binding that holds the atoms or molecules in the solid together. These four types of binding lead to quite distinct attributes of the solids, especially as regards their ability to conduct electricity. (1) Molecular solids are ones in which the individual atoms or molecules are bound by van der Waals forces—residual Coulomb interactions between neutral particles. This is a weak binding, as is illustrated by the fact that H_2, a molecular solid, melts at 14 K. (2) On the other hand, ionic binding is relatively strong binding. A classic example is ordinary table salt, NaCl, in which the $3s$ valence electron in sodium is effectively donated to the adjacent chlorine atom, filling the $3p$ level. The crystal structure is cubic, with sodium and chlorine atoms in alternate positions in the lattice. The binding between adjacent atoms is due to the Coulomb attraction of these oppositely charged atoms. (3) Covalent bonding results from electromagnetic interactions as well. As we saw Section 4.5, such binding, as seen in elements such as silicon and germanium, results from the "sharing" of valence electrons between the individual atoms. Thus in molecular, ionic, and covalent bonding there are not typically "free" electrons to conduct electricity. (4) In metals, on the other hand, the valence electrons are free to propagate throughout the crystal.

Solid	Resistivity (Ω-m)
Ag	1.6×10^{-8}
Cu	1.7×10^{-8}
Al	2.8×10^{-8}
Pb	22×10^{-8}
Ge	0.46
Si	640
S	10^{15}
Diamond	10^{16}

Table 8.1 Resistivities at room temperature of various crystalline solids, including metals, semiconductors, and insulators.

Table 8.1 shows the resistivity ρ for a variety of solids. The resistivity is an intrinsic measure of the resistance to current flow in a particular material at a given temperature. It is the inverse of the conductivity. We often talk about resistance instead of resistivity. The resistance R of a wire, for example, depends not only on the resistivity of the material of which the wire is composed but also on its length L and cross-sectional area A. Clearly, the longer the wire, the larger the resistance, while the bigger the cross-sectional area the more easily electrons can flow through the wire. In fact, the resistance of a wire grows linearly with the length and is inversely proportional to the cross sectional area of the wire:

$$R = \rho \frac{L}{A} \tag{8.27}$$

where the resistivity is the constant of proportionality.

Metals

Our discussion of the Kronig–Penney model and the band structure of crystalline solids shows why the resistivity is small for a metal. First and foremost, the conduction electrons are not localized near an individual atom. Rather, electrons have amplitudes to be anywhere in the metal. Solid-state physicists often call these electrons "free" to emphasize this point, even though these electrons are of course bound in the metal itself. When the valence band is only partially filled, there are plenty of empty energy states available. Thus when an electric field is applied to the metal, it can "accelerate" the electrons and promote them into these once empty energy states. Consequently a current can flow easily. This is the rationale for calling a partially filled valence band the conduction band.

While a metal is a good conductor of heat as well as electricity, heat does not enhance the electrical conductivity of a metal. Rather, the resistivity tends to increase with increasing temperature, generally because of the interactions of the electrons with the ions in the lattice, an interaction that tends to scatter individual electrons. Put another way, as the temperature increases, thermal agitation causes the ions in the lattice to vibrate, reducing the periodicity of the lattice, thus reducing the chances that the electron's wave function satisfies the Bloch condition. And the Bloch condition is the condition that ensures that the electron is free to roam around the solid.

The Kronig–Penney model provides the justification for the free-electron model of a solid that we introduced in Section 7.4. At 0 K, the electron energy levels are filled to

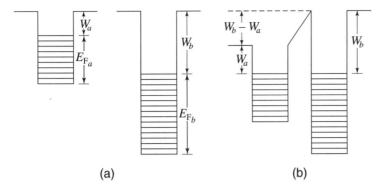

Figure 8.8 (a) Two widely separated metals with differing work functions. (b) When the metals are in "contact," electrons tunnel from metal a to metal b until the Fermi energies are equalized.

a maximum energy that is called the Fermi energy. The difference between the Fermi energy and the top of the potential well confining the electrons in the solid is the work function W that we discussed in Section 1.3. An interesting situation arises if we put two metals (metals a and b) with different work functions in close proximity. As long as the barrier is not too wide or too high, electrons in metal a can tunnel through the barrier to the empty energy levels in metal b, levels that are at a lower energy since the work function for metal b is presumed to be greater than the work function for metal a, as shown in Fig. 8.8. But this tunneling makes the charge on metal b negative, thus raising the potential energy of an electron in metal b and lowering the potential energy of an electron in metal a, which is now positively charged. This tunneling continues until the Fermi energies have equalized. At this point, there is an electric potential difference between the two metals equal to $(W_b - W_a)/e$. This electric potential difference is called the **contact potential**. It is the potential difference between two metals with different work functions placed in "contact" with each other, even though neither metal initially carries any charge. Table 8.2 gives the work function and Fermi energy for a variety of metals.

An interesting application of the contact potential is a thermocouple. Since the Fermi energy varies with temperature [the energy at which $n(E_F) = 1/2$ drops with increasing temperature, as illustrated in Fig. 7.11], it is possible to measure temperatures with two metals in contact. If one of the metals is placed in a thermal bath, the change in Fermi energy means a change in work function for that metal and hence a change in the contact potential. Once calibrated, a thermocouple can serve as an inexpensive thermometer capable of measuring over a broad range of temperatures.

Metal	W (eV)	E_F (eV)
Ag	4.7	5.5
Au	4.8	5.5
Cu	4.1	7.1
K	2.1	2.1
Li	2.3	4.7
Na	2.3	3.1

Table 8.2 Work function and corresponding Fermi energy for a variety of metals.

Insulators and Semiconductors

Let's next examine what happens if the valence band is filled and the conduction band is empty, at least at 0 K. Let's call the size of the energy gap between the top of the valence band and the bottom of the conduction band E_g. To see how many electrons are excited to the conduction band at nonzero temperature, we need to determine the location of the Fermi energy E_F. If we set $E = E_F$ in the Fermi–Dirac distribution function (7.46), we find that $n(E_F) = 1/2$. Thus the Fermi energy cannot be at the top of the valence band since the electron energy levels are all filled there. And it cannot be at the bottom of the conduction band since the energy levels there are empty. Thus the location of the Fermi energy must be somewhere in between, in the gap, even though there are no energy levels there. The symmetry of the situation suggests that the Fermi energy might be half way between the top of the valence band and the bottom of the conduction band, as indicated in Fig. 8.9. And indeed this is the case provided the density of states at the top of the valence band (let's call this $E = 0$) is the same as the density of states at the bottom of the conduction band ($E = E_g$). To see this, first note that the number of electrons excited to the conduction band at temperature T is given by

$$D(E_g)\,dE\,\frac{1}{e^{(E_g-E_F)/k_BT}+1} \tag{8.28}$$

where $D(E_g)$ is the density of states at the bottom of the conduction band. Where do these electrons come from? They must come from the valence band. But each electron excited from the valence band leaves an empty energy state, a hole if you will, behind. Since

$$D(0)\,dE\left(1-\frac{1}{e^{-E_F/k_BT}+1}\right) \tag{8.29}$$

is the number of holes at the top of the valence band, the requirement that the number of holes in the valence band must equal the number of electrons in the conduction band means that

$$D(E_g)\,dE\,\frac{1}{e^{(E_g-E_F)/k_BT}+1} = D(0)\,dE\left(1-\frac{1}{e^{-E_F/k_BT}+1}\right) \tag{8.30}$$

Assuming $D(E_g) = D(0)$, we find

$$\frac{1}{e^{(E_g-E_F)/k_BT}+1} = \frac{e^{-E_F/k_BT}}{e^{-E_F/k_BT}+1} \tag{8.31}$$

$$= \frac{1}{e^{E_F/k_BT}+1} \tag{8.32}$$

which shows that $E_g - E_F = E_F$ or simply

$$E_F = \frac{E_g}{2} \tag{8.33}$$

Therefore the average number of electrons per state at the bottom of the conduction band is given by

$$n(E_g) = \frac{1}{e^{E_g/2k_BT}+1} \tag{8.34}$$

In the special case that $E_g \gg k_BT$, we find

$$n(E_g) \cong e^{-E_g/2k_BT} \tag{8.35}$$

Thus the number of electrons excited across the energy gap into the conduction band depends very sensitively on the size of the gap and on the temperature.

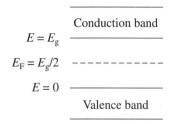

Figure 8.9 The Fermi energy is located at the midpoint of the energy gap between a filled valence band and an empty conduction band, presuming the density of states at the top of the valence band is equal to the density of states at the bottom of the conduction band.

The elements in column IV of the periodic table (see Fig. 7.2) provide a striking example. For diamond the band gap at room temperature is 5.5 eV. Since $k_B T = 1/40$ eV at room temperature, $n(E_g) = e^{-20(5.5)} = 10^{-48}$. Although there are roughly 10^{23} valence electrons per cubic centimeter of the solid, the number excited across the gap into the conduction band is consequently very small. Thus diamond is an excellent insulator. Interestingly, silicon and germanium—two other elements in the same column of the periodic table as diamond—have significantly smaller energy gaps at room temperature (1.14 eV and 0.67 eV, respectively). These substances are good insulators at low temperatures. But at room temperature a band gap of 1 eV means that $n(E_g) = e^{-20(1)} = 2 \times 10^{-9}$. Although the average population per state is still small, the absolute number of electrons excited across the gap by thermal energy is now substantial and these elements conduct electricity reasonably well (but not to the extent that a metal does, as can be seen by examining the resistivities in Table 8.1). Hence these materials are called semiconductors. As we will discuss in the next section, these semiconductors have been the building blocks for the computer/electronics revolution.

EXAMPLE 8.1 Consider a photoelectric effect experiment (see Section 1.3) in which the cathode is composed of sodium and is connected via a copper wire to the anode, which is also composed of copper. Monochromatic light is incident on the cathode. Using the data in Table 8.2, calculate the contact potential between the cathode and the anode. What is the minimum photon energy needed to get a current to flow? Figure 8.8 is a representative energy-level diagram for the two metals when separated and when in contact if metal a is taken to be sodium and metal b is taken to be copper.

SOLUTION The contact potential is the difference in the work functions divided by e, or $(4.1 - 2.3)$ V $= 1.8$ V. As Fig. 8.8b shows, in order for an electron to be liberated from sodium and overcome the contact potential between sodium and copper, it is necessary for the electron to have an energy of $W_a + (W_b - W_a) = W_b$, namely the work function of copper, or 4.1 eV.

EXAMPLE 8.2 Calculate the average number of electrons per state at the bottom of the conduction band for silicon at room temperature and for a temperature of 360 K.

SOLUTION Since $E_g = 1.14$ eV is much greater than $k_B T = 1/40$ eV at room temperature, we can make use of (8.35). Thus at room temperature

$$n(E_g) = e^{-20(1.14)} = 1.25 \times 10^{-10}$$

At $T = 360$ K, $k_B T = 0.031$ eV and therefore

$$n(E_g) = e^{-(1.14)/(0.062)} = 1.03 \times 10^{-8}$$

Thus a change in absolute temperature of 20% produces an increase in the value of $n(E_g)$ by almost 100. Thus the conductivity of semiconductors like silicon is very sensitive to temperature. Unlike metals, for which the conductivity decreases slowly with increasing temperature, the conductivity of a semiconductor increases markedly with temperature.

EXAMPLE 8.3 Use the band structure of crystalline solids to explain why diamond is transparent and silicon is opaque to visible light.

SOLUTION For visible light 400 nm $< \lambda <$ 700 nm, corresponding to photon energies between 1.8 eV and 3.1 eV. The band gap for diamond is 5.5 eV, so photons of visible light do not have sufficient energy to excite electrons in diamond from the valence band to the conduction band. Thus diamond is transparent to electromagnetic radiation in the visible part of the spectrum. On the other hand, the band gap for silicon is 1.1 eV. Thus photons of visible light do have sufficient energy to excite electrons in silicon from the valence band to the conduction band. Consequently, these photons can be absorbed, making silicon opaque to electromagnetic radiation in the visible part of the spectrum.

8.3 The Silicon Revolution

Do you think of yourself as living in revolutionary times? In 1965, shortly after the invention of the integrated circuit, Gordon Moore, one of the founders of Intel, noted that the number of transistors per square inch on an integrated circuit built on a silicon chip was doubling roughly every year. In 1975 Moore proposed a doubling every two years was likely for the foreseeable future. At that time, the number of transistors per square inch had reached a few thousand. Today, that number exceeds one billion. Doubling the number of transistors from 1 thousand to 2 thousand in two years may not seem to be a big deal, but doubling the number from 1 billion to 2 billion in that same period is pretty amazing. This is the characteristic of exponential growth. Of course exponential growth cannot continue forever. Making the circuits on a wafer of silicon smaller and smaller can continue only so far before we reach the level of individual atoms. Nonetheless, at this point there are no indications that Moore's law, as it is called, is breaking down and that the exponential growth phase is coming to an end. Moreover, the consequences of this growth are all around us. The computing power in a common laptop computer exceeds that of a supercomputer a decade ago. Today you can purchase a cell phone that contains a high resolution digital camera and an MP3 player and allows you to surf the Internet, too. And not only has the power of electronics devices increased dramatically, but their cost has decreased at the same remarkable rate. Thus these electronic devices are available to more and more people, truly changing the world in which we live.

The basic building block in this electronics revolution is a semiconductor such as silicon that has been doped with impurities in a very special way. A doped semiconductor is called an extrinsic semiconductor because its properties have been externally modified. As we have noted, silicon resides in column IV of the periodic table. Its electronic structure is $1s^2 2s^2 2p^6 3s^2 3p^2$. In a crystal of silicon, each silicon atom bonds with four adjacent silicon atoms in a tetrahedral structure, as indicated in Fig. 8.10a. In this way, each of the $3s$ and $3p$ electrons participates in covalent bonds with neighboring silicon atoms. It is this tetrahedral covalent bonding that is responsible for the energy gap between the

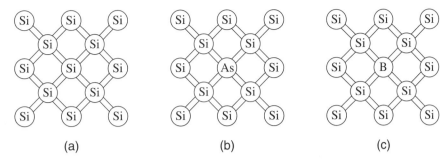

Figure 8.10 (a) A diagram indicating how each silicon atom forms covalent bonds with the four neighboring atoms in crystalline silicon. (b) In an n-type semiconductor a small fraction of silicon atoms are replaced with elements such as arsenic from column V of the periodic table. (c) In a p-type semiconductor, silicon is doped with elements such as boron from column III of the periodic table.

valence band and the conduction band in silicon.[2] If a small fraction (typically 10^{-10} to 10^{-5}) of the silicon atoms are replaced with arsenic atoms, for example, in a process called doping, we generate what is called an **n-type semiconductor**. Since arsenic belongs to column V of the periodic table, it has five valence electrons instead of silicon's four. Four of these electrons can form covalent bonds with the four adjacent silicon atoms, as indicated in Fig. 8.10b, thus simulating a silicon atom. The "extra" electron in arsenic is attracted to the arsenic atom rather than to a silicon atom because of the additional positive charge on the arsenic nucleus. But the binding is quite modest, with a binding energy of approximately 0.05 eV. Thus it takes relatively little energy to dislodge this extra electron and push it into the conduction band. Therefore the arsenic impurity is referred to as a **donor impurity**, since it donates an electron, which of course has negative charge, to carry the current. Hence the name n-type semiconductor. Figure 8.11a shows the corresponding energy-level diagram. The presence of these additional filled energy levels just below the conduction band raises the Fermi energy above the midpoint, with the degree of raising dependent on the concentration of the dopant [as you might surmise by looking back at the dependence on the density of states in the derivation of (8.33)]. In addition to arsenic, common donor atoms for silicon include phosphorus and antimony.

Let's now examine a **p-type semiconductor**. For silicon, a good candidate for doping is boron. Boron is from column III of the periodic table and thus has one fewer electron than is needed to form the four covalent bonds with the adjacent silicon atoms. With a modest amount of energy, again typically 0.05 eV, it is possible for a boron atom to borrow an electron from a silicon atom to form the requisite covalent bonds with its neighboring silicon atoms, as illustrated in Fig. 8.10c. The net effect is to introduce empty energy levels just above the valence band, as indicated in Fig. 8.11b. Thus an impurity like boron

---

[2]The same sort of tetrahedral bonding occurs in diamond. The much larger energy gap between the valence band and the conduction band in diamond is due to the fact that the carbon atoms in diamond are somewhat closer together than are the silicon atoms. The sensitivity of the mechanical and electrical properties of these crystals on the type of bonding is highlighted by comparing graphite, in which the carbon atoms bond in a hexagonal lattice, with diamond. Graphite is a reasonably good conductor with the $2p$ valence electrons partially filling the conduction band, unlike diamond which is, as we have noted, an outstanding insulator. And whereas diamond is one of the hardest known substances, graphite is commonly used as a lubricant.

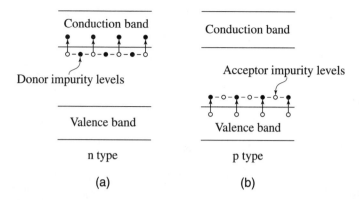

Figure 8.11 An energy-level diagram for an n-type semiconductor in (a) and a p-type semiconductor in (b).

is referred to as an **acceptor impurity**. When boron accepts an electron from silicon, a hole is left behind in the valence band. If an electric field is applied to the crystal, the holes propagate in the direction of the electric field, just like a positive charge. Thus the name p-type semiconductor. Since the energy levels just above the top of the valence band are empty at low temperatures, the Fermi energy is pulled down from the midpoint of the band gap by the presence of these impurities. As for an n-type semiconductor, the larger the percentage of impurities, the more the Fermi energy deviates from the middle of the gap.

Through diffusion of the impurities, it is possible in a small piece of silicon to move in the space of tens of nanometers from silicon with an acceptor impurity to silicon with a donor impurity, thus generating a **p-n junction**. Figure 8.12a shows

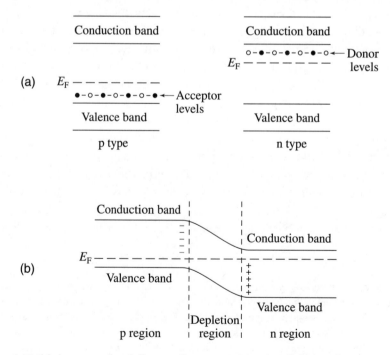

Figure 8.12 (a) An energy-level diagram for an n-type semiconductor and a p-type semiconductor. Notice how the Fermi energy is raised or lowered, respectively, from its position at the middle of the energy gap. (b) An energy-level diagram for an n-type and a p-type semiconductor in contact.

the energy-level diagram before equilibrium is established. The situation is reminiscent of the situation with two conductors with different work functions. In particular, the Fermi energy on the n side is higher than the Fermi energy on the p side. Thus electrons on the n side have empty energy levels on the p side into which they can move, meaning the p side is becoming negatively charged, leaving the n side positively charged with a surplus of holes. This transfer of charge continues until the Fermi energies have equalized, at which point the junction region looks like a parallel-plate capacitor, with a contact potential on the order of one volt. See Fig. 8.12b. The region between the "plates" is referred to as the **depletion region** because it is depleted of mobile charge carriers by the strong electric field (on the order of 10^8 V/m) that exists there.

Even when the Fermi energies are in equilibrium, a p-n junction is a dynamic place. There are currents flowing back and forth across the junction. On the p side, electrons are excited by thermal energy across the band gap into the conduction band. These electrons can then migrate toward the n side, where they find a downhill potential and can slide over to the n side. This current is referred to as the thermal current. We will call it I_0. It is balanced by the recombination current, which arises when electrons on the n side in the conduction band use thermal energy to surmount the potential energy barrier between the n side and the p side. If these electrons migrate to the p side, their likely fate is to fall into one of the holes available in the valence band, hence the name recombination current. In equilibrium, the two currents are of course equal.

But look what happens if we apply a potential difference across the junction. Let us connect a battery to the junction with the negative terminal attached to the p side and the positive terminal to the n side. This is known as reverse biasing. The net effect of the battery is to push up the potential energy of the p side, raising the height of the potential energy barrier between the two semiconductors. What then happens to the currents? The thermal current is not affected by this change, because the thermal current simply requires that the electrons have enough energy to cross the band gap, which is not affected by the applied potential. But the recombination current decreases because electrons on the n side now have to overcome a larger potential energy barrier. The probability of electrons having this energy is dictated by the Boltzmann factor. Thus $I_0 \to I_0 e^{e\varphi/k_B T}$ with the potential φ negative.[3] Thus the recombination current is exponentially suppressed when the junction is reverse biased. If on the other hand we forward bias the p-n junction by connecting the positive terminal of the battery to the p side of the junction and the negative terminal to the n side, the recombination current will be exponentially increased as the barrier that electrons must overcome to travel from the n side to the p side is reduced. Thus as before $I_0 \to I_0 e^{e\varphi/k_B T}$ but this time with a positive φ. The total current in the p-n junction is thus given by

$$I = I_0 \left(e^{e\varphi/k_B T} - 1 \right) \tag{8.36}$$

---

[3]The e in the exponent is, of course, the "other e," namely the magnitude of the charge on an electron. This is only part of the notational morass raised by this equation. In quantum mechanics, it is customary to use the symbol V for potential energy. In introductory treatments of electromagnetism, on the other hand, the same symbol is used for the potential, which is measured in volts in SI units. To avoid confusion, I have used φ for the potential, as is commonly done in more advanced treatments of electrodynamics. The potential energy V is then given by $q\varphi$, where q is the charge.

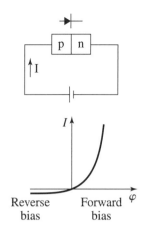

Figure 8.13 Current (I) in a p-n junction as a function of applied voltage, measured in volts (V). Note that the current is quite small (microamps) when the junction is reverse biased but increases exponentially when the junction is forward biased.

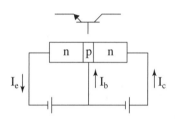

Figure 8.14 An n-p-n transistor.

Figure 8.13 shows a plot of the current for both types of bias. For $\varphi < 0$, the current asymptotically approaches the value I_0 as the recombination current is suppressed. But when $\varphi > 0$, the current grows exponentially. Even though the value of I_0 may be quite small, the exponential growth can make the current very large for sufficiently large values of φ. Thus on the tiny piece of silicon we have constructed a **diode**, a device that transmits current in one direction but not in the other when a potential difference is applied.

To illustrate how a p-n junction enters into the construction of a transistor, consider the following n-p-n configuration, as shown in Fig. 8.14.[4] Here the middle region, the p-type semiconductor, is a very thin region called the base. The n region on the left is referred to as the emitter and the n region on the right the collector. The base is sufficiently thin that electrons flowing from the emitter to the base do not have sufficient time to fill in the holes in the p region. Hence these electrons diffuse into the collector. Typically, the doping in the base is smaller than in the emitter, thus reducing the odds that the electrons in the emitter fall into one of the holes there. If the emitter–base p-n junction is forward biased, a small change in the voltage produces a big change in the current. On the other hand, the base–collector p-n junction is reverse biased with a large bias voltage. The power (the current times the voltage) in the base–collector circuit can be quite large and depends sensitively on the current flow in the emitter–base region. Thus this transistor can serve as a power amplifier, say between a microphone and a speaker.

In addition to serving as the basic component of a transistor, the p-n junction is the key element in a light-emitting diode (LED) and a solar cell. In some materials, such as gallium arsenide, an electron in the conduction band emits visible light when it makes a transition to an empty energy state (a hole) in the valence band in the depletion region. If the junction is forward biased, electrons and holes are replenished as the current flows through the junction. Moreover, filled electron states in the conduction band and empty energy states in the valence band constitutes a population inversion (see the discussion in Section 7.8). Thus a light-emitting diode can also function as a semiconductor laser.

A solar cell is essentially a light-emitting diode operating in reverse. Sunlight can be used to excite an electron from the valence band to the conduction band in the depletion region, leaving a hole in the valence band. The electric field in the depletion region sweeps the electron to the n side and the hole to the p side before the electron can recombine with the hole. In this way a current is generated that can be used to supply power to an external circuit. For photons to reach the depletion region, the p-type layer in the solar cell must be thin so that it is sufficiently transparent to sunlight. The overall efficiency of such a solar cell is typically 15%, which means a solar cell with an area of a square meter produces roughly 100 W of electrical power. The rest of the incident energy from the sun is lost to heat. While solar cells with higher efficiency have been constructed, these alternatives have been more expensive and have not yet proven to be cost effective.

---

[4]The bipolar junction transistor was developed by John Bardeen, Walter Brattain, and William Shockley in 1948. They were awarded the Nobel Prize in physics in 1956. The bipolar junction transistor has the deficiency of being a power hog, which is a major problem when trying to pack a large number of transistors into a small region in an integrated circuit. Alternative transistors in the form of MOSFETs are therefore often used. The transistor is arguably the most important invention of the twentieth century.

8.4 Superconductivity

After our discussion in this chapter of the electrical properties of crystalline solids, including conductors and semiconductors, it is irresistible to say something about superconductors, namely those materials that conduct electricity without any resistance whatsoever. The phenomenon of superconductivity was discovered by Kamerlingh Onnes in 1911 when he noted that the resistance of mercury plummeted to zero when mercury was cooled below a critical temperature $T_C = 4.15$ K. About one third of the pure metals and many alloys are superconductors.

In addition to having zero resistivity, materials that are superconductors expel magnetic fields from the material when cooled below the critical temperature, an effect known as the **Meissner effect**. As the transition temperature is reached, currents circulate on the surface of the superconductor that exactly cancel the magnetic field within the material. This complete expulsion of magnetic field is characteristic of a class of superconductors called Type I superconductors. In these superconductors the magnitude of the critical temperature depends on the strength of the applied field, as indicated in Fig. 8.15. Some Type I superconductors along with the corresponding critical temperature and critical magnetic fields are listed in Table 8.3. For some materials, typically alloys, there are two critical values of the magnetic field, B_{C_1} and B_{C_2}, as illustrated in Fig. 8.16. If the applied magnetic field is less than B_{C_1}, the magnetic field is expelled completely, as is the case for a Type I superconductor. But for magnetic fields $B_{C_1} < B < B_{C_2}$, the magnetic field penetrates the superconductor in narrow filaments called vortex cores. Within these cores the material is a normal conductor, but superconductivity is maintained in the material between the cores. As the field strength increases, the material becomes more closely

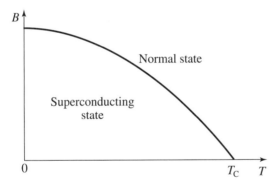

Figure 8.15 The critical magnetic field as a function of temperature in a Type I superconductor.

Material	T_C (K)	$B_C(0)$ (Tesla)
Zn	0.88	0.0053
Al	1.17	0.0105
In	3.41	0.0293
Pb	7.19	0.0803
Nb	9.25	0.198

Table 8.3 Some Type I superconductors. The critical magnetic field is measured as $T \to 0$.

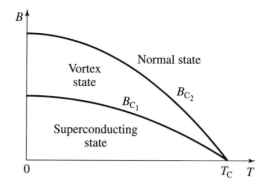

Figure 8.16 The critical magnetic fields B_{C_1} and B_{C_2} as a function of temperature in a Type II superconductor.

Material	T_C (K)	$B_{C_2}(0)$ (Tesla)
PbMoS	14.4	6.0
NbN	16.0	15.3
Nb_3Sn	18.1	24.5
Nb_3Ge	23.2	34.0

Table 8.4 Some Type II Superconductors.

filled with these vortex cores until the cores overlap when the magnetic field strength exceeds B_{C_2} and the superconductivity is destroyed. Since Type II superconductors have significantly larger values for the critical field strength (see Table 8.4), these materials are much better candidates for applications such as carrying large currents, since these currents generate magnetic fields themselves that can destroy the superconducting state. The downside of these Type II materials is that they are often brittle and not easily molded into wires that could be used to carry currents.[5]

Given our discussion of Bose–Einstein condensation in Section 7.7, you may be tempted to surmise that this transition to the superconducting state is a phase transition similar to the one that occurs in liquid helium at 2.17 K, at which point helium atoms form a condensate with superfluid behavior. But something more complicated must be at work here, since the charge carriers in a solid are electrons (or holes). These particles are spin-1/2 particles that obey Fermi–Dirac statistics, not the Bose–Einstein statistics that describe the behavior of the spin-0 helium atoms. There are a couple of hints as to what is occurring. For one, the best conductors, such as copper, gold, and silver, are not superconductors. Materials such as niobium and lead, which are relatively poor conductors at room temperature, have relatively large values of T_C. Recall that resistance in a metal is due to interactions of the electrons with the lattice and really good conductors have minimal lattice interactions. Also, superconductivity in materials that have a number of different isotopes occurs with modestly higher values of T_C for the lighter nuclei. This also suggests that interaction with the lattice is important, since lighter nuclei typically means more vibration of the nuclei at a given temperature.

In 1959 John Barden, Leon Cooper, and Robert Schrieffer proposed a theory that explains the phenomenon of superconductivity (with the exception of the high-T_C

---

[5]Roughly 10% of the electrical energy used in the United States is lost in the transmission process.

superconductors). This BCS theory posits that, despite their mutual Coulomb repulsion, electrons with equal and opposite momenta bind together in these materials to form bound pairs, generally referred to as **Cooper pairs**. One of the electrons in the pair pulls the ion cores toward it through Coulomb attraction. The other electron is then attracted to this displacement of the ion cores. A useful analogy is to consider two people sleeping together on a soft mattress. When one of the people rolls away from the other, that person leaves a depression in the mattress into which the second person may fall, thus "binding" the two people together. In the case of superconductivity, the BCS theory predicts the binding energy of each Cooper pair, referred to as the gap energy, has a value at absolute zero of

$$E_g = 3.53 k_B T_C \tag{8.37}$$

It is clear that thermal energy can dissociate these Cooper pairs if the temperature of the material gets too large. Since the two electrons in a Cooper pair are in a total-spin-0 state [see the spin state in (7.25)], these pairs are themselves bosons. Thus for sufficiently low temperature a phase transition to a condensate can occur. In this condensate, there is an incentive for the pairs to be in the same total-momentum state (see the discussion in Section 7.8), thus preventing the scattering by the lattice that is responsible for resistivity. Presumably, this pairing of electrons also occurs in the high-temperature Type II superconductors that were discovered in the late 1980s. These materials, which have values of T_C ranging from 30 K to 140 K, are layered ceramic materials with planes of copper oxide. However, the mechanism responsible for the electron pairing in these high-temperature superconductors is still not understood. Since nitrogen gas liquefies at 77 K, these high temperature superconductors can be maintained in the superconducting state at a much lower cost than can superconductors that require liquid helium for cooling, since liquid helium is much more expensive than liquid nitrogen. Moreover, these high-temperature superconductors have large values for the critical magnetic field strength, which means that they can be used to transport large currents. The downside of these ceramic superconductors is that they are quite brittle and therefore difficult to manipulate.

8.5 Summary

The main message of this chapter is that the periodic nature of the potential energy for a crystalline solid generates bands of allowed energy levels separated by energy gaps. If the valence band is partially filled at low temperatures, it is referred to as the conduction band. Such solids are metallic in nature and readily conduct electricity. If the valence band is completely filled as the temperature approaches zero and the conduction band is empty, the solid is either an insulator or a semiconductor, depending on the size of the band gap. By doping (adding impurities to) a semiconductor such as silicon, one can produce n-type and p-type semiconductors that can be joined together to form a p-n junction. This junction is a solid-state device that conducts electricity in only one direction (a diode). The p-n junction is the building block of transistors, such as the n-p-n bipolar junction transistor. Our ability to package more and more transistors per unit area in an integrated circuit on a substrate of silicon is responsible for the electronics revolution that is taking place during our lifetimes.

Problems

8.1. Verify that imposition of the Bloch ansatz

$$\psi(x+a) = e^{i\theta}\psi(x)$$

on the two solutions to the Schrödinger equation

$$\psi = A_n \sin k(x-na) + B_n \cos k(x-na)$$
$$(n-1)a < x < na$$

and

$$\psi = A_{n+1} \sin k[x-(n+1)a]$$
$$+ B_{n+1} \cos k[x-(n+1)a]$$
$$na < x < (n+1)a$$

yields the conditions

$$A_{n+1} = e^{i\theta} A_n \qquad B_{n+1} = e^{i\theta} B_n$$

8.2. Calculate the amount the Fermi energy is raised if 1 C of electrons is added to a cubic centimeter of a metal. *Note*: One coulomb is a large amount of charge.

8.3. Semiconductors such as ZnSe, CdTe, GaAs, and Ge are used to manufacture lenses. These materials are particularly useful in the infrared (2 μm to 30 μm), despite the fact that they are highly opaque in the visible region of the spectrum, as shown in Fig. 8.17. For Ge, for example, the energy gap between the valence band and the conduction band is $E_g = 0.67$ eV. Use this fact to explain why Ge is a good material for an infrared lens.

Figure 8.17 Galium arsenide (GaAs) is used to manufacture infrared lenses. (Courtesy II–VI, Inc.)

8.4. An ohmmeter is sometimes used to determine the "direction" of a diode by connecting the ohmmeter to the diode one way and then reversing the ohmmeter leads. If the ohmmeter applies an emf of 0.5 V to the diode in order to determine resistance, what would be the ratio of reverse resistance to forward resistance at 300 K? Assume the diode behaves as an ideal p-n junction. What do you conclude from this calculation if the forward resistance is 30 kΩ and the maximum range of the ohmmeter is 2000 kΩ? *Suggestion*: Start with (8.36).

8.5. Draw energy-level diagrams for the metals silver and copper when separated and in contact. Indicate the contact potential on your diagram. Use the values in Table 8.2 to determine the relative positions of the work functions and the Fermi energies.

8.6. Use the band structure of crystalline solids to explain why a thin piece of sulfur, a good insulator, transmits light that is reddish yellow when illuminated from the back side with white light.

8.7. Use Faraday's law to argue that the magnetic field cannot vary with time in a material that has zero resistivity (and hence infinite conductivity).

8.8. The critical magnetic field in a superconductor has the temperature dependence

$$B_C(T) = B_C(0)\left[1 - \left(\frac{T}{T_C}\right)^2\right]$$

Calculate the maximum current a superconducting niobium wire 1 mm in diameter can carry at 4 K, the temperature at which helium liquefies. See Table 8.3 for the values of T_C and $B_C(0)$ for niobium.

8.9. Use (8.37) to determine the size of the energy gap in eV for niobium.

CHAPTER 9

Nuclear Physics

In 1911, in Ernest Rutherford's laboratory, Hans Geiger and an undergraduate student, Ernest Marsden, discovered the atomic nucleus in an experiment in which alpha particles were projected at a thin gold foil. The relatively high fraction of the alpha particles deflected through large angles was very surprising but could be accounted for by assuming the positive charge in the nucleus was concentrated in a massive central core, or nucleus.[1] In the simplest atom, the hydrogen atom, the nucleus is the proton, which is 1836 times more massive than the electron. But of what does the nucleus of gold consist? The answer to this question was not known until 1932, when James Chadwick discovered the neutron, a neutral particle similar in mass to the proton. In the nucleus of gold, for example, there are 79 protons and 118 neutrons. One can say that 1932 marked the beginning of nuclear physics. Nuclear physics is a field of enormous importance. Understanding which nuclei are stable and why allows us to understand which elements occur in nature. Moreover, as we noted in Section 3.2, the energy scale of nuclear physics is millions of times greater than that of atomic physics. The main focus of this chapter is the so-called curve of binding energy. The nuclear binding energy released in fusion reactions is the energy source that powers our Sun (and the other stars in our universe as well). These fusion reactions are ones that we would very much like to utilize in a controlled fashion here on Earth since this would provide an almost limitless source of power. As we will also discuss in this chapter, nuclear energy can also be released in fission reactions, for peaceful purposes in a nuclear reactor and for less peaceful purposes in the form of nuclear weapons.

---

[1]The discovery of the existence of the nucleus was the death sentence for classical physics. For if the atom consists of electrons orbiting around the positively charged nucleus in a planetary-like system, the electromagnetic radiation from these accelerating electrons would lead to the collapse of the atom in a very short time period. See Problem 9.1. Rutherford later commented that the discovery of large-angle deflection of alpha particles was the most incredible event of his life, "as if you fired a 15-inch shell at a piece of tissue paper and it came back and hit you." This wasn't as much of an exaggeration as you might suppose given the odds for such large-angle scattering in the "plum-pudding" model of the atom that was current at the time, a model in which the positive charge was thought to be uniformly distributed through out the atom (the pudding with the electrons as the raisins).

9.1 Nuclear Notation and Properties

A nucleus is specified by listing the number of protons (Z) and the number of neutrons (N) it contains. Of course, the number of protons is the same as the number of electrons in the corresponding (neutral) atom, or element. The symbol Z is generally referred to as the **atomic number**. If you know the elements well, then, for example, you know that carbon (C) has $Z = 6$ and uranium (U) has $Z = 92$. The number of neutrons in the nucleus of a particular element can vary, however. Elements with the same number of protons but different numbers of neutrons are referred to as **isotopes**. The combined number of neutrons and protons in the nucleus is referred to by the symbol $A = Z + N$, the **atomic mass number**. Thus specifying Z and A is sufficient to determine the composition of a particular nucleus, since $N = A - Z$. The standard notation is of the form $^A_Z X$, although the subscript Z is often dispensed with since the same information is implicitly contained in the symbol X for the atom and putting both the superscript and the subscript makes for a rather cumbersome notation. The simplest atom, hydrogen, has three isotopes, ^{1}H with a nucleus consisting of a single proton, ^{2}H with a nucleus consisting of one proton and one neutron, and ^{3}H with a nucleus consisting of one proton and two neutrons. Of these three isotopes, ^{1}H and ^{2}H are stable. The isotope ^{2}H is called deuterium and the corresponding nucleus is referred to as the deuteron (denoted by the symbol d). The isotope ^{3}H is called tritium and the nucleus, the triton, is often given the symbol t. The element with $Z = 2$ is helium, which has two stable isotopes, ^{3}He and ^{4}He. The nucleus of the most common isotope of helium, ^{4}He, is referred to as the alpha particle and denoted by the symbol α. As a last example, carbon has two stable isotopes: ^{12}C, which has six protons and six neutrons in its nucleus and has a natural abundance of 99%, and ^{13}C, with a nucleus of six protons and seven neutrons. The unstable isotope ^{14}C is used in radioactive carbon dating (which we will discuss in Section 9.3). It should be noted that nuclei having different values for Z and N but the same value for A are referred to as **isobars** (^{14}C and ^{14}N, for example).

Although the proton and neutron differ in their electric charges, perhaps the most striking thing about them is their similarity. Table 9.1 gives the masses of the electron, proton, and neutron. Note that while the proton and the neutron are both significantly more massive than the electron, the masses of the neutron and proton differ by just slightly more than 1 part in 1000. In determining the mass of the atom, the electron plays a very small role. As we will see in the next section, the mass of the nucleus (and hence the atom) is roughly A times the mass of the proton or neutron. Hence the name atomic mass number for A. In fact, the nuclear force treats the proton and neutron in exactly the same way, as if they were identical particles.[2] Thus the constituents of the nucleus, whether they are protons or neutrons, are typically referred to as **nucleons**.

Particle	kg	MeV/c^2
Electron	9.109×10^{-31}	0.511
Proton	1.673×10^{-27}	938.3
Neutron	1.675×10^{-27}	939.6

Table 9.1 Mass of the electron, proton, and neutron

---

[2]Consequently, the nuclear force is said to exhibit **charge independence**.

Most nuclei are roughly spherical in shape. Scattering experiments with electrons as the projectile carried out by Robert Hofstadter in the 1950s showed that the radius of a nucleus composed of A nucleons is given roughly by

$$R = r_0 A^{1/3} \tag{9.1}$$

with $r_0 = 1.2$ fm. This translates into a volume

$$V = \frac{4}{3}\pi R^3 = \frac{4}{3}\pi r_0^3 A = V_0 A \tag{9.2}$$

where V_0 is roughly the same for all nuclei. Consequently, nuclear matter tends to have a uniform density of 2×10^{17} kg/m^3, roughly fourteen orders of magnitude more dense than ordinary matter. This large density is consistent with the fact that the size of an atom is four to five orders of magnitude larger than that of the nucleus while essentially all the mass is concentrated in the nucleus.

Electron scattering experiments effectively measure the charge density of the nucleus since electrons do not experience the nuclear force. The assumption is that the charge density—the distribution of protons in the nucleus—coincides with the distribution of nucleons in general. To map out the distribution of nucleons, neutrons as well as protons, in a scattering experiment, it is necessary to use as the projectile a particle such as the neutron that experiences the nuclear force. But determining a radius of the nucleus in this way is more complicated (in an interesting and important way) than it might first appear. In determining the likelihood that a neutron interacts with the nucleus it is convenient to introduce the concept of the **cross section**. Consider a beam of neutrons of intensity I incident on a slab of matter of thickness dx, cross-sectional area S, and density $\tilde{n}$ (the number of atoms per unit volume). Then $\tilde{n} S dx$ is the number of nuclei in the slab. After traversing the slab, the intensity of the beam is $I + dI$, as indicated in Fig. 9.1. The intensity decreases ($dI < 0$) because of interactions with the nuclei in the slab. If we denote by σ the *effective area* of the nucleus for scattering neutrons, then $\sigma \tilde{n} S dx$ is the effective area covered by all the nuclei in the slab, and the fraction of the neutrons that get scattered is $\sigma \tilde{n} S dx / S$. Note that by choosing the slab to be of infinitesimal thickness dx, we have avoided the issue of one nucleus being partially obscured by one in front of it. Putting everything together,

$$\frac{dI}{I} = -\frac{\sigma \tilde{n} S dx}{S} = -\sigma \tilde{n} dx \tag{9.3}$$

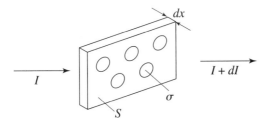

Figure 9.1 A slab of thickness dx and cross-sectional area S is composed of nuclei each with *effective* cross-sectional area σ. A beam of particles with intensity I is incident on slab. The value of σ varies with the energy and the type of projectile in the beam.

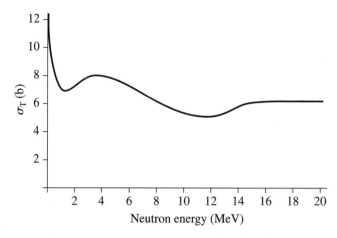

Figure 9.2 The cross section for the scattering of neutrons by uranium. Adapted from R. L. Henkel, L. Cranberg, G. A. Jarvis, R. Nobles, and J. E. Perry, *Phys. Rev.* **94**, 141 (1954).

which integrates to

$$I(x) = I(0)e^{-\sigma \tilde{n} x}$$
$$= I(0)e^{-x/\Lambda} \tag{9.4}$$

where

$$\Lambda = \frac{1}{\sigma \tilde{n}} \tag{9.5}$$

is the referred to as the **attenuation length**, namely the distance the beam must traverse so that its intensity is attenuated to $1/e$ of its initial value. Thus by measuring the attenuation length we can obtain a value for σ.

Figure 9.2 shows a plot of the cross section for neutron scattering by uranium. Notice that the cross section is measured in barns, where 1 barn $= 10^{-28}$ m^2. If we use the radius (9.1) with $A = 238$ and calculate the area πR^2, we obtain a value of 2 barns, which is in rough accord with the measured value for the cross section, at least at high energy. The cross section, however, varies markedly with energy and is much larger for low-energy neutrons. The growth in the cross section at low energies does not mean that the nuclei are getting bigger, but rather they are becoming more effective at scattering neutrons. Heuristically, we can argue that the probability that a neutron interacts with the nucleus increases as the neutron spends more time traversing the nucleus. Since the time the neutron takes to traverse the nucleus is inversely proportional to the velocity of the neutron, this increase in the cross section at low energies is sometimes called the $1/v$ effect. In effect, the cross section is proportional to λR instead of R^2, where $\lambda = h/p$ is the de Broglie wavelength of the neutron.[3]

---

[3] A barn might seem like a very small area, but a cross section of a barn in a nuclear physics experiment is such a large one that one can say it is as easy to scatter particles as hitting the broad side of a barn. This terminology was suggested in 1942 by two Purdue University physicists working on the Manhattan Project (see Section 9.6) and was considered classified information until after World War II.

9.2 The Curve of Binding Energy

The "curve of binding energy" is arguably the most important curve in science, responsible for our very existence and potentially our extinction as well. Moreover, it is the potential solution to many of the world's energy and environmental problems. Let's start with a simple example from atomic physics to illustrate the role that binding energy plays. We know that the constituents of a hydrogen atom are a proton and an electron. Is the mass of a hydrogen atom equal to the mass of the proton plus the mass of the electron? No, because of binding energy. The proton and the electron have, of course, opposite charges and attract each other. Breaking (or ionizing) hydrogen into its constituents requires the input of 13.6 eV of energy. Thus

$$m_H c^2 + 13.6 \text{ eV} = m_p c^2 + m_e c^2 \tag{9.6}$$

where m_H is the mass of the hydrogen atom, m_p is the mass of the proton, and m_e is the mass of the electron. More generally, we can write

$$m_H = m_p + m_e - \text{B.E.}/c^2 \tag{9.7}$$

where the binding energy B.E. = 13.6 eV is 10^{-6} % of the mass of the hydrogen atom, or on the order of 10 parts per billion. We saw in Section 6.3 how one would calculate this binding energy. The reason that hydrogen is stable—that is it doesn't decompose naturally into an electron and a proton—is that the mass of the atom is indeed less than the sum of the masses of the constituents. There is no way for a hydrogen atom on its own to decompose and conserve energy.

The corresponding two-body system in nuclear physics is the deuteron, whose mass m_d can be expressed as

$$m_d = m_p + m_n - \text{B.E.}/c^2 \tag{9.8}$$

where again m_p is the mass of the proton, m_n is the mass of the neutron, and B.E. = 2.2 MeV, which is 0.1% of the mass of the deuteron. Note that the energy scale of nuclear physics is MeV not eV. This is one reason for characterizing the nuclear interactions that holds the neutron and proton together as **strong interactions**. For the ^{4}He nucleus, or alpha particle,

$$m_\alpha = 2m_p + 2m_n - \text{B.E.}/c^2 \tag{9.9}$$

where B.E. = 28 MeV, which is 0.7% of the mass of the alpha particle. If we extend this line of reasoning to a more complicated nucleus, one composed of Z protons and $N = A - Z$ neutrons in the nucleus, we have

$$m_{\text{nucleus}} = Zm_p + (A - Z)m_n - \text{B.E.}/c^2 \tag{9.10}$$

Mass spectroscopy (see Problem 9.10) measures the masses of atoms, not the nuclei themselves. Thus it is most useful to rewrite (9.10) as

$$m_{\text{atom}} = Zm_H + (A - Z)m_n - \text{B.E.}/c^2 \tag{9.11}$$

which includes the masses of the Z electrons that are bound to the atom. Since the nuclear binding energy holding the protons and neutrons together in the nucleus is so much bigger than the atomic binding energy holding the electrons in the atom, the binding energies in (9.10) and (9.11) can be taken to be essentially the same to the accuracy we need.

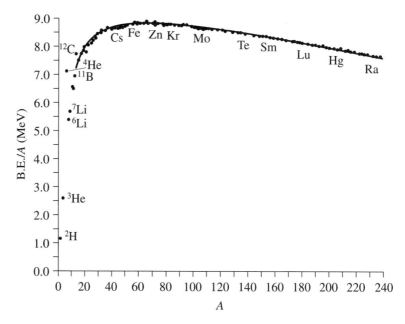

Figure 9.3 The binding energy per nucleon. The smooth curve is a plot of the binding energy (9.12) divided by A.

Figure 9.3 shows the value of B.E./A, the binding energy per nucleon, plotted as a function of A for the naturally occurring isotopes. This solid line, which provides a good fit to the data for all but the smallest values of A, results from writing the binding energy as

$$\text{B.E.} = a_1 A - a_2 A^{2/3} - a_3 \frac{Z^2}{A^{1/3}} - a_4 \frac{(Z - \frac{A}{2})^2}{A} + \frac{a_5}{\sqrt{A}} \begin{pmatrix} 1 \\ 0 \\ -1 \end{pmatrix} \quad (9.12)$$

with the following values for the parameters a_1 through a_5:

$$a_1 = 15.75 \text{ MeV} \quad a_2 = 17.8 \text{ MeV} \quad a_3 = 0.711 \text{ MeV} \quad a_4 = 94.8 \text{ MeV} \quad a_5 = 11.2 \text{ MeV} \quad (9.13)$$

The a_5 term makes a positive contribution to the binding energy for nuclei with an even number of protons and an even number of neutrons and a negative contribution to the binding energy for nuclei with an odd number of protons and an odd number of neutrons.

The equation (9.11) with the binding energy given by (9.12) is generally referred to as the **semiempirical mass formula**. The rationale for this name is that there is a lot of physics behind the five terms in (9.12), physics to which we will now turn. We will use a variety of models, including the liquid drop model, the Fermi gas model, and the shell model, to understand the properties of nuclei. The word model in general and the introduction of a number of different models in particular may be upsetting to you. After all, in physics our goal is to obtain a deep, fundamental understanding, which seems at odds with using a variety of different models. The challenge we are facing is that sorting out the details of nuclear physics is inherently more difficult than it is for atomic and molecular physics, which are governed by well-understood electromagnetic interactions. As we will discuss in the next chapter, the neutrons and protons in the nucleus are themselves composed of more fundamental entities called quarks. The equations (quantum chromodynamics) that govern the interactions between quarks are nonlinear. Consequently, the interaction between, say, a neutron and a proton in the deuteron is different than between a neutron and a proton in the tritium nucleus, which includes an additional neutron. This is in striking contrast to Maxwell's equations, the equations that govern electromagnetic interactions. Maxwell's equations are linear differential equations and we can take full advantage of the principle of superposition. The electromagnetic force between the electron and the proton in a hydrogen atom, for example, is the same as the electromagnetic force between these two particles in a helium atom, despite the fact that the helium atom possesses an additional electron and an additional proton in the nucleus.

The Liquid Drop Model

The nuclear force is a short-range force, acting over a distance scale on the order of a fermi. One often says, speaking loosely, that the nucleons essentially need to be in contact to experience the nuclear force since the nucleons themselves have a size scale on the order of a fermi as well. The nuclear force is attractive unless the nucleons are squeezed close together, in which case the force becomes repulsive. These characteristics of the nuclear force are reminiscent of the characteristics of the force between molecules and leads one to suppose that a nucleus should behave like a droplet of a fairly incompressible fluid, an idea first suggested by Niels Bohr. Because of the relatively small size of the box confining the nucleons in the nucleus, the zero-point energy of nucleons in, say, a solid-like lattice would be so large that the solid phase would not be stable; the solid would immediately melt to form a liquid. The attractive nuclear force tends to pull the individual nucleons inward, forming a droplet. Nucleons on the surface of the droplet, however, do not get the full advantage of the attractive force since they do not have nearest neighbors on all sides. The bigger the surface area of a droplet, the larger the shortfall in binding. As is the case for fluids, the droplets tend to be predominantly spherical in shape since, for a given volume, a sphere has the minimum surface area. In a fluid, the effect that leads to the shrinking of the exposed surface of a fluid droplet is referred to as surface tension. We can say there is an analogous nuclear surface tension as well.

We can now see the physics responsible for the first two terms in the semiempirical mass formula. The first term, the $a_1 A$ term, is called the **volume term**. It arises since, as a first approximation, the binding energy is directly proportional to the number A of nucleons in the nucleus. Each nucleon bonds to its nearest neighbors and the number of such bonds is proportional to the number of nucleons. Moreover, the number of nucleons is proportional to the volume of the nucleus, as illustrated in Fig. 9.4. The second term, known as the **surface term**, in the semiempirical mass formula is essentially a correction term to the first term. The nucleons at the surface of the nucleus do not form as many bonds as do the nucleons in the interior of the nucleus. Thus they do not benefit from the full binding that the interior nucleons experience with neighbors on all sides. Consequently, we must subtract a term proportional to the surface area of the nucleus since we have effectively over counted the binding effectiveness with the volume term. Since the radius of the nucleus is proportional to $A^{1/3}$, the surface area is proportional to $A^{2/3}$. Hence the term $a_2 A^{2/3}$. This terms enters with a minus sign corresponding to a reduction in the net binding energy.

Figure 9.4 The nucleons in a nucleus are closely packed in a roughly spherical shape.

The volume and surface terms in the semiempirical mass formula arise from the short-range nature of the strong nuclear force. Each proton in the nucleus, on the other hand, interacts as well with all the other protons through long-range Coulomb interactions. The effect of this Coulomb repulsion is to reduce the binding energy holding the nucleons together in the nucleus. For example, if we treat the nucleus as a uniform ball of charge $Q = Ze$, we can calculate the energy required to assemble this charge distribution (see Fig. 9.5):

$$\int_0^Q \frac{q}{4\pi\epsilon_0 r} dq = \frac{3}{5} \frac{Q^2}{4\pi\epsilon_0 R} \tag{9.14}$$

where R is the radius of the sphere. If we substitute $R = r_0 A^{1/3}$, we obtain a term proportional to $Z^2/A^{1/3}$, namely the third term, or **Coulomb term**, in the semiempirical mass formula. Like the surface term, this term also enters with a minus sign since the Coulomb repulsion reduces the binding energy. If we substitute in values for the constants and take $r_0 = 1.2 \times 10^{-15}$ m, we obtain $a_3 = 0.72$ MeV (see Problem 9.3), which compares favorably with the value for a_3 in (9.12). The smaller value of a_3 in comparison to a_1 and a_2 is indicative of the fact that, as noted earlier, the nuclear force is significantly stronger than the electromagnetic force.

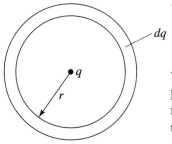

Figure 9.5 The energy required to assemble a uniform sphere of charge can be determined by assembling the sphere in a series of spherical shells, each of charge dq. The work required to attach dq to the sphere of radius r is the product of dq and the potential $q/4\pi\epsilon_0 r$ at the surface of the sphere. The total energy required is the integral as q varies from 0 to Q (and hence r varies from 0 to R).

The Fermi Gas Model

Since the Coulomb term reduces the binding energy, you might expect based on our discussion so far that nuclei would have a lot of neutrons relative to the number of protons. But if you look at Fig. 9.6, a plot of the stable nuclei as a function of Z and N, you see that for small Z the stable nuclei tend to have $Z = N$ and while there is a deviation

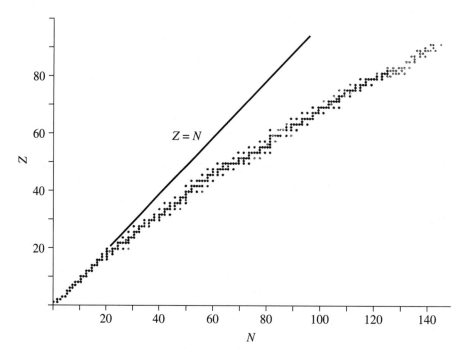

Figure 9.6 The distribution of the naturally occurring nuclei. Note the deviations from the straight line $Z = N$ as A increases.

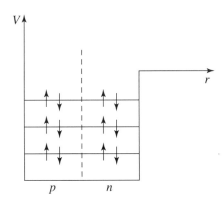

Figure 9.7 A schematic energy-level diagram showing the three lowest energy levels filled by nucleons in the nucleus. The energy levels are presumed to be nondegenerate. For ^{12}C six protons fill the three lowest proton energy levels and six neutrons fill the three lowest neutron energy levels.

toward increasing N relative to Z for larger Z, the deviation is a relatively modest one. We do not see nuclei with a huge excess of neutrons relative to protons, unless you want to include a neutron star in the table of stable nuclei (see Example 9.2). Why is this? For one thing, the neutron is not a stable particle. In free space, a neutron decays into a proton, an electron, and an electron antineutrino with a lifetime (see Section 9.3) of fifteen minutes:

$$n \rightarrow p + e^- + \bar{\nu}_e \tag{9.15}$$

So the real question is not why the nucleus is not primarily composed of neutrons but rather why the neutrons in the nucleus don't decay. To answer this question we need to bring quantum mechanics into our modeling process. Let's assume that we can treat the nucleons as moving independently in the potential energy well that confines these particles in the nucleus. Since the nuclear force treats the neutrons and protons as identical particles, the energy levels for the protons and neutrons will be the same, at least to the extent that we ignore the Coulomb repulsion of the protons. To illustrate, let's consider the example of ^{12}C, as indicated in Fig. 9.7. Since protons and neutrons are fermions, we can put at most two protons and two neutrons in each energy level, thus filling the lowest three proton energy levels and the lowest three neutron energy levels.[4] If one of these neutrons were to decay, it could do so only if there were enough energy released in the decay to put the proton in the fourth energy level, since the three lowest proton energy levels are already filled. Because the neutron is only slightly more massive than the proton, the energy released in the decay (9.15) is not sufficient to do this. Hence the neutrons in the ^{12}C nucleus are stable.

The energy-level diagram for ^{12}C shows why it is natural to have the number of protons equal to the number of neutrons in the nucleus, at least for small Z. The fourth term in the semiempirical mass formula, the **asymmetry term**, is a measure of the reduced binding energy as the nucleus moves away from the $Z = N$ condition. To see how the

---

[4]For the sake of simplicity, we are treating each of the energy levels as nondegenerate, as would be the case if the potential energy well were truly one dimensional.

dependence of this term on Z and A arises, we can apply the Fermi gas model that we used in treating electrons in a conductor in Section 7.4. There we saw that the Fermi energy is given by

$$E_F = \frac{\hbar^2}{2m}\left(\frac{3\pi^2 N}{V}\right)^{2/3} \tag{9.16}$$

where N is the number of identical fermions in the "box" of volume V. Recall that the total energy of the fermions is given by

$$E_{\text{total}} = \frac{3}{5}NE_F \tag{9.17}$$

Since in the nucleus there are two fermion gases, one composed of protons and the other composed of neutrons, the total energy is given by

$$E_{\text{total}} = \frac{3}{5}Z\frac{\hbar^2}{2m}\left(\frac{3\pi^2 Z}{V}\right)^{2/3} + \frac{3}{5}N\frac{\hbar^2}{2m}\left(\frac{3\pi^2 N}{V}\right)^{2/3}$$

$$\sim Z\left(\frac{Z}{V}\right)^{2/3} + N\left(\frac{N}{V}\right)^{2/3} \tag{9.18}$$

where Z is the number of protons and here N is the number of neutrons. We have taken the mass m to be the same for the neutron and the proton. In the last line of (9.18) we have suppressed the constants since we really want to focus on how the result depends on A and Z. Since $V \sim A$ and $N = A - Z$, we see that

$$E_{\text{total}} \sim \frac{1}{A^{2/3}}\left[Z^{5/3} + (A-Z)^{5/3}\right] \tag{9.19}$$

For a fixed number of nucleons, that is, for fixed A, we can find the value of Z (and therefore N) that minimizes the overall energy by setting the derivative of (9.19) with respect to Z equal to zero:

$$\frac{\partial E_{\text{total}}}{\partial Z} \sim \frac{1}{A^{2/3}}\left[Z^{2/3} - (A-Z)^{2/3}\right] = 0 \tag{9.20}$$

which requires $Z = A - Z = N$, that is, the number of protons in the nucleus equals the number of neutrons for the minimum energy state.

We can use expression (9.19) to calculate the excess energy when the nucleus deviates from the $Z = N$ configuration. Call this excess energy δE, the extra energy when

$$Z = A/2 + \delta Z \tag{9.21}$$

This additional energy is given by the *difference* between (9.19) for an arbitrary value of Z and for $Z = N = A/2$:

$$\delta E \sim \frac{1}{A^{2/3}}\left[Z^{5/3} + (A-Z)^{5/3}\right] - \frac{1}{A^{2/3}}\left[\left(\frac{A}{2}\right)^{5/3} + \left(\frac{A}{2}\right)^{5/3}\right] \tag{9.22}$$

Substituting (9.21) for Z, we obtain

$$\delta E \sim \frac{A^{5/3}}{A^{2/3}}\left[\left(1 + \frac{2\delta Z}{A}\right)^{5/3} + \left(1 - \frac{2\delta Z}{A}\right)^{5/3} - 2\right] \tag{9.23}$$

N	Z	Number
Even	Even	166
Odd	Odd	8
Even	Odd	57
Odd	Even	53

Table 9.2 Distribution of stable nuclei

Provided $\delta Z/A \ll 1$, we can use a Taylor series or a binomial expansion to obtain the leading term

$$\delta E \sim \frac{(\delta Z)^2}{A} = \frac{(Z - A/2)^2}{A} \tag{9.24}$$

which is the fourth term in the semiempirical mass formula. It enters with a minus sign in (9.12) since increasing the total energy of the nucleus corresponds to a reduction in the binding energy. Notice that the impact of this term becomes less pronounced for large A because of the presence of A in the denominator. We can think of the neutrons as the nuclear glue that manages to hold the nucleus together despite the increasing Coulomb repulsion of the protons.

We can get some insight into the fifth term, the **pairing term** in (9.12), from the Fermi gas model as well. The distribution of stable nuclei in Table 9.2 shows that there is a large preference for nuclei that have an even number of protons and an even number of neutrons (even–even nuclei). Moreover, very few of the stable nuclei have an odd number of protons and an odd number of neutrons. The pairing term responds to this distribution with a contribution that increases the binding energy for even–even nuclei and reduces it for odd–odd nuclei. We will see in Section 9.4 that the pairing term plays a crucial role in determining which nuclei are suitable candidates to fuel a nuclear reactor (or an atomic bomb). ^{12}C is a nice example of an even–even nucleus (see Fig. 9.7). Of course when two identical fermions are in the same energy level, the spin state must be antisymmetric under exchange since the spatial state is symmetric. As we noted in our discussion of the electrons in the helium atom in Section 7.3, this leads to the particles having a higher probability of overlapping in space than they would if they were in different energy levels. Such an overlap leads to an increase in the nuclear binding energy because of the attractive nature of the nuclear force.

The Shell Model

While we are on the subject of nuclear models, it is good to include the shell model in our discussion. One of the interesting features of nuclear physics that is missed by the semiempirical mass formula (see Fig. 9.8) is the unusual stability of nuclei with certain **magic numbers**

$$Z \text{ and/or } N = 2, 8, 20, 28, 50, 82, 126$$

The most striking example is the alpha particle, the ^{4}He nucleus, which has $Z = N = 2$. If you look at Fig. 9.3, you can see how the binding energy per nucleon for ^{4}He stands out relative to its nearby neighbors. Even more striking is the fact that it takes more than

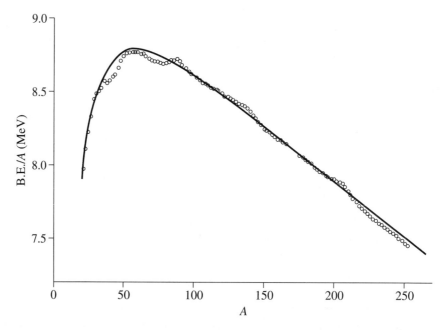

Figure 9.8 Deviations from the curve of binding energy (solid line) due to shell effects that are not included in the semiempirical mass formula. Note the expanded scale of binding energy per nucleon on the vertical axis in comparison with that used in Fig. 9.3. Adapted from W. F. Hornyak, *Nuclear Structure*, Academic Press, New York, 1975.

20 MeV to remove a proton or a neutron from the ^{4}He nucleus.[5] This stability is reminiscent of the stability of noble gases, which we understand in terms of filled shells of electrons in the atom. The situation is more complex with nuclei since we are not able to work out from first principles the contribution of each nucleon to the overall potential energy well confining these nucleons. And simple assumptions about the nature of the overall potential energy fail to account for the magic numbers. This problem was resolved independently by Maria Goeppert-Mayer and J. Hans D. Jensen in 1949. In particular, Goeppert-Mayer and Jensen, who shared the Nobel Prize for their work on nuclear structure, proposed that nucleons were subject to a strong coupling of each nucleon's intrinsic spin and its orbital angular momentum via the nuclear force. This coupling has the opposite sign to the spin–orbit coupling that occurs for electrons in atoms.[6]

---

[5] Another illustration of the unusual stability of nuclei with magic numbers is the fact that tin, which has $Z = 50$, has ten stable isotopes, whereas the average number of isotopes in that region of the periodic table is three or four. And there are six stable nuclei with $N = 50$ and seven stable nuclei with $N = 82$, whereas normally there are only two or three stable nuclei with the same number of neutrons. One area of much interest is whether it might be possible to leapfrog over the heaviest quite unstable elements to an "island of stability" (say $Z = 126$, for example) by colliding nuclei together. Recent reports suggest that an element with $Z = 118$ may have been created in the collision of calcium and californium.

[6] It should be emphasized that whereas the spin–orbit coupling that occurs in atomic physics can be understood as a consequence of the interaction between the magnetic moment of the electron and the magnetic field present in the atom (due to the "motion" of the nucleus in the electron's rest frame), the nuclear spin–orbit coupling arises from the dependence of the nuclear force on the intrinsic spin of the nucleon. We will see a striking example of this dependence of the nuclear force on intrinsic spin when we discuss the interactions between quarks in Chapter 10.

EXAMPLE 9.1 Use (9.12) to calculate the binding energy of the iron nucleus ^{56}Fe. What is the binding energy per nucleon?

SOLUTION Iron has 26 protons and 30 neutrons. Substituting the values $Z = 26$, $N = 30$, and $A = 56$ into (9.12), we obtain

$$\text{B.E.} = 56 a_1 - (56)^{2/3} a_2 - \frac{(26)^2}{(56)^{1/3}} a_3 - \frac{(26-28)^2}{56} a_4 + \frac{a_5}{\sqrt{56}}$$

$$= [(56)(15.75) - (14.64)(17.8) - (176.69)(0.711)$$
$$\quad - (0.0714)(94.8) + (0.1336)(11.2)] \text{ MeV}$$

$$= 490.5 \text{ MeV}$$

Note we have included the pairing term a_5 since iron is an even–even nucleus. Thus the binding energy per nucleon is

$$\frac{\text{B.E.}}{A} = \frac{490.5 \text{ MeV}}{56} = 8.76 \text{ MeV}$$

EXAMPLE 9.2 In the expression (9.12) for the binding energy in the semiempirical mass formula we have taken into account strong and electromagnetic interactions but not gravitational interactions. Show that gravity adds a term of the form $a_6 A^n$. Obtain values for a_6 and n.

SOLUTION In (9.14) we calculated the energy required to assemble a ball of charge Q with radius R. Since the gravitational force and the electrostatic force exhibit the same $1/r^2$ behavior, we can carry over the result of (9.14) to gravity with the replacements $Q^2 \to M^2$, where M is the mass of the nucleus, and $(1/4\pi\epsilon_0) \to G$, where G is the gravitational coupling constant. Thus the additional term that must be added to (9.12) arises from the energy

$$\frac{3}{5} \frac{GM^2}{R} = \frac{3}{5} \frac{GA^2 m^2}{r_0 A^{1/3}}$$

where m is the mass of a nucleon. This additional term is of the form

$$a_6 A^{5/3}$$

with

$$a_6 = \frac{3}{5} \frac{Gm^2}{r_0}$$
$$= \frac{3}{5} \frac{(6.67 \times 10^{-11})(1.67 \times 10^{-27})^2}{1.2 \times 10^{-15}} \text{ J}$$
$$= 5.8 \times 10^{-37} \text{ MeV}$$

Since gravity is an attractive force, this term *adds* to the binding energy and thus enters (9.12) with a positive sign. Note that the value for a_6 is roughly 38 orders of magnitude smaller than the terms due to strong interactions in (9.12). Nonetheless, this term can play a significant role if the number of nucleons is sufficiently large. In fact, we can use (9.12) with this additional gravity term to estimate the minimum mass required for a neutron star. See Problem 9.16.

9.3 Radioactivity

The heavier naturally occurring nuclei included in Fig. 9.3 are typically unstable. For example, ^{238}U has a lifetime of 6.45×10^9 yr, while ^{226}Ra has a lifetime of 2300 yr. The lifetime of ^{238}U is sufficiently large that we can understand its presence on Earth as a residue of the material out of which the Earth was formed roughly 5 billion years ago. But what about radium? Shouldn't all the radium have decayed away by now? Most of the naturally occurring radioactive nuclei with $A \geq 210$ can be grouped into three series, the $4n$, $4n+2$, and $4n+3$ series, with long-lived parent nuclei ^{232}Th, ^{238}U, ^{235}U, respectively.[7] Nuclei such as ^{226}Ra are referred to as daughter nuclei, since these nuclei are byproducts of the decay of the parents.

Let's first focus on the parent nuclei. How do we determine how likely an unstable nucleus such as ^{235}U is to decay? If you line up a collection of ^{235}U nuclei, they are all identical. Some of these nuclei may have been created in a supernova explosion that occurred billions of years ago and some of them may have been produced yesterday in a nuclear reactor. They are all the same. None of them has wrinkles or gray hair that would allow us to see signs of aging. Consequently, when the decay rate for these nuclei is calculated using quantum mechanics, we find that R, the probability of decay per unit time, is independent of time.[8] The odds that any particular unstable nucleus decays in the next second is independent of how long that nucleus has been in existence. Thus if we have a collection of N of these nuclei, the number of these nuclei will be equal to $N + dN$ a time dt later where

$$dN = -NR\,dt \tag{9.25}$$

since the rate of decay per unit time multiplied by the time interval is the rate for decay in that period of time. We can also write this equation simply as

$$\frac{dN}{dt} = -NR \tag{9.26}$$

Equation (9.25) is easily integrated by separating the variables:

$$\frac{dN}{N} = -R\,dt \tag{9.27}$$

Integrating the number of nuclei from $N(0)$ to $N(t)$ as time varies from 0 to t, we obtain

$$\int_{N(0)}^{N(t)} \frac{dN}{N} = -\int_0^t R\,dt \tag{9.28}$$

$$\ln \frac{N(t)}{N(0)} = -Rt \tag{9.29}$$

Thus

$$N(t) = N(0)e^{-Rt}$$
$$= N(0)e^{-t/\tau} \tag{9.30}$$

---

[7]The rationale for naming the series in this way will become apparent shortly.

[8]The proof is often referred to as Fermi's Golden Rule, since Enrico Fermi established this result in his lectures on quantum mechanics. This result applies to unstable atoms as well as nuclei.

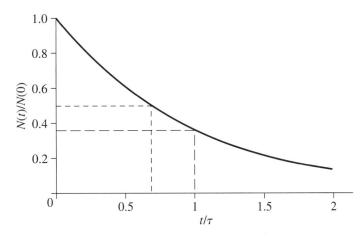

Figure 9.9 The number of unstable nuclei decreases exponentially with time. At time $t = 0$ the number of nuclei is $N(0)$. The half-life $t_{1/2}$ (short dashed line) is the time for the number to fall to $N(0)/2$ while the lifetime τ (long dashed line) is the time for the number to fall to $N(0)/e$.

where in the second line we have defined $\tau = 1/R$. The quantity τ is referred to as the **lifetime** of the unstable nucleus. It is the time for the number of nuclei to fall to $1/e$ of the initial number. See Fig. 9.9. While particle physicists (see Chapter 10) typically use the lifetime to characterize the decays of unstable particles, nuclear physicists ordinarily refer to the **half-life**, namely the time for the number of nuclei to fall to one half the initial number:

$$N(t_{1/2}) = \frac{1}{2} N(0) \tag{9.31}$$

or equivalently

$$\frac{1}{2} = e^{-t_{1/2}/\tau} \tag{9.32}$$

Thus

$$t_{1/2} = \tau \ln 2 = 0.693\, \tau \tag{9.33}$$

If we try to work out the number of daughter nuclei that are present at time t, we must take into account the formation of these daughter nuclei from the decay of the parents as well as the decay of the daughter nuclei themselves. Consequently, we have the two differential equations

$$\frac{dN_P}{dt} = -R_P N_P \tag{9.34}$$

for the parent and

$$\frac{dN_D}{dt} = R_P N_P - R_D N_D \tag{9.35}$$

for the daughter, where we have inserted subscripts to distinguish the number of parent nuclei N_P from the number of daughter nuclei N_D. The first equation leads to

$$N_P(t) = N(0) e^{-R_P t} \tag{9.36}$$

where $N(0)$ is the number of parent nuclei present at $t = 0$ and therefore (9.35) becomes

$$\frac{dN_D}{dt} = R_P N(0) e^{-R_P t} - R_D N_D \tag{9.37}$$

To solve this differential equation, multiply by $e^{R_D t} dt$, leading to

$$e^{R_D t} dN_D + R_D N_D e^{R_D t} dt = R_P N(0) e^{(R_D - R_P)t} dt \tag{9.38}$$

or

$$d\left(N_D e^{R_D t}\right) = R_P N(0) e^{(R_D - R_P)t} dt \tag{9.39}$$

$$N_D e^{R_D t} = \frac{N(0) R_P}{R_D - R_P} e^{(R_D - R_P)t} + C \tag{9.40}$$

where C is a constant of integration. At $t = 0$, $N_D = 0$. Therefore

$$C = -\frac{R_P N(0)}{R_D - R_P} \tag{9.41}$$

and hence

$$N_D(t) = \frac{N(0) R_P}{R_D - R_P} \left(e^{-R_P t} - e^{-R_D t}\right) \tag{9.42}$$

If the parent is long-lived compared to the daughter, $R_P = 1/\tau_P \ll R_D = 1/\tau_D$

$$N_D(t) = \frac{N(0) R_P}{R_D} \left(e^{-R_P t} - e^{-R_D t}\right) \tag{9.43}$$

For $t \gg \tau_D$, then $e^{-R_D t} \ll 1$ and since $N_P(t) = N(0) e^{-R_P t}$, (9.43) reduces to

$$N_P(t) R_P = N_D(t) R_D \tag{9.44}$$

once equilibrium is established. Thus for times that are short compared with the parent's lifetime, the number of parent nuclei stays essentially constant. Since the daughter nuclei are presumed to decay quickly after they are formed, their population is dictated by the resupply from the decay of the parents. This is a special case of the more general result for a long-lived parent P and a series consisting of a short-lived daughter D, a short-lived granddaughter GD, and so on in equilibrium

$$N_P(t) R_P = N_D(t) R_D = N_{GD}(t) R_{GD} = \cdots \tag{9.45}$$

There are three common modes for unstable nuclei to decay, typically referred to as α, β, and γ decay. These Greek letters were assigned initially to these decay modes by Rutherford because the nature of the decay processes was not clear. Now we know that alpha decay involves the emission of a ^{4}He nucleus, or alpha particle; β decay involves the emission of an electron (or possibly the electron's antiparticle, the positron); and γ decay involves the emission of a high-energy photon. The reason the photon tends to be so energetic, at least in nuclear decays, comes from the fact that the potential energy well confining the nucleons in the nucleus is so much smaller than the size of the atom itself, leading to a much larger separation between the energy levels than is the case for the energy levels of an atom (see Example 3.1).

The parent nuclei in the three naturally occurring radioactive series all decay by emission of an alpha particle. Since an alpha particle has $A = 4$ and $Z = 2$, emission

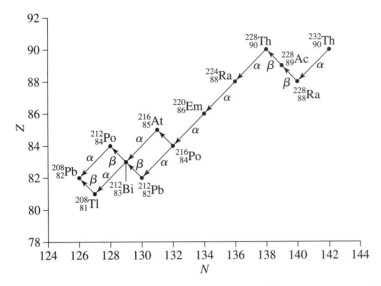

Figure 9.10 The elements in the $4n$ series, which starts with ^{232}Th and ends with ^{208}Pb.

of an alpha particle changes the atomic mass number of the nucleus by four units and the charge of the nucleus by two units. A beta decay, on the other hand, does not affect the atomic mass number but changes the charge on the nucleus by one unit ($Z \to Z+1$ when the beta particle is an electron). A gamma decay does not, of course, change A or Z. Thus each of the elements in the $4n$ series with parent ^{232}Th has an atomic mass number divisible by 4. See Fig. 9.10. On the other hand, the elements in the series with ^{238}U as the parent all have $A = 4n + 2$ and those in the series with ^{235}U as the parent have $A = 4n + 3$, where n is an integer. There is, in addition, a $4n + 1$ series. This series, however, does not occur naturally, since the parent nucleus ^{237}Np has a half-life of 2.1×10^6 yr. Thus the parent nuclei have all decayed away in the roughly five billion years the Earth has been in existence (see Problem 9.24). This series can be generated artificially, say in a nuclear reactor.

Alpha Particle Decay

One of the striking features of the various radioactive series in Fig. 9.10 is the large difference in alpha particle decay rates. For example, in the $4n$ series, the parent ^{232}Th has a half-life of 1.4×10^{10} yr with a kinetic energy of the emitted α particle of 4.1 MeV, while ^{212}Po, one of the daughters, has a half-life of 0.30×10^{-9} s with a kinetic energy of the emitted alpha particle of 9.0 MeV. Although the kinetic energy of the alpha particle is only twice as large in the decay of this isotope of polonium relative to that for thorium, the half-life is 7×10^{-25} times shorter!

In the early days of the development of quantum mechanics, George Gamow applied the tunneling approximation that we worked out in Section 4.7 to calculate the lifetime of these alpha particle emitters. His approach involved an interesting mix of classical physics and quantum mechanics. In particular, Gamow reasoned that the emitter (with Z protons and N neutrons) could be thought of as a core nucleus with $Z - 2$ protons and $N - 2$ neutrons and an alpha particle. The potential energy of interaction of the alpha particle and the core is shown in Fig. 9.11. Within the core, the strong interactions

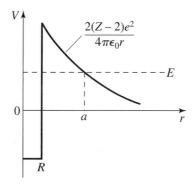

Figure 9.11 The potential energy experienced by an alpha particle in the nucleus, including an attractive nuclear part and a repulsive Coulomb part. An alpha particle tunneling through the Coulomb potential exits the nucleus with energy E.

between the alpha particle and the core generate a potential energy well of radius R that confines the alpha particle. Outside the core ($r > R$), the potential energy is given by the Coulomb term $2(Z-2)e^2/4\pi\epsilon_0 r$. This Coulomb interaction effectively generates a barrier through which the alpha particle can tunnel. The alpha particle can be thought of as bouncing back and forth in the potential well, making repeated attempts to tunnel through the barrier. The probability of transmission through the barrier is given by

$$T = e^{-G} \tag{9.46}$$

where

$$G = 2\int_R^a \frac{\sqrt{2m[V(r)-E]}}{\hbar} \, dr \tag{9.47}$$

with the potential energy shown in Fig. 9.11. The value of a is determined by setting

$$\frac{2(Z-2)e^2}{4\pi\epsilon_0 a} = E \tag{9.48}$$

where E is the kinetic energy of the emitted alpha particle, since for $r > a$ the energy of the alpha particle is larger than the potential energy and tunneling is no longer involved. The lifetime for the decay can be approximated as

$$\tau = \frac{2R}{v} e^G \tag{9.49}$$

where v is the speed of the alpha particle in the well and therefore $2R/v$ is the time between encounters with the barrier. The number of encounters before tunneling takes place is taken to be e^G, namely the inverse of the transmission probability. If we substitute the expression (9.48) for the kinetic energy E in (9.47), we obtain

$$G = \frac{4}{\hbar}\sqrt{\frac{m(Z-2)e^2}{4\pi\epsilon_0}} \int_R^a \left(\frac{1}{r} - \frac{1}{a}\right)^{1/2} dr \tag{9.50}$$

Given that

$$\int_R^a \left(\frac{1}{r} - \frac{1}{a}\right)^{1/2} dr = \sqrt{a}\left[\cos^{-1}\sqrt{\frac{R}{a}} - \left(\frac{R}{a} - \frac{R^2}{a^2}\right)^{1/2}\right] \tag{9.51}$$

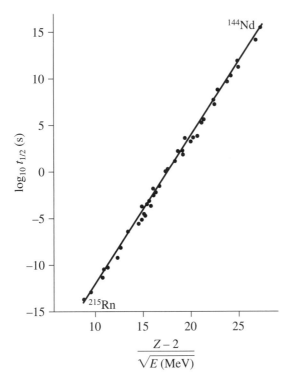

Figure 9.12 The logarithm of the half-life of a large number of alpha particle emitters as a function of $(Z-2)/\sqrt{E}$.

and $R \ll a$, we can approximate G as

$$G \cong \frac{2\pi}{\hbar}\sqrt{\frac{m(Z-2)e^2 a}{4\pi\epsilon_0}} \qquad (9.52)$$

Given the dependence of a on the kinetic energy E and charge $Z-2$ on the core, we find the numerical value for G can be written as

$$G \cong \frac{4(Z-2)}{\sqrt{E(\text{MeV})}} \qquad (9.53)$$

provided the energy is expressed in MeV. Since G occurs in the exponent in (9.49), the lifetime depends very sensitively on the values for Z of the nucleus and the kinetic energy E of the emitted alpha particle. Figure 9.12 shows a plot of $\log_{10} t_{1/2}$ of alpha particle emitters versus $(Z-2)/\sqrt{E}$. Notice the agreement over 30 orders of magnitude.

Beta and Gamma Decay

Our discussion of alpha decay may lead you to think that beta decay involves the emission of an electron that is contained within the nucleus. This is definitely not the case. In some ways, beta decay has more in common with gamma decay, in that the photon that is emitted in gamma decay is not present in the nucleus before the emission but rather is created when the nucleus makes a transition from one energy level to one with a lower energy. Similarly, the electron (or positron) and associated neutrino that are emitted in beta decay are created as the nucleus makes the transition from one state to another. Of course, to conserve charge the nucleus in the final state will be a different nucleus than the one in the initial state. The prototype beta decay is the decay of the neutron, as shown in (9.15), with a lifetime of 900 s. Other interesting examples of beta decay include the decay of tritium

$$t \to {}^3\text{He} + e^- + \bar{\nu}_e \tag{9.54}$$

for which $t_{1/2} = 12.3$ yr, the decay of carbon-14

$$^{14}\text{C} \to {}^{14}\text{N} + e^- + \bar{\nu}_e \tag{9.55}$$

for which $t_{1/2} = 5730$ yr, and

$$^{239}\text{Np} \to {}^{239}\text{Pu} + e^- + \bar{\nu}_e \tag{9.56}$$

for which $t_{1/2} = 24$ d. These examples all suggest a nucleus that is unstable because it has too many neutrons is making a transition to one with one fewer neutron and one more proton by a neutron within the nucleus decaying. It is better, however, to treat the initial and final state nuclei as units and not try to understand the beta decay as simply neutron decay since, for example, it is not possible to account for the wide variation in lifetimes in this way. Moreover, there are some beta decays that involve the emission of a positron, such as

$$^{239}\text{Np} \to {}^{239}\text{U} + e^+ + \nu_e \tag{9.57}$$

Alternatively, the neptunium nucleus can absorb an electron from the atom, undergoing the reaction

$$^{239}\text{Np} + e^- \to {}^{239}\text{U} + \nu_e \tag{9.58}$$

in a process referred to as **electron capture**. Each of the reactions (9.54) through (9.58) is an example of a **weak interaction**. These interactions are given this name because the strength of the interaction is many orders of magnitude weaker than the strong interactions that bind the nucleons together in the nucleus. In Chapter 10 we will examine these fundamental interactions in more detail.

EXAMPLE 9.3 The relative natural abundance on Earth of the two isotopes ^{235}U and ^{238}U is roughly $1/138$. If we assume that the two elements were initially formed with the same abundance, then how much time has elapsed since these elements were created? *Note*: The half-life of ^{235}U is 7.04×10^8 yr and the half-life of ^{238}U is 4.47×10^9 yr.

SOLUTION From the half-lives we know that $\tau_{235\text{U}} = 1.02 \times 10^9$ yr and $\tau_{238\text{U}} = 6.45 \times 10^9$ yr. Since

$$N(t) = N(0)e^{-t/\tau}$$

therefore $N_{235\text{U}} = N(0)e^{-t/\tau_{235\text{U}}}$ and $N_{238\text{U}} = N(0)e^{-t/\tau_{238\text{U}}}$. Hence

$$\frac{1}{138} = \frac{e^{-t/\tau_{235\text{U}}}}{e^{-t/\tau_{238\text{U}}}}$$

Substituting in the lifetimes, we find $t = 6.0 \times 10^9$ yr. See the discussion in Section 9.5 on the formation of the elements.

EXAMPLE 9.4 In the atmosphere, cosmic ray protons collide with the nuclei of air molecules and produce neutrons. These neutrons are absorbed by ^{14}N nuclei in the reaction

$$n + {}^{14}N \to {}^{14}C + p$$

In the steady state, the ratio ^{14}C/all Carbon $= 1.3 \times 10^{-12}$. ^{14}C then beta decays to ^{14}N in the reaction

$$^{14}C \to {}^{14}N + e^- + \bar{\nu}_e$$

with $t_{1/2} = 5730$ yr. Consequently, all plants are radioactive, but the radioactivity diminishes after the plant dies since the ^{14}C is no longer replenished by photosynthesis. If the activity of wood from an ancient building is 12 decays/minute per gram of C and the activity of living wood is 15 decays/minute per gram of C, how old is the building?

SOLUTION
$$\frac{dN}{dt} = \frac{d}{dt}\left(N(0)e^{-t/\tau}\right) = -\frac{N(0)}{\tau}e^{-t/\tau}$$

Thus

$$\ln\frac{-dN/dt}{N(0)/\tau} = -t/\tau$$

The initial activity is $N(0)/\tau$ and the activity at time t is $-dN/dt$. Therefore

$$t = \tau \ln\frac{N(0)/\tau}{-dN/dt} = (8270 \text{ yr})\ln(15/12) = 1845 \text{ yr}$$

where $\tau = t_{1/2}/(0.693) = 8270$ yr.

EXAMPLE 9.5 The half-life of $^{228}_{88}$Ra, the first daughter of the thorium series, is 5.75 yr. The parent, $^{232}_{90}$Th, has a half-life of 1.41×10^{10} yr. In 200 g of thorium ore, how much radium is there?

SOLUTION The number of thorium nuclei in the rock is

$$\left(\frac{200}{232}\right)6.02 \times 10^{23} = 5.2 \times 10^{23}$$

Because $t \gg \tau_{Ra}$, we can use (9.44):

$$\frac{N_{Th}}{\tau_{Th}} = \frac{N_{Ra}}{\tau_{Ra}}$$

Hence the number of radium atoms

$$N_{Ra} = N_{Th}\frac{\tau_{Ra}}{\tau_{Th}}$$

$$= (5.2 \times 10^{23})\left(\frac{8.3 \text{ yr}}{2 \times 10^{10} \text{ yr}}\right) = 2.15 \times 10^{14}$$

corresponding to a mass of

$$228 \text{ g}\left(\frac{2.15 \times 10^{14}}{6.02 \times 10^{23}}\right) = 8.2 \times 10^{-8} \text{ g}$$

9.4 Nuclear Fission

Given our discussion of radioactivity, we expect that a nucleus will be potentially unstable whenever the sum of the masses of the final state products is less than the mass of the nucleus. Total energy, including mass energy and kinetic energy, is, of course, conserved in this decay. The difference in mass energy is released in the form of kinetic energy, say in an alpha particle decay, or in the form of electromagnetic energy of the emitted photon and kinetic energy of the recoiling nucleus in gamma decay. But this raises a question: why are any of the nuclei heavier than iron stable, given that the curve of binding energy peaks at iron? It appears that a heavy nucleus such as uranium should find it energetically advantageous to split into two lighter nuclei, ones that are closer to the peak in the curve of binding energy. Since the binding energy peaks at roughly 8.5 MeV per nucleon while the binding energy for the heavy nucleus is roughly 7.5 MeV per nucleon, for a nucleus with 200 nucleons roughly 200 MeV of energy should be released. This is a factor of 10^8 times greater than the few eV of energy released in a typical chemical reaction, or a factor of 10^6 per unit mass.

Where does this energy release come from? It is primarily the reduction in Coulomb energy that occurs in dividing a nucleus with Z protons into two fragments, each with significantly fewer protons than the initial nucleus. These two fragments then repel each other and move apart, gaining approximately 200 MeV of kinetic energy. As was noted earlier, most nuclei are spherical in shape. A sphere is noteworthy for having the smallest surface area per unit volume. Thus if a nucleus is to split, or fission, the surface area must increase (see Fig. 9.13). But as we saw in the semiempirical mass formula, there is a penalty in binding energy for having more nucleons at the surface, which is after all why the nuclear surface tension tends to pull the nucleons into a spherical shape. Thus there is a surface tension energy barrier to the fission process, as illustrated in Fig. 9.14. The

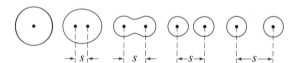

Figure 9.13 A schematic diagram of the steps involved in nuclear fission. Note the increase in surface area as the process takes place.

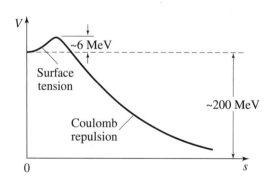

Figure 9.14 A rough plot of the potential energy experienced by a large fissioning nucleus. Note the surface tension barrier and the effect of Coulomb repulsion once the barrier is passed.

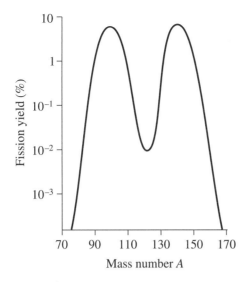

Figure 9.15 The distribution of nuclei generated in nuclear fission of ^{235}U by low-energy neutrons.

barrier is roughly 6 MeV in height. In spontaneous fission the mass of the nucleus that must tunnel through this barrier is much greater than the mass of the alpha particle that tunnels through the Coulomb barrier in alpha particle decay. For a nucleus such as ^{238}U the half-life for such spontaneous fission events is on the order of 10^{16} yr. Consequently, only one in every 2 million ^{238}U nuclei decays by spontaneous fission as opposed to alpha particle decay.

Alternatively, it is possible to induce fission reactions with neutrons. If a nucleus such as ^{235}U absorbs a neutron, even a slow neutron with very little kinetic energy, temporarily becoming an excited ^{236}U nucleus, the energy of excitation is larger than the height of the surface tension barrier. Thus such an excited nucleus can rapidly fission:

$$^{236}\text{U} \rightarrow X + Y + 2.5\,n \qquad (9.59)$$

The two fragments that are produced tend to be asymmetric, as indicated in Fig. 9.15. Since the fission process has dramatically reduced the number of nucleons in each fragment, these fragments are relatively rich in neutrons in comparison with the stable nuclei shown in Fig. 9.6. Thus the fission fragments tend to be quite radioactive with lots of beta decay. Moreover, a typical fission reaction releases 2 or 3 additional neutrons [on average 2.5 neutrons, as indicated in (9.59)], opening up the possibility of a **chain reaction**. If each fission event initiates a single additional fission event, the reaction is self-sustaining and has a reproduction constant $K = 1$, which is called the **critical condition**. In this case the reaction runs at a constant rate. If $K > 1$, the reaction is supercritical. For example, if two of the neutrons produced in each fission event were each to induce an additional fission reaction (corresponding to a reproduction constant $K = 2$), we would potentially have 2^n fission reactions after n generations. Since each of these reactions releases 200 MeV of energy, the amount of energy released can be very large indeed. Moreover, since the time between subsequent events is as short as 10^{-8} s, the release of energy can be very rapid as well.

Why should we focus on ^{235}U which is, as we noted earlier, not a very abundant isotope of uranium? If we calculate the excitation energy after the more abundant isotope ^{238}U absorbs a low-energy neutron turning into ^{239}U, we see that the excitation energy is

$$Q = \left(m_n + m_{238_U} - m_{239_U}\right) c^2 = 5 \text{ MeV} \tag{9.60}$$

as can be checked using the semiempirical mass formula (9.12). Thus the excitation energy is 1 MeV below the top of the surface tension barrier. On the other hand, if we repeat the same calculation for ^{235}U, we find

$$Q = \left(m_n + m_{235_U} - m_{236_U}\right) c^2 = 6.6 \text{ MeV} \tag{9.61}$$

or 0.6 MeV above the barrier. Why the difference? The ^{238}U nucleus is an even–even nucleus ($Z = 92$ and $N = 146$). Thus it is more tightly bound than the ^{235}U nucleus, which is an even–odd nucleus ($Z = 92$ and $N = 143$). Consequently, when we calculate the excitation energy Q, we get a smaller value for ^{238}U as the target than for ^{235}U. Of course, neutrons with kinetic energy greater than 1 MeV would readily produce fission if absorbed by ^{238}U. The problem with ^{238}U as the fuel for a chain reaction is that neutrons produced in the fission reactions, which are typically quite energetic, can lose energy quickly through inelastic collisions and fall below the 1 MeV threshold. These neutrons are then absorbed by the ^{238}U nucleus through the reaction

$$n + {}^{238}\text{U} \to {}^{239}\text{U}^* \to {}^{239}\text{U} + \gamma \tag{9.62}$$

Rather than fissioning, the ^{239}U nucleus then undergoes a beta decay

$$^{239}\text{U} \to {}^{239}\text{Np} + e^- + \bar{\nu}_e \tag{9.63}$$

Neutrons that follow this pathway therefore do not generate a subsequent fission event.[9]

Nuclear Reactors

As we saw in the previous section, the natural abundance on Earth of ^{235}U is only 0.7%. Moreover, as we have just discussed, neutrons that interact with the more abundant isotope ^{238}U tend to be captured by the nucleus without producing a fission event. Thus it is a challenge to generate a self-sustaining chain reaction in a nuclear reactor using natural uranium. Figure 9.16 shows the cross section for inducing fission in ^{235}U as a function of neutron kinetic energy. The cross section rises rapidly as the neutron's kinetic energy decreases, reaching a value of 10^3 barns for thermal neutrons, that is neutrons with kinetic energy characteristic of neutrons at room temperature, namely 1/40 eV. Moreover, the cross section for ^{238}U to absorb a low-energy neutron is substantially less than that for ^{235}U. This is a consequence of the pairing interaction that we discussed earlier. In general, nuclei with an odd number of neutrons have a larger cross section for absorbing a low-energy neutron than do nuclei with an even number of neutrons. In order to take advantage of these effects, it is necessary to slow down the neutrons

---

[9]The neptunium produced in the reaction (9.63) undergoes a subsequent beta decay via the reaction $^{239}\text{Np} \to {}^{239}\text{Pu} + e^- + \bar{\nu}_e$. This particular isotope of plutonium is an even–odd nucleus, like ^{235}U. When it absorbs a neutron, it too has sufficient energy to surmount the surface tension barrier and fission rapidly. See Example 9.7. Otherwise, the fate of ^{239}Pu is to alpha decay to ^{235}U with a half-life of 24,000 yr.

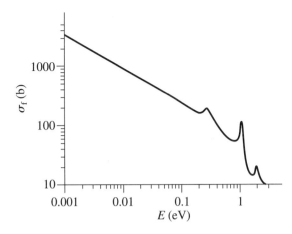

Figure 9.16 The cross section for neutron-induced fission of ^{235}U as a function of neutron kinetic energy for low-energy neutrons.

produced in a fission event by letting them interact with a moderator. A potentially good moderator is H_2O, since a neutron colliding with a proton can lose much of its kinetic energy to the proton very much as what happens when two billiard balls of equal mass collide. But it turns out the cross section for a proton in the water molecule to absorb a neutron (becoming a deuteron) is sufficiently large that ordinary water is not a suitable moderator for natural uranium ore. One (relatively expensive) alternative is to use heavy water as the moderator, since deuterons have a much smaller cross section for absorbing neutrons than do protons. Another alternative is to use graphite as the moderator. Although the mass of a carbon nucleus is obviously larger than that of a proton, graphite is widely available (although special consideration must be given to providing graphite of very high purity, since the impurities in graphite, especially boron, have large cross sections for absorbing neutrons). The very first nuclear reactor built in 1942 by a group headed by Enrico Fermi at the University of Chicago used graphite as the moderator. A third option is to use ordinary water as the moderator but enrich the amount of ^{235}U to roughly 3%.[10]

Most nuclear reactors in the United States use enriched uranium and water as the moderator. In a pressurized water reactor (see Fig. 9.17), the water in the primary loop is kept under pressure to keep it from boiling. The heat is transferred through a heat exchanger to a secondary water loop, where steam is generated to drive a turbine that generates electricity. The water in the secondary loop is isolated from the primary so that the secondary water and steam are not contaminated with radioactivity.

A nuclear reactor runs with a reproduction constant that is very close to one to provide for a self-sustaining chain reaction. Control rods composed of materials such as cadmium, which has a high cross section for neutron absorption, are used to guarantee that the reactor does not go supercritical ($K > 1$). Because a small fraction of the neutrons that contribute to the fission process are delayed neutrons that are emitted more slowly than the prompt neutrons that predominate in the reaction (9.59), there is adequate time to respond to changes in K.

---

[10]This enrichment process presents a technical challenge, one that we will discuss in more detail in Section 9.6.

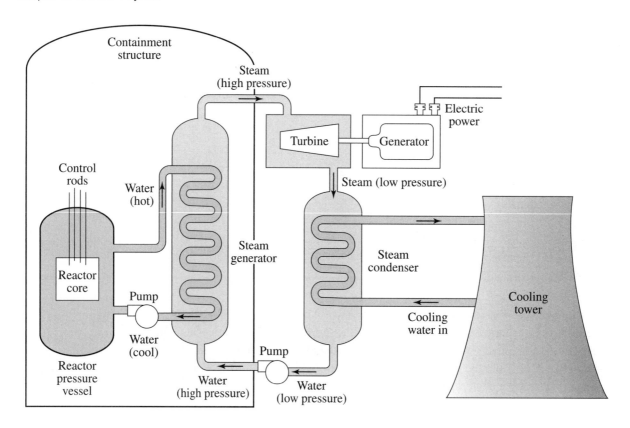

Figure 9.17 A schematic diagram of a pressurized water nuclear reactor.

The primary concerns with reactor safety involve a loss of coolant that might cause the reactor to overheat, melting the fuel elements and perhaps generating a nonnuclear explosion. There are also concerns with the long-term disposal of the radioactive waste products that are generated in the reactor. Moreover, there is the risk that enriched uranium or the plutonium that is produced in the reactor could be the target of terrorists seeking the capacity to make a bomb. Nonetheless, nuclear energy has the considerable advantage of not generating greenhouse gases as do conventional power plants that use fossil fuels such as coal and oil. And since a nuclear reactor can create (or breed) its own fuel through the production of plutonium [see the discussion following (9.62)], there is not the same concern about running out of fuel for a reactor based on nuclear fission as there is for a power plant that uses fossil fuels.

EXAMPLE 9.6 Calculate the energy released if ^{238}U were to fission into two equal-size fragments. Ignore the effect of the pairing term in (9.12).

SOLUTION We want to evaluate

$$[m_{\text{nuc}}(Z, A) - 2m_{\text{nuc}}(Z/2, A/2)]\, c^2$$
$$= a_2 \left[A^{2/3} - 2\,(A/2)^{2/3} \right] + a_3 \left[\frac{Z^2}{A^{1/3}} - 2\frac{(Z/2)^2}{(A/2)^{1/3}} \right]$$

where $m_{\text{nuc}}(Z, A)$ is the mass of a nucleus with atomic number Z and atomic mass number A for $Z = 92$ and $A = 238$. Notice that only the surface term and the Coulomb term contribute to this difference in mass energy. Therefore

$$[m_{\text{nuc}}(92, 238) - 2m_{\text{nuc}}(46, 119)]c^2 = (17.8 \text{ MeV})\left[(238)^{2/3} - 2(119)^{2/3}\right]$$

$$+ (0.71 \text{ MeV})\left[\frac{(92)^2}{(238)^{1/3}} - 2\frac{(46)^2}{(119)^{1/3}}\right]$$

$$= -178 \text{ MeV} + 359 \text{ MeV}$$

$$= 181 \text{ MeV}$$

Thus the reduction in Coulomb energy is a whopping 359 MeV, but the net energy released in the fission is "only" 181 MeV since the increase in surface area required 178 MeV.

EXAMPLE 9.7 Calculate the energy of excitation when ^{239}Pu absorbs a low-energy neutron.

SOLUTION

$$Q = (m_n + m_{^{239}\text{Pu}} - m_{^{240}\text{Pu}})c^2 = \text{B.E.}_{^{240}\text{Pu}} - \text{B.E.}_{^{239}\text{Pu}}$$

where we can make use of the expression (9.12) for the binding energies. Thus

$$Q = a_1(240 - 239) - a_2[(240)^{2/3} - (239)^{2/3}] - a_3(94)^2\left[\frac{1}{(240)^{1/3}} - \frac{1}{(239)^{1/3}}\right]$$

$$- a_4\left[\frac{(94-120)^2}{240} - \frac{(94-119.5)^2}{239}\right] + \frac{a_5}{\sqrt{240}}$$

Substituting in the values for a_1 through a_5 from (9.13), we obtain

$$Q = 6.9 \text{ MeV}$$

which is higher than the 6 MeV surface tension barrier. Thus neutron-induced fission can proceed quite rapidly for this isotope of plutonium.

9.5 Nuclear Fusion

Let's now turn our attention to the other end of the curve of binding energy. As we move from the lighter elements toward the heavier ones, the binding energy per nucleon increases. Thus if two lighter elements fuse together, there will be a release of energy. For example,

$$d + d \rightarrow {}^3\text{He} + n + 3.3 \text{ MeV} \tag{9.64}$$

$$d + d \rightarrow t + p + 4.0 \text{ MeV} \tag{9.65}$$

Each of these reactions releases roughly 1 MeV per nucleon involved in the reaction, comparable to the energy released per nucleon in fission reactions. Moreover, the fuel for these reactions is deuterium, which is present at the level one part per 6000 in the hydrogen atoms on Earth. In fact, there is enough deuterium in the Earth's oceans to supply the world's energy needs for the foreseeable future. And while the tritium produced in the reaction (9.65) is unstable and undergoes beta decay through the reaction (9.54), the half-life is only 12.3 years. Thus, the issue of taking care of the radioactive wastes for millennia, which is a concern with waste products from a fission reactor, is not such a problem for a fusion reactor.

So what's the problem with fusion? Why don't the lighter elements naturally fuse into the heavier ones, at least up to iron? Reactions such as (9.64) and (9.65) are strong interactions, that is, interactions mediated by the nuclear force. As we remarked in Section 9.1, for these reactions to occur the deuterons must essentially touch each other. But each of the deuterons has a charge $+e$ and thus it is necessary to overcome the Coulomb repulsion to get them sufficiently close together. In order to reach a separation of R, it is necessary that the kinetic energy of the deuterons be on the order of $e^2/4\pi\epsilon_0 R$. For $R = 1$ fm, this energy is 10^6 eV. Or better said, if we set this energy equal to $k_B T$, where k_B is the Boltzmann constant, we obtain a temperature of 10^{10} K. At these high temperatures, the deuterium atoms are ionized. The resulting gas of charged particles is referred to as a **plasma**. Of course, it isn't possible to heat the plasma to 10^{10} K in a crucible with a bunsen burner. In fact, confining the deuterons while they are heated to such high temperatures presents an enormous challenge. The main strategy has been to use magnetic fields to confine the particles. The plasma can be heated through the application of RF (radio frequency) waves and the injection of energetic neutral particles into the plasma. In order to reach a state in which the energy generated by the fusion reactions exceeds the energy used to heat the plasma it is necessary that the product of the ion density and the containment time be greater than 10^{16} s/cm^3 for the deuteron–deuteron reactions.[11] So far, while it has been possible to reach ignition temperatures in a plasma of sufficiently large density, it has not been possible to confine the plasma for a sufficiently long time to make a fusion power reactor feasible. Because of turbulence the plasma tends to leak out. So far the only way to minimize this leakage has been to make the volume containing the plasma very large, hence making a potential reactor too expensive to be economically viable, at least for the time being.

Nonetheless, fusion reactions are of immense importance to us, for these are the reactions that power our Sun as well as the other stars in the universe. In stars like our Sun that have an abundance of hydrogen, the fusion reactions are

$$p + p \to d + e^+ + \nu_e + 0.4 \text{ MeV} \tag{9.66}$$

$$p + d \to {}^3\text{He} + \gamma + 5.5 \text{ MeV} \tag{9.67}$$

$${}^3\text{He} + {}^3\text{He} \to \alpha + p + p + 12.9 \text{ MeV} \tag{9.68}$$

The net effect of these reactions [2 of (9.66) + 2 of (9.67) + (9.68)] is

$$4p \to \alpha + 2e^+ + 2\nu_e + 24.7 \text{ MeV} \tag{9.69}$$

---

[11] This condition on the product of the density and the containment time is referred to as the Lawson criterion.

which is effectively converting hydrogen into helium. Since the two positrons produced in these reactions annihilate with electrons in the star, the total energy released is roughly 27 MeV. The first step in this proton–proton cycle, as it is called, is a weak interaction, like the beta decays that we discussed in Section 9.3. Thus the cross section for this reaction is very small in comparison to the cross section for the reactions (9.64) and (9.65) that involve the strong interactions. Nonetheless, the enormous gravity of the Sun makes it possible to achieve the high density and long confinement times that are required for fusion to occur even though the temperature in the interior of the Sun is "only" 1.5×10^7 K.[12]

Nucleosynthesis: The Formation of the Elements

After a star has expended most of its hydrogen through nuclear fusion, it starts to collapse under its own gravity. This compression raises the temperature to the point where additional fusion reactions can occur. Next up is fusing helium into carbon.[13] Once the necessary supply of helium required to stave off gravitational collapse is exhausted, the process of contraction and ignition of additional fusion reactions repeats itself, thus forming elements up to iron, essentially at the peak of the curve of binding energy. Elements heavier than iron are produced by neutron capture and subsequent beta decay. In this way, elements up to bismuth are generated. As stars come to the end of their life cycle having exhausted their nuclear fuel, they may undergo a catastrophic explosion (a supernova), which releases an enormous flux of neutrons. Again neutron capture can produce heavier elements, including, for example, uranium. In general, the abundance of these elements is inversely proportional to their cross sections for absorbing neutrons. If this neutron capture cross section is large, then the corresponding nucleus is more likely to absorb a neutron, leaving less of that particular isotope in the mix.

A likely residue of a supernova explosion is a neutron star or a black hole. Two noteworthy examples of supernova explosions include the one that generated the Crab Nebula in 1054 and the one that occurred in 1987. The former, which left behind a neutron star, or pulsar, was sufficiently bright that it was seen on Earth in broad daylight

---

[12]It is not strictly necessary to give the particles sufficient energy to overcome the Coulomb repulsion since the protons can tunnel through the Coulomb barrier. Moreover, at any particular temperature T, there is always a distribution of particle energies in which some of the particles will be much more energetic than the average.

[13]Three helium nuclei have the requisite number of protons and neutrons to fuse together to make ^{12}C. But the odds of three particles colliding simultaneously are truly minuscule. Rather, two ^{4}He nuclei first collide to make ^{8}Be. This nucleus is unstable and decays back to two ^{4}He nuclei in less than 10^{-16} s unless it captures another ^{4}He nucleus, thereby making ^{12}C. The odds of this happening would be very low except for the fact that ^{12}C has an excited state 7.65 MeV above the ground state. Because of this excited state there is a resonance in the cross section if the temperature is 10^8 K (see the discussion of resonant scattering in Section 10.3). This resonance makes the cross section sufficiently large for the formation of ^{12}C to proceed. Once the excited ^{12}C is formed, it primarily decays back into ^{8}Be and ^{4}He, but a small fraction of the decays (4 in 10,000) are ones in which two photons are emitted in the transition from the excited state to the ^{12}C ground state. Since life as we know it would not be possible without carbon (and heavier atoms as well), we should be especially thankful for the existence of this excited state in the carbon nucleus and for the nuclear physics that makes it possible. In fact the existence of this excited state at 7.6 MeV in ^{12}C was predicted by Fred Hoyle before its discovery solely on the basis that without it Fred Hoyle would not have been in existence.

for a brief period. The one in 1987 occurred in the large Magellanic Cloud, roughly 168,000 light years from Earth, and was visible with the naked eye, although not in the daytime. It was the closest supernova since the one observed in 1604. Interestingly, supernova 1987A was also observed by three neutrino observatories with the detection of a total of 24 neutrino events. It is believed that 99% of the 10^{46} J of energy released in the supernova explosion was radiated away in the form of neutrinos. So far no residual neutron star has been detected from supernova 1987A. It is possible that the progenitor star, a blue supergiant with a mass of 18 solar masses, has collapsed to a black hole.

9.6 Nuclear Weapons: History and Physics

As we noted at the beginning of this chapter, one can say that the birth of nuclear physics dates from 1932 with Chadwick's discovery of the neutron. The 1930s also marked the beginning of a tumultuous time in world history. Many people perceived that the outbreak of war with Germany was imminent. The physicist Leo Szilard, for one, foresaw as early as 1933 the possibility of using a chain reaction to generate an explosion of unusual destructiveness.[14] The actual discovery of fission was made by the chemists Otto Hahn and Fritz Strassmann in 1938, when they succeeded in chemically separating barium from a sample of uranium ore that had been irradiated with neutrons. The presence of barium was a mystery until the physicists Lise Meitner and Otto Frisch (her nephew) explained how absorption of a neutron could induce splitting of a uranium nucleus.[15] Shortly thereafter, Niels Bohr and John Wheeler gave an explanation of fission in terms of excitation of the liquid drop model of the nucleus. Then, in September 1939, Germany invaded Poland, starting World War II. Scientists on both sides of the conflict began in earnest to investigate the possibility of using nuclear fission to make a bomb.

In contemplating construction of a bomb, the crucial issue that needed to be resolved was how much fissile material is required. As we noted earlier, the critical condition necessary to generate a self-sustaining chain reaction is that at least one of the neutrons produced in each fission event induce another nucleus to fission. One common way for the neutrons to be lost from the chain reaction is to simply escape from the volume containing the fissile material. According to (9.4), the probability that a neutron travels a distance x in matter without interacting decreases exponentially with the distance traveled. But neutrons that reach the surface of the volume of fissile material can escape without generating another fission event. In short, the larger the volume of fissile material, the more likely that the neutrons produced in each fission event will generate more such events. The amount of fissile material required to reach the critical condition is called the **critical mass**. The critical mass for pure ^{235}U is 53 kg, corresponding to a sphere 18 cm in diameter (50% larger than a softball). For ^{239}Pu the critical mass is roughly one third this amount. Of course, if the goal is to liberate a lot of energy in a short period of time—an explosion—then we want as many of the neutrons that are released in each fission event to initiate another fission event as possible. One way to reduce the critical mass is

---

[14] See Richard Rhodes, *The Making of the Atomic Bomb*, Simon & Schuster, New York, 1986, pp. 13–28. This book is highly recommended.

[15] Meitner, a Jewish colleague and collaborator with Hahn at the Kaiser-Wilhelm Institute, had been forced to flee Nazi Germany months before the discovery by Hahn and Strassmann.

to "tamp" the fissile material with a shell of ^{238}U. The tamper holds the fissile material in place, allowing fission to proceed for a longer period before the explosion reduces the density of the fissile material. Moreover, the tamper reflects some of the escaping neutrons back into the fissile material, where they can induce additional fission reactions.

At the beginning of World War II, the value for the critical mass for ^{235}U was not known because the cross section for fast-neutron induced fission of ^{235}U had not yet been measured. Both sides in the conflict made estimates of the critical mass. Rudolph Peierls and Otto Frisch in Britain obtained an estimate of 1 kg and the Germans under the direction of Werner Heisenberg obtained an estimate of 500 kg. Thus one side was an order of magnitude too low and the other an order of magnitude too high in their estimates. Since making a bomb required enrichment to at least 85% ^{235}U, this difference in the early estimates of the critical mass almost certainly had a profound impact on the degree to which each side was willing to undertake the necessary investment of money and manpower to make an atomic bomb.

Since ^{235}U and ^{238}U are the same chemically, separating the two isotopes requires a physical mechanism that takes advantage of the slight difference in mass of the two isotopes. A number of different strategies were adopted, including the use of high-speed centrifuges, gaseous diffusion, and electromagnetic separation. In the final enrichment stage, gaseous diffusion of uranium hexafluoride (UF$_6$) was used. It eventually became the standard form of enrichment. Gaseous diffusion takes advantage of the fact that at temperature T, the average kinetic energy of the molecules of the gas is given by

$$\frac{1}{2}mv^2 = \frac{3}{2}k_B T \qquad (9.70)$$

where k_B is the Boltzmann constant. Thus the molecules with ^{235}U nuclei move slightly faster than those with ^{238}U nuclei. If a barrier between two containers (or rooms) has a series of small holes in it, more uranium hexafluoride molecules with ^{235}U nuclei will diffuse into the next room since these molecules will strike the holes more often than the slightly slower molecules with ^{238}U nuclei. Repeating this process multiple times can lead to sufficient enrichment.[16] Also, in early 1941, the transuranic element plutonium was identified by the chemist Glenn Seaborg. This discovery increased the importance of developing an operational nuclear reactor since plutonium was a byproduct of the fission process of uranium and could be separated chemically from the other fission byproducts. Thus a second potentially easier pathway toward the development of an atomic bomb seemed possible. Consequently, a full-scale production reactor was built at a site on the Columbia River at Hanford, Washington.

The efforts to develop the atomic bomb were given the code name the Manhattan Project. General Leslie Groves, the military officer charged with overseeing the effort, chose the theoretical physicist J. Robert Oppenheimer to head the bomb research and development laboratory, despite Oppenheimer's not having had any substantial prior administrative experience. It was an inspired choice. At Oppenheimer's suggestion, a

---

[16]Gaseous diffusion of uranium hexafluoride was carried out at Oak Ridge, TN. The building was four stories high and a half mile long, the largest building in the world at that time. It consisted of 3122 enrichment stages. Finding an acceptable porous barrier was very difficult since the holes had to be less than one millionth of an inch in diameter, uniform in size, and able to withstand high-pressure uranium hexafluoride gas, which is extremely corrosive, especially to the grease in the pumps used to circulate the gas. A new seal material (Teflon) was developed for the pumps.

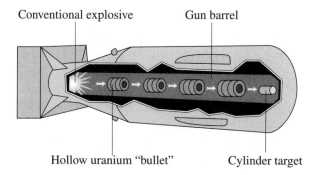

Figure 9.18 The gun-assembly method used in a uranium bomb. Two sub-critical masses of ^{235}U are brought together to produce a supercritical mass.

top-secret facility was established in Los Alamos, New Mexico. By the end of the war, as many as 6,000 people, including the physicists Hans Bethe, Enrico Fermi, and Richard Feynman, the chemist George Kistiakowski, and the mathematicians Stan Ulam and John von Neumann, were participating in this engineering project. Building a bomb using ^{235}U was relatively straightforward. Two subcritical fragments could be brought together quickly by firing one of the fragments into the other using the gun assembly illustrated in Fig. 9.18. The major concern was whether enough ^{235}U could be separated in time to make a bomb feasible. In the case of plutonium, there were two concerns. The Hanford plant started producing plutonium for use in bomb making in early 1945, so availability was also an issue. But there was a more pressing problem with plutonium, one that had come to the surface in 1944. The gun-type assembly method does not work for plutonium. The rate for spontaneous fission of plutonium is too high. When two subcritical fragments of plutonium approach each other, the neutrons released in these spontaneous fission reactions initiate other fission events before the critical mass is reached, leading to what is termed predetonation and a nuclear fizzle that keeps the subcritical fragments apart. An alternative suggested by Kistiakoskwi and Nedermeyer was to implode a subcritical sphere of plutonium by surrounding it with chemical explosives.[17] See Fig. 9.19. This

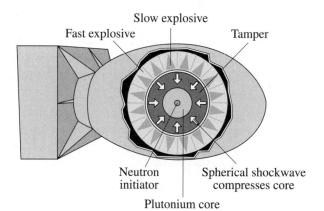

Figure 9.19 The implosion method for a plutonium bomb. Chemical explosives compress a subcritical mass of plutonium until it becomes supercritical.

---

[17]The critical mass is proportional to $1/\rho^2$, where ρ is the density. Notice, in particular, how the density enters (9.4).

Figure 9.20 J. Robert Oppenheimer and General Leslie Groves at the Trinity Test Site after the explosion of the first atomic bomb. (Digital Photo Archive, Department of Energy (DOE), reproduced with permission from AIP Emilio Segre Visual Archives)

was a challenging engineering problem that required the creation of a chemical explosive lens with elements that were detonated within a millisecond of each other, something that had never been done before. Initially, there was a lot of skepticism that the implosion technique could succeed and the period leading up to a test of an implosion-design plutonium bomb was a period of deep concern about the fate of the entire project. On July 16, 1945, the first atomic bomb, a plutonium bomb, was exploded at a site near Alamagordo, NM, with a yield equivalent to the explosion of roughly 20 kilotons of TNT, marking the beginning of the atomic age.[18] See Fig. 9.20.

The fighting in Europe had ended in May, 1945, with the surrender of Germany. Nonetheless, work on developing an atomic bomb had continued with the goal of using the weapon to end the war with Japan.[19] On August 6, 1945, a uranium bomb (Little Boy) was detonated over Hiroshima, Japan, killing approximately 70,000 people instantly with a death toll reaching 140,000 by the end of the year. Little Boy's yield was 15 kilotons. Three days later, on August 9, a plutonium bomb (Fat Man) was dropped on Nagasaki,

---

[18] 1 kiloton of TNT $= 4.3 \times 10^{12}$ J.

[19] For many scientists, the decision to contribute to the building of the atomic bomb was motivated by the fear that Germany would develop the bomb first. Once Germany surrendered, the decision to continue to work on the bomb and subsequently use the bomb on Japan generated some controversy among the scientists.

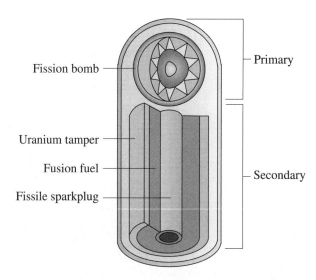

Figure 9.21 A schematic diagram of the Teller–Ulam design for a hydrogen bomb.

killing 40,000 people instantly with the death toll reaching 70,000 by the end of the year. Fat Man's yield was 22 kilotons of TNT. Five days later Japan surrendered, ending World War II.

An atomic bomb is, of course, a nuclear bomb. The energy released in the explosion arises from the difference in nuclear mass energy. While the efforts to develop an atomic bomb were ongoing, there was thought being given, particularly by the physicist Edward Teller, to the next generation of nuclear weapons, ones that derive their energy from fusion as well as fission. We have seen that getting nuclei to fuse requires that these nuclei have high kinetic energies, or high temperatures. Thus a fusion bomb is termed a thermonuclear weapon. It is possible to generate such high temperatures by using a fission bomb to trigger the fusion. A bomb based on the fusion of nuclei is also referred to as a hydrogen bomb because the primary fuel for the fusion part of the bomb consists of the hydrogen isotope deuterium. The primary reactions are

$$d + d \to {}^3\text{He} + n + 3.3 \text{ MeV} \tag{9.71}$$

$$d + d \to t + p + 4.0 \text{ MeV} \tag{9.72}$$

the same reactions that would be used in a fusion power plant. In a bomb, a small amount of tritium is used as a trigger through the reaction

$$t + d \to \alpha + n + 17.6 \text{ MeV} \tag{9.73}$$

Energy is also released in the reaction

$$t + t \to \alpha + 2n + 11.3 \text{ MeV} \tag{9.74}$$

since the triton is also produced in the reaction (9.72). Notice that since an alpha particle with its high binding energy is produced in both these reactions, the energy per nucleon released in these reactions is sizable.

In addition to developing a fission bomb as the trigger, a number of technical issues had to be resolved in order to make the hydrogen bomb a practical weapon. The mathematician Stanislaw Ulam had the insight that the fission bomb and fusion fuel could be in separate parts of the bomb and that the radiation from the fission explosion could be used to compress the fusion material. See Fig. 9.21. Another issue arises from the fact that hydrogen at room temperature is a gas, a gas that doesn't liquefy unless it is cooled to 20 K. Liquid or solid densities are necessary for the fuel for a bomb. The first hydrogen bomb tested by the United States was in fact a cryogenic bomb, since the deuterium was cooled to the point that it liquefied. But such a bomb, which weighed more than 80 tons including the refrigeration equipment, was not a useful weapon. This problem was resolved by the realization that solid densities could be achieved by combining hydrogen with lithium, forming lithium hydride, a gray powder.[20] Thus lithium deuteride has the advantage of a room-temperature solid composed of deuterium, one whose packing material, namely lithium, can contribute to the energy released in the explosion through the reactions

$$n + {}^6\text{Li} \rightarrow \alpha + t + 4.9 \text{ MeV} \tag{9.75}$$

$$d + {}^6\text{Li} \rightarrow 2\alpha + 22 \text{ MeV} \tag{9.76}$$

In addition, (9.75) provides tritium for the primary d-d reaction (9.72).

In a fission bomb, the explosion itself tends to end the nuclear reactions as the density of the fuel for the bomb decreases as the material blows itself apart. In a fusion bomb, the same reduction in density takes place, of course, but the energy released in the explosion also provides additional heating that drives additional fusion reactions, including ones such as (9.75) and (9.76) that require even higher temperatures than those in (9.71) and (9.72) because of the charge on the lithium nucleus. Thus the yield from a thermonuclear weapon can be quite large. Moreover, if the weapon is encased in ^{238}U as a tamper, then the energetic neutrons produced in the fusion reactions can also generate fission reactions in the tamper, boosting the total yield even more. Nuclear tests carried out in the 1950s had yields roughly 1,000 times greater than the fission bombs dropped on Japan. Such bombs would obviously do an enormous amount of damage. In fact, the net effect is one of overkill, since a bomb with a much smaller yield already destroys everything at ground zero. As the technology for targeting nuclear weapons improved, it was deemed more effective to have, say, ten nuclear weapons, each with a yield equivalent of, say, 500 kilotons of TNT targeted on a variety of sites, rather than a single bomb with a yield of 5 megatons. Eventually, through the strategy of mutual assured destruction (MAD), the United States and the Soviet Union each had 25,000 nuclear weapons targeted at the other side.[21] One of the major issues facing humanity today is the proliferation of nuclear weapons to other countries and the fear that nuclear weapons might end up in the possession of terrorists.

---

[20] The first potentially practical hydrogen bomb was, in fact, exploded by the Soviet Union in 1953. Spy planes flying over the Soviet Union picked up significant traces of lithium in the radioactive fallout.

[21] While we no longer are in a perpetual state of cold war since the collapse of the Soviet Union, we have not had a significant reduction in the number of nuclear warheads that are distributed around the world.

9.7 Summary

The main subject of this chapter is the curve of binding energy. A nucleus composed of Z protons and N neutrons has an atomic mass number $A = Z + N$. The mass of the nucleus is less than the sum of the masses of its constituent nucleons by the binding energy

$$m_{\text{nucleus}} = Z m_p + (A - Z) m_n - \text{B.E.}/c^2 \qquad (9.77)$$

In the semiempirical mass formula the binding energy is expressed in terms of five parameters a_1 through a_5 as

$$\text{B.E.} = a_1 A - a_2 A^{2/3} - a_3 \frac{Z^2}{A^{1/3}} - a_4 \frac{(Z - \frac{A}{2})^2}{A} + \frac{a_5}{\sqrt{A}} \begin{pmatrix} 1 \\ 0 \\ -1 \end{pmatrix} \qquad (9.78)$$

where the a_1 term is the volume term, since the volume of the nucleus is directly proportional to the number A of nucleons it contains (and hence the radius is proportional to $A^{1/3}$); the a_2 term is the surface term since nucleons on the surface of the nucleus do not participate as fully in the nuclear binding as do nucleons in the interior of the nucleus; the a_3 term, the Coulomb term, arises from the Coulomb repulsion of the protons in the nucleus; the a_4 term, the asymmetry term, arises from the tendency of nuclei, especially ones with low A, to have $Z = N$; and the a_5 term is the pairing term, which results from the tendency of nuclei to have even numbers of protons and even numbers of neutrons. Key features of the nuclear force that are encompassed in the semiempirical mass formula include its attractive nature, its short range, its large strength relative to the electromagnetic force, and its charge independence—that is, the nuclear force treats protons and neutrons as identical particles.

Nuclei that are unstable decay according to the exponential decay law

$$N(t) = N(0) e^{-Rt} = N(0) e^{-t/\tau} \qquad (9.79)$$

where R is the probability of decay per unit time and $\tau = 1/R$ is called the lifetime. This equation results from the fact that R for a nucleus (or other microscopic system) is independent of time (part of Fermi's Golden Rule). Predominant decay modes of nuclei are called alpha, beta, and gamma decay and correspond to the emission of a ^{4}He nucleus, an electron (or positron) and a corresponding neutrino, and a high-energy photon, respectively.

One of the striking features of the curve of binding energy is that it peaks in the vicinity of the element iron. Thus heavy elements can release roughly 200 MeV of energy by fissioning into two smaller fragments, fragments that are closer to the peak of the curve of binding energy. Such fission events can be induced for elements such as uranium and plutonium by the absorption of a neutron. Since the neutron is neutral, it is not repelled by the large charge on the target nucleus. Nuclear fission can be used to run a reactor or to fuel an atomic bomb.

Alternatively, energy can be obtained from the curve of binding energy through fusion, in which two lighter nuclei are brought together to make a heavier nucleus. The binding energy per nucleon increases as more nucleons are added to the nucleus, at least up to iron. Therefore, if two lighter nuclei can be brought close enough together for these fusion reactions to occur, energy on the order of MeV per nucleon can be released.

Such reactions are referred to as thermonuclear reactions since the nuclei must be heated to high temperatures to overcome the Coulomb repulsion of the fusing nuclei. Nuclear fusion provides the energy that permits stars like our Sun to stave off gravitational collapse.

Problems

9.1. A charge e with acceleration a radiates electromagnetic waves with power

$$\frac{dW}{dt} = \frac{2e^2 a^2}{(4\pi\epsilon_0)3c^3}$$

according to classical electromagnetic theory. Estimate the lifetime of a classical hydrogen atom in which an electron starts orbiting the proton at radius 0.5 Å. *Suggestion*: Use the fact that

$$F = \frac{e^2}{4\pi\epsilon_0 r^2} = ma = \frac{mv^2}{r}$$

to show that the total energy E (kinetic energy plus potential energy) of an electron orbiting at radius r is

$$E = -\frac{e^2}{8\pi\epsilon_0 r}$$

Therefore

$$\frac{dE}{dt} = \frac{e^2}{8\pi\epsilon_0 r^2}\frac{dr}{dt} = -\frac{dW}{dt} = -\frac{2e^2}{(4\pi\epsilon_0)3c^3}\left(\frac{v^2}{r}\right)^2$$

Simplify and then integrate the differential equation.

9.2. Two nuclei are considered mirror nuclei if interchanging the neutrons and protons turns one nucleus into the other. An example is ^{11}B, which consists of five protons and six neutrons, and ^{11}C, which consists of six protons and five neutrons. Determine the difference in mass of these two nuclei assuming that the nuclear binding energy is the same for both and the difference in mass of these nuclei is due to electrostatic energy differences as well as the difference in mass of the nucleon constituents. Assume both nuclei are uniformly charged spheres of the same radius. What value of the radius is required to account for the difference in mass of these two nuclei? The observed atomic mass difference between ^{11}C and ^{11}B is 1.980 MeV/c^2. *Note*: The success of this approach is evidence for the charge independence of the nuclear force, that is, the fact that the nuclear force does not distinguish protons from neutrons.

9.3. Use the expression

$$\frac{3}{5}\frac{Q^2}{4\pi\epsilon_0 R}$$

for the Coulomb energy for a sphere of charge Q of radius R to determine a numerical value for a_3 in the semiempirical mass formula (9.12). Compare your result with the value determined for that parameter in (9.13).

9.4. Show that if only Coulomb energy were involved the fragments in a fission process would have equal numbers of protons to maximize the energy released.

9.5. What is the Fermi energy for protons and what is the Fermi energy for neutrons in the ^{238}U nucleus?

9.6. Retain the constants in the derivation of the asymmetry term (9.24). How does your result compare with the value for a_4 in (9.12)?

9.7. Show that the asymmetry term (9.24) can be written as

$$\frac{(Z-N)^2}{4A}$$

9.8. Determination of the line of stability: Use the expression (9.10) with the binding energy given by (9.12) to determine the value of Z/A that minimizes $m_{\text{nucleus}}c^2$ for fixed A. Show that the answer can be expressed in the form

$$\frac{Z}{A} = \frac{96.1}{189.6 + 1.4A^{2/3}}$$

Suggestion: Differentiate the expression for $m_{\text{nucleus}}c^2$ with respect to Z, keeping A fixed. In this way we determine the mass of the most stable isobar. Take $m_n c^2 = m_p c^2 + 1.3$ MeV.

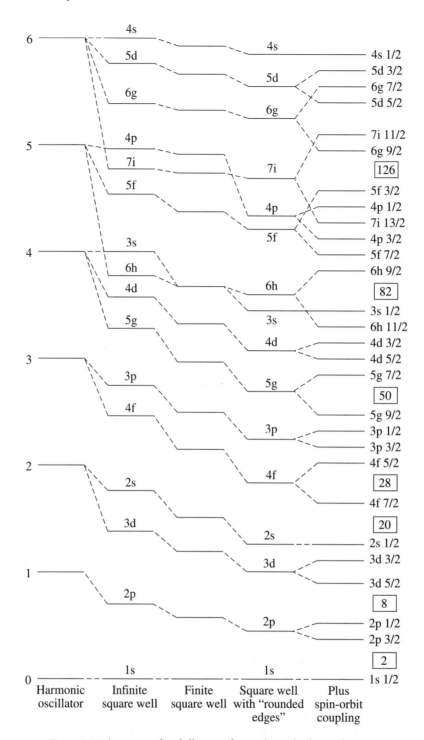

Figure 9.22 An energy-level diagram for nucleons in the nucleus.

9.9. Verify that the energy-level diagram [from B. T. Feld, *Ann. Rev. Nuclear Sci.* **2**, 239 (1955)] shown in Fig. 9.22 yields the magic numbers 2, 8, 20, 28, 50, and 82. Note the last column shows the levels with spin–orbit coupling included. There is an error in the energy-level assignments between the magic numbers 82 and 126. See if you can find it. *Reminder*: The orbital angular momentum states are labeled by the letters s ($l = 0$), p ($l = 1$), d ($l = 2$), f ($l = 3$), g ($l = 4$), and h ($l = 5$). The fractions to the right of each energy level indicate a value for the quantum number j that results from the addition of the orbital angular momentum **L** and the spin

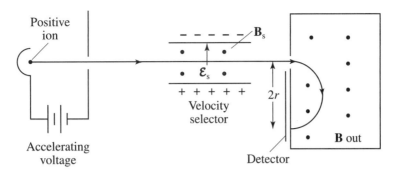

Figure 9.23 A mass spectrometer.

angular momentum **S**. For each value of j there are $2j + 1$ degenerate energy levels.

9.10. Figure 9.23 shows the key features of a mass spectrometer. Positive ions of charge q are accelerated by a potential difference and then enter a region with perpendicular electric and magnetic fields ($\mathcal{E}_s$ and B_s, respectively) that serves as a velocity selector. Upon entering a magnetic field of strength B that is out of the page, the ions travel along a semicircular path of radius r. Measuring the radius of this path permits a determination of the mass of the ion. Show that the mass M of the ion can be determined from the relation

$$M = \frac{rBB_s q}{\mathcal{E}_s}$$

9.11. In practice, it is easiest to measure mass differences with a mass spectrometer. The standard mass is that of ^{12}C, which is defined to have a mass of 12.000 u. (1 u = 931.5 MeV/c^2.) Use the semiempirical mass formula to determine the mass of ^{16}O in u's. The mass of the neutron is 1.0086654 ± 0.0000004 u and the mass of ^{1}H is 1.0078252 ± 0.0000001 u. The experimental value for the mass of ^{16}O is 15.994915 ± 0.000001 u.

9.12. What degree of accuracy in mass spectroscopy would be required to detect the effects of the binding energy of an outer electron on the mass of an atom?

9.13. Calculate the fractional difference in radius for ^{235}U and ^{238}U in the mass spectrometer of Problem 9.10. This illustrates the key idea behind electromagnetic separation of isotopes.

9.14. Show the exponential decay law $N(t) = N(0)e^{-t/\tau}$ can be written as $N(t) = N(0)2^{-t/t_{1/2}}$, where $t_{1/2} = \tau \ln 2$.

9.15. An unusual depletion of ^{235}U in a natural ore deposit was discovered in Gabon, Africa in 1972. It is believed that this depletion is evidence for a natural fission reactor that operated in the distant past. Assume that the minimum concentration of ^{235}U for a reactor moderated with ordinary water is 3% and determine how long ago the reactor could have been operational. *Note*: Included in the evidence for such a natural fission reactor is the discovery in the immediate vicinity of fission products, including isotopes of neodymium, that are characteristic of those found in spent fuel from a nuclear power plant.

9.16. We can think of a neutron star as a giant nucleus. Use the expression (9.12) for the binding energy with the addition of the gravitation term from Example 9.2 to estimate the minimum mass required for positive binding energy for very large A. For simplicity, take the number of protons in the "nucleus" to be zero. *Note*: More detailed calculations yield a *minimum* mass for a neutron star to be on the order of 0.1 solar mass.

9.17. The **activity** of a radioactive sample is the number of decays per second of the sample. The unit of activity is the curie (1 Ci = 3.7×10^{10} disintegrations per second). The curie was initially defined as the number of decays per second for 1 g of radium (^{226}Ra), an alpha particle emitter. What are the lifetime and the half-life of radium? *Note*: A curie is a very strong radioactive source. Common source strengths include the millicurie (mCi) and the microcurie (μCi). In SI units, the unit of radioactivity is the Becquerel (1 Bq = 1 disintegration per second). Thus 1 Ci = 3.7×10^{10} Bq.

9.18. In 1947 an Arab shepherd found religious documents in a cave on the shore of the Dead Sea. These Dead Sea Scrolls were made of parchment and papyrus, so carbon dating could be used to determine their age. The scrolls, which were eventually found in 11 caves and included almost all the books of the Hebrew Bible, were determined to range in age from 2000 to 2300 years. What fractional activity level in comparison with living papyrus was found?

9.19. In 1991 a couple hiking discovered a body (the Ice Man) sticking out of the ice in a glacier in the Alps near the border between Austria and Italy. The body was somewhat unusual in color and size. Beside the body was a copper axe. Using radioactive carbon dating techniques on a piece of tissue, scientists determined that the level of activity was 53% of its value for a living person. How old was the body?

9.20. A 5,000 year-old wooden coffin containing evidence of human mummification in Egypt was found in 2003. This is the earliest evidence for Egyptian mummification. What level of activity in comparison with living wood would be expected for the coffin?

9.21. (*a*) In Section 9.3 it was noted that the half-life for the beta decay of tritium ($t \to {}^3\text{He} + e^- + \bar{\nu}_e$) is 12.3 yr. What is the lifetime? (*b*) The half-life for the beta decay of carbon-14 (${}^{14}\text{C} \to {}^{14}\text{N} + e^- + \bar{\nu}_e$) is 5730 yr. What is the lifetime?

9.22. The half-life of ${}^{238}\text{U}$ is 4.47×10^9 yr and the half-life of ${}^{235}\text{U}$ is 7.04×10^8 yr. How many decays per second are there from 1 g of uranium? Assume the abundance of ${}^{235}\text{U}$ is 0.7%.

9.23. The half-life for the decay ${}^{239}\text{Pu} \to {}^{235}\text{U} + \alpha$ is 24,000 yr. What is the lifetime τ?

9.24. What fraction of the neptunium nuclei ($t_{1/2} = 2.14 \times 10^6$ yr) will survive for 5×10^9 yr?

CHAPTER 10

Particle Physics

The goal of Chapters 6 through 9 has been to give us insight into the role that quantum mechanics plays in describing the world around us, including the basic physics of atomic, solid-state, and nuclear systems. In this last chapter we turn our attention to particle physics. To the extent that a long-standing goal of physics is to understand all of nature in terms of a few underlying principles or laws, particle physics is of central importance. The fundamental interactions—the strong, electomagnetic, weak, and gravitational interactions—are all described by field theories, since fields are the natural way to ensure that these interactions are consistent with special relativity. If you suddenly move a charge, for example, the disturbance in the electromagnetic field that is generated by this motion propagates outward with speed c in vacuum. Consequently, another charge, situated a distance d away, cannot know the first charge has been displaced until a time d/c later. Quantum field theory, which results from the union of field theory and quantum mechanics, is consequently an inherently relativistic enterprise. In Chapter 1 we utilized results that come from a quantum treatment of the electromagnetic field in our discussion of the behavior of light. In this chapter we will look more carefully at the quantum nature of electromagnetic interactions as well as the strong and weak interactions.

10.1 Quantum Electrodynamics

> Thirty-one years ago, Dick Feynman told me about his "sum over histories" version of quantum mechanics. "The electron does anything it likes," he said. "It just goes in any direction at any speed, forward or backward in time, however it likes, and then you add up the amplitudes and it gives you the wave function."
> I said to him, "You're crazy."
> But he wasn't.
>
> —*Freeman Dyson*

As we have seen in the earlier chapters, in nonrelativistic quantum mechanics we can obtain exact solutions for the energy eigenvalues and eigenfunctions for the free particle, the harmonic oscillator, and the hydrogen atom. When we turn to relativistic quantum field theory, we can solve exactly only the free particle. Consequently, it is necessary to resort to approximation techniques to take into account interactions between the particles. One standard approximation technique is referred to as **perturbation theory**. We will outline some of the key features of this perturbative approach within the framework of nonrelativistic quantum mechanics, saving the details for a next course in quantum mechanics.[1]

We start by dividing the Hamiltonian into two parts

$$H = H_0 + H_1 \tag{10.1}$$

where H_0 is presumed to consist of the major portion of the Hamiltonian and H_1 is the smaller portion. (In quantum field theory, H_0 is the Hamiltonian of free noninteracting particles and H_1 is the energy of interaction between the particles.) We are assuming that we can solve for the eigenfunctions and eigenvalues of H_0, namely

$$H_0 \psi_n^{(0)} = E_n^{(0)} \psi_n^{(0)} \tag{10.2}$$

Of course these eigenfunctions form a basis, so we can write

$$\Psi = \sum_n c_n \psi_n^{(0)} \tag{10.3}$$

If H_1 were zero, we would know that the exact time dependence is given $e^{-iE_n^{(0)}t/\hbar}$. With this in mind, we express the time dependence in the form

$$\Psi = \sum_n c_n(t) e^{-iE_n^{(0)}t/\hbar} \psi_n^{(0)} \tag{10.4}$$

Thus if we can find $c_n(t)$, we will know how the probability amplitudes evolve in time. The requirement that $\Psi(t)$ satisfies

$$H\Psi = i\hbar \frac{\partial \Psi}{\partial t} \tag{10.5}$$

then leads to the coupled differential equations

$$\sum_n \left[c_n(t) e^{i(E_f^{(0)} - E_n^{(0)})t/\hbar} \int \psi_f^{(0)*} H_1 \psi_n^{(0)} d^3r \right] = i\hbar \frac{dc_f}{dt} \tag{10.6}$$

that must be solved to determine the amplitudes $c_f(t)$, the amplitudes to find the system in an eigenstate of H_0 with energy $E_f^{(0)}$ at time t.

So far, we have not made any approximations. If we assume that at time $t = 0$ the system is in the state ψ_i, namely $c_n(0) = \delta_{ni}$ in (10.4), then it is possible to express the solution as an infinite series:

$$c_f(t) = \delta_{fi} - \frac{i}{\hbar} \int_0^t e^{i(E_f^{(0)} - E_i^{(0)})t'/\hbar} (H_1)_{fi} dt' + \cdots \tag{10.7}$$

where

$$(H_1)_{fi} = \int \psi_f^{(0)*} H_1 \psi_i^{(0)} d^3r \tag{10.8}$$

with the integral in (10.8) is understood to be over all space. You don't need to worry about the details that lead to (10.7). The important point here is that the amplitudes $c_f(t)$ are expressed as a series of terms. The higher order terms, the ones not shown in (10.7), involve more powers of H_1. For example in the third term in the series H_1 acts twice,

---

[1] See John S. Townsend, *A Modern Approach to Quantum Mechanics*, University Science Books, Sausalito, CA, 2000.

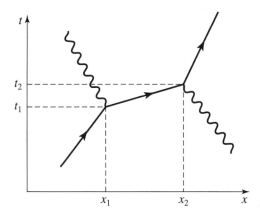

Figure 10.1 A space–time diagram for Compton scattering.

in the fourth term H_1 acts three times, and so on. In quantum field theory, the terms in the series can be neatly expressed in terms of diagrams called **Feynman diagrams**. Ideally, with the appropriate dictionary one can translate the diagram into the probability amplitude from which we can calculate the probability that the system evolves from an initial state ψ_i to a final state ψ_f.

The quantum field theory in which we calculate those amplitudes for processes that involve solely electromagnetic interactions is referred to as quantum electrodynamics, or QED for short. As an illustration, let's examine Compton scattering. In Section 1.3 we derived from conservation of momentum and energy the Compton formula

$$\lambda' - \lambda = \frac{h}{mc}(1 - \cos\theta) \tag{10.9}$$

Equation (10.9) tells us how the wavelength of the outgoing photon varies with the scattering angle θ from an electron at rest. What the Compton formula doesn't tell us is the probability with which this occurs. This probability is calculated with the aid of Feynman diagrams such as the one shown in Fig. 10.1, which is a short-hand way of representing a term in the series (10.7) in which H_1 acts twice. In this space–time diagram the straight lines indicate electrons and the curvy lines photons. The initial and final states consist of an electron and a photon. To make things clear, we restrict ourselves to one dimension. In this diagram the initial electron emits a photon at position x_1 at time t_1. The electron then travels to position x_2, where it absorbs the incident photon at time t_2. One entry in the dictionary that translates this diagram into a complex number includes a factor of e at each vertex, which seems natural since the photon couples to electric charge. Another dictionary entry associates the outgoing photon that is emitted at x_1 with a factor $e^{ik'x_1}$, where $k' = 2\pi/\lambda'$. The incident photon, with wavelength λ, which is absorbed at x_2, is associated with a factor e^{-ikx_2}. The dictionary for the electron lines, including the internal line between x_1 and x_2, requires the use of solutions to the relativistic Dirac equation. One can view this diagram as the solution to the relativistic wave equations, namely those of Maxwell and Dirac in a similar way that the wave function is a solution to the Schrödinger equation.[2]

---

[2] The δ_{fi} term in the expansion (10.7) would correspond to world lines for the photon and the electron that are straight lines, indicating that the final state of the particles is the same as the initial state.

Recall that in calculating the amplitude for a particle to hit a particular point in the detection plane in a double-slit experiment, we added the amplitudes for each possible path the particle can take between the source and the detector. Similarly, when calculating the total amplitude for Compton scattering, we must add the amplitudes for every way the process in question can occur. In practice, this means integrating over all possible values of x_1, t_1, x_2, and t_2 that occur in the diagram, that is, we must add the amplitudes for all possible values of the intermediate positions and times. This is the sum-over-histories generalization of the sum-over-paths approach that we used in Chapter 1.

One of the strange things about diagrams such as the one shown in Fig. 10.1 is that the fundamental subcomponent of this diagram, which is shown in Fig. 10.2, doesn't seem to make physical sense. As is shown in Example 10.1, a single electron in free space cannot emit a photon and conserve energy and momentum. But this is what occurs at position x_1 in the diagram in Fig. 10.1. However, the electron that is traveling between x_1 and x_2 is a "virtual electron" and not a real one to the extent that this electron, which is represented by an internal line in the diagram, is not one of the particles that is actually measured and observed in the initial or final state. We can argue that the Heisenberg energy–time uncertainty relation $\Delta E \, \Delta t \geq \hbar/2$ allows for such processes to the extent that the evolutionary time

$$\Delta t = t_2 - t_1 \tag{10.10}$$

is finite, since therefore there is an inherent uncertainty in the energy of the system that involves these virtual particles. On the other hand, the time interval involved in the overall process of Compton scattering is assumed to be essentially infinite, since the incoming electron and photon are understood to be produced in the distant past and the outgoing electron and photon in this scattering are presumed to be observed in the distant future. Thus the time interval for the overall event pictured in Fig. 10.1 is sufficiently large that we should not expect to observe violations of conservation of energy in Compton scattering itself.

Since we are to include in our calculation diagrams that arise for all values of t_1 and t_2, we must include diagrams in which t_2 precedes as well as follows t_1, as shown in Fig. 10.3. There is an alternative way to see the need for such time orderings that emphasizes the inherently relativistic nature of our calculation. Since the calculation of

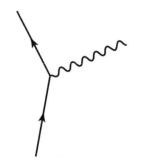

Figure 10.2 The fundamental subunit in Feynman diagrams in quantum electrodynamics involves the emission (or absorption) of a photon by a charged spin-1/2 particle such as the electron.

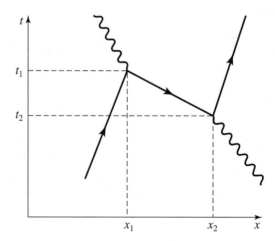

Figure 10.3 All time orderings, including the one in this figure, are required to calculate the probability amplitude for Compton scattering.

the probability should be consistent with the relativity principle, we should be able to calculate the probability for Compton scattering in a frame of reference that is in motion at speed V relative to any other inertial reference frame. If we examine the amplitude that arises from solving these relativistic equations, we find the amplitude corresponding to Fig. 10.1 is nonzero when the quantity

$$c^2(\Delta\tau)^2 = c^2(\Delta t)^2 - (\Delta x)^2 \tag{10.11}$$

is negative as well as positive. A negative separation is called space-like since it results from $\Delta x > c\Delta t$, where Δt is defined by (10.10) and

$$\Delta x = x_2 - x_1 \tag{10.12}$$

Under the Lorentz transformation

$$x' = \gamma (x - Vt) \tag{10.13}$$

$$t' = \gamma \left(t - \frac{Vx}{c^2}\right) \tag{10.14}$$

where

$$\gamma = \frac{1}{\sqrt{1 - V^2/c^2}} \tag{10.15}$$

we find that

$$\Delta t' = \gamma \left(\Delta t - \frac{V\Delta x}{c^2}\right)$$

$$= \gamma \Delta t \left(1 - \frac{V\Delta x}{c^2 \Delta t}\right) \tag{10.16}$$

Thus it is possible that $\Delta t' < 0$ even if $\Delta t > 0$ provided $\Delta x > c\Delta t$. The time ordering of events is not preserved under a Lorentz transformation if the events are space-like separated. For the amplitude shown in Fig. 10.1, this effect occurs only on the microscopic level, when $\Delta\tau$ in (10.11) is on the order of $\hbar/mc^2$, where m is the mass of the electron.[3]

What are we to make of the event that occurs at x_2 at time t_2 in Fig. 10.3? One possibility, as indicated in the Feynman quote that precedes this section, is to suggest that the electron that emitted the outgoing photon at time t_1 propagates backwards in time to t_2, at which point it interacts with the incident photon. But there is a better way to think about this process, one that eliminates the need to consider particles propagating backwards in time. We can argue that the first event at t_2 involves the creation of the outgoing electron and a positively charged particle, so as to conserve charge. This particle, which has the same mass as the electron, is the positron, the antiparticle of the electron. The positron then propagates forward in time to t_1, at which point it annihilates with the incident electron with the emission of the outgoing photon. In Feynman diagrams, the positron is distinguished by having the arrow pointing opposite to the direction of increasing time. In fact, since the time-orderings shown in Fig. 10.1 and Fig. 10.3 are both required by

---

[3]This result provides a caution against interpreting the Feynman diagrams too literally. After all, if the events at x_1, t_1 and x_2, t_2 were really connected by an electron traveling between them, then necessarily $c^2(\Delta\tau)^2$ would be positive since an electron must travel at a speed less than c. The internal line in this Feynman diagram represents a "virtual" electron, not a real one that we actually observe.

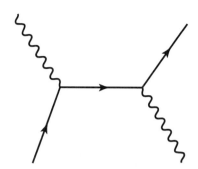

Figure 10.4 A Feynman diagram for Compton scattering. This diagram is understood to include the probability amplitudes shown in Fig. 10.1 and Fig. 10.3.

special relativity, it is natural to combine them in the single Feynman diagram in Fig. 10.4. This diagram is then understood to be shorthand expression for both diagrams.

If we are to take seriously the idea that relativistic quantum mechanics requires the existence of an antiparticle for the electron, it should be possible to observe this particle as a final-state particle. In fact, the positron was first observed in 1932 in a cloud chamber by Carl Anderson, just a few years after Dirac introduced his relativistic wave equation. Now a single photon (with momentum p_0) cannot materialize into a particle–antiparticle pair (with momenta p_- and p_+) despite what the diagram shown in Fig. 10.3 seems to suggest. Such a process cannot conserve energy and momentum since to conserve momentum

$$\mathbf{p_0} = \mathbf{p_-} + \mathbf{p_+} \tag{10.17}$$

and to conserve energy we must have

$$p_0 c = \sqrt{p_-^2 c^2 + m^2 c^4} + \sqrt{p_+^2 c^2 + m^2 c^4} \tag{10.18}$$

The minimum values for p_- and p_+ occur when the electron and positron head off in the same direction as the incident photon, namely

$$p_0 = p_- + p_+ \tag{10.19}$$

but it is evident that (10.18) and (10.19) cannot be satisfied simultaneously. Thus a photon cannot materialize into an electron and positron in free space no matter how large its energy. However, if some other charged particle is present, say the nucleus of a heavy atom such as lead, then it is possible to transfer all the momentum of the photon to the nucleus while transferring very little of the photon's energy. The kinetic energy of the nucleus is then

$$\frac{p_0^2}{2M} = \frac{(p_0 c)^2}{2Mc^2} \tag{10.20}$$

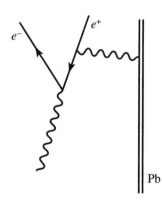

Figure 10.5 A Feynman diagram used to determine the cross section for the creation of an electron–positron pair in the vicinity of a heavy nucleus such as lead.

In the case of lead, this leads to a kinetic energy transferred to the nucleus of 3 eV. Thus the threshold energy for production of an electron–positron pair is very close to $2mc^2 = 1.02$ MeV. Figure 10.5 shows a Feynman diagram corresponding to the creation of an electron–positron pair in the interaction of a photon with a nucleus. In addition to pair creation, photons can also be absorbed by the atom via the photoelectric effect, or simply scattered in a process called Thomson scattering. The total cross section for

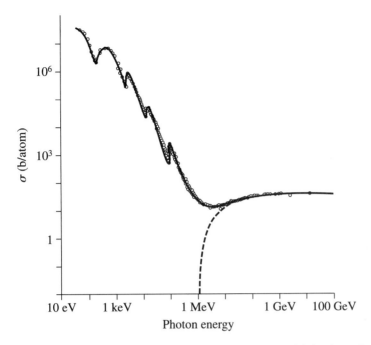

Figure 10.6 The total cross section for the interaction of photons with lead as a function of photon energy. The sharp dips in the cross section with decreasing photon energy occur when the photon energy drops below the binding energy of electrons in a particular shell. The dashed line shows the contribution to the cross section of electron–positron pair creation in the field of the nucleus. Adapted from J. H. Hubbel, H. A. Gimm, and I. Overbo, *J. Phys. Chem. Ref. Data* #9, 1023 (1980).

photons incident on lead is shown in Fig. 10.6. The growth in the total cross section for photon energies above 1 MeV is due to electron–positron pair production.

Returning to Compton scattering, diagrams like the one shown in Fig. 10.4 lead to a differential cross section (the effective area for scattering into solid angle $d\Omega$)

$$d\sigma = \pi \alpha^2 \left(\frac{\hbar}{mc}\right)^2 \left(\frac{\lambda}{\lambda'}\right)^2 \left(\frac{\lambda}{\lambda'} + \frac{\lambda'}{\lambda} - 2 + 4\cos^2\theta\right) d\Omega \qquad (10.21)$$

where λ is the wavelength of the incident photon and λ' is the wavelength of the outgoing, or scattered, photon. The lowest order Feynman diagram for Compton scattering is proportional to e^2, since the diagram has two vertices. In practice, each factor of e^2 enters in the combination

$$\alpha = \frac{e^2}{4\pi\epsilon_0 \hbar c} = \frac{1}{137.0360} \qquad (10.22)$$

where α is a dimensionless constant called the **fine-structure constant**. Thus the cross section (10.21), which is obtained from the amplitude multiplied by its complex conjugate, is proportional to α^2.

As another example, consider electron–electron scattering. Fig. 10.7a is a lowest order diagram contributing to this process. As indicated in Fig. 10.7b, there are other diagrams, other amplitudes, that should be included. Each of these additional diagrams has a larger number of vertices than the lowest order diagram. And although we haven't specified in detail how to convert these diagrams into probability amplitudes, we have noted that at each vertex where a photon couples to an electron (or a positron), the charge e appears in the amplitude. Thus diagrams such as the ones in Fig. 10.7b have four vertices, four

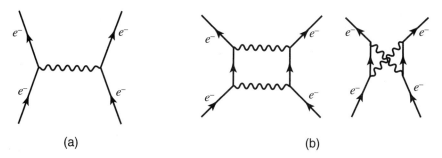

Figure 10.7 (a) The lowest order Feynman diagram for electron–electron scattering. (b) Some of the higher order diagrams that contribute to this process.

factors of e, and are therefore proportional to e^4, in comparison with the lowest order diagram that has two factors of e and is therefore proportional to e^2. Thus an amplitude that is proportional to α^2 is roughly 1% as large as one proportional to α. Therefore if we are interested in calculating the cross section for scattering to a certain accuracy, it makes sense to neglect the contribution of higher order diagrams at a certain point. Nonetheless, these higher order amplitudes in QED can be evaluated in principle and their contribution included if needed, although the higher order terms become progressively more difficult to evaluate.

In general, summing all the diagrams that contribute to a particular process generates an expansion with an infinite number of terms, very much in the spirit of the expansion

$$z = 1 + x + \frac{x^2}{2!} + \frac{x^3}{3!} + \cdots \tag{10.23}$$

However, unlike the expansion (10.23), which you probably recognize as yielding $z = e^x$, no one has succeeded in carrying out the sum of the Feynman diagrams for processes in QED. It is possible that at some point we will figure out how to calculate the overall amplitude without such an expansion, in which case the diagrams may become an historical artifact. But as things stand now, the diagrams are an essential tool.

As a striking example of how successful this approach has been, consider the magnetic moment of the electron. In Section 6.5 we noted that the magnetic moment of the electron is given by

$$\boldsymbol{\mu} = g \left(\frac{-e}{2m}\right) \mathbf{S} \tag{10.24}$$

The value of g has been determined theoretically and experimentally with great accuracy. Current results are

$$\frac{g-2}{2} = (1,159,652,201.4 \pm 27.1) \times 10^{-12} \quad \text{theory}$$
$$= (1,159,652,180.85 \pm 0.76) \times 10^{-12} \quad \text{experiment} \tag{10.25}$$

The theory calculation has a somewhat higher uncertainty, in part because the theory calculation depends on experimental values for α, which has its own uncertainty, but also because the theory calculation does not include all the higher order diagrams. It is possible to estimate how much error neglecting these amplitudes contributes to the final answer. It is likely that the theorists will respond to the more accurate experimental values with some more detailed calculations of their own. In any case, it is agreement like that shown in (10.25) that led Feynman to describe quantum electrodynamics as "the jewel of physics—our proudest possession."[4]

EXAMPLE 10.1 Show that a free electron cannot emit a photon and conserve energy and momentum.

SOLUTION It is easiest to do this calculation in the rest frame of the electron. Then if a decay into a photon and an electron were possible, momentum conservation dictates that the two particles have equal and opposite momenta. Consequently, conservation of energy yields

$$mc^2 = \sqrt{p^2c^2 + m^2c^4} + pc$$

The only solution to this equation is $p = 0$, i.e., no photon.

EXAMPLE 10.2 Which of the Feynman diagrams in Fig. 10.8 is a legitimate Feynman diagram in QED for electron–positron annihilation?

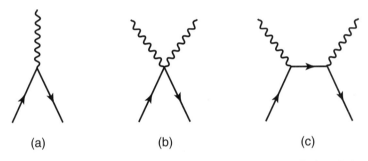

Figure 10.8 Possible Feynman diagrams for electron–positron annihilation. Only one of these diagrams is a valid one.

SOLUTION Figure 10.8a has the correct structure since it involves the fundamental subunit shown in Fig. 10.2. However, in this process none of the particles is presumed to be a virtual particle, since there are no internal lines in the diagram. And an electron and a positron cannot annihilate into a single photon and conserve energy and momentum, as is easily seen by analyzing the process in the center-of-mass frame of the electron–positron pair. Thus at least two photons must be produced in the annihilation process. Figure 10.8b, which is referred to as a "seagull" diagram (you can see why) has the two photons being produced at a single point in space time. Since the particles annihilating are spin-1/2 particles, this is not an allowed diagram in that it is not composed of the fundamental subunit shown in Fig. 10.8a (or Fig. 10.2). It would, however, be a valid Feynman diagram if the charged particles participating in the annihilation were spin-0 particles. Figure 10.8c is an appropriate Feynman diagram.

---

[4]R.P. Feynman, *QED—The Strange Theory of Light and Matter*, Princeton University Press, Princeton, NJ, 1985.

EXAMPLE 10.3 What are the lowest order Feynman diagrams for electron–positron scattering?

SOLUTION For electron–positron scattering, there are two lowest order diagrams. See Fig. 10.9. One diagram is similar to the diagram in Fig. 10.7a, with a positron replacing the electron (note the switch in the direction of one of the arrows). However, there is an additional diagram in which the electron and positron annihilate into a virtual photon, which then generates the final electron–positron pair.

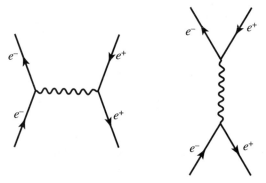

Figure 10.9 Lowest order Feynman diagrams for electron–positron scattering.

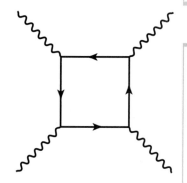

Figure 10.10 The lowest order Feynman diagram for the scattering of light by light.

EXAMPLE 10.4 In classical physics, light does not interact with light since light itself is uncharged. Construct a Feynman diagram that would contribute to the cross section for light by light scattering (Delbrück scattering).

SOLUTION Since photons are neutral and therefore do not couple to each other, the only way for light to scatter light is to have virtual charged particles involved, as illustrated in Fig. 10.10. Since this diagram is proportional to α^2, the cross section for light-by-light scattering is proportional to α^4. The size of the cross section is consequently very small in comparison to that for photon–electron scattering. See the dependence of (10.21) on α.

10.2 Elementary Particles

One thing contributing to the remarkable success of QED is the elementary nature of electrons, positrons, and photons. These are point particles as far as we can tell. They do not have any internal structure. Protons and neutrons—particles that experience the strong nuclear force—have a size of roughly one fermi, as was noted in Chapter 9. If particles have a size, it is natural to ask what's inside. In the case of atoms, it is of course electrons and nuclei. In the case of nuclei, it is the nucleons. And in the case of the nucleons themselves, it is quarks. As far as we know, the quarks themselves are

Leptons	Quarks	Gauge Bosons
e, ν_e	u, d	photon (γ)
μ, ν_μ	c, s	gluons (g)
τ, ν_τ	t, b	$W^\pm, Z^0$

Table 10.1 The elementary particles consist of leptons and quarks, which are spin-1/2 particles, and the gauge bosons, the mediators of the interactions, which are spin-1 particles.

elementary, that is, they don't have any size or internal structure. But no one has ever isolated a quark outside a proton or a neutron, so it is hard to know for sure.[5]

Particle physicists call strongly interacting particles **hadrons** (which is derived from the Greek word *hadros*, meaning strong). In addition to protons and neutrons, there are many, many subatomic (or better, subnuclear) particles that experience the strong nuclear force. Some examples are given in the next section. We now know that each of these particles is composed of **quarks**, which have intrinsic spin $s = 1/2$. The hadrons with half-integral intrinsic spin ($s = 1/2, 3/2, \ldots$) are composed of three quarks (or three antiquarks in the case of the antiparticles) and are called **baryons**. The hadrons with integral intrinsic spin ($s = 0, 1, \ldots$) are composed of a quark and an antiquark and are called **mesons**. And to make things more complicated still, the quarks themselves come in six different types, or **flavors**, namely, up (u), down (d), strange (s), charm (c), bottom (b), and top (t). As illustrated in Table 10.1, it is convenient to group these quarks in pairs, u and d, c and s, and t and b. The first quark in each pair (u, c, and t) has charge $+2e/3$, while the second quark (d, s, and b) has charge $-e/3$. The proton is composed of two u quarks and one d quark. The neutron is composed of two d quarks and one u quark. You can see this gives the correct charge for each of these nucleons. In the beta decay reaction

$$n \to p + e^- + \bar{\nu}_e \tag{10.26}$$

that we examined in Chapter 9, one of the d quarks in the neutron is converted into a u quark with the emission of an electron and an electron antineutrino, namely

$$d \to u + e^- + \bar{\nu}_e \tag{10.27}$$

The electron and the neutrino, like the quarks, are spin-1/2 elementary particles, particles that do not have any internal structure. But unlike the quarks, the electron and the neutrino do not participate in the strong interactions. They are generally referred to as **leptons**, namely fundamental spin-1/2 particles that do not have strong interactions. Since electrons and quarks are charged, they of course participate in electromagnetic

---

[5]Whether we can assess if a particle has some internal structure depends on the wavelength of the object that we use as a probe. If the wavelength of the probe—whether the probe consists of photons in a standard optical microscope or electrons in an electron microscope—is large compared with the size of the particle, the particle will appear to be point like. As was noted in Section 9.1, it was scattering experiments carried out in the 1950s with high-energy electrons from an accelerator that determined the size of nuclei. Similarly, in the late 1960s scattering experiments carried out at the Stanford Linear Accelerator Center with much higher energy electrons gave clear evidence for the quark substructure of nucleons.

interactions. Neutrinos, on the other hand, are neutral; they don't have magnetic moments or other electromagnetic properties. Apart from gravity, neutrinos participate only in the weak interactions. Consequently, emission of a neutrino in a decay is a telltale sign of a weak interaction. The electron and the electron neutrino appear in the first row in Table 10.1, along side the u and d quark.

Given our discussion in the preceding section of the natural role that antiparticles play in relativistic quantum mechanics, it should not be surprising that there are antiparticles for each of the quarks and leptons. The positron, the antiparticle of the electron, has the same mass but opposite charge of an electron. But what distinguishes a neutrino from an antineutrino? One difference is a quantum number that we call **lepton number**. We assign the electron and its neutrino lepton number $+1$. The antiparticles, the positron and the antineutrino, then have lepton number -1. The beta decay reaction (10.26) conserves lepton number, since the neutron (or the proton, for that matter) is not a lepton (it has lepton number 0), and the final state has lepton number 0 as well since the electron has lepton number $+1$ and the antineutrino has lepton number -1. Thus the lepton number of the final state is the same as the lepton number of the initial state.

To see the significance of lepton number for neutrinos, consider a reaction that does not occur even though energy, linear momentum, angular momentum, and charge are conserved:

$$\bar{\nu}_e + n \not\to p + e^- \tag{10.28}$$

What keeps this reaction from occurring is the fact that it does not conserve lepton number, since the lepton number of the initial state is -1 and the lepton number of the final state is $+1$. On the other hand,

$$\bar{\nu}_e + p \to n + e^+ \tag{10.29}$$

does conserve lepton number and does occur.

You may be wondering why the neutrinos and antineutrinos have a subscript e. The reason is that there are two other generations of leptons, as noted in Table 10.1. The muon behaves very much like a heavy electron with a mass $m_\mu = 105.66$ MeV/c^2. But when the muon goes to decay, it does not do so radiatively: $\mu^- \not\to e^- + \gamma$. What is observed is consistent with two types of neutrinos, namely ν_e and ν_μ. The primary decay mode of the negatively charged muon is

$$\mu^- \to e^- + \bar{\nu}_e + \nu_\mu \tag{10.30}$$

with a lifetime of 2.2×10^{-6} s.[6] This reaction has muon lepton number $+1$ for the initial and final states in addition to electron lepton number 0 for the initial and final states. In short, there is a separate conservation law for electron lepton number and for muon lepton number. Thus a beam of muon antineutrinos incident on a liquid hydrogen target can initiate the reaction

$$\bar{\nu}_\mu + p \to n + \mu^+ \tag{10.31}$$

---

[6]The experimental upper limit on the branching ratio for $\mu^- \to e^- + \gamma$ to $\mu^- \to e^- + \bar{\nu}_e + \nu_\mu$ is 1.2×10^{-11}.

provided the antineutrino has sufficient energy to create the muon, but not the reaction

$$\bar{\nu}_\mu + p \not\to n + e^+ \tag{10.32}$$

even though the positron is much less massive than the muon.[7]

A similar story holds for the τ lepton, which was discovered in 1975 and has mass $m_\tau = 1777$ MeV/c². Why we have three generations of leptons is still somewhat of a mystery. After the discovery of the muon, I. I. Rabi famously remarked, "Who ordered that?" This mystery has been repeated at the level of the quarks, since they too occur in three generations.

The third column in Table 10.1 shows the **gauge bosons**, the fundamental spin-1 particles that mediate the strong, electromagnetic, and weak interactions. Section 10.1 on QED is devoted to electromagnetic interactions. All charged particles have electromagnetic interactions. Figure 10.7a is the prototype Feynman diagram illustrating how two electrons interact through the exchange of a (virtual) photon. As Section 10.4 explains, the strong interactions between the quarks are mediated by the exchange of gluons, as illustrated in Fig. 10.17. And the weak interactions between the quarks and leptons are mediated by the exchange of W and Z bosons, as illustrated in Fig. 10.21 and Fig. 10.22 from Section 10.5. The fundamental role that symmetry principles play in dictating the nature of these interactions is outlined in Section 10.8. This section also explains why the spin-1 particles that mediate the interactions are referred to as gauge bosons.

EXAMPLE 10.5 Use conservation of lepton number to explain why the following reactions occur or do not occur. (i) $\nu_e + p \not\to n + e^+$, (ii) $\bar{\nu}_\mu + n \not\to p + \mu^-$, (iii) $\mu^+ \to e^+ + \bar{\nu}_\mu + \nu_e$.

SOLUTION Only reaction (iii) conserves lepton number. Reaction (iii) is like (10.30) with each particle replaced by its antiparticle. Thus this is the predominant decay mode for the μ^+. Reaction (i) violates electron lepton number conservation and reaction (ii) violates muon lepton number conservation.

EXAMPLE 10.6 Construct a Feynman diagram leading to the production of a $\mu^- - \mu^+$ pair in $e^- - e^+$ annihilation. What is the threshold energy in the center-of-mass frame required for this reaction to occur?

SOLUTION The lowest order diagram for this process is shown in Fig. 10.11. Since the rest mass energy of the muon is 105.66 MeV, the center-of-mass total energy required for this reaction is 211.32 MeV.

Figure 10.11 The lowest order Feynman diagram leading to the production of a muon–antimuon pair in electron–positron annihilation.

---

[7]As we will see in Section 10.6, the recent evidence of neutrino oscillations indicates that the individual lepton numbers are not absolutely conserved. In fact, unlike conservation of charge, which is an absolute conservation law that is connected to an underlying symmetry principle (see Section 10.8), conservation of lepton number is a somewhat ad hoc conservation law that has been introduced to account for the observed reactions.

10.3 Hadrons

In 1934, long before anyone knew about quarks and gauge bosons, Hideki Yukawa suggested that the strong force was mediated by the exchange of a particle with a mass of roughly 100 to 200 MeV. Although we now have a more complete understanding of how the strong force arises from the interactions of quarks and gluons, it is instructive to retrace Yukawa's reasoning using the Feynman diagram in Fig. 10.12, a diagram that would arise from the exchange of such a particle between two nucleons. The fact that the nuclear force has a short range R, on the order of a fermi, dictates very roughly what the mass of the exchanged particle should be. Using the Heisenberg energy–time uncertainty relation $\Delta E \Delta t \simeq \hbar$ with $\Delta t \simeq R/c$ (since the speed of the exchanged particle would typically be on the order of the speed of light) and $\Delta E \simeq mc^2$, where m is the mass of the exchanged particle (since emitting the particle requires a violation of conservation of energy by roughly mc^2), we find

$$R \simeq \frac{\hbar}{mc} \qquad (10.33)$$

Substituting in the observed range of the nuclear force, we find $mc^2 \simeq 200$ MeV. This argument is a heuristic one at best (Yukawa did better by actually solving a wave equation that led to this result), but it does show the deep connection between the mass of the exchanged particle and the range of the interaction. For electromagnetism, the mass of the exchanged particle, the photon, is zero, and indeed the range of that interaction is infinite.

A particle called the pion, or π meson, with a mass fitting Yukawa's prediction was discovered in 1947 in the debris from cosmic ray collisions. Pions come in three different charges: π^+, π^-, and π^0. We now know the π^+ is a bound state of a u quark and an anti-d quark ($\bar{d}$), the π^- is the antiparticle of the π^+ and is therefore a bound state of an anti-u quark ($\bar{u}$) and a d quark, and the π^0 is a linear superposition of a $u\bar{u}$ and a $d\bar{d}$. The pion was termed a meson initially because its mass was intermediate between the heavier proton and neutron, which is why they were termed baryons, meaning heavyweight, and the lighter electron, which was termed a lepton, meaning lightweight. Subsequently, quark–antiquark bound states with masses heavier than the proton and the neutron have been discovered, but they are referred to as mesons nonetheless.

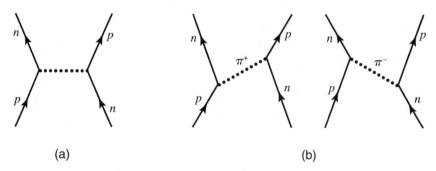

Figure 10.12 (a) A Feynman diagram for proton–neutron scattering involving exchange of a meson called the pion. As for Compton scattering, the two time orderings shown in (b) are understood to be included in the diagram shown in (a). Consequently, there must be positively and negatively charged pions. In addition, neutral pions are required to account for the strong interactions in, say, neutron–neutron scattering.

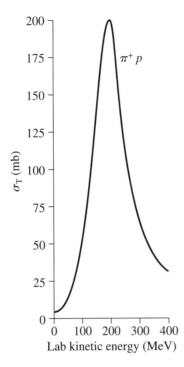

Figure 10.13 The cross section for π^+-p scattering for protons at rest as a function of the kinetic energy of the pion. The cross section shows a resonance at a center-of-mass energy of 1232 MeV.

In the early 1960s particle physics came of age as an experimental science. Instead of relying on data from cosmic ray collisions, physicists built particle accelerators with sufficient energy to create particles such as the pion in the laboratory. As scattering experiments were performed at progressively higher energies, many new "particles" were created in these collisions. An interesting example is shown in Fig. 10.13, in which the scattering cross section for positively charged pions incident on a liquid hydrogen target is plotted as a function of the kinetic energy of the incident pion. The large bump in the cross section that occurs at a center-of-mass energy of 1232 MeV corresponds to a resonance in which a new particle, the Δ^{++} baryon, is created:

$$\pi^+ + p \rightarrow \Delta^{++} \rightarrow \pi^+ + p \tag{10.34}$$

Since the π^+ and the proton each have charge e, the Δ^{++} must have charge $+2e$, hence the superscript. Recall that in terms of quarks, the proton is a uud and the π^+ is a $u\bar{d}$. In this collision, the d and the $\bar{d}$ annihilate, resulting in the state uuu. Since each u quark has charge $+2e/3$, a state composed of 3 u quarks does have charge $+2e$. The Δ^{++} also has intrinsic spin $s = 3/2$. The Δ baryon has four different charge states: Δ^{++}, Δ^+, Δ^0, and Δ^-. For example, the Δ^0 is created in π^--p collisions, again with a center-of-mass energy of roughly 1232 MeV.

One of the striking things that is observed in the data in Fig. 10.13 is that the bump in the cross section occurs over a broad range of center-of-mass energies, corresponding to a spread, or uncertainty, $\Delta E = 125$ MeV in the energy at which the Δ^{++} is produced. Using the energy–time uncertainty relation (see Example 5.6), this spread in energy corresponds to a lifetime τ for the Δ^{++} of 5×10^{-24} s, after which it decays back into a p and a π^+. At first it may seem surprising that we can measure a lifetime this short, but

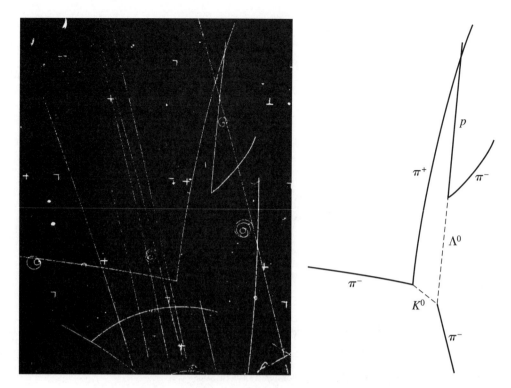

Figure 10.14 A hydrogen bubble chamber picture showing the reaction $\pi^- + p \to \Lambda^0 + K^0$. Unlike the π^- and p, the Λ^0 and K^0 are neutral and do not leave a track of ions where bubbles start to form. The macroscopic distance between the formation and the decay of the Λ^0 and K^0 is indicative of the long lifetime of these hadrons. Image courtesy Lawrence Berkeley National Laboratory.

we are making the measurement indirectly by observing the width ΔE of the resonance. And while 5×10^{-24} s may seem like a short time on a macroscopic scale, it is a typical hadronic lifetime. It is of the same order of magnitude as the time for light to travel a distance of a fermi, which is roughly the size of a strongly interacting particle.

Figure 10.14 shows a different sort of event that can occur when a π^- collides with a proton at a higher center-of-mass energy. In this event, a different baryon, the Λ^0 baryon, is created together with a new type of meson, the K^0 meson, or kaon. The Λ^0 was given its name because the tracks left by the p and π^- into which it decays look as if they form the Greek letter Λ. The K^0 meson decays into a π^+ and π^-. The lifetimes of the Λ^0 and the K^0 are roughly 10^{-10} s. Like the Δ^{++}, the Λ^0 and the K^0 are hadrons. The cross section for the reaction

$$\pi^- + p \to \Lambda^0 + K^0 \tag{10.35}$$

that produces them is typically 20–30 mb, the same size as other strong interaction cross sections. And like the Δ^{++}, the Λ^0 and the K^0 decay into ordinary hadrons, but with lifetimes that are thirteen orders of magnitude longer than that of the Δ^{++}. We now understand that this strange behavior is generated by the production of a new type of quark in the collision, the strange quark s, and its antiparticle $\bar{s}$. The Λ^0 baryon is a uds bound state while the K^0 meson is a $d\bar{s}$ bound state. The strange quark s has charge $-e/3$, just like the d quark. The Λ^0 baryon is the lightest baryon containing a strange quark, while the K^0 meson is the lightest meson containing a strange quark. We say the strange

Meson	Quark Content	Mass (MeV/c^2)
π^+	$u\bar{d}$	140
π^-	$d\bar{u}$	140
π^0	$(u\bar{u}-d\bar{d})/\sqrt{2}$	135
K^+	$u\bar{s}$	494
K^-	$s\bar{u}$	494
K^0	$d\bar{s}$	498
$\bar{K}^0$	$s\bar{d}$	498
η^0	$(2s\bar{s}-u\bar{u}-d\bar{d})/\sqrt{6}$	547

Table 10.2 The low-mass spin-0 mesons.

Baryon	Quark Content	Mass (MeV/c^2)
p	uud	938.3
n	udd	939.6
Λ^0	uds	1116
Σ^+	uus	1189
Σ^0	uds	1193
Σ^-	dds	1197
Ξ^0	uss	1315
Ξ^-	dss	1321

Table 10.3 The low-mass spin-1/2 baryons.

quark has a new quantum number called strangeness ($S = -1$) and the antistrange quark has $S = +1$. Strangeness is conserved in the production of these particles, but the decay violates strangeness conservation. Consequently, the decay is governed by the weak interactions, not the strong interactions that govern the decay of the Δ^{++}.

When reactions like (10.35) were first observed, the rationale for the strange behavior in which hadrons were produced strongly but decayed weakly into other hadrons was a mystery. Now we understand that the creation of a new flavor of quark is involved. The strong interactions conserve flavor, but the weak interactions do not. The quark has the symbol s for strange because of the at first seemingly strange behavior in which it was involved. While we are on the subject of terminology, it might be good to say a few words about names and symbols for u and d quarks. The u stands for "up" and the d stands for "down." The switch in quark content in going from a proton to a neutron is the change of a u quark to a d quark. We remarked in Chapter 9 that the strong interactions treat the proton and the neutron as identical particles. Heisenberg had the idea that the nuclear force treats them as two different spin states of a new type of spin, which we now call isospin. Isospin is *not* angular momentum, but rather it is connected to rotations in an abstract space in which a $180°$ rotation changes a proton into a neutron. The nucleon is an isospin-1/2 particle: the proton is isospin up and the neutron is isospin down. But once we understand the underlying quark structure of these nucleons, we see that the switch from isospin up to isospin down is the switch from a u to a d. This is the rationale for referring to u and d as up and down quarks, respectively. Table 10.2 and Table 10.3 give the quark content and masses of the lightest spin-0 mesons and spin-1/2 baryons, respectively. You can see that the mesons that contain a strange quark are more massive than those composed purely of up and down quarks. A similar pattern occurs for the baryons. Hadrons that contain c quarks or b quarks become progressively more massive. For example, the D^+ meson, a $c\bar{d}$ bound state, has a mass of 1869 MeV/c^2 and the B^+ meson, a $u\bar{b}$ bound state, has a mass of 5279 MeV/c^2.[8]

There are other ways to increase the mass of the hadrons apart from adjusting their quark content. As is the case for electrons in atoms, increasing the orbital angular momenta of the quarks increases their energy and consequently increases the mass of the hadrons. Moreover, the nuclear force is extremely spin dependent. For example, the ρ^+ meson is composed of a u quark and a $\bar{d}$ antiquark, just like the π^+ meson. The ρ^+ meson,

---

[8] Because the effective mass of the t quark is very large, the t quark decays so quickly that mesons and baryons containing the t quark do not have a chance to form.

a spin-1 particle, has a mass of 770 MeV/c^2 while the π^+ meson, a spin-0 particle, has a mass of 140 MeV/c^2. The change in going from the pion to the ρ meson is simply "flipping a spin," in that the quark and antiquark in the pion are in the single spin-0 state (the singlet spin state)

$$\frac{1}{\sqrt{2}}[\chi_+(1)\chi_-(2) - \chi_-(1)\chi_+(2)] \tag{10.36}$$

while in the ρ meson they are in one of the three spin-1 states (the triplet spin states)

$$\chi_+(1)\chi_+(2) \tag{10.37}$$

$$\frac{1}{\sqrt{2}}[\chi_+(1)\chi_-(2) + \chi_-(1)\chi_+(2)] \tag{10.38}$$

$$\chi_-(1)\chi_-(2) \tag{10.39}$$

Thus changing from the total spin-0 to the total spin-1 state increases the mass of the meson by a whopping 630 MeV/c^2. This is in marked contrast to the 5.8×10^{-6} eV energy difference between the singlet and triplet energy states in the ground state of hydrogen, the hyperfine energy splitting that arises from the interaction of the magnetic moments (and hence the spins) of the electron and proton in the atom.

EXAMPLE 10.7 The Ω^- is a spin-3/2 baryon with strangeness $S = -3$. Of what quarks is it composed?

SOLUTION Recall that the strange quark s has charge $-e/3$ and strangeness $S = -1$. A baryon is composed of three quarks. Thus it must be that $\Omega^- = sss$ in order to make the charge and strangeness of the Ω^- work out. The Ω^- is indeed a very strange baryon!

10.4 Quantum Chromodynamics

The Δ^{++} is the most uppity baryon possible, since it is composed of 3 u quarks. Moreover, since it has intrinsic spin 3/2, one of the four possible spin states for this baryon is for each quark to be in a spin-up state [$\chi_+(1)\chi_+(2)\chi_+(3)$]. Another possible spin state is for each of the quarks to be in the spin-down state [$\chi_-(1)\chi_-(2)\chi_-(3)$]. Each of these spin states (as well as the other two spin-3/2 spin states) is symmetric under exchange. Moreover, the Δ^{++} is the lowest mass baryon with spin equal to 3/2. Thus the quarks are in the spatial ground state ($l = 0$), which is also symmetric under exchange of any two of the quarks. Therefore, the three identical fermions appear to be bound in a state that is symmetric under exchange of any two of the u quarks. But as was noted in Section 7.2, one of the fundamental tenets of relativistic quantum mechanics is the spin-statistics theorem that states that the overall "wave function" for identical fermions must be antisymmetric under exchange. The resolution of this conundrum is to assert that the quarks have an additional quantum number called **color**. There are three possible colors—red, green, and blue. The antisymmetry of the overall state is restored by positing that the color state of the three quarks in the Δ^{++} is given by

$$\Psi_A(1, 2, 3) = \frac{1}{\sqrt{6}} [u_R(1)u_G(2)u_B(3) + u_G(1)u_B(2)u_R(3)$$
$$+ u_B(1)u_R(2)u_G(3) - u_B(1)u_G(2)u_R(3)$$
$$- u_G(1)u_R(2)u_B(3) - u_R(1)u_B(2)u_G(3)] \quad (10.40)$$

where the subscripts R, G, and B refer to red, green, and blue, respectively. Please do not interpret color as the color that we see in everyday life. Color for quarks is a quantum number like electric charge. There are two types of electric charge called positive and negative, initially named by Ben Franklin. Similarly, there are three types of color charge, called red, green, and blue. Nonetheless, color does turn out to be a particularly apt name for this additional "charge."

One striking confirmation of the existence of a color as a physical entity that affects observable quantities is the value of the ratio of the cross section for electron–positron annihilation into hadrons to the cross section for electron–positron annihilation into μ^+–μ^-:

$$R = \frac{\sigma(e^- + e^+ \to \text{hadrons})}{\sigma(e^- + e^+ \to \mu^- + \mu^+)} \quad (10.41)$$

Figure 10.15 shows this ratio, which is commonly called R, in terms of Feynman diagrams. The Feynman diagram in the denominator is the one that appears in Fig. 10.11. The three Feynman diagrams in the numerator are the diagrams needed to calculate the cross section for $e^- + e^+ \to u + \bar{u}$, $e^- + e^+ \to d + \bar{d}$, and $e^- + e^+ \to s + \bar{s}$. Although free quarks have never been observed, we know that the quarks are the fundamental building blocks for all hadrons. Since there is a 100% probability that the quarks and antiquarks produced in electron–positron annihilation will turn into hadrons, the parts of the diagrams that correspond to this conversion of quarks to hadrons will simply generate a multiplier of unity in each case. What is neat about the ratio R is that although you may not know how to calculate any one of the diagrams in detail, you do know how to calculate the ratio provided we assume that the center-of-mass energy is sufficiently high that we can ignore the differences in mass of the final-state particles. Since the processes

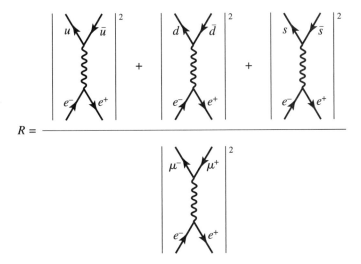

Figure 10.15 The Feynman diagrams that contribute to the ratio R.

in question are electromagnetic, the difference between the diagrams in the numerator and the one in the denominator is the coupling constant that occurs at the vertex where the virtual photon couples to the outgoing particles (quarks and antiquarks in the numerator and μ^+ and μ^- in the denominator). Since the u quark has charge $+2e/3$ and the s and d quarks have charge $-e/3$ while the muon has charge e, at first the ratio looks as if it should be

$$R = \left(\frac{2}{3}\right)^2 + \left(\frac{1}{3}\right)^2 + \left(\frac{1}{3}\right)^2 = \frac{2}{3} \tag{10.42}$$

Figure 10.16 shows the experimental results for the quantity R. For center-of-mass energies less than 3.7 GeV, the value of R is approaching a value of 2, not 2/3. The error in our calculation is that we neglected color. For each one of the Feynman diagrams in the numerator, there are really three diagrams, say for the production of a red–antired u–$\bar{u}$ pair, a blue–antiblue u–$\bar{u}$ pair, and a green–antigreen u–$\bar{u}$ pair. Since these color states are distinguishable final states, we should add the probability of producing each set of colored quark–antiquark pairs for each quark flavor to determine the total cross section for producing hadrons. The net effect is to multiply the value in (10.42) by 3.

You can't help but notice that there is a sizable jump in the value for R starting after the ψ' resonance. In this region there is sufficient energy to create two mesons, one containing a c quark and one containing a $\bar{c}$ quark. At higher energies baryons containing c and $\bar{c}$ quarks are produced as well. Since the c quark has charge $+2e/3$, the additional hadrons that are introduced by exceeding this threshold energy add $3 \times (2/3)^2 = 4/3$ to the value for R. Thus R asymptotically should approach the value of $2 + 4/3$, or $3\ 1/3$, until the threshold for producing the b–$\bar{b}$ pair at a center-of-mass energy of approximately

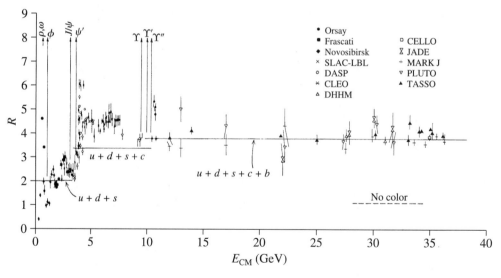

Figure 10.16 The ratio $R = \sigma(e^- + e^+ \to \text{hadrons})/\sigma(e^- + e^+ \to \mu^- + \mu^+)$ as a function of the center-of-mass energy (in GeV) of the electron–positron system. The arrows labeled by ρ and ω indicate the location of spin-1 mesons composed of $u\bar{u}$ and $d\bar{d}$ quarks. These mesons show up as resonances in $\sigma(e^- + e^+ \to \text{hadrons})$. The ϕ meson is an $s\bar{s}$ bound state. Similarly the J/ψ at 3.1 GeV and ψ' at 3.7 GeV are $c\bar{c}$ bound states and the Υ resonances, which occur at a center-of-mass energy of roughly 10 GeV, are $b\bar{b}$ bound states. The J/ψ, the ψ', and the Υ resonances are very narrow spikes in the $e^- + e^+ \to \text{hadrons}$ cross section. See Fig. 10.29 for a more detailed look at the cross section in the 3.1 GeV region. Adapted from F. Halzen and A. D. Martin, *Quarks and Leptons*, John Wiley & Sons, New York, 1984, p. 229.

10 GeV, at which point $3 \times (1/3)^2 = 1/3$ is added to the value of R. The t quark does not contribute to the production of hadrons in the data in Fig. 10.16, since the mass of the t quark is very large.[9]

Just as electric charge is the source of the electromagnetic field that mediates electromagnetic interactions between charged particles through the exchange of photons, color is the source of the strong interactions between the quarks. The quarks interact by exchanging quanta called **gluons**. For example, absorption or emission of a gluon can turn a red quark into a blue one, as shown in Fig. 10.17. The gluons can be thought of as arising from the states that are generated by combining a color with an anticolor. Thus the reaction

$$u_R \to u_B + g_{R\bar{B}} \tag{10.43}$$

corresponds to a red u quark turning into a blue u quark with the emission of a red–antiblue gluon $g_{R\bar{B}}$. Although the quark changes color in (10.43), color is conserved since the red color of the quark in the initial state is the same as the sum of the colors (red + blue + antiblue) in the final state. You might at first think there should be nine different gluons that arise from the nine different possible ways of combing three colors and three anticolors together. But one of these ways, the one corresponding to the linear combination of color–anticolor states

$$\frac{1}{\sqrt{3}} \left(R\bar{R} + G\bar{G} + B\bar{B} \right) \tag{10.44}$$

does not occur as a mediator. This state is referred to as a **color singlet**. It is the single state that arises from the combination of three colors and three anticolors that has no net

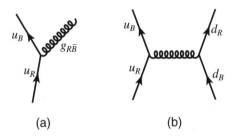

Figure 10.17 (a) A red up quark emits a red–antiblue gluon and becomes a blue up quark. In QCD there are seven additional quark-gluon vertices since there are eight different color–anticolor combinations for the gluons. (b) The Feynman diagram showing the interaction of a red up quark and a blue down quark through the exchange of a gluon.

---

[9]Mentioning the mass of a quark makes for an awkward discussion since, as we will discuss later in this section, there is strong reason to believe that individual quarks cannot be isolated. Thus the mass of a quark cannot be directly measured. Nonetheless, "effective" masses of the heavier quarks can be determined from the masses of hadrons that contain these quarks. The t quark is sufficiently heavy (roughly equal to that of a gold nucleus) that it decays into lighter quarks so quickly that hadrons containing the t quark do not have time to form.

Figure 10.18 Since the gluons themselves carry color, they couple to themselves as well as to quarks. In QCD there are vertices in which three gluons and four gluons couple to each other.

color, just as the combination of spin states for two spin-1/2 particles

$$\frac{1}{\sqrt{2}}[\chi_+(1)\chi_-(2) - \chi_-(1)\chi_+(2)] \tag{10.45}$$

is a spin-singlet state that has no net spin angular momentum, as shown in Example 7.3.

One of the striking differences between quantum chromodynamics (QCD) and quantum electrodynamics (QED) is that because the gluons carry color, they can couple to each other. QCD dictates that two types of coupling between gluons are possible, one with three gluons at a vertex and the other with four. See Fig. 10.18. Because the gluons themselves carry color, QCD is much more difficult to solve than is QED. Moreover, the coupling between quarks and gluons or between gluons themselves is stronger than the coupling between electrons and photons. The analogue of α, the fine structure constant, for the strong interactions can be as large as 15 at low energies.[10]

What we believe happens is that a color "dipole" consisting of a quark with one color, say red, and an antiquark of the opposite color, antired, produces a color field that is quite different from that produced by an electric dipole in which two opposite electric charges are separated by some distance. Figure 10.19 shows the fields in the two cases. In the electromagnetic dipole the electric fields spread out over space. But for sufficiently large separations of the quark and the antiquark in the color dipole, the color field forms a flux tube. As the separation between the quark and the antiquark increases, so does the length of the flux tube and the correponding energy. An energy linearly proportional to separation distance leads to a linearly rising potential energy. If the tube doesn't rupture, it would take an infinite amount of energy to separate the quark and the antiquark to infinity. Hence the quark and the antiquark are confined, forming a meson. This is the rationale for calling the mediators of the color interaction gluons, since this interaction glues the quarks and antiquarks together into mesons (or three quarks together into a baryon). However, if the quark and the antiquark are produced with a sufficiently large energy, say in electron–positron annihilation, the flux tube can rupture as the quark and the antiquark produced in the annihilation separate, leading to the creation of an additional quark–antiquark pair. The net effect, as illustrated in Fig. 10.20, is to produce two quark–antiquark pairs, namely two mesons.

Why are the mesons able to escape the grasp of the strong interactions as their separation increases? The mesons are all in the very special superposition of

---

[10]These coupling constants are not truly constant. Their values depend mildly (logarithmically) on the energy of the exchanged particle. At sufficiently high energies, the coupling constant for the strong interactions becomes small enough that it is feasible to use Feynman diagrams to calculate probability amplitudes for the strong interactions.

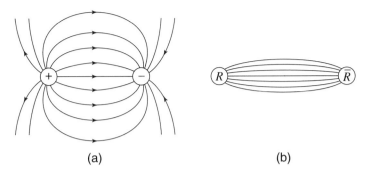

Figure 10.19 An electric dipole consists of a positive and a negative charge separated by a distance d. The electric field produced by such a dipole is shown in (a). A quark and an antiquark separated by a distance d form a color dipole as well as an electric dipole. The form of the color fields at largish separation of the quark and the antiquark is shown in (b).

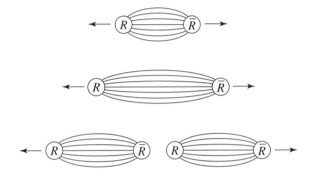

Figure 10.20 As the separation between the quark and the antiquark increases, it becomes energetically advantageous for the flux tube to rupture with the creation of an additional quark–antiquark pair, thus producing two mesons. The mesons are not purely $R\bar{R}$ but are rather in the color singlet (10.44).

color–anticolor states shown in (10.44). As we noted in our discussion of the color states of gluons, this superposition is a color singlet. It has no net color. Just as two neutral atoms interact electromagnetically much more weakly than do two ions, so too do two mesons each of which is "colorless." Similarly, the antisymmetric color state

$$\frac{1}{\sqrt{6}}(RGB + GBR + BRG - BGR - GRB - RBG) \qquad (10.46)$$

of three quarks that we argued was necessary to ensure that the three u quarks in the Δ^{++} were in an antisymmetric state under exchange is also a color singlet. The linear combinations (10.44) and (10.46) are the only color singlet states that can be made from the three color charges. Therefore only quark–antiquark states (mesons) or three quark states (baryons) have the possibility of being colorless.[11]

Finally, let's now return to the issue of the force between two nucleons. In our discussion of nuclear physics in the preceding chapter, we characterized this force as the strong force. Now we see that the nuclear force between color neutral hadrons is the relatively weak residue of the underlying interactions of quarks and gluons just as the relatively

---

[11] The state $\frac{1}{\sqrt{6}}(\bar{R}\bar{G}\bar{B} + \bar{G}\bar{B}\bar{R} + \bar{B}\bar{R}\bar{G} - \bar{B}\bar{G}\bar{R} - \bar{G}\bar{R}\bar{B} - \bar{R}\bar{B}\bar{G})$ in which each of the colors in (10.46) is replaced by its anticolor is also a color singlet. Thus antibaryons (such as $\bar{p}$ and $\bar{n}$) consisting of three antiquarks can also exist.

weak van der Waals force between neutral atoms or molecules is a residue of the underlying Coulomb interactions between the charged particles that constitute these atoms or molecules. Although the residual nuclear force between nucleons is indeed strong relative to the Coulomb interactions between the nucleons themselves, it is not as strong nor of such long range as the underlying interaction between the quarks themselves.

10.5 Quantum Flavor Dynamics

In QED, the quantum theory of electromagnetic interactions, electric charge is conserved. In QCD, the quantum theory of strong interactions, color charge is conserved. What about weak interactions? What do they conserve? The short answer, as we will see in this section and the following two sections, is not as much.

First, it would be good to have a nice three-letter acronym for the weak force. A case can be made for calling it QFD, short for quantum flavor dynamics, for the weak interactions are first and foremost interactions that couple to flavor. We commented earlier that there are six different flavors of quarks, namely u, d, s, c, b, and t. Similarly there are six different types, or flavors, of leptons e, ν_e, μ, ν_μ, τ, and ν_τ. All quarks and leptons participate in the weak interactions. But flavor, unlike charge and color, is not necessarily conserved. The weak interaction

$$\nu_e + n \to e^- + p \tag{10.47}$$

can be expressed in terms of quarks as

$$\nu_e + d \to e^- + u \tag{10.48}$$

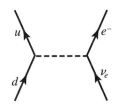

Figure 10.21 A typical Feynman diagram for the weak interactions. The mediator of this interaction is the W boson.

We see that there are two flavor changes in this reaction, namely, $d \to u$ and $\nu_e \to e^-$. The lowest order Feynman diagram for this process is shown in Fig. 10.21. The mediator for this process is referred to as the W boson (or intermediate vector boson, which is a way of saying that the W boson has intrinsic spin one). The two time-orderings implicit in this Feynman diagram mean there are two W bosons, the W^+ and the W^-. The masses of the W bosons are 80 GeV/c^2. You may be surprised that a proton can turn into a neutron and emit a particle with a mass of 80 GeV/c^2. But if you look back at the discussion on the connection of the range of a force to the mass of the exchanged particle, you see that what this is really telling us is that the weak interaction is a very short-range one. For particles to interact through the weak interactions they must be very close together. Thus the effective area for interaction is very small. This is the reason the cross section for a reaction such as that shown in (10.47) is roughly 13 orders of magnitude smaller than a strong interaction cross section such as that for pion–nucleon scattering. For example, a 1 MeV neutrino is likely to travel through 35 light years of water without interacting.

Not only are weak interactions mediated by the charged bosons $W^\pm$, but there is a neutral mediator as well, the Z^0 boson. Since the Z^0 is electrically neutral, it couples directly to all the different flavors of quarks and leptons without a flavor change. Thus reactions such as

$$\nu_e + p \to \nu_e + p \tag{10.49}$$

are possible. In terms of the quarks, the corresponding Feynman diagrams are given in Fig. 10.22. The mass of the Z^0 is 91 GeV/c^2, somewhat larger than that of the $W^\pm$. Thus the cross section for these neutral current interactions, as they are called, is comparable with the cross section for charged current interactions, those mediated by

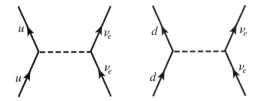

Figure 10.22 Weak interactions are also mediated by the Z^0 boson.

the $W^\pm$. Because reactions such as (10.49) do not entail the conversion of a neutrino into a charged lepton that can leave a track of ions as it moves through a liquid or a gas, these neutral current interactions were not noticed until they were predicted to exist by theory, a theory that unified electromagnetic and weak interactions.

In our discussion of QED we made much of the fact that the coupling of the photon to a charged particle is dictated by the charge the particle possesses. In the case of the strong interactions, the coupling constant for gluons to quarks is significantly larger than is the coupling of photons to charged particles such as the electron. Before our discussion of the mass of the mediators of the weak interactions, you might have supposed that the weak interactions have a much smaller coupling constant than e. In the Glashow–Weinberg–Salam theory that unifies weak and electromagnetic interactions, the coupling constant for a W coupling to a quark or a lepton is $e/\sin\theta_w$, where the angle $\theta_w = 28.7°$ is determined from experiment. Thus the charged current coupling constant is actually somewhat larger than e. The reason that the coupling constant is expressed in terms of the angle θ_w, which is called the **Weinberg angle**, is that in this unification of electromagnetic and weak interactions the photon and the Z^0 are different linear combinations of the two other basis states, the so-called W^0 and B, namely

$$\gamma = B\cos\theta_w + W^0 \sin\theta_w \quad (10.50)$$

and

$$Z^0 = -B\sin\theta_w + W^0 \cos\theta_w \quad (10.51)$$

Using cosine and sine of an angle to express this superposition is convenient, since it assures us that the sum of the probabilities is one ($\cos^2\theta_w + \sin^2\theta_w = 1$) for each superposition. Moreover, if we think of the W^0 and B states as orthogonal states, then the linear combinations that correspond to the photon and the Z^0 are orthogonal, too. See Problem 5.3. The mass of the photon is, of course, zero, while the theory predicts that the mass of the Z^0 is related to the mass of the W via $M_W = M_Z \cos\theta_w$, in agreement with experiment. Thus the two superpositions in (10.50) and (10.51) generate two very different particles. The different masses for the photon and the Z^0 are generated by a mechanism called spontaneous symmetry breaking (see Section 10.8).

Because of their large masses, a high-energy accelerator is required to create the W and Z bosons. High-energy experimental physics is an enormous enterprise, involving multinational collaborations and large teams of scientists. Figure 10.23 shows the UA1 detector at CERN that was used to make the first detection of the W produced in proton–antiproton collisions with a center-of-mass energy of 540 GeV. Typically, the quarks in the proton share one half of the proton's kinetic energy, with the gluons sharing the other half. Thus each quark in the proton (and each antiquark in the antiproton) has an energy of roughly 45 GeV, so the center-of-mass energy in a quark–antiquark collision is sufficient to make a W boson. The production and subsequent decay of a W boson is

Figure 10.23 The UA1 detector at CERN was used to discover the W and Z bosons. (Courtesy CERN)

governed, of course, by the weak interactions. In proton–antiproton collisions, the vast majority of the particles produced are in purely hadronic events. To pick out the W from this enormous background of events, the experimentalists looked for leptons produced with very high momenta transverse to the beam direction. Such events could arise only from the leptonic decay modes of the W. Carlo Rubbia and Simon van der Meer shared the Nobel Prize in physics in 1984 for their efforts leading to the discovery of the W and Z just one year earlier, in 1983.

EXAMPLE 10.8 Construct Feynman diagrams for the processes $\nu_e + \mu^- \rightarrow \nu_\mu + e^-$ and $\mu^- \rightarrow \nu_\mu + e^- + \bar{\nu}_e$.

SOLUTION Fig. 10.24 shows the Feynman diagrams. Notice that the electron and muon lepton numbers are separately conserved at each vertex.

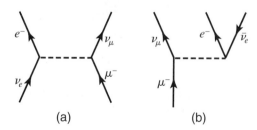

Figure 10.24 Lowest order Feynman diagrams for $\nu_e + \mu^- \rightarrow \nu_\mu + e^-$ and $\mu^- \rightarrow \nu_\mu + e^- + \bar{\nu}_e$.

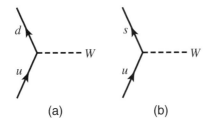

Figure 10.25 Emission or absorption of a W boson changes a u to d, as shown in (a), or a u to s, as shown in (b).

10.6 Mixing Angles

In Section 10.3 we noted that low-mass hadrons containing a strange or antistrange quark like the Λ^0 baryon and the K^0 meson decay weakly into hadrons composed of up and down quarks and antiquarks. Consequently, the weak interactions mediated by the $W^\pm$ bosons not only change the flavor of a u quark to a d quark, as shown in Fig. 10.25a, they also allow for the change of flavor from u to s, as shown in Fig. 10.25b. Alternatively, these vertices can be turned on their sides, as shown in Fig. 10.26, thus permitting a u and a $\bar{d}$ or a u and an $\bar{s}$ to annihilate into a W^+ as happens in the decays $\pi^+ \to \mu^+ + \nu_\mu$ and $K^+ \to \mu^+ + \nu_\mu$, respectively. See Fig. 10.27. It turns out that the coupling to strange quarks is somewhat weaker than the coupling to down quarks. The physicist Nicola Cabbibo recognized that a natural way to express the differences in the strength of the interaction is to say that the u quark couples to the following linear combination:

$$d' = d \cos \theta_c + s \sin \theta_c \qquad (10.52)$$

where θ_c is called the **Cabibbo angle**. Thus there is a probability $\cos^2 \theta_c$ for a u to turn into a d and a probability $\sin^2 \theta_c$ for a u to turn into an s. We call this particular linear combination of d and s quarks d' because it is relatively close to the d, since $\theta_c = 13.1°$ from experiment.[12]

The Feynman diagram for the decay $K^0 \to \mu^+ + \mu^-$ shown in Fig. 10.28a raises an interesting problem. The rate for the decay obtained by "squaring" the amplitude corresponding to this Feynman diagram is significantly larger than the observed decay rate. It was suggested by Glashow, Illiopoulos, and Maiani in 1970 that what was needed was another amplitude, shown in Fig. 10.28b, that could cancel most of the contribution of the diagram in Fig. 10.28a when the two amplitudes were added together. What was needed was another quark, the c quark, with charge $+2e/3$ like the u quark, that coupled to the following linear combination of d and s quarks:

$$s' = -d \sin \theta_c + s \cos \theta_c \qquad (10.53)$$

where θ_c is the same Cabibbo angle as in (10.52). The amplitude in Fig. 10.28a is then proportional to $\cos \theta_c \sin \theta_c$, while the amplitude in Fig. 10.28b is proportional to $-\sin \theta_c \cos \theta_c$. Thus provided the c quark is not too dissimilar in mass from the u quark,

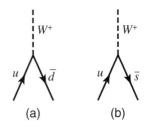

Figure 10.26 A u and a $\bar{d}$ in (a) and a u and an $\bar{s}$ in (b) annihilate into a W^+.

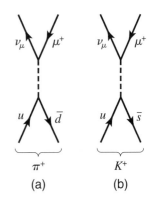

Figure 10.27 Feynman diagrams for π^+ and K^+ decay.

---

[12]Strictly, the d' is a linear combination of the three quarks with charge $-e/3$, namely d, s, and b. However the coupling to the b quark is quite small. We ignore it here for simplicity. Cabibbo did not know about the b quark, since it wasn't discovered until much later.

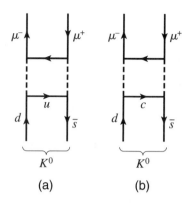

Figure 10.28 The contribution of the Feynman diagram with a virtual u shown in (a) to the decay $K^0 \to \mu^- + \mu^+$ is partially canceled by the diagram with a virtual c quark shown in (b).

there would be sufficient cancellation between the two amplitudes to suppress the rate for the decay $K^0 \to \mu^+ + \mu^-$ to the point where the calculated rate is not in conflict with the experimental results. And such a quark does indeed exist. It was discovered subsequently, in 1974. As was discussed in Section 10.4, evidence for this quark is seen in the rise in the value of the ratio R in electron–positron annihilation above 3.7 GeV. Because of (10.53), weak decays of baryons and mesons that contain a c quark tend to be preferentially into hadrons that contain a strange quark.[13]

One of the noteworthy features of Fig. 10.16 that we have not discussed is the sharp spikes in the cross section at certain energies. For example, in the vicinity of a center-of-mass energy of 3.1 GeV, the cross section for producing hadrons increases sharply, as shown in Fig. 10.29. At SLAC, this resonance was referred to by the letter ψ. The ψ is a bound state of a charm–anticharm quark pair. It was simultaneously observed at Brookhaven National Laboratory (BNL) and called the J particle there. Consequently, this particle is often referred to as the J/ψ. Just as an electron–positron bound state is called positronium, the charm–anticharm quark bound state is called charmonium.[14] The production of charmonium in e^-–e^+ annihilation produces a resonance, very much in the way the production of the Δ^{++} generates a resonance in π^+–p scattering. Because the sum of the masses of the lightest pair of mesons containing a c and a $\bar{c}$ quark exceeds the mass of the J/ψ particle—the ground state of charmonium—the J/ψ cannot decay into hadrons containing charm quarks. Consequently, the decay proceeds through an intermediate state of three gluons, as indicated in Fig. 10.30. As was noted in Section 10.4,

---

[13] The c quark is said to be the charm quark. The rationale for calling this quark the charm quark arose from the as yet unexplained symmetry in the quark and lepton flavor worlds. Before 1974 two generations of leptons had been discovered (e and ν_e constituted one generation and μ and ν_μ constituted a second, more massive generation). In the quark sector, there was one generation consisting of u and d quarks and a single more massive strange quark s. Glashow and Bjorken had suggested in 1964 that it would be charming if there were two generations of quarks to match the two generations of leptons. Now of course we know there are three generations of quarks and three generations of leptons, so the symmetry between quarks and leptons has been maintained.

[14] The discovery of the J/ψ created much excitement among particle physicists. At one point a theoretical preprint was circulated with two panda bears with complementary spots stuck together on the cover, reflecting the pandemonium that was common at the time in the high-energy physics community.

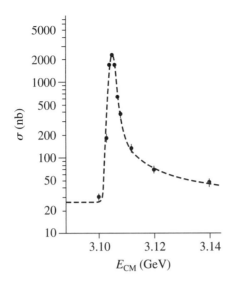

Figure 10.29 The cross section for $e^- + e^+ \to$ hadrons in the vicinity of a center-of-mass energy of 3.1 GeV. The width of the J/ψ resonance is 50 keV, which is much less than the spread of several MeV in the beam energies of the electron and the positron. Results are from the Stanford Positron-Electron Accelerating Ring (SPEAR) at the Stanford Linear Accelerator Center (SLAC). Adapted from J. E. Augustin et al., *Phys. Rev. Lett.* **33**, 1406 (1974).

the strength of the coupling of the gluon to a quark decreases as the energy carried by the gluon increases. Consequently, because of the relatively large mass of the J/ψ, each of the three gluons in the intermediate state in Fig. 10.30 carries enough energy that the coupling between these gluons and the quarks is relatively small—thus the long lifetime and corresponding narrow width of the J/ψ. In addition to the J/ψ, there are additional excited states of charmonium. The first excited state, the ψ', has a mass of 3.7 GeV/c^2. Once created in e^--e^+ annihilation, the primary decay mode of the ψ' is to the lower mass J/ψ with the emission of two pions ($\psi' \to J/\psi + \pi^+ + \pi^-$). See Fig. 10.31. After the discovery of charmonium particle physicists found themselves solving the Schrödinger equation to try to determine the form of the potential energy responsible for binding the c and $\bar{c}$ together. Given the discussion in Section 10.4 on the linearly rising potential energy that is generated in attempting to pull a quark and an antiquark apart, it will not come as a surprise to learn that a potential energy of the form $V = ar$ works very well for all but the smallest values of r, where r is the separation between the c and $\bar{c}$ quarks.[15]

One final comment on Fig. 10.16 is in order. In addition to the J/ψ and ψ', there are spikes labeled ω, ρ, ϕ, and Υ, all of which are resonances that occur in e^--e^+ annihilation through the formation of quark–antiquark bound states. The ρ and ω correspond to two different linear combinations of $u\bar{u}$ and $d\bar{d}$, the ϕ is an $s\bar{s}$ bound state, while the Υ, Υ', and Υ'' correspond to different energy states of the $b\bar{b}$ system, or bottomonium.

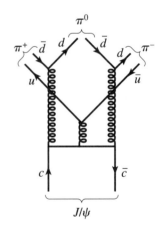

Figure 10.30 A decay mode of the J/ψ meson in which the c and $\bar{c}$ quarks annihilate through a three-gluon channel.

---

[15] Thus Problem 4.15 illustrates how numerical techniques can be used to determine the energy levels of charmonium.

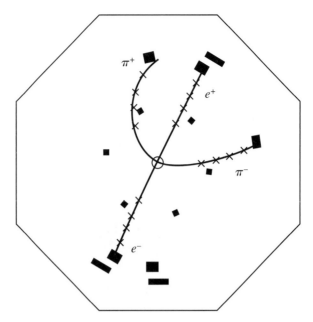

Figure 10.31 A spark chamber picture of the decay $\psi' \to J/\psi + \pi^+ + \pi^-$. The magnetic field and the SPEAR beam pipe are perpendicular to the plane of the figure. The 150-MeV pions bend more in the magnetic field than do the 1.5-GeV electron and positron emitted in the decay of the ψ. Adapted from G. S. Abrams et al., *Phys. Rev. Lett.* **34**, 1181 (1975). Given the difficulty that ensued in finding the "right" name for the J/ψ particle, many SLAC physicists could not resist wearing T-shirts showing this picture, indicating how nature seemed to come down on the matter.

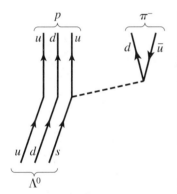

Figure 10.32 Feynman diagram for the decay $\Lambda^0 \to p + \pi^-$.

EXAMPLE 10.9 Construct a Feynman diagram for the decay $\Lambda^0 \to p + \pi^-$.

SOLUTION Fig. 10.32 shows the Feynman diagram. The Λ^0 baryon is a *uds* bound state. The strange quark decays into an up quark by emitting a W^- boson, which subsequently decays into a down quark and an anti-up quark. Strictly, either of the d quarks in the final state could be grouped with the $\bar{u}$ to make a π^- meson.

Neutrino Oscillations

This mixing, as it is called, of d and s quarks in the coupling to the W boson is apparently repeated in the lepton sector as well and is the resolution of a long-standing mystery. Let's start with a description of the mystery. Recall from our discussion of nuclear fusion in Section 9.5 that the initial reaction in a series of reactions [see (9.66) through (9.68)] responsible for powering the Sun is the weak interaction

$$p + p \to d + e^+ + \nu_e + 0.4 \text{ MeV} \qquad (10.54)$$

where here d is the deuteron. The power output of the Sun is well known and consequently so should be the number of nuclear reactions taking place in the Sun's interior. Each conversion of four protons into an alpha particle releases two electron neutrinos, as summarized in (9.69). Consequently, we expect a huge flux of neutrinos (roughly 100 billion per square centimeter per second) from the Sun to be incident on the Earth (see Problem 10.13). Although the cross section for neutrino interactions is very small, it is possible to predict, using the theory of weak interactions and our understanding of the workings of the interior of the Sun, how many neutrinos should be detected in, say, an underground detector that is sufficiently isolated from cosmic ray background. The first experiment to detect solar neutrinos was carried out by R. Davis in a gold mine in South Dakota with a vat about the size of a swimming pool filled with C_2Cl_4 (cleaning fluid). Electron neutrinos interact with chlorine atoms via the reaction

$$\nu_e + Cl \rightarrow e^- + Ar \tag{10.55}$$

In particular, ^{37}Cl is converted to ^{37}Ar, which is radioactive with a half-life of 35 days. Davis isolated the argon and measured its radioactivity. In the initial experiment, a few argon atoms each month were expected. In this experiment and in other more recent experiments, the number of neutrinos detected is roughly one-third to one-half of that predicted by theory, generating what has been called **the solar neutrino problem**.

The resolution of the solar neutrino problem resides in the sort of mixing that occurs between d and s quarks in (10.52) and (10.53). To make this discussion as straightforward as possible, let's focus on two neutrino types, ν_e and ν_μ. Assume that these neutrinos are the analogues of the d' and s', that is,

$$\nu_e = \nu_1 \cos\theta + \nu_2 \sin\theta \tag{10.56}$$

and

$$\nu_\mu = -\nu_1 \sin\theta + \nu_2 \cos\theta \tag{10.57}$$

where θ is yet another mixing angle, different in value from the Cabibbo angle (or, for that matter, the Weinberg angle). The ν_1 and ν_2 are presumed to be the energy eigenstates (like s and d). If the corresponding energy eigenvalues E_1 and E_2 are different, then the superpositions in (10.56) and (10.57) are time dependent:

$$\nu_e(t) = \nu_1(0)e^{-iE_1t/\hbar} \cos\theta + \nu_2(0)e^{-iE_2t/\hbar} \sin\theta \tag{10.58}$$

and

$$\nu_\mu(t) = -\nu_1(0)e^{-iE_1t/\hbar} \sin\theta + \nu_2(0)e^{-iE_2t/\hbar} \cos\theta \tag{10.59}$$

where $\nu_e(0)$ is the amplitude of an electron neutrino at $t = 0$ and $\nu_\mu(0)$ is the amplitude of a muon neutrino at $t = 0$. If we presume that $\nu_e(0) = 1$ and $\nu_\mu(0) = 0$, as would be the case for the neutrinos produced by the Sun, then

$$\nu_1(0) = \nu_e \cos\theta \quad \text{and} \quad \nu_2(0) = \nu_e \sin\theta \tag{10.60}$$

and hence

$$\nu_e(t) = \nu_1(0)e^{-iE_1t/\hbar} \cos\theta + \nu_2(0)e^{-iE_2t/\hbar} \sin\theta$$
$$= \left(\cos^2\theta e^{-iE_1t/\hbar} + \sin^2\theta e^{-iE_2t/\hbar}\right)\nu_e(0) \tag{10.61}$$

The probability of the neutrino being an electron neutrino at time t is therefore given by

$$\text{Prob}(\nu_e \rightarrow \nu_e) = \cos^4\theta + \sin^4\theta + \cos^2\theta \sin^2\theta \left[e^{i(E_2-E_1)t/\hbar} + e^{i(E_2-E_1)t/\hbar}\right]$$
$$= 1 - \sin^2(2\theta) \sin^2\left[(E_2 - E_1)t/2\hbar\right] \tag{10.62}$$

Neutrinos are generally ultrarelativistic. Presuming that $pc \gg mc^2$, we can use the binomial expansion to approximate the energy

$$E_i = \sqrt{p^2c^2 + m_i^2 c^4} = pc\sqrt{1 + \frac{m_i^2 c^2}{p^2}}$$

$$\cong pc + \frac{m_i^2 c^3}{2p} \qquad (10.63)$$

and therefore

$$E_2 - E_1 \cong \frac{\left(m_2^2 - m_1^2\right) c^3}{2p} \qquad (10.64)$$

Thus measurements of the rate of neutrino oscillations permit us to determine the differences in the square of the masses of the neutrinos.

Figure 10.33 shows the experimental setup for the Super-Kamiokande experiment, one of the recent experiments that has provided strong evidence for neutrino oscillation. The Super-Kamiokande experiment consists of 50,000 metric tons of ultra pure water in a tank surrounded by 11,000 photomultipliers, each 50 cm in diameter. The

Figure 10.33 The Super-Kamiokande detector partially filled with water. The leptons produced by neutrino interactions in the water emit Cerenkov radiation that is detected by the photomultipliers lining the tank. [Courtesy Kamioka Observatory, ICRR (Institute for Cosmic Ray Research), The University of Tokyo]

detector is located underground in an active zinc mine in the Japanese Alps. Not only is Super-Kamiokande capable of detecting solar neutrinos, but it can also observe the neutrinos that are generated in cosmic ray collisions in the atmosphere. Cosmic ray protons striking the nuclei of air molecules produce pions. The predominant pion decay modes

$$\pi^+ \to \mu^+ + \nu_\mu \tag{10.65}$$

and

$$\pi^- \to \mu^- + \bar{\nu}_\mu \tag{10.66}$$

generate a flux of muon neutrinos. Cosmic rays strike the atmosphere on all sides of the Earth and the probability that a muon neutrino scatters in traveling up through the Earth is small, as indicated in Fig. 10.34. Without neutrino oscillation there should be equal fluxes of energetic muon neutrinos entering the tank of water from above and from below. But in fact the number from below is less than the number from above because the neutrinos from below travel farther and therefore have more time (distance) to oscillate into tau neutrinos. The energy of the incident neutrinos is not large enough to produce a tau lepton and thus these neutrinos do not produce a charged lepton in their passage through the water and are not observed. Figure 10.35 shows the probability of a muon neutrino remaining a muon neutrino as a function of distance traveled. Similarly, electron neutrinos from the Sun that oscillate into muon neutrinos do not have sufficient energy to create a charged muon and thus cannot interact through W exchange with the material in the target.

The results from Super-Kamiokande as well as a number of other experiments show $(m_2^2 - m_1^2)c^4 = 8 \times 10^{-5}$ eV^2 and $\sin^2(2\theta_{12}) = 0.86$, with uncertainties on the order of 5%. The data also indicate that $(m_3^2 - m_2^2)c^4$ is on the order of 2×10^{-3} eV^2 and $\sin^2(2\theta_{23}) = 1.0$, with a larger uncertainty. An upper limit on the mass for the electron neutrino of 2.2 eV comes from measuring the energies of the most energetic electrons emitted in the beta decay of tritium, while comparable upper limits on the mass of muon and tau

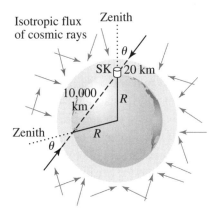

Figure 10.34 Cosmic rays strike the atmosphere of the Earth from all sides, producing charged pions which decay predominately into muons and muon neutrinos. Consequently, without neutrino oscillations the flux of muon neutrinos entering an underground detector should not show any dependence on the angle θ.

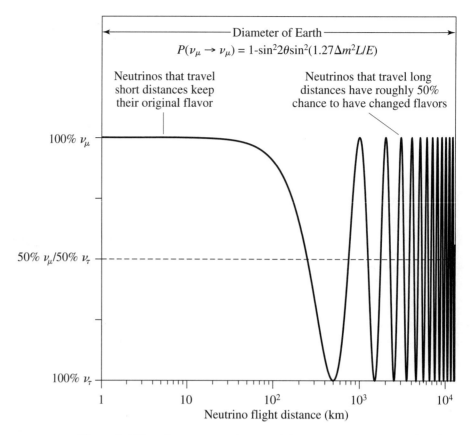

Figure 10.35 The probability of a muon neutrino remaining a muon neutrino as a function of distance traveled by the neutrino. The probability for a muon neutrino to remain a muon neutrino is the same as the result (10.62) with ν_μ replacing ν_e and $E_2 - E_1$ replaced with $E_3 - E_2 = (m_3^2 - m_2^2)c^3/2p = \Delta m^2 c^3/2p$ since ν_μ oscillates mainly into ν_τ. The time t in (10.62) is set equal to L/c, where L is the distance traveled by the neutrino. The numerical value in the argument of the sine function arises from presuming that Δm^2 is measured in eV^2, L in meters, and E in MeV. Note that the horizontal axis is a log scale. The first minimum in the oscillation occurs for neutrinos that travel five hundred kilometers. (Courtesy M. Messier)

neutrinos are in the MeV range. The natural presumption is that all three masses are near zero, but the neutrino oscillation results are sensitive only to differences in mass.[16]

Neutrinos are everywhere. Not only do stars emit a large flux of neutrinos, as has been noted, but there is a cosmic background radiation of neutrinos that consists of roughly 300 neutrinos per cubic centimeter throughout the universe. This background radiation of neutrinos is somewhat cooler than that for photons, because the neutrinos decoupled from matter very shortly after the Big Bang because of their weak interaction with matter. These neutrinos contribute to the energy density of the universe in a form that is called hot dark matter—hot because the neutrinos are moving at ultrarelativistic speeds and

---

[16] The observation of neutrino oscillations was startling to many physicists. Neutrino oscillations cannot take place unless the neutrinos have different masses, i.e., different energy eigenvalues. The presumption had been that the masses were all zero, since there were no facts to the contrary and it was difficult to understand if neutrinos had a mass why it would be so small.

dark because they are nearly undetectable apart from gravitational effects. A limit on the sum of the masses of all three flavors of neutrinos of less than 1 eV can be inferred from the limits placed on the amount of hot dark matter in the universe. A larger amount of hot dark matter would generate a distribution of galaxies in the universe as a whole that would be in disagreement with what is observed.

10.7 Symmetries and Conservation Laws

In our discussion of particle physics, we have made use of a variety of conservation laws, including conservation of energy, linear momentum, angular momentum, charge, and color. As we will discuss in this section and the next, there is a very deep connection between conservation laws and an associated underlying symmetry in the laws of physics. We will also discuss the discrete space–time symmetries of parity, charge conjugation, and time reversal and the discovery that the weak interactions do not possess the same discrete space–time symmetries as do the strong and electromagnetic interactions.

To see the connection between symmetries and conservation laws, start with (5.73), namely

$$\frac{d\langle A \rangle}{dt} = \frac{i}{\hbar} \int_{-\infty}^{\infty} \Psi^*[H, A_{\text{op}}]\Psi \, dx + \int_{-\infty}^{\infty} \Psi^* \frac{\partial A_{\text{op}}}{\partial t} \Psi \, dx \qquad (10.67)$$

where $[H, A_{\text{op}}] = H A_{\text{op}} - A_{\text{op}} H$ is the commutator of the Hamiltonian with the operator corresponding to the observable A. Recall that we derived this result by taking the time derivative of

$$\langle A \rangle = \int_{-\infty}^{\infty} \Psi^* A_{\text{op}} \Psi \, dx \qquad (10.68)$$

making use of the fact that the time dependence of the wave function is governed by the Schrödinger equation

$$H\Psi(x, t) = i\hbar \frac{\partial \Psi(x, t)}{\partial t} \qquad (10.69)$$

where H is the Hamiltonian. A conserved quantity is one for which $\langle A \rangle$ is independent of time.

Linear Momentum and Energy

Let's first take a look at conservation of linear momentum in one dimension, for which

$$H = \frac{p_{x_{\text{op}}}^2}{2m} + V(x_{\text{op}}) \qquad (10.70)$$

Since

$$\frac{d\langle p_x \rangle}{dt} = \frac{i}{\hbar} \int_{-\infty}^{\infty} \Psi^*[H, p_{x_{\text{op}}}]\Psi \, dx \qquad (10.71)$$

for linear momentum to be conserved, we require

$$[H, p_{x_{\text{op}}}] = \left[\frac{p_{x_{\text{op}}}^2}{2m} + V(x_{\text{op}}), p_{x_{\text{op}}}\right]$$
$$= 0 \qquad (10.72)$$

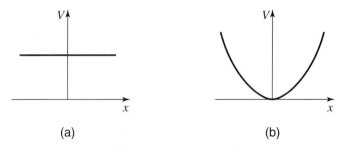

Figure 10.36 (a) A constant potential energy is translationally invariant. (b) The potential energy of the harmonic oscillator does not exhibit translational symmetry.

The linear momentum operator clearly commutes with itself and

$$[x_{\text{op}}, p_{x_{\text{op}}}] = i\hbar \qquad (10.73)$$

Therefore, linear momentum will be conserved whenever $V(x)$ is independent of x, that is, $V(x) = V_0$, where V_0 is a constant, as illustrated in Fig. 10.36a.

A symmetry operation is an operation that leaves the system invariant, or unchanged. In this case, we say the system has translational invariance since the potential energy satisfies $V(x + a) = V(x)$ for any a. Put another way, $V(x)$ is independent of where we position the origin of our coordinate system. Contrast this constant potential energy with that of the harmonic oscillator $V(x) = m\omega^2 x^2/2$, shown in Fig. 10.36b. For the harmonic oscillator, there is no translational invariance. The potential energy varies with the position x and the origin in Fig. 10.36b is that special point at which the potential energy is a minimum. Applying (10.71) to the harmonic oscillator, we obtain

$$\frac{d\langle p_x \rangle}{dt} = -m\omega^2 \langle x \rangle \qquad (10.74)$$

which is just Hooke's law expressed in terms of expectation values. Thus when we do not have translational symmetry, linear momentum is not conserved. Our results are just what we might expect from classical physics, just phrased somewhat differently, since if the potential energy is independent of x, then the force $F = -\partial V/\partial x$ vanishes, indicating the momentum is conserved.

Similarly, if the Hamiltonian is invariant under time translations, that is H is independent of time, then of course $\partial H/\partial t = 0$. Consequently, (10.67) tells us that energy is conserved since the Hamiltonian commutes with itself.

This translational invariance is actually built quite deeply into the way we think about the world. For example, when we measure the lifetime of an excited state of hydrogen, we naturally expect that we will get the same answer if we were to perform the experiment in a different part of the room or indeed in a different lab, perhaps one in another country. Why do we believe this? One can say it is due to the homogeneity of space, the fact that the physics is translationally invariant. The confirmation is how deeply we believe in and have experimental evidence for conservation of linear momentum.

Angular Momentum

We now turn to conservation of angular momentum. If the potential energy is rotationally invariant, that is $V = V(r)$, independent of θ and ϕ, then

$$\frac{d\langle L_z \rangle}{dt} = \frac{i}{\hbar} \int \Psi^* [H, L_{z_{\text{op}}}] \Psi \, d^3r = 0 \qquad (10.75)$$

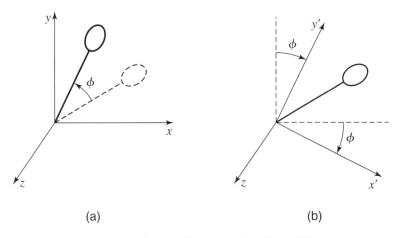

Figure 10.37 A schematic diagram illustrating the effect of (a) rotating an experiment by angle ϕ counterclockwise about the z axis and (b) keeping the experiment fixed and rotating the coordinate axes by angle ϕ clockwise about the z axis.

since the Hamiltonian and the operator corresponding to the z component of the orbital angular momentum

$$L_{z_{\text{op}}} = \frac{\hbar}{i} \frac{\partial}{\partial \phi} \qquad (10.76)$$

commute.[17] This is consistent with what we know from classical physics for a system with central forces since the torque $\mathbf{r} \times \mathbf{F}$ vanishes when the force $\mathbf{F}$ is parallel (or antiparallel) with the radius $\mathbf{r}$, as will be the case when $V = V(r)$.

Testing rotational invariance in the lab would require us to take the experimental equipment and rotate it counterclockwise, say by an angle ϕ about the z axis, and perform the experiment again, as indicated in Fig. 10.37a. Taking a complex piece of experimental apparatus, say one that is carefully aligned, rotating it, and performing the measurements again is an experiment that we seldom do. We do, however, often do something comparable, namely we may choose to analyze the experiment from a different coordinate system, say one that is rotated clockwise by angle ϕ about the z axis, as indicated in Fig. 10.37b. We learn from the very beginning in physics that we have the freedom to orient our coordinate axes in any way we choose. Why are we so comfortable doing this? Why do we believe so naturally in the isotropy of space? The answer is all the evidence that we have for conservation of angular momentum.

Parity

Just as it seems common sense that we have the freedom to translate or rotate our coordinate axes at our discretion, one would suppose that we could equally well invert our coordinate axes in analyzing a particular experiment. Inverting the coordinate axes turns a right-handed coordinate system into a left-handed one, as indicated in Fig. 10.38.

---

[17]Similarly, $L_{x_{\text{op}}}$ and $L_{y_{\text{op}}}$ depend only on angles as well and commute with the Hamiltonian when the potential energy is a function of r only.

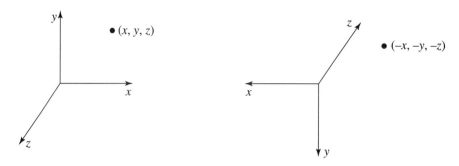

Figure 10.38 Inversion of the coordinate axes.

The coordinates of the position vector **r** are consequently inverted:

$$\mathbf{r} = (x, y, z) \to (-x, -y, -z) = -\mathbf{r} \tag{10.77}$$

Inversion symmetry is closely connected to mirror symmetry. A **mirror reflection**, one in which just one of the axes is inverted, say

$$(x, y, z) \to (x, -y, z) \tag{10.78}$$

can be obtained from an inversion of coordinates followed by a rotation of 180° about the y axis. Since the laws of physics are rotationally invariant, these laws should also look the same in a mirror if it doesn't matter whether we use a right-handed or a left-handed coordinate system in our analysis.

As we noted in Section 5.1, the quantum mechanical operator that inverts the coordinates of the wave function is the parity operator

$$\Pi \psi(x, y, z) = \psi(-x, -y, -z) \tag{10.79}$$

Since

$$\frac{d\langle \Pi \rangle}{dt} = \frac{i}{\hbar} \int_{-\infty}^{\infty} \Psi^* [H, \Pi] \Psi \, d^3r \tag{10.80}$$

if the Hamiltonian commutes with the parity operator, then the expectation value of the parity does not change with time and parity is conserved. The underlying symmetry associated with conservation of parity is the invariance of the laws of physics to switching from a right-handed to a left-handed coordinate system.

There is a lot of evidence that the strong and electromagnetic interactions conserve parity. In trying to sort out a puzzle involving the decays of the K mesons, T. D. Lee and C. N. Yang proposed in 1956 that perhaps parity is not conserved in the weak interactions. Moreover, they noted that whether parity was or was not conserved had never been tested by experiment. Subsequently, C–S. Wu examined the beta decay

$$^{60}\text{Co} \to {}^{60}\text{Ni} + e^- + \bar{\nu}_e \tag{10.81}$$

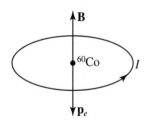

Figure 10.39 Schematic diagram of C-S. Wu's experiment on the beta decay of ^{60}Co nuclei, which are located at the center of a current-carrying wire. The electrons are preferentially emitted opposite to the direction of the magnetic field.

for cobalt nuclei that had been cooled to low temperature in a strong magnetic field. A very schematic diagram of the experiment is shown in Fig. 10.39. Because ^{60}Co has a large intrinsic spin and consequently a large magnetic moment, the nuclei align themselves in the magnetic field before the decay. In the decay more electrons are emitted antiparallel to the magnetic field than parallel to the magnetic field. While this result sounds fairly innocuous, it is clear evidence for nonconservation of parity. After all, the direction of the magnetic field is determined using the "right-hand rule." You wrap your right hand around the current loop in the direction of the current and your thumb then points in the

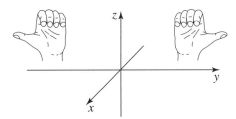

Figure 10.40 A hand and its mirror image.

direction of the magnetic field. Thus by analyzing the ^{60}Co beta decay, we can determine unambiguously what is right handed by noting that the electrons are emitted preferentially opposite to the direction of your thumb if you use the right-hand rule. If on the other hand, had you used your left hand to determine the magnetic field direction, you would then say the electrons are preferentially emitted parallel to the direction of your thumb on that hand. In this sense, the experiment seems to say that the weak interactions are left handed. But whether something is left or right handed depends on whether you use a right-handed coordinate system, as the mirror image of a hand in Fig. 10.40 shows. Thus the result of this experiment is evidence for nonconservation of parity.

This description of the beta decay of ^{60}Co may raise questions about conservation of parity in the electromagnetic interactions. After all, don't electromagnetic interactions depend on the right-hand rule? That is how we determined the direction of the magnetic field from the current carrying wire. But in electromagnetism the fields are used to determine the direction of forces on charged particles. If you examine the Lorentz force law $\mathbf{F} = q\mathbf{E} + q\mathbf{v} \times \mathbf{B}$, you see that to determine the direction of the magnetic force, which involves a cross product, we use the right-hand rule twice, once to determine the direction of the magnetic field and once to determine the direction of the force. Applying the right-hand rule twice gives the same result as applying the left-hand rule twice. Thus as long as we are consistent, whether we use the right-hand rule or the left-hand rule for electromagnetism, we obtain the same results. There is no distinction between a left-handed versus a right-handed coordinate system. Consequently, all electromagnetic processes conserve parity. And although the beta decay of ^{60}Co is just one manifestation of nonconservation of parity in the weak interactions, all weak processes violate conservation of parity. It is a hallmark of the weak interactions.

Charge Conjugation and Time Reversal

For a brief period of time after the discovery of nonconservation of parity, physicists thought there was a way to restore the symmetry that had been lost to nonconservation of parity. In particular, they hoped that even if the weak interactions were not invariant under inversion of coordinates, they might be invariant under the combined operations of inversion and charge conjugation. In charge conjugation, every particle is replaced with its antiparticle. The electromagnetic interaction between an antiproton and a positron in an antihydrogen atom, for example, is identical to that between a proton and an electron in a hydrogen atom. The energy levels, the decay rates, all the physics that is due to electromagnetic interactions does not permit us to distinguish the particle world and the antiparticle world. The physics is the same for both, at least as long as we restrict ourselves to electromagnetic interactions. Since electromagnetic interactions are invariant under

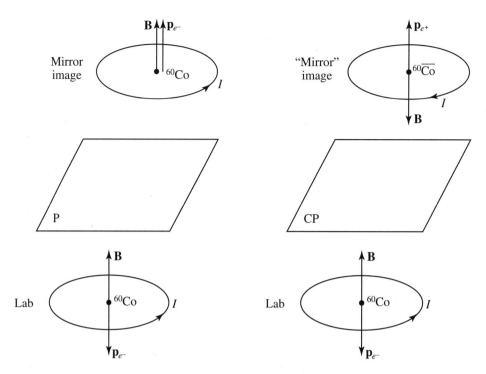

Figure 10.41 A mirror that combines the operations of inversion and charge conjugation (a CP mirror) restores the symmetry in the beta decay of ^{60}Co. In the lab, electrons are emitted preferentially opposite to the magnetic field, which is also true for the positrons observed in the CP mirror.

inversion of the coordinates as well, these interactions are invariant under the combined action of charge conjugation and inversion, an action that is typically referred to as CP.

Let's go back to the Wu experiment and consider the combined actions of inversion of the coordinates and the replacement of each particle by its antiparticle. Suppose there is a mirror on the ceiling of the lab, as indicated in Fig. 10.41. The direction of current in the lab is the same as in the mirror, but the direction of the emitted electrons is reversed in the mirror. As we have noted, the physics is not symmetric in the lab and in the mirror image of the lab, that is in a right-handed and a left-handed coordinate system. But suppose the mirror is not an ordinary mirror, but rather one that combines the operations of charge conjugation and inversion (a CP mirror). In the CP mirror, the decay is

$$^{60}\overline{\text{Co}} \rightarrow\, ^{60}\overline{\text{Ni}} + e^+ + \nu_e \tag{10.82}$$

While the positrons emitted in this beta decay travel in the opposite direction to that observed in the lab, the magnetic field in the CP mirror is reversed as well because the direction of current flow in the CP mirror is opposite to that in the lab since the charge carriers are the antiparticles of the ones in the lab. But with both the direction of the positrons and the magnetic field reversed in the CP mirror, more positrons come out antiparallel to the magnetic field, just as the electrons do in the lab. Thus the symmetry has been restored.

Until 1964 there was reason to believe that the weak interactions were invariant under CP. But in that year in an analysis of the decays of the K^0 mesons it was discovered that there is a small violation, at the level of a couple of parts per thousand, in CP conservation. This CP nonconservation is to be contrasted with nonconservation of parity, which one can say is maximal in the sense that if a system starts in an eigenstate of parity with eigenvalue $+1$, it is equally likely to evolve into a state that has parity $+1$ or one that has parity -1, all other things being equal. Although the violation of CP conservation is a small one, the effect can be seen because of quantum interference effects in the time evolution of the K^0–$\bar{K}^0$ system—interference effects that are very similar to those responsible for neutrino oscillations. One of the interesting implications of nonconservation of CP is that time reversal must also be violated, since it can be shown that any theory that respects both quantum mechanics and special relativity conserves CPT, where T stands for time reversal. Thus if CP is violated and CPT is conserved, T must be violated as well.

One of the virtues of CP nonconservation may be our very existence. We live in a universe that started as a Big Bang in which particle and antiparticle creation and annihilation occurred at the same rate, as is clear from Feynman diagrams such as that shown in Fig. 10.11. But the universe now is dominated by what we call matter, not antimatter. How did this asymmetry arise? As was noted by A. Sakharov, a necessary condition for the predominance of matter over antimatter is CP nonconservation in the laws of physics. Thus our universe would be a very different place without CP nonconservation. Interestingly, it can also be shown that CP nonconservation requires three generations of quarks/leptons, since the generalization of the mixing angle to three generations demands a matrix with four parameters, one of which must be a complex phase associated with CP nonconservation. While we do not know yet at a fundamental level why we have three generations of quarks/leptons, it seems that without three generations life as we know it could not have developed.

EXAMPLE 10.10 Show that total momentum is conserved for two particles that interact through a potential energy $V(x_1 - x_2)$. What symmetry does the potential energy possess?

SOLUTION First note that for this potential energy the system is not translationally invariant for $x_1 \to x_1 + a$, that is if we change the position of one of the particles but not the other, the potential energy varies. On the other hand, if we make the simultaneous transformation $x_1 \to x_1 + a$ and $x_2 \to x_2 + a$, then the distance separating the particles does not change and neither does the potential energy. The total momentum operator is

$$P_{x_{op}} = \frac{\hbar}{i}\frac{\partial}{\partial x_1} + \frac{\hbar}{i}\frac{\partial}{\partial x_2}$$

The commutator of the total momentum operator and the Hamiltonian reduces to

$$\left[P_{x_{op}}, V(x_1 - x_2)\right]\psi = \left(\frac{\hbar}{i}\frac{\partial V(x_1 - x_2)}{\partial x_1} + \frac{\hbar}{i}\frac{\partial V(x_1 - x_2)}{\partial x_2}\right)\psi = 0$$

where in the last step we have taken advantage of the fact that

$$\frac{\partial}{\partial x_1}V(x_1 - x_2) = -\frac{\partial}{\partial x_2}V(x_1 - x_2)$$

It is interesting to note the similarity in this derivation with the comparable result in classical physics, where the force on particle 1 is $F_1 = -\partial V/\partial x_1$ and the force on particle 2 is $F_2 = -\partial V/\partial x_2$. For the potential energy $V(x_1 - x_2)$, $F_1 = -F_2$, a statement of Newton's third law, and therefore $F_1 + F_2 = 0$, indicating the total momentum is a constant.

10.8 The Standard Model

One fundamental conservation law that was not included in our discussion in the previous section is conservation of electric charge. Given the connection that we have established between a conservation law and symmetry, it is natural to ask what symmetry is associated with conservation of charge. It turns out there are actually two closely intertwined symmetries, namely **gauge invariance** in Maxwell's equations and **overall phase invariance** in quantum mechanics.

Let's start with Maxwell's equations. One of Maxwell's equations, Gauss's law for magnetism, in differential form requires the magnetic field **B** to satisfy

$$\nabla \cdot \mathbf{B} = 0 \tag{10.83}$$

A way to be sure the divergence of the magnetic field vanishes is to write

$$\mathbf{B} = \nabla \times \mathbf{A} \tag{10.84}$$

since the divergence of a curl is zero. The vector field **A** is known as the **vector potential**. When we substitute $\mathbf{B} = \nabla \times \mathbf{A}$ into Faraday's law in differential form

$$\nabla \times \mathbf{E} = -\frac{\partial \mathbf{B}}{\partial t} \tag{10.85}$$

we obtain

$$\nabla \times \left(\mathbf{E} + \frac{\partial \mathbf{A}}{\partial t} \right) = 0 \tag{10.86}$$

This equation is satisfied provided

$$\mathbf{E} + \frac{\partial \mathbf{A}}{\partial t} = -\nabla \varphi \tag{10.87}$$

since the curl of a gradient vanishes. Consequently,

$$\mathbf{E} = -\nabla \varphi - \frac{\partial \mathbf{A}}{\partial t} \tag{10.88}$$

The scalar field φ is called the **scalar potential**.

Within classical physics, **A** and φ are auxiliary fields that are introduced to facilitate the calculation of **E** and **B**. The scalar potential is particularly useful in electrostatics, since it is often easier to calculate the scalar φ for some charge distribution and then take its gradient than to calculate the vector **E** directly. The fields **A** and φ are not considered to be physical fields since it is possible to modify them according to

$$\mathbf{A} \to \mathbf{A} + \nabla \chi \quad \text{and} \quad \varphi \to \varphi - \frac{\partial \chi}{\partial t} \tag{10.89}$$

where χ is an arbitrary scalar field, without affecting the electric field **E** and the magnetic field **B**. See Problem 10.16. A simple illustration is the freedom we have to set the value of the scalar potential at some point (say the point at infinity or perhaps the point we call ground in a circuit) to zero. We could just as easily define it as 100 V without affecting any of the physics. It is only potential differences that matter. This shift in the zero of the scalar potential φ is a special case of the more general **gauge transformation** shown in (10.89). Consequently, the fields **A** and φ are called **gauge fields**. And the fact that **E** and **B** are not affected by a gauge transformation is referred to as gauge invariance.

While providing a rigorous proof of the connection between gauge invariance and conservation of charge is beyond the scope of this text, following E. Wigner we can give a heuristic argument that shows they are connected. Start by assuming that charge is not conserved and a charge is created at a point in space at which the scalar potential has the value φ. The energy to create the charge must be independent of the value of φ by gauge invariance. If φ is spatially dependent, then according to (10.88), there is an electric field present that will accelerate the particle to a region where the scalar potential has a different value, say φ'. If the charge is now presumed to disappear, the energy gained in this annihilation must be independent of the value of φ', again because of gauge invariance. But the particle has gained kinetic energy $q(\varphi - \varphi')$ in this movement, energy that could, for example, be used to run a perpetual motion machine. If we presume that this is crazy, that energy is conserved, then charge must be conserved as well.

Now we turn to what at first looks like an unrelated symmetry that is deeply embedded in quantum mechanics, namely *overall* phase invariance. If you take any wave function and change the overall phase

$$\psi(\mathbf{r}, t) \to e^{i\delta} \psi(\mathbf{r}, t) \tag{10.90}$$

there will be no change in the probability density $\psi^*(\mathbf{r}, t)\psi(\mathbf{r}, t)$ or in any of the probabilities $|c_n|^2 = \left| \int \psi_n^* \psi \, d^3 r \right|^2$ that we calculate from this wave function. In making the case for overall phase invariance, it is important to note that the phase δ is independent of space and time (and consequently the phase change can be described as a *global* phase change). Otherwise not only will the $|c_n|^2$ be affected, but in addition the change generated in (10.90) will modify the fundamental equations, such as the Schrödinger equation, that involve derivatives of the wave function. If we were to consider what is termed a *local* phase change on the wave function for a particle with charge q of the form

$$\psi(\mathbf{r}, t) \to e^{iq\chi(\mathbf{r},t)/\hbar} \psi(\mathbf{r}, t) \tag{10.91}$$

where $\chi(\mathbf{r}, t)$ is an arbitrary scalar field, then a partial derivative of the wave function, say with respect to x, becomes

$$\frac{\partial \psi(\mathbf{r}, t)}{\partial x} \to e^{iq\chi(\mathbf{r},t)/\hbar} \left(\frac{\partial \psi(\mathbf{r}, t)}{\partial x} + \frac{iq}{\hbar} \frac{\partial \chi(\mathbf{r}, t)}{\partial x} \psi(\mathbf{r}, t) \right) \tag{10.92}$$

and a partial derivative with respect to t becomes

$$\frac{\partial \psi(\mathbf{r}, t)}{\partial t} \to e^{iq\chi(\mathbf{r},t)/\hbar} \left(\frac{\partial \psi(\mathbf{r}, t)}{\partial t} + \frac{iq}{\hbar} \frac{\partial \chi(\mathbf{r}, t)}{\partial t} \psi(\mathbf{r}, t) \right) \tag{10.93}$$

The whole idea of phase invariance is that the overall phase not have observable consequences. But in (10.92) and in (10.93) we see the presence of an extra term involving the derivative of the phase. Such a term should not be present if quantum mechanics is to be locally phase invariant. The equations themselves should not be modified by making such a phase change. Now the argument for expanding global phase invariance to **local phase invariance** may not be an easy sell for nonrelativistic quantum mechanics (and therefore for the Schrödinger equation), but in relativistic quantum mechanics it is natural to ask why we should require that the overall phase change that we are making, say at a particular time t, should be the same everywhere in the universe, since it would take a time d/c for the information about the phase change to propagate a distance d to some other point in space.

The way to ensure local phase invariance is to require that every derivative be accompanied by an extra field. For example, you can check that if each partial derivative with respect to x is modified according to

$$\frac{\hbar}{i}\frac{\partial}{\partial x} \to \frac{\hbar}{i}\frac{\partial}{\partial x} - qA_x \tag{10.94}$$

then the extra term that arises in (10.92) will be canceled out and the equations themselves will suffer only an overall phase change of $e^{iq\chi(\mathbf{r},t)/\hbar}$ provided that simultaneously with the change (10.91) we change A_x via

$$A_x \to A_x + \frac{\partial \chi}{\partial x} \tag{10.95}$$

To accomplish the same trick for the partial derivatives with respect to y and z, we need additional fields A_y and A_z. The three fields A_x, A_y, and A_z form a vector field $\mathbf{A}$. The spatial derivatives then all enter in the form

$$\frac{\hbar}{i}\nabla \to \frac{\hbar}{i}\nabla - q\mathbf{A} \tag{10.96}$$

with

$$\mathbf{A} \to \mathbf{A} + \nabla \chi \tag{10.97}$$

the generalization of (10.95) to three dimensions. Similarly, if the partial derivatives with respect to t are modified according to

$$i\hbar\frac{\partial}{\partial t} \to i\hbar\frac{\partial}{\partial t} - q\varphi \tag{10.98}$$

where φ is a scalar field, then the extra term in (10.93) will also be canceled out provided

$$\varphi \to \varphi - \frac{\partial \chi}{\partial t} \tag{10.99}$$

Thus the overall physics will not be affected provided we introduce a vector field $\mathbf{A}$ and a scalar field φ according to (10.96) and (10.98), respectively, and require that

$$\mathbf{A} \to \mathbf{A} + \nabla \chi \quad \text{and} \quad \varphi \to \varphi - \frac{\partial \chi}{\partial t} \tag{10.100}$$

when

$$\psi(\mathbf{r}, t) \to e^{iq\chi(\mathbf{r},t)/\hbar}\psi(\mathbf{r}, t) \tag{10.101}$$

What are these fields $\mathbf{A}$ and φ? They are the same auxiliary fields that we introduced in our discussion of gauge invariance in Maxwell's equations. But now see that they are not merely auxiliary fields. Rather $\mathbf{A}$ and φ are essential fields that undergo the gauge transformation (10.100) when the local phase change (10.101) takes place. Thus local phase invariance for charged particles *requires* the introduction of electromagnetic fields. For example, implementing local gauge invariance for the Schrödinger equation for a free particle with charge q by making the changes (10.96) and (10.98) leads to

$$\frac{1}{2m}\left(\frac{\hbar}{i}\nabla - q\mathbf{A}\right)^2 \psi = i\hbar\frac{\partial \psi}{\partial t} - q\varphi\psi \tag{10.102}$$

or

$$\frac{1}{2m}\left(\frac{\hbar}{i}\nabla - q\mathbf{A}\right)^2 \psi + q\varphi\psi = i\hbar\frac{\partial \psi}{\partial t} \tag{10.103}$$

which is the Schrödinger equation for a particle with charge q interacting with an electromagnetic field. The $q\varphi$ term is, of course, the potential energy of the particle [$-e^2/(4\pi\epsilon_0 r)$ for the hydrogen atom, for example], while the terms involving **A** are needed to include the effects of magnetic fields in the Hamiltonian. Thus not only is there a symmetry called local gauge symmetry (or gauge invariance) connected with conservation of charge, but this symmetry requires the existence of electromagnetic fields, fields that satisfy Maxwell's equation.[18]

What about conservation of color? A similar local gauge invariance applies to the color charges of the quarks. In this case, the analogue of (10.101) is a generalized rotation that arises from the principle that the strong interactions between quarks are invariant under a transformation that turns, say, a red quark into an arbitrary linear combination of a red, a green, and a blue quark, subject to the constraint that the transformation not affect the probability that a quark exists. Requiring that the laws of physics be invariant under such transformations that differ at different points in space and time dictates the existence of eight color gauge fields. These are the fields that give rise to the eight types of gluons that mediate the strong interactions. The equations satisfied by the gauge fields and the quarks are the equations of quantum chromodynamics. It is this local gauge invariance argument that permits us to deduce these equations despite the fact that the strong interactions are short-range interactions that do not have a classical limit, unlike Maxwell's equations whose existence was deduced by Gauss, Faraday, Ampere, and, of course, Maxwell long before the advent of quantum mechanics.

The model in which the strong and electroweak interactions are mediated by gauge fields is called the **Standard Model**. It has been spectacularly successful in its agreement with experiment. While the theoretical predictions of the Standard Model have not been carried out across the board to the extent that they have in QED [see (10.25)], there are no confirmed disagreements between theory and experiment. One key element of the standard model has not yet been discovered, however. There is a scalar field, called the Higgs field, that is required to break spontaneously the gauge symmetry in electroweak interactions. This Higgs field is responsible not only for the large masses of the $W^\pm$ and Z^0 but also for the masses of all the quarks and leptons. It should also manifest itself in the existence of at least one spin-0 particle called the Higgs boson. As the Large Hadron Collider (LHC) begins operation at CERN, physicists are optimistic that the Higgs boson will be found, thereby completing the Standard Model.

Spontaneous symmetry breaking in quantum field theory is a complex phenomenon that is beyond the scope of this text. A couple of the interesting features of spontaneous symmetry breaking are captured, however, in a simple physical example. Consider a flexible plastic needle that is placed under stress in the vertical direction by uniform inward pressure. Even though the system initially has rotational, or azimuthal symmetry about an axis passing through the needle, under this pressure the needle bows laterally as shown in Fig. 10.42. The direction of the bowing, say to the left or to the right to pick two of the infinite number of possibilities, cannot be predicted. Nonetheless, such bowing is essential if the system is to find its lowest energy state. Thus we say that the ground state of the system "breaks" the (azimuthal) symmetry that the system possesses. Note

---

[18] Sometimes (10.101) is referred to as a local gauge transformation of the first kind while (10.100) is referred to as a gauge tranformation of the second kind. But it is probably best to think of them as a unit, namely a local gauge transformation.

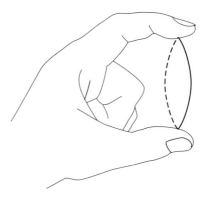

Figure 10.42 Spontaneous symmetry breaking in a flexible plastic needle.

that if we perturb the system by applying an additional force to the needle perpendicular to the plane of the figure, the needle spins easily about the axis. On the other hand, pushing radially inward or outward on the needle causes the needle to oscillate. Thus the dynamics are very different in response to these two perturbations. In quantum field theory, the "easy" azimuthal excitation leads to a massless particle, while the "harder" radial excitation leads to a particle with nonzero mass.

One of the disconcerting aspects to the Standard Model is the large number of parameters that need to be specified in the model, including such parameters as the value for the fine structure constant α, the masses of the leptons and the quarks, the various mixing angles, and so forth. Since the strong and electroweak interactions are local gauge theories, there has been hope that these theories can be be joined together in a grand unified theory (GUT). Such a grand unified theory would logically tie together the lepton and quark sectors, presumably through interactions that would permit a quark to turn into a lepton. Such interactions would naturally lead to decays such as

$$p \to e^+ + \pi^0 \tag{10.104}$$

and would be mediated by very massive spin-1 particles. Searches for proton decay have been carried out with the same detectors that are used as neutrino detectors. Reactions such as (10.104) should produce a very clear signal. The π^0 decays with a lifetime of 10^{-16} s into two 67.5-MeV photons. And the positron would annihilate with one of the many electrons in the detector, producing two 0.5-MeV photons. Such a reaction would be hard to miss! So far, experiments such as the Super-Kamiokande experiment described in Section 10.6 have put stringent limits on the lifetime of the proton. For the decay mode (10.104), $\tau > 10^{33}$ yr. Since the age of the universe is "only" 1.4×10^{10} yr, a limit of 10^{33} yr may seem so large as to be irrelevant to everyday life. But recall that

$$N(t) = N(0)e^{-t/\tau} \tag{10.105}$$

and for $t \ll \tau$

$$N(t) \cong N(0)\left(1 - \frac{t}{\tau}\right) \tag{10.106}$$

Thus we would expect on average one decay per year for every 10^{33} protons, corresponding to a mass of roughly 1000 tonnes (1 tonne $= 10^3$ kg).[19] Experiments such as the Super-Kamiokande experiment should enable us to put stringent limits on the lifetime of the proton and thus put limits on the mass of any heavy particles that mediate interactions that convert quarks into leptons.[20] But because of the background due to 1 GeV neutrinos that are produced in cosmic ray showers in the upper atmosphere, the best even a detector as large as that of Super-Kamiokande can do is a limit of 10^{34} yr. Thus there is a relatively narrow window through which we may be able to obtain evidence in support of a grand unified theory.

10.9 Summary

The elementary particles consist of quarks, leptons, and gauge bosons. The quarks and leptons are spin-1/2 particles while the gauge bosons are spin-1 particles. Strongly interacting particles (hadrons) are grouped into baryons (hadrons with half-integral intrinsic spin) and mesons (hadrons with integral intrinsic spin). Baryons are composed of three quarks, and mesons are quark–antiquark pairs. There are six different flavors of quarks (u, d, s, c, b, and t) and six different types, or flavors, of leptons (e, ν_e, μ, ν_μ, τ, and ν_τ) as well. Moreover, the quarks come in three colors (called red, green, and blue). Just as electric charge is the source of electromagnetic interactions, color is the source of strong interactions. The leptons do not have color and thus do not participate in the strong nuclear force. Electromagnetic interactions are mediated by photons, strong interactions are mediated by gluons, and the weak interactions are mediated by the W and Z bosons. In the Standard Model, each of these interactions is what is termed a local gauge theory, a quantum field theory in which the fields that mediate the interactions arise from a symmetry principle such as local phase invariance. The Standard Model includes a unification of the weak and electromagnetic interactions, similar in spirit to the unification that takes place between electric and magnetic interactions in Maxwell's equations. The search is on for a theory that unifies all the interactions. Such a theory would presumably yield connections between the many parameters that are now required to specify the Standard Model.

You may have noticed that we have left gravity out of our discussion of fundamental interactions and quantum mechanics. On the scale of particle physics, gravity is so much weaker than the strong and electroweak interactions that its impact on particle processes is negligible. On the other hand, if we wish to develop a theory that truly unifies all the interactions, then gravity should be included. While in general relativity we have a very successful and beautiful theory of gravity on the classical level, so far there is no accepted

---

[19] Steve Weinberg has remarked that we know in our bones that the lifetime of the proton is greater than 10^{22} yr, since otherwise we would be lethal to ourselves from reactions such as (10.104).

[20] To legislate against reactions such as (10.104) ever occurring, it was deemed necessary to invoke a new conservation law, namely conservation of baryon number. Within the quark model, we would say that each quark has baryon number 1/3 and consequently the proton and the neutron, for example, each has baryon number +1. In neutron decay ($n \to p + e^- + \bar{\nu}_e$), baryon number is conserved. The proton, on the other hand, doesn't decay because it is the lightest baryon.

theory of gravity that includes quantum mechanics, with the possible exception of string theory. But so far, no one has been able to put string theory to the test required of any viable theory of physics, that it yield predictions that can be tested against experiment and observation. If this book has succeeded, it has given you a sense of how broadly and successfully quantum mechanics has met this test.

Problems

10.1. Draw a Feynman diagram that would contribute to bremsstrahlung (braking radiation), that is the emission of a photon when a charged particle such as an electron collides with a heavy nucleus.

10.2. What is the Feynman diagram for the process $e^- + e^+ \to \gamma + \gamma$?

10.3. The π^0 decays into two photons with a lifetime of 10^{-16} s. Construct a Feynman diagram for this decay.

10.4. Table 10.4 lists the ten lowest mass spin-3/2 baryons, the decuplet. Check that the quark content is in agreement with the charges for each of these baryons. Explain why there is a Δ^{++} baryon and no Δ^{--} baryon. Estimate the effective mass difference between the strange quark and the up or down quark from the differences in masses among these baryons.

Spin-3/2 baryons	Quark content	Mass (MeV/c^2)
$\Delta^{++}, \Delta^+, \Delta^0, \Delta^-$	uuu, uud, udd, ddd	1232
$\Sigma^{*+}, \Sigma^{*0}, \Sigma^{*-}$	uus, uds, dds	1385
Ξ^{*0}, Ξ^{*-}	uss, dss	1533
Ω^-	sss	1672

Table 10.4 The decuplet

10.5. Each of the baryons in the decuplet except the Ω^- has a lifetime of roughly 10^{-23} s, i.e., a strong decay. What is a possible decay mode for the Σ^{*+}? Repeat for Ξ^{*0}. Suggestion: Take into account the meson masses in Table 10.2 and the baryon masses in Table 10.3 in identifying decay modes that conserve charge, strangeness, energy, etc.

10.6. Why is the ratio R defined as

$$R = \frac{\sigma(e^- + e^+ \to \text{hadrons})}{\sigma(e^- + e^+ \to \mu^- + \mu^+)}$$

instead of

$$R = \frac{\sigma(e^- + e^+ \to \text{hadrons})}{\sigma(e^- + e^+ \to e^- + e^+)}$$

Hint: Which Feynman diagrams contribute to the process $e^- + e^+ \to e^- + e^+$?

10.7. What is a possible final state that would permit the creation via the strong interactions of the Ω^- baryon in the reaction in which a K^- collides with a proton? Suggestion: Be sure to conserve strangeness. Can you see the advantage of initiating this reaction with a K^- meson that has strangeness $S = -1$?

10.8. Construct a Feynman diagram for the decay $K^0 \to \pi^- + e^+ + \nu_e$. Suggestion: Examine the decay process at the level of the constituent quarks as in Example 10.9.

10.9. Derive (10.62). Note: The identity

$$(\cos^2 \theta + \sin^2 \theta)^2 = 1$$

is useful here.

10.10. Determine the eigenvalues of the charge conjugation operator. Hint: Apply the operator twice.

10.11. Determine the threshold energy for a photon to produce an electron–positron pair in the vicinity of an electron. Note: At threshold, the electron and the electron–positron pair are at rest in the center-of-mass frame.

10.12. The local newspaper reports the discovery of a new "elementary particle" X^{++}. This particle is reported to decay into two positively charged pions, i.e.,

$$X^{++} \to \pi^+ + \pi^+$$

Explain why there is reason to be skeptical of this discovery.

10.13. The solar constant is the amount of energy per unit area from the Sun incident on the outer surface of Earth's atmosphere perpendicular to the incident rays. The magnitude of the solar constant is 1370 W/m^2. Assuming the flux of neutrinos from the Sun comes predominantly from the fusion reactions that convert hydrogen into helium, determine the number of neutrinos per m^2 per second incident on the Earth from the Sun.

10.14. Which of the following reactions are forbidden? Why? (a) $\pi^0 \to \gamma$ (b) $\pi^0 \to \gamma + e^- + \bar{\nu}_e$ (c) $p + \bar{n} \to \pi^0 + \pi^+$ (d) $e^+ + n \to p + \bar{\nu}_e$ (e) $e^- + p \to n + \gamma$.

10.15. Determine the radius of the track left by a proton with kinetic energy equal to its mass energy ($K = mc^2$) in a 1-T magnetic field.

10.16. Show that under the gauge transformation

$$\mathbf{A} \to \mathbf{A} + \nabla \chi \quad \text{and} \quad \varphi \to \varphi - \frac{\partial \chi}{\partial t}$$

the electric field $\mathbf{E}$ and the magnetic field $\mathbf{B}$ are unchanged. Or, in other words, show that $\mathbf{E}$ and $\mathbf{B}$ are gauge invariant.

APPENDIX A

Special Relativity

A.1 The Relativity Principle

Let's start our discussion of the relativity principle with Newton's second law, $\mathbf{F} = d\mathbf{p}/dt$. Newton's law holds in a nonaccelerating reference frame, commonly called an **inertial reference frame**. In such a frame the sum of the forces acting on a particle is equal to the time derivative of the momentum $\mathbf{p}$ of the particle. In particular, in an inertial reference frame a particle with no forces acting on it moves in a straight line with a constant velocity $\mathbf{v}$. Let us call this frame S. Now consider another inertial reference frame S', one that is moving with a uniform velocity V in the positive x direction relative to the frame S. If the origins of these two coordinate frames coincide at time $t = 0$, then the coordinates x, y, and z in S and the coordinates x', y', and z' in S' are related by

$$x' = x - Vt$$
$$y' = y$$
$$z' = z \tag{A.1}$$

as indicated in Fig. A.1. The transformation of coordinates specified in (A.1) is referred to as a Galilean transformation. Before the development of the special theory of relativity, it was presumed that time was a universal quantity independent of the choice of reference frame. Thus the requirement that the time in S and the time in S' are the same was an unstated implicit assumption in a Galilean transformation.

Under a Galilean transformation, the components of the velocity of a particle as observed in the two frames are given by

$$v'_x = \frac{dx'}{dt} = \frac{dx}{dt} - V = v_x - V$$
$$v'_y = \frac{dy'}{dt} = \frac{dy}{dt} = v_y$$
$$v'_z = \frac{dz'}{dt} = \frac{dz}{dt} = v_z \tag{A.2}$$

that is, the x component of the velocity of the particle as observed in S' is simply the x component of the velocity of the particle as observed in S minus the relative velocity V

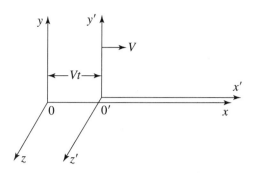

Figure A.1 A reference frame S' moves at speed V along the x axis relative to the frame S.

of the two frames, while the y and z components of the velocity are the same in the two frames. We can write the velocity transformation equation in the vector form

$$\mathbf{v}' = \mathbf{v} - \mathbf{V} \tag{A.3}$$

where $\mathbf{V} = V\mathbf{i}$. Since $\mathbf{V}$ is assumed to be independent of time, we see that

$$\mathbf{a}' = \frac{d\mathbf{v}'}{dt} = \frac{d\mathbf{v}}{dt} = \mathbf{a} \tag{A.4}$$

But if $\mathbf{a}' = \mathbf{a}$, then $\mathbf{F} = m\mathbf{a} = m\mathbf{a}' = \mathbf{F}'$. Thus observers in the two frames will deduce that the same net force acts on the particle. The key idea of the relativity principle is that the laws of physics are the same in any inertial reference frame. It was Einstein who attached Galileo's name to the transformation (A.1) since Galileo had initially proposed that it would not be possible for an observer in an inertial reference frame to determine that the frame was in motion by performing an experiment within that reference frame. You have probably had a visceral experience of the relativity principle if you have been seated in an airplane on the tarmac waiting for permission to taxi. If you look out your window and see another plane directly beside you in motion, you can't be certain whether you have started to move and are now in motion relative to the ground (presuming that the acceleration of your plane has been so gentle that you did not feel it) or the other plane is in motion, unless you look at some other object such as the ground.

A.2 The Postulates of Special Relativity

Einstein believed strongly in the relativity principle and felt that it should apply to *all* laws of physics, not just the laws of mechanics. This is, in fact, the fundamental postulate of special relativity:

> The laws of physics are the same in any inertial reference frame.

In his 1905 paper on the electrodynamics of moving bodies Einstein focused on the relativity that is inherent in electromagnetism. For example, he noted how a current is induced in a wire moving in the vicinity of a magnet. In one frame of reference, the one at rest relative to the wire, the time-varying magnetic field induces an electric field that accelerates the charges in the wire, thus generating a current. In another frame of reference, the one at rest with the magnet, the charge carriers in the wire experience a Lorentz force $q\mathbf{v} \times \mathbf{B}$ that causes a current to flow in the direction of the wire. What really matters physically is the *relative* motion between the wire and the magnet.

Einstein also gave a lot of thought to the nature of electromagnetic waves. He came to the conclusion that electromagnetic waves can propagate in a vacuum. A time-varying electric field generates a time-varying magnetic field, according to Ampere's law as modified by Maxwell. This time-varying magnetic field then generates an electric field, according to Faraday's law. These time-varying electromagnetic fields thus feed on each other, forming electromagnetic waves. Therefore, electromagnetic waves do not require a medium (the "ether") in which to propagate. But without an ether, there is no rationale for supposing that there is a special frame of reference (one at rest relative to the ether) in which light propagates with speed c. Consequently, following the logic of the relativity principle, Einstein was led to propose a second postulate:

The speed of light in vacuum is the same in any inertial reference frame.

This postulate, which is clearly at odds with the way velocities add in Galilean relativity [see (A.3)], requires a new understanding of the roles that space and time play in physics, as well as a major revision in the fundamental laws of mechanics.

EXAMPLE A.1 Make a Galilean transformation to a frame S' moving in the positive x direction at speed c on the electromagnetic wave $\mathcal{E} = \mathcal{E}_0 \cos(kx - \omega t)$.

SOLUTION

$$\mathcal{E} = \mathcal{E}_0 \cos(kx - \omega t) = \mathcal{E}_0 \cos[k(x - ct)] \tag{A.5}$$

since

$$\frac{\omega}{k} = \frac{2\pi \nu}{2\pi/\lambda} = \lambda \nu = c \tag{A.6}$$

Thus under the Galilean transformation $x' = x - ct$ we see that

$$\mathcal{E} = \mathcal{E}_0 \cos kx' \tag{A.7}$$

Looking back on the path that led him to propose the postulate on the constancy of the speed of light, Einstein later said "After ten years of reflection such a principle resulted from a paradox upon which I had already hit at the age of sixteen: If I pursue a beam of light with the velocity c, I should observe such a beam as an electromagnetic field constant in time, periodic in space. However, there seems to exist no such thing, neither on the basis of experience, nor according to Maxwell's equations. . . ."

Time Dilation

The requirement that the speed of light is the same for all observers in inertial reference frames has a number of surprising consequences.[1] As Einstein noted, it finally came to him that "time was suspect." To understand time itself, consider as a simple time-keeping instrument a "light clock" consisting of two mirrors separated by a distance L_0, as shown

---

[1] The approach that I follow in this appendix is very much influenced by my experiences teaching special relativity with my colleague T. M. Helliwell. His *Special Relativity* (University Science Books, Sausalito, CA, 2010) gives a more thorough discussion of these topics.

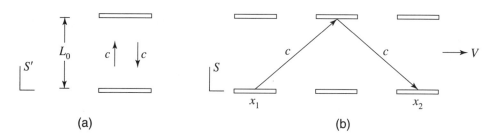

(a)   (b)

Figure A.2 (a) A light clock in its rest frame, which is taken to be the reference frame S'. (b) The light clock as viewed from the reference frame S.

in Fig. A.2a. The time it takes light to make a round trip between the mirrors is given by

$$\Delta t' = \frac{2L_0}{c} \quad (A.8)$$

Thus if the distance between the mirrors is 15 cm, the time required for the light to complete a round-trip is one nanosecond. Put another way, we might say the clock "ticks" every nanosecond. We call this time $\Delta t'$ since we suppose initially that this clock is at rest in the reference frame S'.

We want to compare this time interval with the time between ticks of the clock as observed in the reference frame S in which the clock is moving in the positive x direction at speed V. In S the light starts its round-trip at a particular position along the x axis, say at x_1, and completes the round-trip at the position x_2, as indicated in Fig. A.2b. To determine the time between ticks that observers in S measure, we need to suppose that there are two synchronized clocks (perhaps light clocks) at rest in S, one at x_1 and one at x_2. The light leaves the bottom mirror at time t_1 and completes a round-trip at time t_2 as measured in S. Since the light travels at speed c in S as well as in S', the time interval between ticks in S is given by the transverse distance $2L_0$ traveled by the light divided by the transverse velocity of the light, namely $\sqrt{c^2 - V^2}$, as indicated in Fig. A.3. Thus

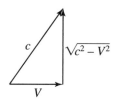

Figure A.3 In S the light travels at speed c along a diagonal path to the mirror, which is moving horizontally at speed V. Thus the transverse velocity of the light is $\sqrt{c^2 - V^2}$.

$$\Delta t = t_2 - t_1 = \frac{2L_0}{\sqrt{c^2 - V^2}}$$
$$= \frac{1}{\sqrt{1 - V^2/c^2}} \frac{2L_0}{c} \quad (A.9)$$

Comparing this result with (A.8), we see that

$$\Delta t = \frac{1}{\sqrt{1 - V^2/c^2}} \Delta t'$$
$$= \gamma \Delta t' \quad (A.10)$$

where

$$\gamma = \frac{1}{\sqrt{1 - V^2/c^2}} \quad (A.11)$$

Since $\gamma \geq 1$, the time interval between ticks is longer in S than in S'. In other words, the time between ticks on the moving clock is greater than the time between ticks in the rest frame of the clock. For example, if the reference frame S' is moving along the positive x axis with speed $V = 4/5\,c$ relative to S, then $\sqrt{1 - V^2/c^2} = 3/5$ and the time between ticks of the moving clock is 5/3 ns, as compared with 1 ns in the clock's

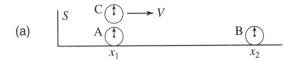

Figure A.4 (a) Two clocks A and B are at rest in S, one at position x_1 and the other at position x_2. Each clock reads $t = 0$ when a third clock C, also reading $t = 0$, moves past clock A at speed V. (b) Later, when clock C reaches x_2, it reads less than clock B by $\sqrt{c^2 - V^2}$.

rest frame. Equivalently, since

$$\Delta t' = \sqrt{1 - V^2/c^2}\, \Delta t \qquad (A.12)$$

if 1 ns elapses on the clocks at rest in S, then 3/5 ns elapses on the moving clock, the one at rest in S'. In short, we say that *moving clocks run slow by a factor of* $\sqrt{1 - V^2/c^2}$, as illustrated in Fig. A.4. We call this effect **time dilation**.

The time registered by a clock in its rest frame is called the **proper time** and is typically denoted by the symbol τ. Thus in terms of proper time (A.10) becomes

$$\Delta t = \frac{\Delta \tau}{\sqrt{1 - V^2/c^2}} = \gamma \Delta \tau \qquad (A.13)$$

This result applies to all clocks, not just light clocks. For after all, if mechanical clocks and biological clocks as well as light clocks run at the same rate in one inertial reference frame, they must run at the same rate in all inertial reference frames. Otherwise, these inertial reference frames would not be equivalent, violating the relativity principle. Thus we see how the two postulates of special relativity have called into question the very nature of time itself.

Length Contraction

Time dilation is closely related to another relativistic effect known as length contraction. Consider a stick of length L_0 in its rest frame. We call this length the **proper length** of the stick. Suppose a clock moving at speed V moves past the stick, as shown in Fig. A.5a.

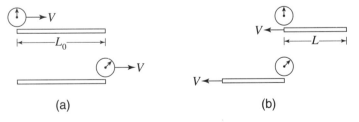

Figure A.5 (a) A clock moves at speed V past a stick of rest length L_0. The elapsed time on the clock is $L_0\sqrt{1 - V^2/c^2}/V$, since the moving clock runs slow by $\sqrt{1 - V^2/c^2}$. (b) In the rest frame of the clock, the stick has length L and the time required for the stick to move past the clock is L/V.

In the rest frame of the stick, the time for the clock to move from one end of the stick to the other is L_0/V. But since the clock is moving, it runs slow by $\sqrt{1 - V^2/c^2}$ and thus the time that has elapsed on the clock is $L_0\sqrt{1 - V^2/c^2}/V$. In the rest frame of the clock (see Fig. A.5b), the time for the stick to pass by the clock is simply L/V, where L is the length of the stick as measured in the clock's rest frame. Observers in the rest frame of the stick and the rest frame of the clock must agree on the reading of the clock when the clock and the right end of the stick coincide. After all, one might imagine that there is a hook on the right end of the stick that stops the clock at that moment. All observers would then see the same reading on the clock. Consequently,

$$\frac{L_0\sqrt{1 - V^2/c^2}}{V} = \frac{L}{V} \tag{A.14}$$

and therefore

$$L = L_0\sqrt{1 - V^2/c^2} \tag{A.15}$$

Thus the length of a stick is contracted by a factor of $\sqrt{1 - V^2/c^2}$ along its direction of motion. This effect is often referred to as Lorentz–Fitzgerald contraction, or **Lorentz contraction** for short.[2]

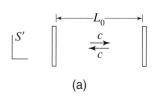

(a)

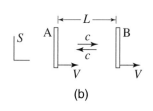

(b)

Figure A.6 (a) A light clock oriented along the x' axis in its rest frame S'. (b) In reference frame S, the clock moves along the x axis at speed V.

An alternative way to see that both length contraction and time dilation are required for self consistency is to suppose that the stick is positioned along the x axis between two mirrors, forming a "longitudinal" light clock, as shown in Fig. A.6. In the rest frame of the clock, which we take to be S', the time required for the clock to tick is $\Delta t' = 2L_0/c$ just as for the transverse light clock, since the distance between the mirrors in the clock's rest frame is L_0. Since the longitudinal light clock and the transverse light clock run at the same rate in the frame S', they must also run at the same rate as observed in the reference frame S to be consistent with the relativity principle. As before, in S we call the distance between the two mirrors L. First note that the distance $c\Delta t_1$ the light travels between the mirrors on the outward leg of the trip from mirror A to mirror B is given by

$$c\Delta t_1 = L + V\Delta t_1 \tag{A.16}$$

since mirror B starts a distance L from mirror A and travels an additional distance $V\Delta t_1$ in the time interval Δt_1 it takes light to travel between the mirrors. Therefore

$$\Delta t_1 = \frac{L}{c - V} \tag{A.17}$$

On the return trip, from mirror B to mirror A, the light need travel only a distance $c\Delta t_2$ given by

$$c\Delta t_2 = L - V\Delta t_2 \tag{A.18}$$

---

[2]Length contraction may call into question the implicit assumption made in our discussion of the light clock in the previous section that transverse lengths are not affected by the motion. But suppose that there are horizontal razor blades at the location of the mirrors of the clock at rest in S'. If the transverse length between the mirrors as observed in S were shorter than L_0, for example, then this moving clock would cut the wood casing of an identical clock at rest in S *inside* the location of the mirrors on that clock. But this would imply that if there were razor blades at the location of the mirrors of the clock in S, the cut marks on the clock in S' would be *outside* the location of the mirrors. But the relativity principle assures us that the situation is symmetrical. Thus the transverse length L_0 must be the same in the two frames of reference.

since mirror A has moved toward the light in the time interval Δt_2 it takes light to complete the trip. Consequently,

$$\Delta t_2 = \frac{L}{c+V} \tag{A.19}$$

The total amount of time that has transpired is then

$$\Delta t = \Delta t_1 + \Delta t_2$$
$$= \frac{L}{c-V} + \frac{L}{c+V}$$
$$= \frac{2L}{c(1-V^2/c^2)} \tag{A.20}$$

Since this time interval must be the same as (A.9), we see that

$$\frac{2L}{c(1-V^2/c^2)} = \frac{2L_0}{c\sqrt{1-V^2/c^2}} \tag{A.21}$$

and therefore

$$L = L_0\sqrt{1-V^2/c^2} = \frac{L_0}{\gamma} \tag{A.22}$$

Simultaneity

Perhaps the most counterintuitive consequence of the postulates of special relativity is the fact that the order in time in which events occur is not absolute but rather can differ between reference frames. In particular, it is not hard to see that events such as the synchronization of clocks that are simultaneous in one inertial reference frame need not be simultaneous in another inertial reference frame.

To examine the issue of synchronization of clocks, consider two clocks separated by a distance L_0 in the rest frame of the clocks, as shown in Fig. A.7a. A straightforward way to synchronize these clocks is to set off a light flash in the clocks' rest frame exactly halfway between the two clocks. It takes time $L_0/2c$ for the light to reach either clock, at which point each clock could be set to a particular time, say $t' = 0$, as shown in Fig. A.7b. But we see immediately that the clocks will not be synchronized as observed from the frame S, as indicated in Fig. A.7c. Following the line of reasoning that led to (A.19), in S the light signal reaches the clock at A first, in a time interval

$$\Delta t_1 = \frac{L_0\sqrt{1-V^2/c^2}}{2(c+V)} \tag{A.23}$$

since the distance between the clocks in this frame is $L_0\sqrt{1-V^2/c^2}$. Again, as in (A.17), the light signal reaches the clock at B after the longer time interval

$$\Delta t_2 = \frac{L_0\sqrt{1-V^2/c^2}}{2(c-V)} \tag{A.24}$$

Thus the clock at A leads the clock at B by

$$\Delta t = \Delta t_2 - \Delta t_1 = \frac{L_0\sqrt{1-V^2/c^2}}{2(c-V)} - \frac{L_0\sqrt{1-V^2/c^2}}{2(c+V)}$$
$$= L_0\sqrt{1-V^2/c^2}\frac{V}{c^2-V^2} \tag{A.25}$$

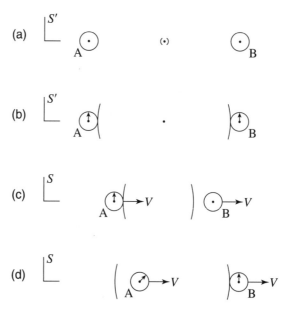

Figure A.7 (a) Two clocks are synchronized in their rest frame (S') by setting off a flash of light midway between the clocks. (b) In S' the light flash arrives at the same time at both clocks, which are then set to 0. (c) In reference frame S, the trailing clock receives the light signal first and the clock is set to 0. (d) At a later time the light flash arrives at the leading clock, which is then set to zero as well. Consequently, the leading clock lags the trailing clock.

But this is the time interval as determined in S. Since the clocks in S' run slow relative to the clocks in S, the total time that has transpired on the clock at A when the clock at B is set to zero is therefore

$$\Delta t' = \sqrt{1 - V^2/c^2}\,\Delta t = \frac{L_0 V}{c^2} \tag{A.26}$$

From the vantage point of observers in S, the clock at A leads the clock at B by the amount of time shown in (A.26), as indicated in Fig. A.7d. Since the clock at A trails the clock at B as the clocks move past observers in S, we often say trailing, or chasing, clocks lead. Alternatively, we might say that the clock at B, the leading clock as the clocks move past observers in S, lags the clock at A by this amount of time, thus, "leading clocks lag," which is a useful mnemonic. It should be emphasized that the distance that occurs in (A.26) is the separation of the two clocks in their rest frame. Thus we see that events that occur simultaneously in one frame of reference—say the arrival of the light flash at the two clocks in the clocks' rest frame—are not simultaneous as observed from a reference frame in which the clocks are in motion. And in fact the order in time in which these events occur is frame dependent. After all, observers in a reference frame in which the clocks move in the negative x direction instead of the positive x direction would see that the light signal that is used to synchronize the clocks arrives first at clock B (which is now the trailing clock) and then later at A.

The fact that clocks that are synchronized in one inertial reference frame are not synchronized in a reference frame that is in uniform motion relative to this frame allows us to understand many of the seemingly paradoxical results of special relativity. For example, observers in S see clocks at rest in S' running slow. But to be consistent with the relativity principle observers in S' must see clocks at rest in S running slow as well. How can this be? First, note that when time $\Delta \tau$ transpires on a clock at rest in S', this

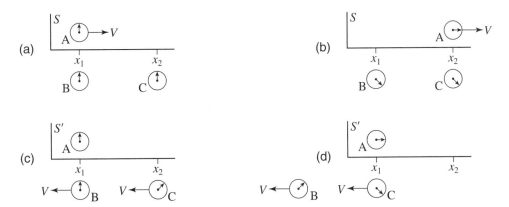

Figure A.8 (a) In reference frame S clock A moves past position x_1 with speed V. Clock A reads the same as clock B, which is synchronized with clock C in S. (b) Clock A moves past clock C at position x_2. The elapsed time on A is less than on clocks B and C since the moving clock runs slow. (c) and (d) show the same events as observed in frame S', which is the rest frame for clock A. (c) Clock B lags clock C (since "leading clocks lag") when clocks A and B coincide. (d) Clock C coincides with clock A, at which point the readings of clocks A and C for this event are the same as those shown in (b). Notice that clocks B and C run slow relative to clock A in the reference frame S'.

clock moves by an amount $\Delta x = V \Delta t$ as observed in S. How is this time interval Δt measured? It is helpful to think of each point in S as having with it an associated clock and therefore an associated time. As the clock at rest in S' passes a point in S, say x_1, the S clock at this point reads time t_1. When the S' clock reaches position x_2, the S clock at this point reads t_2. See Fig. A.8. Time dilation means $\Delta t = t_2 - t_1 = \gamma \Delta \tau$, as we have seen. But notice that in making this measurement in S we have had to use two different clocks, clocks that presumably have been synchronized in their rest frame S so they initially read the "same" time (in S). But as viewed from S', the clock at x_2 is observed to have a "head start" relative to the clock at x_1, since this clock is a trailing clock. Thus even though it runs slow to an observer in S', observers in S and in S' agree on the reading of this clock when the clock at x_2 and the clock in S' coincide.

EXAMPLE A.2 The Stanford Linear Accelerator Center (SLAC) houses a 3-km long linear accelerator, "the world's straightest object." See Fig. A.9. (*a*) Determine the time it takes for an electron moving at the uniform speed of $V = 3/5\,c$ to traverse the length of the beam line.[3] Assume there are synchronized clocks located at the beginning and end of the beam line and that the clock at the beginning of the beam line is set to $t = 0$ when the electron departs. What does the clock at the end of the beam line read when the electron arrives? (*b*) Suppose that the electron carries with it a (small) clock that is also set to zero at the beginning of the trip. What does this clock read when the electron reaches the end of the beam line?

Figure A.9 An aerial view of SLAC. (Courtesy SLAC National Accelerator Laboratory)

---

[3] As its name suggests, the purpose of SLAC is to *accelerate* electrons. The maximum beam energy at SLAC is 50 GeV, corresponding to a value of v/c of 0.999 999 999 95 for the electron. It is possible to use special relativity to analyze accelerating objects, but care must be exercised. Thus we consider here a more straightforward example in which the electron moves at a constant velocity.

SOLUTION (*a*) In the accelerator rest frame, the distance traversed by the electron is 3 km at a uniform speed of $3/5\,c$. Thus

$$t = \frac{3 \text{ km}}{(3/5)c} = 5 \text{ km}/c$$

(*b*) Moving clocks run slow by a factor of $\sqrt{1 - V^2/c^2} = 4/5$. Thus the reading on the clock traveling with the electron is

$$t' = \sqrt{1 - V^2/c^2}\, t = \left(\frac{4}{5}\right)(5 \text{ km}/c) = 4 \text{ km}/c$$

Note: In more conventional units, $1 \text{ km}/c = 3.33 \times 10^{-6}$ s $= 3.33\ \mu$s.

EXAMPLE A.3 Repeat Example A.2 in the electron's rest frame. In this reference frame the beam line speeds past the electron at $V = 3/5\,c$. When the end of the beam line reaches the electron, (*a*) what does the electron's clock read and (*b*) what does the clock at the end of the beam line read?

SOLUTION (*a*) In the electron's rest frame the beam is Lorentz contracted from its laboratory rest frame length L_0 to L, where

$$L = \sqrt{1 - V^2/c^2}\, L_0 = \left(\frac{4}{5}\right)(3 \text{ km}) = \frac{12}{5} \text{ km}$$

Thus the electron's clock reads

$$t' = \frac{L}{V} = \frac{(12/5) \text{ km}}{(3/5)c} = 4 \text{ km}/c$$

in agreement with the result obtained in the laboratory rest frame.
(*b*) In the electron's rest frame, the laboratory clocks run slow by $\sqrt{1 - V^2/c^2}$. Therefore in this reference frame the time that transpires on the laboratory clocks is

$$\Delta t_{\text{lab}} = \sqrt{1 - V^2/c^2}\, t' = \left(\frac{4}{5}\right) 4 \text{ km}/c = \frac{16}{5} \text{ km}/c$$

Notice that in each reference frame, observers see the clocks in the other reference frame running slow. But in the electron's rest frame, the clocks at the beginning and end of the beam line are *not* synchronized. The separation between these clocks in their rest frame is 3 km. Thus the chasing clock, the one at the end of the beam line, leads the one at the beginning of the beam line by

$$\frac{V L_0}{c^2} = \left(\frac{3c}{5}\right)\left(\frac{3 \text{ km}}{c^2}\right) = \frac{9}{5} \text{ km}/c$$

Therefore the reading of the clock at the end of the beam line when this clock reaches the position of the electron is

$$t = \Delta t_{\text{lab}} + \frac{VL_0}{c^2} = \left(\frac{16}{5} + \frac{9}{5}\right) \text{ km}/c = \frac{25}{5} \text{ km}/c = 5 \text{ km}/c$$

which is the same as the reading on the clock as determined in the laboratory rest frame!

A point in space and time at which something occurs is referred to as **an event**. Events exist apart from the reference frame that is used to describe them. For example, one such event is the arrival of the electron at the end of the beam line. Observers in the electron's rest frame and in the lab would both agree that such an event has occurred. Moreover, one could suppose that the electron's arrival at the end of the beam line stops two clocks, one in the electron's rest frame and one in the lab's rest frame. These clocks, which could have digital readouts, are located at the event. Observers in the electron's rest frame and in the lab would then agree on the reading of each of these clocks, even though the two clocks do not agree with each other.

Experimental Confirmation

In the more than one hundred years since Einstein proposed the principles of special relativity, the theory has been tested against experiment many times and has never been found to be wanting. In his 1905 paper Einstein himself pointed out the time dilation would cause a clock at the equator to run slow relative to a clock at the North Pole by "a very small amount." Of course, in 1905 there were no clocks that could test this prediction. This situation changed with the development of atomic clocks. Initial tests of time dilation with atomic clocks carried out in the 1970s with the clocks carried by jet airplanes circumnavigating the Earth were in good agreement with theory. Nowadays, such effects are more commonplace to the extent that the Global Positioning System (GPS) (in which 24 satellites with atomic clocks orbit the earth twice a day) must take both time dilation and simultaneity into account in order to achieve the desired accuracy. Ignoring such relativistic effects would limit the accuracy of the GPS system to hundreds of meters.[4]

An especially noteworthy example of a natural clock is the lifetime of an unstable particle such as the muon. In its rest frame, the muon has a mean lifetime τ of 2.2 μs. Without time dilation, the largest distance on average a muon could travel before it decays into an electron and two neutrinos would be roughly $c\tau = (3 \times 10^8 \text{ m/s})(2.2 \times 10^{-6} \text{ s}) = 660$ m. But at sea level, perhaps as much as ten kilometers from the point where the muons are created as a byproduct of collisions between cosmic rays and nuclei of the molecules in the upper atmosphere, muons are raining down upon us at the rate of 180 m^{-2}s^{-1}.

---

[4]In addition, the GPS system must account for the fact that clocks at high altitude tick more rapidly than those at the Earth's surface, a consequence of the general theory of relativity (another of Einstein's "discoveries").

In our reference frame at rest on the Earth's surface, the muons are typically produced with a speed such that $\gamma = 1/\sqrt{1 - v^2/c^2} \approx 20$. Thus taking time dilation into account, before decaying the muon can on average travel a distance

$$d = vt = v\frac{\tau}{\sqrt{1 - v^2/c^2}} = v\gamma\tau \tag{A.27}$$

which is roughly $(20)(660)$ m $= 13$ km. Alternatively, in the muon's rest frame, the Earth's atmosphere is Lorentz contracted by a factor of 20, giving the surface of the Earth adequate time to reach the muon before it decays.

Such time dilation (or Lorentz contraction) effects are truly commonplace in experimental high-energy physics. Almost all the so-called elementary particles are unstable. Since these particles are typically produced with velocities approaching the speed of light at a high-energy particle accelerator, they can travel much larger distances in the lab frame than would be possible without time dilation. For example, a 500-GeV K^+ meson with a lifetime $\tau = 10^{-8}$ s (and hence $c\tau = 3$ m) can travel on average 3 km before it decays. In other words, the building housing the detectors that might be used to analyze the decay of a 500-GeV K^+ meson produced at the Fermilab accelerator (located outside Chicago) would be located roughly 3 km from the point at which the K^+ meson is produced. Looking to the future, a group of physicists is proposing to build a muon collider in which positively and negatively charged muons would be accelerated to energies up to 4 TeV and then would make a head-on collision in the laboratory rest frame. As we have noted, in their rest frame muons live only for a couple of microseconds. But a 4-TeV muon (for which $v/c = 0.999\,999\,999\,7$) can live for almost a tenth of a second and travel more than 25,000 km in the laboratory rest frame before decaying.

A.3 The Lorentz Transformation

Before the advent of special relativity, we would think of each point in a coordinate system as being labeled by a particular value of the coordinates x, y, and z. Now we see that we should include a time coordinate t as well. That is, at each point in space in an inertial reference frame we might imagine there is a clock. And all these clocks should be synchronized and tick at the same rate in that particular frame of reference if this time coordinate is to be physically meaningful. The transformation of coordinates to an inertial reference frame S' that moves at speed V along the positive x axis relative to the reference frame S is given by the Lorentz transformation

$$\begin{aligned} x' &= \gamma(x - Vt) \\ y' &= y \\ z' &= z \\ t' &= \gamma\left(t - \frac{Vx}{c^2}\right) \end{aligned} \tag{A.28}$$

where as before

$$\gamma = \frac{1}{\sqrt{1 - V^2/c^2}} \tag{A.29}$$

This Lorentz transformation is the replacement for the Galilean transformation (A.1). And a replacement is definitely needed, since the Galilean transformation leads to a velocity of $c + V$ in S for light propagating along the x' axis with velocity c in S'. Notice how the first three equations in (A.28) reduce to the Galilean transformation when $V/c \ll 1$.

The inverse Lorentz transformation is given by

$$x = \gamma(x' + Vt')$$
$$y = y'$$
$$z = z'$$
$$t = \gamma\left(t' + \frac{Vx'}{c^2}\right) \tag{A.30}$$

since in S' the reference frame S moves at speed V along the negative x' axis. Thus one goes from (A.28) to (A.30) by interchanging the primed and unprimed coordinates and switching V to $-V$.

We can check that the Lorentz transformation (A.28) reproduces the results of the previous section. For example, consider a clock at rest at in S'. In S in the time interval Δt the location of the clock moves along the x axis by $\Delta x = V \Delta t$. Then from the Lorentz transformation (A.28) the time interval $\Delta t'$ that elapses on the clock is given by

$$\Delta t' = \gamma\left(\Delta t - \frac{V \Delta x}{c^2}\right) = \gamma \Delta t \left(1 - \frac{V^2}{c^2}\right) = \sqrt{1 - V^2/c^2}\, \Delta t \tag{A.31}$$

Thus

$$\Delta t = \frac{\Delta t'}{\sqrt{1 - V^2/c^2}} \tag{A.32}$$

which is the time-dilation result (A.10).

Similarly, a stick of length L_0 in the rest frame S' is measured to have length L in S. In S', $L_0 = x'_2 - x'_1$, where the left end of the stick is located at the position x'_1 and the right end of the stick is located at x'_2. In S the stick moves to the right at speed V. In order for observers in S to determine the length of this moving stick we need to define carefully how the measurement is carried out. We suppose that an observer in S at a particular location, say x_1, sees the left end of the stick at a particular time, say t_1, located directly in front of this observer. This single observer might simultaneously try to look to see the location of the right end of the stick, but this is not a satisfactory approach since it takes time for the light from the right end of this stick to reach the observer and in that additional time the location of the stick has changed from the location it had when the observer noted the location of the left end of the stick. A better approach is to assume that there are many observers in S located along the x axis. If one of the observers determines that the left end of the stick is located at a particular point x_1 at time t_1, then one of the other observers in S located at x_2 observes that the right end of the stick is at this particular point in space at this *same moment in time* in frame S. From the Lorentz transformation (A.28)

$$\begin{aligned} L_0 &= x'_2 - x'_1 \\ &= \gamma\left[x_2 - x_1 - V(t_2 - t_1)\right] \\ &= \gamma(x_2 - x_1) \\ &= \gamma L \end{aligned} \tag{A.33}$$

where we have set $t_2 = t_1$ since we are measuring the locations of the end of the stick at the same time in S. Thus

$$L = L_0\sqrt{1 - V^2/c^2} \tag{A.34}$$

which is the length contraction result (A.22). This measurement of the length of the stick in S gives an operational definition of what it means to say the length of the stick is Lorentz contracted when observed in S. Also note that the transverse coordinates y and z are not affected by a Lorentz transformation, consistent with our assumption that the distance between the mirrors in the transverse light clock did not change as viewed in the frame S moving with speed V along the x axis.

Finally, consider two clocks at rest in S' separated by a distance L_0. If we observe the clocks from S at the same time ($t_2 = t_1$), we see that

$$\begin{aligned} t_2' - t_1' &= \gamma\left[t_2 - t_1 - \frac{V}{c^2}(x_2 - x_1)\right] \\ &= -\gamma \frac{V}{c^2} L_0 \sqrt{1 - V^2/c^2} \\ &= -\frac{V L_0}{c^2} \end{aligned} \tag{A.35}$$

where the minus sign indicates that the leading clock, the one at x_2, lags the trailing clock, the one at x_1, assuming $x_2 > x_1$. This result coincides with (A.26).

EXAMPLE A.4 Use the inverse Lorentz transformation (A.30) to derive time dilation.

SOLUTION Consider a clock at rest in S', say at $x' = 0$. In the time interval $\Delta t'$

$$\Delta t = \gamma\left(\Delta t' + \frac{V \Delta x'}{c^2}\right)$$

Since the clock is at rest in S', $\Delta x' = 0$ and consequently

$$\Delta t = \gamma \Delta t'$$

or

$$\Delta t' = \frac{1}{\gamma}\Delta t = \sqrt{1 - V^2/c^2}\,\Delta t$$

showing that moving clocks run slow by a factor of $\sqrt{1 - V^2/c^2}$.

It is straightforward to derive Lorentz contraction and the leading-clocks-lag rule from (A.30) as well.

Relativistic Velocity Transformation

The Lorentz transformation makes it easy to relate the spacetime coordinates of an event in one frame to the spacetime coordinates of an event in another frame. As a particle moves from one position at a particular time to another position at a later time in a particular frame of reference it traces out a sequence of events. From the Lorentz transformation

(A.28) we see that

$$dx' = \gamma(dx - V dt)$$
$$dy' = dy$$
$$dz' = dz$$
$$dt' = \gamma \left(dt - \frac{V dx}{c^2} \right) \tag{A.36}$$

Consequently, we can relate the components of the velocity of a particle in S' to the components in S:

$$v'_x = \frac{dx'}{dt'} = \frac{\gamma (dx/dt - V)}{\gamma [1 - V(dx/dt)/c^2]} = \frac{v_x - V}{1 - V v_x/c^2}$$

$$v'_y = \frac{dy'}{dt'} = \frac{dy/dt}{\gamma[1 - V(dx/dt)/c^2]} = \frac{v_y \sqrt{1 - V^2/c^2}}{1 - V v_x/c^2}$$

$$v'_z = \frac{dz'}{dt'} = \frac{dz/dt}{\gamma[1 - V(dx/dt)/c^2]} = \frac{v_z \sqrt{1 - V^2/c^2}}{1 - V v_x/c^2} \tag{A.37}$$

This is clearly a more complex transformation than that of the spatial coordinates x, y, and z in the Lorentz transformation (A.28). Notice in particular that if $v_x = c$, then

$$v'_x = \frac{c - V}{1 - V/c} = c \tag{A.38}$$

as well, consistent with the fact that the speed of light is equal to c in the two reference frames.

As for the Lorentz transformation, the inverse velocity transformation can be obtained from (A.37) by interchanging the primed and unprimed coordinates and replacing V with $-V$:

$$v_x = \frac{v'_x + V}{1 + V v'_x/c^2}$$

$$v_y = \frac{v'_y \sqrt{1 - V^2/c^2}}{1 + V v'_x/c^2}$$

$$v_z = \frac{v'_z \sqrt{1 - V^2/c^2}}{1 + V v'_x/c^2} \tag{A.39}$$

EXAMPLE A.5 A K meson decays into a neutrino and a muon. In the rest frame of the K meson, which we take to be S', the velocity of the muon along the x' axis is $v'_x = 0.85c$. What is the velocity of the muon as measured in the reference frame S in which the K meson before decaying has a speed $V = 0.5c$ along the x axis?

SOLUTION

$$v_x = \frac{v'_x + V}{1 + V v'_x/c^2} = \frac{0.85c + 0.5c}{1 + (0.5)(0.85)} = \frac{1.35c}{1.425} = 0.95c$$

Notice how the term in the numerator of the velocity transformation equation yielded the result $1.35c$ that one would have expected from Galilean relativity. The correction due to the term in the denominator ensures that the velocity of the muon as measured in S is indeed less than c.

A.4 Four-vectors

It is convenient to rewrite the Lorentz transformation (A.28) in terms of the four variables, $x_0 = ct$, $x_1 = x$, $x_2 = y$, $x_3 = z$ as follows:

$$x'_0 = \gamma \left(x_0 - \frac{V}{c} x_1 \right)$$
$$x'_1 = \gamma \left(x_1 - \frac{V}{c} x_0 \right)$$
$$x'_2 = x_2$$
$$x'_3 = x_3 \tag{A.40}$$

We say that these components are part of what is termed the position four-vector x_μ, where $\mu = 0, 1, 2, 3$. You can verify that under a Lorentz transformation

$$x_0^2 - x_1^2 - x_2^2 - x_3^2 = {x'}_0^2 - {x'}_1^2 - {x'}_2^2 - {x'}_3^2 \tag{A.41}$$

We characterize this quantity as a **Lorentz invariant**, since (A.41) shows it has the same value in all inertial reference frames. If we express this invariance in terms of the more customary x, y, z, and t coordinates, we see that

$$c^2 t^2 - (x^2 + y^2 + z^2) = c^2 t'^2 - (x'^2 + y'^2 + z'^2) \tag{A.42}$$

If a light flash goes off at the moment the origins of the coordinate axes for the reference frames S and S' coincide, then the location of the light front at time t in S is determined by the condition $x^2 + y^2 + z^2 = c^2 t^2$, or $c^2 t^2 - (x^2 + y^2 + z^2) = 0$. Similarly, in S' the light also expands spherically outward and the location of the light front is determined by the condition $x'^2 + y'^2 + z'^2 = c^2 t'^2$, or $c^2 t'^2 - (x'^2 + y'^2 + z'^2) = 0$. Thus the Lorentz-invariant condition (A.41) [or (A.42)] is a natural generalization of the condition that light travels at speed c in all reference frames.

A Lorentz invariance such as (A.42) is reminiscent of the invariance of $x^2 + y^2 + z^2$ under rotations. For example,

$$x' = \cos\phi\, x + \sin\phi\, y$$
$$y' = -\sin\phi\, x + \cos\phi\, y$$
$$z' = z \tag{A.43}$$

corresponds to a *rotation* of the coordinate axes by angle ϕ counterclockwise about the z axis, as indicated in Fig. A.10. You can check that $x^2 + y^2 + z^2 = x'^2 + y'^2 + z'^2$ under this rotation. Just as all three-vectors follow the same transformation rules under rotations, all four-vectors follow the same rules as those that govern the position four-vector under Lorentz transformations.[5]

---

[5]In the position four-vector time and space are joined together in a new type of space often referred to as Minkowski space or, more simply, spacetime. Ordinary three-dimensional space is often referred to as a Euclidean space in that the sum of the squares of the components of any vector is unchanged by a rotation of coordinates. Minkowski space is a non-Euclidean space in that the sum of the squares of the coordinates ($c^2 t^2 + x^2 + y^2 + z^2$) is *not* an invariant. Rather the combination of coordinates that is invariant consists of $c^2 t^2 - (x^2 + y^2 + z^2)$, which is the way to express the invariance favored by particle physicists, or the negative of this expression, namely $x^2 + y^2 + z^2 - c^2 t^2$, which is favored by general relativists. The Lorentz transformation (A.28) can be thought of as a generalized "rotation" in Minkowski space, mixing space and time together. See Problem A.10.

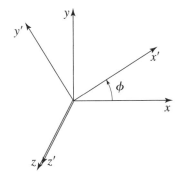

Figure A.10 Two coordinate axes. One is rotated relative to the other by angle ϕ about the z axis.

A.5 Momentum and Energy

Let us use the velocity transformation (A.37) to analyze an elastic collision between two particles of equal mass m. In the frame S', which in this case is the center-of-mass frame, the two particles, each with speed V, make a head-on collision, as indicated in Fig. A.11a. After the collision, the two particles rebound with speed V perpendicular to the initial direction. In the frame S (shown in Fig. A.11b), which is moving in the negative x direction at speed V relative to S', the two particles after the collision have speed V in the positive x direction. And the "target" particle in this collision is at rest in S. Using (A.39), we see that the speed of the incident particle is given by

$$v_x = \frac{2V}{1 + V^2/c^2} \tag{A.44}$$

At first it seems disconcerting that

$$mv_x = m\frac{2V}{1 + V^2/c^2} \neq 2mV \tag{A.45}$$

since the left-hand side of this equation should be the x component of the linear momentum before the collisions and the right-hand side the x component of the linear momentum

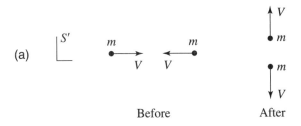

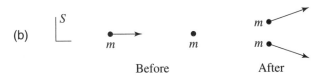

Figure A.11 (a) Two particles of equal mass m and speed V make a head-on elastic collision in reference frame S'. After the collision the particles rebound along the y' axis. (b) The same collision as viewed from reference frame S.

after the collision. Our expectation is that linear momentum should be conserved in this collision. After all, it was an analysis of collisions such as the one outlined here that led Newton to introduce the concept of linear momentum. The resolution of this seeming paradox is that the linear momentum of a particle is *not* equal to $m\mathbf{v}$. Rather, the correct definition of momentum for a particle of mass m with velocity $\mathbf{v}$ is

$$\mathbf{p} = \frac{m\mathbf{v}}{\sqrt{1 - v^2/c^2}} \tag{A.46}$$

If you apply this definition of momentum to the collision in the frame S, you can verify that linear momentum is indeed conserved in the collision. Of course, this definition of linear momentum corresponds to the nonrelativistic expression $\mathbf{p} = m\mathbf{v}$ when $v/c \ll 1$.

One way to "derive" the correct expression for the linear momentum is assert that the three-momentum $\mathbf{p} = (p_x, p_y, p_z)$ should be components of a four-vector p_μ just as the position coordinates $\mathbf{r} = (x, y, z)$ are three of the four components of the position four-vector x_μ. To construct a candidate "momentum" four-vector, we first create a four-vector velocity by taking the derivative of the components of the position four-vector not with respect to time (which is itself a component of a four-vector) but with respect to the proper time τ. For a particle that is moving, that is one whose spatial coordinates shift by Δx, Δy, and Δz in time Δt, we define the change in proper time through the Lorentz invariant

$$c^2 \Delta \tau^2 = c^2 \Delta t^2 - \Delta x^2 - \Delta y^2 - \Delta z^2 \tag{A.47}$$

Thus in the rest frame of the particle, where $\Delta x = \Delta y = \Delta z = 0$, $\Delta \tau = \Delta t$. As was noted in Section A.2, the proper time is the time as kept on a clock in its rest frame. All observers, independent of their frame of reference, "see" the same reading for this clock. Thus in the language of (four-) vectors, we can characterize the proper time as a Lorentz scalar and the four-vector velocity

$$u_\mu = \frac{dx_\mu}{d\tau} \tag{A.48}$$

is the derivative of the four-vector position with respect to this Lorentz scalar. Once we have a four-vector velocity, we then define the four-vector momentum by

$$p_\mu = m u_\mu = m \frac{dx_\mu}{d\tau} \tag{A.49}$$

But we know from (A.10) that

$$\frac{dt}{d\tau} = \gamma \tag{A.50}$$

Consequently,

$$p_0 = mc \frac{dt}{d\tau} = \gamma mc = \frac{mc}{\sqrt{1 - v^2/c^2}} \tag{A.51}$$

The spatial components of the four-vector momentum can be expressed in terms of the ordinary velocity by using the chain rule. The x component, for example, is given by

$$p_x = m \frac{dx}{d\tau} = m \frac{dx}{dt} \frac{dt}{d\tau} = m v_x \frac{dt}{d\tau} \tag{A.52}$$

and therefore the spatial components of p_μ are the spatial components of the linear momentum

$$\mathbf{p} = \gamma m \mathbf{v} = \frac{m\mathbf{v}}{\sqrt{1 - v^2/c^2}} \qquad (A.53)$$

In sum,

$$p_\mu = (p_0, p_x, p_y, p_z) = (\gamma mc, \gamma m v_x, \gamma m v_y, \gamma m v_z) \qquad (A.54)$$

But what is the meaning of p_0? This component of the four-momentum is just E/c, provided we identify

$$E = \gamma m c^2 = \frac{mc^2}{\sqrt{1 - v^2/c^2}} \qquad (A.55)$$

When the particle is at rest, this energy consists entirely of mass energy, namely

$$E = mc^2 \quad \text{when} \quad v = 0 \qquad (A.56)$$

perhaps the most famous equation in physics. When the particle is in motion, then $\gamma > 1$ and the particle has a larger energy, consisting of both mass energy and kinetic energy K. This gives us a relativistically correct way to define kinetic energy:

$$K = E - mc^2 = (\gamma - 1)mc^2 \qquad (A.57)$$

When $v/c \ll 1$, the nonrelativistic limit can be obtained by carrying out a binomial or Taylor series expansion of γ:

$$K = \left(\frac{1}{\sqrt{1 - v^2/c^2}} - 1\right) mc^2 = \left(1 + \frac{v^2}{2c^2} + \cdots - 1\right) mc^2 = \frac{1}{2}mv^2 + \cdots \qquad (A.58)$$

Thus the relativistic expression for the kinetic energy reduces to the nonrelativistic $mv^2/2$ in the limit that $v \ll c$. On the other hand, as $v \to c$, $E = \gamma mc^2 \to \infty$. Consequently, the velocity of a particle of mass m is restricted to be less than c since it would take an infinite amount of energy to get the particle up to the speed of light. Only a particle with $m = 0$ can travel at c.

Before Einstein, it was presumed that in collisions there were three conservation laws—conservation of mass, conservation of linear momentum, and conservation of energy—primarily because measurements of masses in chemical reactions were not precise enough to detect a change in mass, although such changes do occur. A significantly larger change in mass occurs in many of the nuclear reactions that were first observed in the twentieth century. It is the conversion of mass energy to kinetic energy that fuels a nuclear power plant and the atomic bomb. These topics are discussed at length in Chapter 9. Moreover, in Chapter 10 reactions involving particle–antiparticle annihilation in which the rest mass of the particles disappears entirely are discussed. We now know the correct conservation laws are simply conservation of momentum and conservation of energy, where energy is understood to include mass energy as well as kinetic energy. In fact, in special relativity conservation of momentum and conservation of energy really collapse to a single conservation law, namely conservation of four-momentum.

It is common to refer to the momentum four-vector as the energy–momentum four-vector, since its components include the energy as well as the components of the linear momentum of the particle. Like any four-vector, the momentum four-vector behaves just like the position four-vector under a Lorentz tranformation. Therefore

$$p'_0 = \gamma\left(p_0 - \frac{V}{c}p_1\right)$$
$$p'_1 = \gamma\left(p_1 - \frac{V}{c}p_0\right)$$
$$p'_2 = p_2$$
$$p'_3 = p_3 \tag{A.59}$$

relating the energy and the components of the linear momentum in the two reference frames S and S'. Consequently, the quantity

$$p_0^2 - p_1^2 - p_2^2 - p_3^2 = \frac{E^2}{c^2} - p_x^2 - p_y^2 - p_z^2 \tag{A.60}$$

has the same value in all inertial reference frames. Like (A.41), it is a Lorentz invariant. In the rest frame of the particle $\mathbf{p} = 0$ and $E = mc^2$. Thus

$$\frac{E^2}{c^2} - p_x^2 - p_y^2 - p_z^2 = m^2c^2 \tag{A.61}$$

and therefore

$$E = \sqrt{p^2c^2 + m^2c^4} \tag{A.62}$$

where $p^2 = p_x^2 + p_y^2 + p_z^2$. This expression connecting the energy E of a particle with mass m and momentum p is one of the truly fundamental and important results of special relativity. Although our derivation was carried out for a particle with a nonzero rest mass m, this result (A.62) is a quite general one. For a particle for which $pc \gg mc^2$,

$$E = pc\sqrt{1 + \frac{m^2c^4}{p^2c^2}} = pc\left(1 + \frac{m^2c^4}{2p^2c^2} + \cdots\right) \qquad pc \gg mc^2 \tag{A.63}$$

Thus for an ultrarelativistic particle, $E \cong pc$. For the special case of a massless particle, like a photon, (A.62) reduces to

$$E = pc \quad \text{when} \quad m = 0 \tag{A.64}$$

The scattering of a photon from an electron at rest in the Compton effect (see Section 1.3) provides a nice confirmation of (A.64).

EXAMPLE A.6 Use the Lorentz transformation of the momentum four-vector to determine the energy of a photon observed on Earth that is emitted by a source that recedes from the Earth at speed V, as illustrated in Fig. A.12. Call the energy of the photon in the rest frame of the source E_{em}. Therefore the magnitude of the momentum of the photon in this frame is E_{em}/c.

SOLUTION Denote the rest frame of the source as S'. The four-momentum of the photon in S' is

$$p'_\mu = (p'_0, p'_1, p'_2, p'_3) = (E_{\text{em}}/c, -E_{\text{em}}/c, 0, 0)$$

since we are assuming that S' moves in the x direction as observed in S. Note that $p'_1 = -E_{\text{em}}/c$ since the photon is emitted in the negative x' direction in S'. Using

Earth

Figure A.12 A source of light receding from the Earth at speed V.

the inverse Lorentz transformation of (A.59), namely

$$p_0 = \gamma \left(p'_0 + \frac{V}{c} p'_1 \right)$$

we see that

$$\frac{E_{\text{obs}}}{c} = \gamma \left(\frac{E_{\text{em}}}{c} - \frac{V}{c} \frac{E_{\text{em}}}{c} \right) = \gamma \left(1 - \frac{V}{c} \right) \frac{E_{\text{em}}}{c}$$

where the energy of the photon as observed on Earth is denoted by E_{obs}. Therefore

$$E_{\text{obs}} = \frac{1 - V/c}{\sqrt{1 - V^2/c^2}} E_{\text{em}} = \sqrt{\frac{1 - V/c}{1 + V/c}} E_{\text{em}}$$

As discussed in Section 1.3, the energy E of a photon is directly proportional to its frequency ν, namely $E = h\nu$, where h is Planck's constant. Thus we see that the frequency ν_{obs} of the light as observed on Earth is given by

$$\nu_{\text{obs}} = \sqrt{\frac{1 - V/c}{1 + V/c}} \nu_{\text{em}}$$

where ν_{em} is the frequency of the light in the rest frame of the source. Thus when the source recedes from the Earth, the frequency the light as observed on Earth is lower than that in the rest frame of the source. This is the relativistic Doppler effect. Since the wavelength $\lambda = c/\nu$, a decrease in frequency corresponds to an increase in wavelength. Such a shift in wavelength is referred to as a **redshift**, since red light has a longer wavelength than blue light. The observed redshift of light emitted from atoms in distant galaxies provides compelling evidence for an expanding universe and the Big Bang theory.[6]

EXAMPLE A.7 The mass of the π^0 meson is 135 MeV/c^2. The predominant decay mode of the π^0 is into two photons ($\pi^0 \to \gamma + \gamma$). If the π^0 is at rest before the decay, determine the momentum of each of the photons.

SOLUTION In order to conserve momentum, the photons must be emitted back to back with the same magnitude of the momentum for each. Consequently, each photon has the same energy E and hence

$$m_{\pi^0} c^2 = 2E$$

Thus $E = 67.5$ MeV. Then since $E = pc$ for a photon, $p = 67.5$ MeV/c.

---

[6]There are roughly 100 galaxies, most in our own local group of galaxies, that exhibit a blueshift in the radiation they emit. Consequently, these nearby galaxies are moving toward us rather than away from us.

EXAMPLE A.8 Use Lorentz invariance arguments to determine the threshold kinetic energy of the incident proton for the production of an antiproton via the reaction

$$p + p \to p + p + p + \bar{p}$$

in which the target proton is at rest in the lab.

SOLUTION Conservation of four-momentum in the collision requires that

$$p_{1_\mu} + p_{2_\mu} = p_{3_\mu} + p_{4_\mu} + p_{5_\mu} + p_{6_\mu}$$

where p_{1_μ} is the four-momentum of the incident proton, p_{2_μ} is the four-momentum of the target proton, etc. Taking advantage of the Lorentz invariant "length" of these four-vectors, we see that

$$\left(\frac{E_1 + E_2}{c}\right)^2 - (\mathbf{p}_1 + \mathbf{p}_2)^2 = \left(\frac{E_3 + E_4 + E_5 + E_6}{c}\right)^2 - (\mathbf{p}_3 + \mathbf{p}_4 + \mathbf{p}_5 + \mathbf{p}_6)^2$$

Since both sides of this equation are Lorentz invariants, their values are frame independent. Evaluating the left-hand side in the laboratory frame and the right-hand side in the center-of-mass frame when the final-state particles are at rest, i.e., at threshold, we obtain

$$\left(\frac{E_1 + mc^2}{c}\right)^2 - p_1^2 = \left(\frac{4mc^2}{c}\right)^2$$

Consequently

$$E_1^2 + 2E_1 mc^2 + m^2 c^4 - p_1^2 c^2 = 16\, m^2 c^4$$

Since

$$E_1^2 - p_1^2 c^2 = m^2 c^4$$

we find

$$2E_1 mc^2 = 14\, m^2 c^4$$

Figure A.13 The Berkeley Bevatron, Ernest Orlando Lawrence Berkeley National Laboratory, reproduced with permission from AIP Emilio Segre Visual Archives.

and thus
$$K_1 = E_1 - mc^2 = 7mc^2 - mc^2 = 6mc^2$$

The Berkeley Bevatron (see Fig. A.13), one of the first "high-energy" accelerators built in the 1950s, accelerated protons to kinetic energies of approximately 6 GeV with the expectation that when these protons interacted with the protons in a copper target antiprotons would be created.

Problems

A.1. The tau lepton has a proper lifetime of 0.3 ps. What speed must a tau lepton have in order to travel, on average, 1 mm before decaying?

A.2. An astronaut has a resting heartbeat of 60 beats/minute. The astronaut flies past the Earth in a spaceship traveling at $3/5\,c$. How does the astronaut's heartbeat rate as measured by an observer in the ship compare with that measured by observers on Earth?

A.3. The radius of our galaxy is approximately 3×10^{20} m. How fast must a spaceship travel so that the crossing the galaxy takes 40 years as measured on board the ship? How long would the trip take as measured on Earth?

A.4. Upon arrival at the star Alpha Centauri, which is 4 light years away, an astronaut sends a radio message back to planet Earth. The message is received 9 years after the astronaut departed Earth. How fast did the astronaut travel? How much time had elapsed for the astronaut when the message was sent?

A.5. The Global Positioning System (GPS) relies on very accurate atomic clocks aboard a network of 24 satellites, each of which orbits the Earth in 12 hours. To provide a resolution better than 1 meter on Earth, the clocks must not gain or lose more than 3 ns in 12 hours. That is, the clocks must be accurate to
$$3 \times 10^{-9} \text{ s}/(12 \text{ hr}) = 7 \times 10^{-14}$$
The satellites move at a speed $v = 3.9$ km/s in circular orbits. Is it necessary for GPS receivers on Earth to account for special relativistic effects?

A.6. A rod of length L_0 is oriented so it makes an angle θ_0 with respect to the x' axis. Observers in the reference frame S see the rod moving with speed V along the x axis. Show that the angle θ the rod makes with the x axis is given by $\tan\theta = \gamma \tan\theta_0$. *Suggestion*: Take the lower end of the rod to be located at the origin of the S' reference frame.

A.7. The most famous paradox in special relativity is the twin paradox. Consider two identical twins, Al and Bert. Bert decides to make a round-trip to a nearby star. Since moving clocks run slow, the time interval dt on Al's clock and the time $d\tau$ on Bert's clock are related by $d\tau = dt\sqrt{1 - v(t)^2/c^2}$, where $v(t)$ is the speed of Bert relative to Al. Thus the total elapsed time on the Bert's clock is given by
$$\tau = \int_0^t \sqrt{1 - v(t)^2/c^2}\, dt$$
Consequently Bert returns from the trip younger than Al. The paradox arises in asserting that from the Bert's perspective it is Al who did the traveling and thus Al should be younger than Bert because of time dilation. What is the fallacy in Bert's reasoning?

A.8. Show with the aid of the Lorentz transformation (A.28) that the quantity $c^2t^2 - x^2 - y^2 - z^2$ is an invariant, namely
$$c^2t^2 - (x^2 + y^2 + z^2) = c^2t'^2 - (x'^2 + y'^2 + z'^2)$$

A.9. Use the inverse Lorentz transformation (A.30) to derive length contraction and the leading-clock-lags rule.

A.10. Prove that the Lorentz transformation (A.28) can be cast in the form
$$x' = \cosh\phi\, x - \sinh\phi\, ct$$
$$ct' = -\sinh\phi\, x + \cosh\phi\, ct$$
$$y' = y$$
$$z' = z$$

with $\tanh\phi = V/c$. *Suggestions*: You may wish to make use of the identity

$$\cosh^2\phi - \sinh^2\phi = 1$$

A.11. In reference frame S event B occurs 4 μs after event A. Events A and B are separated by 1.5 km along the x axis. With what speed must an observer move in the positive x direction so that events A and B occur simultaneously?

A.12. In reference frame S event A takes place at $x_A = 500$ m. A second event, event B, occurs 5 μs later at $x_B = 1500$ m. With what speed must an observer move in the positive x direction so that the events occur at the same point in space in that observer's frame?

A.13. Two protons move at speed v, one along the x axis and one along the y axis. What is the relative speed of the two protons? Evaluate your answer for the special case $v = 3/5\,c$.

A.14. In the laboratory rest frame two particles are approaching in opposite directions, each with speed of $0.9c$. Find the velocity of one of the particles as determined in the rest frame of the other particle.

A.15. Show in the collision pictured in Fig. A.11b that linear momentum is conserved.

A.16. A proton has an energy of 1 TeV. How fast is the proton moving? For simplicity, take $mc^2 = 1$ GeV.

A.17. Two protons, each with an energy of 1 TeV, make a head-on collision at the Large Hadron Collider (LHC). What is the velocity of one of the protons as measured in the rest frame of the other proton? Take $mc^2 = 1$ GeV.

A.18. At the SLAC National Accelerator Laboratory electrons are accelerated to an energy of 50 GeV. How fast are these electrons moving?

A.19. An electron has a kinetic energy six times greater than its mass energy. Find (a) the electron's total energy and (b) its speed.

A.20. A particle with mass M at rest decays into two particles, one with mass m_1 and the other with mass m_2. Use conservation of energy and momentum to show that

$$E_1 = \frac{(M^2 + m_1^2 - m_2^2)\,c^2}{2M}$$

A.21. Find the kinetic energy of a proton moving with half the speed of light. Express your result in terms of mc^2, the rest mass energy of the proton.

A.22. A photon of energy $E = 15$ GeV moving to the right is absorbed by a particle of mass m moving to the left with speed $v = \frac{3}{5}c$. After absorption, the particle has mass m_f and is moving to the right with speed $\frac{3}{5}c$. Using energy units, find m and m_f.

A.23. At what speed must a body travel so that its total energy is twice its rest energy mc^2?

A.24. Show that

$$\frac{v}{c} = \frac{pc}{E}$$

for a particle with mass m.

A.25. How large an error is made in using the nonrelativistic expression $K = mv^2/2$ for the kinetic energy if $v/c = 0.1$?

A.26. A K^0 meson traveling at $0.9c$ decays into two pions. Take the mass of the K^0 meson as 500 MeV/c^2 and the mass of each pion to be 140 MeV/c^2. One of the pions travels backwards and the other forwards relative to the direction of motion of the K^0. Determine the velocities of the two pions. *Suggestion*: Determine the velocities of the pions in the center-of-mass frame of the K^0 and then use the velocity transformation to determine the velocities in the lab.

A.27. A π^0 traveling at speed v decays into two photons. One of the photons emerges parallel to the direction of travel of the π^0 and the other antiparallel. If the energy of one of the photons is three times the energy of the other, determine the speed of the π^0. *Suggestion*: Express the energy of the π^0 in the form $\sqrt{p_\pi^2 c^2 + m_\pi^2 c^4}$. Use conservation of energy and momentum to determine p_π and then use Problem A.24 to determine v.

A.28. Figure A.14 shows the spectrum of emission lines of atomic hydrogen excited in a discharge tube on Earth (a) and as measured with a spectrograph attached to a telescope that is pointed at a distant galaxy (b). Is the galaxy moving toward us or away from us? Determine the approximate speed of the galaxy relative to us.

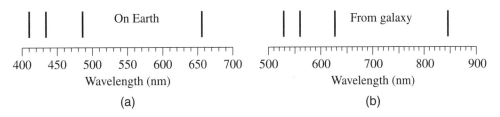

Figure A.14 Emission lines from atomic hydrogen.

A.29. The redshift parameter z is defined by the relationship

$$z = \frac{\lambda_{obs} - \lambda_{em}}{\lambda_{em}}$$

or

$$1 + z = \frac{\lambda_{obs}}{\lambda_{em}}$$

where λ_{em} is the emitted wavelength and λ_{obs} is the observed wavelength. The largest confirmed galactic redshift is $z \sim 7$. How fast is this galaxy receding from us?

A.30. A photon collides with an electron at rest. Determine the threshold energy for creation of an electron–positron pair in the reaction

$$\gamma + e^- \rightarrow e^- + e^- + e^+$$

A.31. An electron and positron make a head-on collision with equal and opposite velocities. What is the threshold energy required to produce a proton–antiproton pair in the reaction

$$e^- + e^+ \rightarrow p + \bar{p}$$

APPENDIX B

Power-Series Solutions

Our goal in this appendix is to show how to solve the differential equations that result from the energy eigenvalue equation for the harmonic oscillator, the eigenvalue equation for the magnitude squared of the orbital angular momentum, and the radial equation for the hydrogen atom using a power-series solution. In the case of the harmonic oscillator and angular momentum, there are in addition elegant operator techniques for obtaining the eigenvalues and eigenstates from the fundamental commutation relations. You are certain to see those techniques in your next course in quantum mechanics.

B.1 The Simple Harmonic Oscillator

It is possible to solve the energy eigenvalue equation

$$-\frac{\hbar^2}{2m}\frac{d^2\psi}{dx^2} + \frac{1}{2}m\omega^2 x^2 \psi = E\psi \tag{B.1}$$

for the harmonic oscillator with a power series. First, we simplify the differential equation by switching to the dimensionless variable

$$y = \frac{x}{a} = \sqrt{\frac{m\omega}{\hbar}}x \tag{B.2}$$

where

$$a = \sqrt{\frac{\hbar}{m\omega}} \tag{B.3}$$

has the dimensions of length. With this change of variables the differential equation (B.1) becomes

$$\frac{d^2\psi}{dy^2} + \left(\epsilon - y^2\right)\psi = 0 \tag{B.4}$$

where we have set

$$\epsilon = \frac{2E}{\hbar\omega} \tag{B.5}$$

If we attempt to solve this differential equation with a power series solution, we obtain a three-term recursion relation. A way around this difficulty is to factor out explicitly the large y behavior of the wave function. For large y the differential equation becomes

$$\frac{d^2\psi}{dy^2} - y^2\psi = 0 \tag{B.6}$$

which has solutions

$$\psi = Ay^n e^{-y^2/2} + By^n e^{y^2/2} \tag{B.7}$$

In fact, in the limit of large y, we can take any power of y (or polynomial in y) multiplied by the decreasing exponential as an asymptotic solution to (B.4).

With this in mind, we make the change of variables

$$\psi(y) = H(y)e^{-y^2/2} \tag{B.8}$$

where $H(y)$ is a polynomial in y that we need to determine to ensure that (B.8) is an exact solution to the differential equation (B.4). Making the substitution (B.8) into (B.4), we find that $H(y)$ satisfies the differential equation

$$\frac{d^2H}{dy^2} - 2y\frac{dH}{dy} + (\epsilon - 1)H = 0 \tag{B.9}$$

At first glance it looks as if we are headed in the wrong direction in that (B.9) has both first and second derivative terms while (B.4) has only a second derivative term. But if we substitute

$$H(y) = \sum_{k=0}^{\infty} a_k y^k \tag{B.10}$$

into (B.9) we obtain

$$\sum_{k=2}^{\infty} k(k-1)a_k y^{k-2} - 2\sum_{k=1}^{\infty} k a_k y^k + (\epsilon - 1)\sum_{k=0}^{\infty} a_k y^k = 0 \tag{B.11}$$

Letting $k' = k - 2$ in the first term yields for this term

$$\sum_{k'=0}^{\infty} (k'+2)(k'+1)a_{k'+2} y^{k'} \tag{B.12}$$

We now rename the dummy index k' as k, which allows us to write (B.11) as

$$\sum_{k=0}^{\infty} [(k+2)(k+1)a_{k+2} - 2k a_k + (\epsilon - 1)a_k] y^k = 0 \tag{B.13}$$

Since the y^k are linearly independent functions, the only way for (B.13) to be satisfied is for the coefficient of each factor of y^k to vanish. In this way, we obtain the two-term recursion relation

$$\frac{a_{k+2}}{a_k} = \frac{2k+1-\epsilon}{(k+2)(k+1)} \tag{B.14}$$

In general, we are left with the terms in an infinite power series, which for large k behaves as

$$\frac{a_{k+2}}{a_k} \to \frac{2}{k} \tag{B.15}$$

This is the same behavior for large k that the function

$$e^{y^2} = \sum_{n=0}^{\infty} \frac{y^{2n}}{n!} = \sum_{k=0}^{\infty} b_k y^k \tag{B.16}$$

exhibits, since for this function $b_k = 1/(k/2)!$ and therefore

$$\frac{b_{k+2}}{b_k} = \frac{(k/2)!}{[(k/2)+1]!)} = \frac{1}{(k/2)+1} \to \frac{2}{k} \tag{B.17}$$

Since the large y behavior is determined by the behavior for large k, the series solution will generate the exponential behavior $e^{-y^2/2}e^{y^2} = e^{y^2/2}$ for $\psi(y)$ that we tried to discard when choosing a viable form for the solution to (B.4). The only way to obtain a normalizable wave function is for the series to terminate, in which case $H(y)$ is a finite polynomial in y known as a Hermite polynomial. Examining the recursion relation (B.14), we see that the series terminates when it reaches a value of k such that $k = n$ provided

$$\epsilon = 2n + 1 \tag{B.18}$$

where n is an integer. Consequently,

$$E_n = \left(n + \frac{1}{2}\right)\hbar\omega \qquad n = 0, 1, 2, \ldots \tag{B.19}$$

The three lowest energy (unnormalized) eigenfunctions are

$$\psi_0 = a_0 e^{-y^2/2} \tag{B.20}$$

$$\psi_1 = a_1 y e^{-y^2/2} \tag{B.21}$$

$$\psi_2 = a_0 \left(2y^2 - 1\right) e^{-y^2/2} \tag{B.22}$$

Each energy eigenfunction with eigenvalue $E_n = (n+1/2)\hbar\omega$ is a polynomial of degree n multiplying the decaying exponential $e^{-y^2/2}$. Up to an overall multiplicative constant, these polynomials are the Hermite polynomials.

B.2 Orbital Angular Momentum

As discussed in Section 6.2, eigenfunctions of the operator

$$\mathbf{L}_{\text{op}}^2 = -\hbar^2 \left[\frac{1}{\sin\theta}\frac{\partial}{\partial\theta}\left(\sin\theta\frac{\partial}{\partial\theta}\right) + \frac{1}{\sin^2\theta}\frac{\partial^2}{\partial\phi^2}\right] \tag{B.23}$$

can be written in the form

$$Y(\theta,\phi) = \Theta(\theta)\Phi(\phi) \tag{B.24}$$

where the ϕ dependence is determined by the fact that the eigenfunctions of $\mathbf{L}_{\text{op}}^2$ can be chosen to be simultaneous eigenfunctions of $L_{z\text{op}}$ since $\mathbf{L}_{\text{op}}^2$ and $L_{z\text{op}}$ commute. Consequently, the eigenvalue equation

$$\mathbf{L}_{\text{op}}^2 Y(\theta,\phi) = \lambda\hbar^2 Y(\theta,\phi) \tag{B.25}$$

reduces to

$$-\hbar^2 \left[\frac{1}{\sin\theta}\frac{d}{d\theta}\left(\sin\theta\frac{d\Theta}{d\theta}\right) - \frac{m_l^2}{\sin^2\theta}\Theta\right] = \lambda\hbar^2\Theta \tag{B.26}$$

This differential equation can be put in a more standard form by changing the independent variable to $z = \cos\theta$, in which case (B.26) becomes

$$\frac{d}{dz}\left[(1-z^2)\frac{d\Theta}{dz}\right] + \left(\lambda - \frac{m_l^2}{1-z^2}\right)\Theta = 0 \tag{B.27}$$

We will restrict our attention to solving this differential equation for the special case $m_l = 0$, since it can be shown that the functions for nonzero m_l can be obtained by differentiating the solutions with $m_l = 0$.[1] Consequently, the eigenvalue problem for $\mathbf{L}^2_{\text{op}}$ reduces to solving Legendre's differential equation

$$(1-z^2)\frac{d^2\Theta}{dz^2} - 2z\frac{d\Theta}{dz} + \lambda\Theta = 0 \tag{B.28}$$

A power-series solution of the form

$$\Theta(z) = \sum_{k=0}^{\infty} a_k z^k \tag{B.29}$$

yields

$$\sum_{k=2}^{\infty} k(k-1)a_k z^{k-2} - \sum_{k=0}^{\infty} k(k-1)a_k z^k - 2\sum_{k=0}^{\infty} k a_k z^k + \lambda \sum_{k=0}^{\infty} a_k z^k = 0 \tag{B.30}$$

Making the change of variables $k' = k - 2$ in the first summation and then renaming the dummy index k' as k, we find

$$\sum_{k=0}^{\infty} \{(k+2)(k+1)a_{k+2} - [k(k+1) - \lambda]a_k\} z^k = 0 \tag{B.31}$$

Since the functions z^k are linearly independent, this equation requires that

$$\frac{a_{k+2}}{a_k} = \frac{k(k+1) - \lambda}{(k+2)(k+1)} \tag{B.32}$$

Notice that as $k \to \infty$

$$\frac{a_{k+2}}{a_k} \to 1 \tag{B.33}$$

and the series diverges at $z = \pm 1$ unless the series terminates.[2] The series terminates with a polynomial of degree l provided $\lambda = l(l+1)$, where $l = 0, 1, 2, \ldots$ The polynomials are called Legendre polynomials and are denoted by $P_l(z)$:

$$P_0 = 1 \tag{B.34}$$

$$P_1 = z \tag{B.35}$$

$$P_2 = \tfrac{1}{2}(3z^2 - 1) \tag{B.36}$$

You can see these Legendre polynomials appearing in the $Y_{l,0}(\theta,\phi)$ given in Section 6.2.

---

[1]For positive m_l

$$\Theta_{m_l} = \left(1 - z^2\right)^{m_l/2} \frac{d^{m_l}\Theta_{m_l=0}}{dz^{m_l}}$$

Once you have worked out the solutions with positive m_l, the solutions with negative m_l are then obtained from those with positive m_l by letting $m_l \to -m_l$.

[2]It turns out this is a very "modest" logarithmic singularity at $z = \pm 1$, at least if you rate degrees of divergence. But any divergence is too much for a well-behaved eigenfunction.

It can similarly be shown that for $m_l \neq 0$ the differential equation

$$\frac{d}{dz}\left[(1-z^2)\frac{d\Theta}{dz}\right] + \left[l(l+1) - \frac{m_l^2}{1-z^2}\right]\Theta = 0 \tag{B.37}$$

which is known as the associated Legendre equation, has well-behaved solutions only if $|m_l| \leq l$. We have shown in Section 6.2 that m_l takes on integral values. Therefore, the allowed values of m_l range from $-l$ to l in integral steps.

B.3 The Hydrogen Atom

Our approach to determining the allowed energies and radial wave functions for the hydrogen atom is very similar to the procedure that we used for the harmonic oscillator. We start with (6.63), namely

$$-\frac{\hbar^2}{2m}\frac{d^2u}{dr^2} + \left[\frac{l(l+1)\hbar^2}{2mr^2} - \frac{Ze^2}{4\pi\epsilon_0 r}\right]u = Eu \tag{B.38}$$

We are searching for bound-state solutions, namely those with energy $E < 0$. We define a dimensionless variable

$$\rho = \frac{\sqrt{-8mE}}{\hbar}r \tag{B.39}$$

Expressed in terms of ρ, (B.38) becomes

$$\frac{d^2u}{d\rho^2} - \frac{l(l+1)}{\rho^2}u + \left(\frac{\lambda}{\rho} - \frac{1}{4}\right)u = 0 \tag{B.40}$$

where

$$\lambda = \frac{Ze^2}{4\pi\epsilon_0 \hbar}\sqrt{\frac{m}{-2E}} \tag{B.41}$$

is a dimensionless constant. If we attempt to solve this differential equation with a power series solution, we obtain a three-term recursion relation. However, for large ρ, the equation simplifies to

$$\frac{d^2u}{d\rho^2} - \frac{1}{4}u = 0 \tag{B.42}$$

which has solutions

$$u = Ae^{-\rho/2} + Be^{\rho/2} \tag{B.43}$$

We discard the solution that increases exponentially with ρ because such a solution does not lead to a normalizable wave function. However, as for the harmonic oscillator, the exponentially increasing solution will resurrect itself unless the values for λ are carefully chosen.

We begin by making the change of variables

$$u = \rho^{l+1}e^{-\rho/2}F(\rho) \tag{B.44}$$

where the ρ^{l+1} behavior is determined by examining the differential equation for small ρ, as we saw in Section 6.3. When (B.44) is substituted into (B.40), we obtain the following differential equation for $F(\rho)$:

$$\frac{d^2 F}{d\rho^2} + \left(\frac{2l+2}{\rho} - 1\right) \frac{dF}{d\rho} + \left(\frac{\lambda}{\rho} - \frac{l+1}{\rho}\right) F = 0 \tag{B.45}$$

Although this differential equation looks more complicated than our starting point (B.40), it is now straightforward to obtain a power-series solution of the form

$$F(\rho) = \sum_{k=0}^{\infty} c_k \rho^k \tag{B.46}$$

Substituting (B.46) into (B.45), we obtain

$$\sum_{k=0}^{\infty} k(k-1) c_k \rho^{k-2} + (2l+2) \sum_{k=1}^{\infty} k c_k \rho^{k-2} + \sum_{k=0}^{\infty} [-k + \lambda - (l+1)] c_k \rho^{k-1} = 0 \tag{B.47}$$

Making the change of variables $k' = k - 1$ in the first two terms and then renaming the dummy index k' as k, we can express (B.47) as

$$\sum_{k=0}^{\infty} \{[k(k+1) + (2l+2)(k+1)] c_{k+1} + [-k + \lambda - (l+1)] c_k\} \rho^{k-1} = 0 \tag{B.48}$$

leading to

$$\frac{c_{k+1}}{c_k} = \frac{k + l + 1 - \lambda}{(k+1)(k+2l+2)} \tag{B.49}$$

Note that as $k \to \infty$

$$\frac{c_{k+1}}{c_k} \to \frac{1}{k} \tag{B.50}$$

which is the same asymptotic behavior as e^ρ. Thus unless the series terminates, the function $u(\rho)$ will grow exponentially as $e^{\rho/2}$. To avoid this fate, we require

$$\lambda = 1 + l + n_r \tag{B.51}$$

where the integer

$$n_r = 0, 1, 2 \ldots \tag{B.52}$$

determines the value of k at which the series terminates. The function F will then be a polynomial of degree n_r known as an associated Laguerre polynomial. You can see these polynomials appearing in the radial functions for the hydrogen atom given in Section 6.3. Finally, notice that restricting the value of λ to the integer $n = 1 + l + n_r$ means that the energy is quantized according to

$$E_n = -\frac{m Z^2 e^4}{2(4\pi \epsilon_0)^2 \hbar^2 n^2} \qquad n = 1, 2, 3, \ldots \tag{B.53}$$

Moreover, we see that the allowed values of l range from 0 to $n - 1$, depending on the order n_r of the associated Laguerre polynomial.

Answers to Odd-Numbered Problems

Chapter 1

1.1 (b) $\omega = kc/n$

1.3 (b) $2.1 \ \mu$m

1.5 2,000 km for $d_{\text{eye}} = 5$ mm

1.7 1.7×10^{17} photons

1.9 (a) 4.22 eV (b) 8.6×10^5 m/s (c) 169 W/m^2

1.11 3.95 eV

1.13 $h = 6.6 \times 10^{-34}$ J·s, $W = 2.3$ eV

1.15 $E' = mc^2/3$, $K = 2mc^2/3$

1.17 No

1.19 $z_1 = e^{i\pi/6}, z_1^* z_1 = 1, z_2 = e^{i\pi/3}, z_2^* z_2 = 1$

1.21 (a) 10^8/s (b) 3.6×10^{-11} W

1.23 (a) 0.2 (b) 0.98

1.25 $2r^2 [1 + \cos(4\pi nd/\lambda)]$, $d_{\min} = \lambda/4n$

1.27 (a) $\sin^2 \left[\frac{2\pi(l_1 - l_2)}{\lambda} \right]$ (b) $l_1 = l_2 + \frac{\lambda}{4}$ (c) 0.25

1.29 $2\sqrt{2}/3$

1.31 (a) $\phi = (2\pi/\lambda)(d_1 - d_2)$ (b) $r^2 \left(\frac{3}{4} + \frac{1}{\sqrt{2}} \cos \phi \right)$

1.33 r^2 for $ka \ll 1$

1.35 r^2

1.37 $3r^2$

1.41 $2d\sin\theta = \left(m + \frac{1}{2}\right)\lambda$

Chapter 2

2.1 1 km/s

2.3 15

2.5 (a) 2.5 pm (b) 31 μm

2.7 1.2×10^{-9} m, 3.9×10^{-11} m, 8.7×10^{-13} m

2.9 For $m \cong 1$ kg, $v \lesssim 6.63 \times 10^{-34}$ m/s, $t \gtrsim 4.5 \times 10^{34}$ s

2.11 (a) 60 keV (b) 2.5 keV

2.13 1.8×10^{-10} m

2.19 $\frac{\hbar k}{m}\left(|A|^2 - |B|^2\right)$

2.21 (b) $m = 1$ g, $t \sim 2 \times 10^{19}$ s and for an electron, $t \sim 2 \times 10^{-16}$ s

2.23 (a) $\Delta p \simeq \frac{\hbar}{\Delta x} = 3.5 \times 10^{-30}$ kg·m/s (b) 3 nm

2.25 (a) $v_{\text{ph}} = \sqrt{\frac{Tk}{\rho}}$, $v_g = \frac{3}{2}\sqrt{\frac{Tk}{\rho}}$

2.29 (b) $kc^2/\sqrt{k^2c^2 + \frac{m^2c^4}{\hbar^2}}$

2.31 $N = \sqrt{\frac{105}{L^7}}$, $\langle x \rangle = \frac{5}{8}L$

2.33 $\langle x \rangle = \frac{1}{2}L$, $\langle p_x \rangle = 0$, $\Delta x = L\sqrt{\frac{1}{12} - \frac{1}{2\pi^2}}$, $\Delta p_x = \frac{\hbar\pi}{L}$, $\Delta x \Delta p_x = \hbar\pi\sqrt{\frac{1}{12} - \frac{1}{2\pi^2}}$

2.35 $\langle x \rangle = 0$, $\Delta x = \frac{1}{\sqrt{2\kappa}}$

Chapter 3

3.3 (a) $\frac{1}{\sqrt{L}}$ (b) $\frac{L}{2\sqrt{3}}$

3.5 $|c_1|^2 = \frac{1}{4}$, $|c_2|^2 = \frac{3}{4}$, $\langle E \rangle = \frac{13}{4}E_1$, $\Delta E = \frac{3\sqrt{3}}{4}E_1$

3.9 (a) $\frac{8}{n^2\pi^2}$ for $n = 1, 3, 5, \ldots$, 0 for $n = 2, 4, 6, \ldots$
(b) $\Psi(x,t) = \sum_{n=1}^{\infty} \frac{\sqrt{2}}{n\pi}(1 - \cos n\pi)e^{-iE_n t/\hbar}\sqrt{\frac{2}{L}}\sin\frac{n\pi x}{L}$

3.11 $840/\pi^6 = 0.874$

Chapter 4

4.7 $2A\cos(ka/2) = Ce^{-\kappa a/2}$ and $C = \dfrac{e^{\kappa a/2}}{\sqrt{2}}\left[\dfrac{1}{\cos^2(ka/2)}\left(\dfrac{a}{4} + \dfrac{\sin ka}{4k}\right) + \dfrac{1}{2\kappa}\right]^{-1/2}$

4.9 (c) $V_0 = 52$ MeV

4.11 (b) $\dfrac{3}{4}\sqrt{\dfrac{\pi}{b^5}}$

4.13 $\sqrt{\dfrac{\hbar}{2m\omega}}$

4.15 (c) 0.81

4.17 $\Psi(x,t) = e^{-iE_0 t/\hbar}\left[\dfrac{1}{\sqrt{2}}\psi_0(x) + \dfrac{1}{\sqrt{2}}e^{-i(E_1-E_0)t/\hbar}\psi_1(x)\right]$

4.25 Same as (4.120) and (4.121)

4.29 (c) $R = 1$

4.31 (c) $T = \dfrac{1}{1+k^2a^2/4}$

Chapter 5

5.5 E_1 for -1 and equal probabilities of E_0 and E_2 for $+1$

5.7 $45°$

Chapter 6

6.1 $E_{n_x,n_y,n_z} = \dfrac{\hbar^2\pi^2}{2m}\left(\dfrac{n_x^2}{a^2} + \dfrac{n_y^2}{b^2} + \dfrac{n_z^2}{c^2}\right)$, $E_{1,1,2}$, nondegenerate

6.3 $(m\omega/\pi\hbar)^{3/4} e^{-m\omega r^2/2\hbar}$, $l=0$

6.11 $+\hbar$

6.15 (a) $E_{l,m_l} = \left\{\dfrac{[l(l+1)-m_l^2]\hbar^2}{2I_1} + \dfrac{m_l^2\hbar^2}{2I_3}\right\}$ (b) $\psi(\theta,\phi,t) = \dfrac{1}{\sqrt{2}}Y_{0,0} + \dfrac{e^{-iE_{1,1}t/\hbar}}{\sqrt{2}}Y_{1,1}$
(c) $E_{1,1}/2$

6.17 (a) $H = \dfrac{L_{z\,\text{op}}^2}{2I}$, $\psi_{m_l} = \dfrac{e^{im_l\phi}}{\sqrt{2\pi}}$, $E_{m_l} = \dfrac{\hbar^2 m_l^2}{2I}$
(b) 0 with probability 2/3, $\hbar$ with probability 1/6, $-\hbar$ with probability 1/6
(c) $\dfrac{1}{\sqrt{3\pi}}\left(1 - e^{-i\hbar t/2I}\cos\phi\right)$ (d) 1/3 (e) $\Delta E = \dfrac{\sqrt{2}}{3}\dfrac{\hbar^2}{2I}$

6.19 2.5×10^{74}

6.21 $-mA^2/2n^2\hbar^2$

6.23 $l = 1$, 1/3 probability that $L_z = 0$

6.25 $r\left[\sin\theta\cos(\phi - \omega_0 t) + \sin\theta\sin(\phi - \omega_0 t) + \cos\theta\right]e^{-r/2b_0}$

6.27 $S_x = \frac{\hbar}{2}\ \frac{1}{\sqrt{2}}\begin{pmatrix}1\\1\end{pmatrix},\ S_x = -\frac{\hbar}{2}\ \frac{1}{\sqrt{2}}\begin{pmatrix}1\\-1\end{pmatrix}$

Chapter 7

7.3 $\frac{1}{4} \pm \frac{16}{9\pi^2}$

7.5 4.4×10^5 eV

7.7 (a) $(3\pi^2)^{1/3}$ (b) $\frac{(3\pi^2)^{1/3}}{4}\hbar c \left(\frac{N}{V}\right)^{4/3}$

7.9 $E_F/2$

7.11 3.5×10^{-11} m²/N

7.17 (a) 1682 K (b) 1.26 cm

7.19 (a) 50 nm (b) 10^6

7.21 1/10

7.23 11.5

7.25 3.2×10^{13}/cm³

7.27 0.65

Chapter 8

8.9 2.82×10^{-3} eV

Chapter 9

9.1 1.3×10^{-11} s

9.3 0.72 MeV

9.5 28 MeV for p, 38 MeV for n

9.9 Should be 5f $j = 5/2$

9.11 15.999 u

9.13 0.013

9.15 1.7×10^9 yr

9.17 $\tau = 2.3 \times 10^3$ yr, $t_{1/2} = 1.6 \times 10^3$ yr

9.19 5250 yr

9.21 (a) 17.8 yr (b) 8270 yr

9.23 34,600 yr

Chapter 10

10.7 $K^- + p \to \Omega^- + K^0 + K^+$

10.11 $4\,mc^2$

10.13 6.3×10^{14} m^{-2}s^{-1}

10.15 $\sqrt{3}mc/qB \simeq 3\sqrt{3}$ m

Appendix A

A.1 $0.996\,c$

A.3 $v/c = 1 - 2 \times 10^{-7}$, $t = 6.4 \times 10^4$ yr

A.5 Yes

A.11 $v/c = 4/5$

A.13 $0.77\,c$

A.17 $v/c = 1 - 1.25 \times 10^{-13}$

A.19 (a) $7\,mc^2$ (b) $v/c = 0.990$

A.21 $0.15\,mc^2$

A.23 $v/c = \sqrt{3}/2$

A.25 1%

A.27 $v/c = 1/2$

A.29 $v/c \sim \frac{63}{65} = 0.97$

A.31 $m_p c^2$

Index

Acceptor impurity, 270
Activity, 315
Alpha decay, 292–295
Amplitude
   of a wave, 1
   probability, 19
Anderson, C., 322
Angular frequency, 2
Angular momentum
   intrinsic spin, 196–200
   orbital, 179–186
Anticoincidence experiment, 18
Antiparticles, 194, 355
   antineutrino, 328
   antiquarks, 327
   positron, 194, 321
Aspect, A., 17, 23
Associated Laguerre polynomials, 189, 398
Associated Legendre equation, 397
Asymmetry term, 285–287
Atom interferometry, 52–57
Atomic bomb, 306–310
   gun-type assembly, 308
   implosion method, 309
Atomic mass number, 278
Atomic number, 278
Attenuation length, 280
Avogadro's number, 232

Balmer series, 190
Band structure of solids, 257–263
   Kronig–Penney model, 257–261
   qualitative analysis, 262–263
Bardeen, J., 274
Barn (unit), 280
Baryon, 327
Baryon conservation, 363

BCS theory, 275
Beam splitter, 18
Bell, J. S., 171
Beta decay, 292, 295–296
Bethe, H., 308
Bevatron, 389
Big Bang, 240, 350, 357, 387
Binding energy
   atomic, 281
   nuclear, 281–288
Black hole, 306
Blackbody radiation, 235–239
Bloch ansatz, 258
Bohr magneton, 195
Bohr model, 187, 207
Bohr radius, 191
Bohr, N., 22, 283, 306
Boltzmann constant, 228
Boltzmann distribution
   for a two-state system, 247
Boltzmann factor, 227
Boltzmann, L., 227
Born interpretation, 64
Bose, S. N., 242
Bose–Einstein condensation, 242
Bose–Einstein distribution function, 242
Bose–Einstein statistics, 229
Boson, 214
   Higgs, 361
   W and Z, 363
Bottom quark, 327
Bottomonium, 345
Bragg relation, 59
Bragg, W. L., 59
Bremmstrahlung, 364

Cabbibo, N., 343
Cabibbo angle, 343
Cavity radiation, 235–239
   energy density, 238
   formula, 248
   Stefan–Boltzmann law, 239
   Wien's law, 239
Centrifugal barrier, 188
CERN, 341
Chadwick, J., 277
Chain reaction, 299
Chandrasekhar, S., 225
Charge conjugation invariance, 355
Charge independence, 278
Charm quark, 327, 344
Charmonium, 344
Chemical potential, 242
COBE satellite, 240
Color, 334
Color singlet, 337, 339
Colorless, 339
Commutation relations
   intrinsic spin, 198, 199
   orbital angular momentum, 182, 184
   position–momentum, 163
Commutator, 160
Commuting operators, 160–162
Completeness, 99
Complex conjugate, 19
Compressibility, 224
Compton formula, 15
Compton scattering, 13–16, 319–322
Compton wavelength, 15
Compton, A. H., 13
Conservation of
   angular momentum, 352–353
   charge, 358
   color, 361
   linear momentum, 351
   parity, 354
      in electromagnetic interactions, 355
Conservation of probability, 66–67
Contact potential, 265
Cooper pairs, 275
Cooper, L., 274
Cornell, E., 245
Correspondence principle, 79, 195, 207
Cosmic background radiation, 240–241
Coulomb term, 284
CP
   conservation, 356
   nonconservation, 356–357
Crab nebula, 226, 305
Critical condition, 299
Critical mass, 306

Critical temperature
   Bose–Einstein condensation, 244
      in liquid helium, 247
Cross section, 279
Crystalline solids
   covalent, 263
   ionic, 263
   metallic, 263
   molecular, 263
Cubic cavity, 236
Curie, 315
Curve of binding energy, 281–288

Dark matter, 350
Davis, R., 347
Davisson, C. J., 60
de Broglie wavelength, 53, 280
Decoherence, 173
Degeneracy, 161, 179
   in the hydrogen atom, 190
Degeneracy pressure, 224
Delta baryon, 331
Density of states
   for nonrelativistic particles, 223
   for photons, 237
   in an energy band, 266
Deuterium, 304
Diffraction
   crystal, 58–62
   gratings, 36–38, 42
   single-slit, 7, 48
Diode, 272
Dipole
   color, 338
   electric, 338
Dirac delta function, 130
Dirac delta-function potential
   double well, 133–135
   single well, 131–132
Dirac, P. A. M., 51
Dispersion relation, 75
Donor impurity, 269
Doppler effect, 387
Double-slit experiment
   helium atoms, 52–56
   light, 32–36
Down quark, 327, 333

Ehrenfest's theorem, 79–81, 167
Eigenvalue, 103
Eigenvalue equation, 103
Einstein A and B coefficients, 248
Einstein, A., 1, 9, 22, 43, 170, 242, 368–369
Einstein–Podolsky–Rosen paradox, 170–171
Electromagnetic spectrum, 3

Electron–positron scattering, 326
Electroweak interactions, 361, 362
Elementary particles, 326–329
Energy
    eigenfunction, 105
    eigenvalue, 105
    expectation value, 102
    operator, 104
Energy–momentum four-vector, 385
Entangled state, 170
Euler identity, 3, 20
Event, 377
Exchange force, 253
Exchange operator, 212
Expectation values, 77
    time dependence, 166, 351

Fermat's principle, 40
Fermi energy, 222–224, 231, 266
Fermi gas model, 282, 284, 286
Fermi temperature, 232
Fermi velocity, 224
Fermi's golden rule, 290
Fermi, E., 290, 301, 308
Fermi–Dirac distribution function, 230
Fermi–Dirac statistics, 229
Fermion, 214
Feynman diagrams, 319–326
Feynman, R., 38, 51, 308, 317
Field emission of electrons, 143
Fine structure of hydrogen, 200
Fine-structure constant, 323
Finite square well, 113–118
Fission, 298–302
Flavors
    leptons, 340
    quarks, 327
Flux tube, 338
Forward bias, 271
Four-momentum, 384–385
Four-vectors, 382
Fowler–Nordheim relation, 144
Free-electron model, 221
Fresnel, A., 43
Frisch, O., 306, 307
Fusion, 303–305

g factor, 197, 324
Galilean transformation, 367
Gamma decay, 292
Gamow, G., 293
Gaseous diffusion, 307
Gauge bosons, 329
Gauge invariance, 358–361
Geiger, H., 277
Gerlach, W., 201

Germer, D. A., 60
Glashow–Weinberg–Salam
    theory, 341
Glauber, R., 28
Global phase invariance, 359
Global positioning system (GPS),
    377, 389
Gluons, 337
Goeppert-Mayer, M., 288
Goudsmit, S., 197
Grand unified theory (GUT), 362
Gratings, 36–38, 42
Group velocity, 73
Groves, L., 307

h-bar, 62
Hadron, 327
Hadrons, 330–334
Hahn, O., 306
Half-life, 291
Hamiltonian, 104
Harmonic oscillator
    in quantum mechanics, 123–130
    power-series solution, 393–395
Heisenberg uncertainty principle, 70
Heisenberg, W., 51
Helium, 215–217
Hermite polynomials, 126, 395
Hermitian operator, 155
    properties, 155–158
Higgs boson, 361
Higgs field, 361
Hilbert space, 175
Hofstadter, R., 279
Hoyle, F., 305
Hydrogen atom, 187–193
    Balmer series, 190
    Bohr radius, 191
    degeneracy, 190
    energy eigenvalues, 189
    Lyman series, 190
    Paschen series, 190
    power-series solution, 397–398
    radial equation, 188
Hydrogen bomb, 310–311
Hyperfine structure, 200

Ice man, 316
Ideal gas constant, 232
Ideal gas law, 232
Identical particles
    in quantum mechanics, 212–214
    indistinguishability, 211, 213, 228–230, 242,
        247, 251
Inertial reference frame, 367
Integrated circuit, 268, 272, 275

Interference
    constructive, 5, 34
    destructive, 5, 34
    fringes, 4, 27
    single-atom, 52–57
    single-photon, 23–28
    thin-film, 31–32
Interferometer
    Mach–Zehnder, 23
    Michelson, 47
Intermediate vector boson, 340
Intrinsic spin, 196–200
    electron, 197
    photon, 198, 237
Invariance
    charge conjugation, 355
    gauge, 358
    inversion of coordinates, 354
    Lorentz, 382
    phase, 359
    rotational, 353
    translational, 352
Isobar, 278
Isospin, 333
Isotope, 278
    effect in superconductivity, 274

Jensen, J. H. D., 288

Kaon, 332, 343
Ketterle, W., 245
Kiln, 235
Kinetic energy, 385
    nonrelativistic, 58
    operator, 104
    threshold, 388
Kistiakowski, G., 308
Kronecker delta, 99, 157
Kronig–Penney model, 257

Lambda baryon, 332, 343
Laser, 247–251
    four-level system, 250
    He-Ne, 251
    population inversion, 250
    pumping, 250
    three-level system, 250
Law of atmospheres, 233
Least action principle, 51
Least time principle, 38–43
LED, 272
Lee, T. D., 354
Legendre polynomials, 396
Length contraction, 371–373
Lepton number, 328
    electron, 328
    muon, 328
Leptons, 327
    flavors, 340
Lifetime, 291
Light-by-light scattering, 326
Liquid drop model, 282–283
Local phase invariance, 359–360
Lorentz contraction, 372
Lorentz force, 355
Lorentz invariant, 382, 384, 386
Lorentz transformation, 321, 378–379
Lorentz, H., 195
Lyman series, 190

Mach–Zehnder interferometer, 23
Magic numbers, 287–288
Magnetic dipole moment, 194–197
Marsden, E., 277
Mass energy, 281, 385
Matrix mechanics, 159
Maxwell's equations, 3, 20
Maxwell, J. C., 43
Maxwell–Boltzmann distribution function, 230
Maxwell–Boltzmann statistics, 229
Measurement
    in Stern–Gerlach experiment, 202–203
    problem, 170, 173
Meitner, L., 306
Meson, 327
Metals, 264–265
Michelson interferometer, 47
Microstates, 226
Millikan, R., 11
Mirror reflection, 354
Mixing angles, 343–350
    Cabibbo angle, 343
    neutrino oscillations, 347
    Weinberg angle, 341
Moderator, 301
Molecular binding, 132–135
Momentum
    conservation, 351–352
    eigenfunction, 104
    operator, 104, 179
Moore's law, 268
Moore, G., 268
Multielectron atoms, 215–218
Multiparticle systems, 211–212
Muon, 328

n-type semiconductor, 269
Natural linewidth, 169
Neutrino, 328
Neutrino oscillations, 346–350
Neutron star, 226, 289, 306
Newton's rings, 4

Newton, I., 7, 43
Noncommuting operators, 162–164
Nonconservation of parity, 354–355
Normalization, 65
    Gaussian wave function, 126–128
    particle in a box, 93
    radial function, 183
    spherical harmonics, 183
Nuclear fission, 298–302
Nuclear force
    charge independence, 278
    range, 283–284, 330
    spin dependence, 288, 333–334
Nuclear fusion, 225, 303–305
Nuclear models
    Fermi gas, 284–287
    liquid drop, 283–284
    shell, 287–288
Nuclear reactors, 300–302
Nucleon, 278

Observables, 155
Operator
    energy, 104
    exchange, 212
    intrinsic spin, 198
    linear momentum, 104, 179
    orbital angular momentum, 181
    parity, 154
    position, 104
Oppenheimer, J. R., 307
Optical path length, 41
Orbital angular momentum
    conservation, 352–353
    eigenfunctions, 183–184
    eigenvalues, 183
    operators, 181–182
Orthonormal basis, 156–157
Orthonormal set, 99
Overall phase invariance, 359

p-n junction, 270–272, 275
p-type semiconductor, 269
Pairing term, 287
Parity, 153–155, 353–355
    conservation, 353–354
    nonconservation, 354–355
Particle in a box, 91–97
    allowed energies, 93
    energy eigenfunctions, 93
    time dependence, 95–97
Particle–antiparticle creation, 322
Paschen series, 190
Pauli principle, 214

Pauli spin matrices, 209
Peierls, R., 307
Penzias, A., 240
Period, 2
Periodic table, 218
Perturbation theory, 318
Phase, 2
    difference, 35
    overall, 35, 359
Phase factor
    overall, 30, 74, 96
    relative, 30, 96
Phase velocity, 72
Photoelectric effect, 9–12
Photons, 10
    intrinsic spin, 198
Pion, 330
Planck function, 238
Planck's constant, 10
Planck, M., 235, 238
Plasma, 304
Population inversion, 250
Positron, 194, 321
Positronium, 194
Power-series solutions
    harmonic oscillator, 393–395
    hydrogen atom, 397–398
    orbital angular momentum, 395–396
Probability amplitude, 19
Probability current, 66
Proper length, 371
Proper time, 371, 384
Proton decay, 362–363

QED, 319
QFD, 340
Quantum chromodynamics, 334–340, 361
Quantum electrodynamics, 317–325, 358–361
Quantum field theory, 317, 358–361
Quantum flavor dynamics, 340–342
Quantum statistics, 226–232
Quarks, 327
    colors, 334
    flavors, 327, 340

Rabi, I. I., 329
Radiation
    absorption, 248
    spontaneous emission, 248
    stimulated emission, 248–249
Radioactivity, 290–296
    carbon dating, 297
Ramsauer–Townsend effect, 150
Range of interaction
    and mass of exchanged particle, 330

Ratio R, 335
Redshift, 387
Reduced mass, 188
Reflection coefficient, 138
Relativity principle, 367–369
Resistance, 264
Resistivity, 264
Reverse bias, 271
Rotational symmetry, 352–353
Rubbia, C., 342
Rutherford, E., 277
Röntgen, W., 58

Sakharov, A., 357
Scalar potential, 358
Scanning tunneling microscope, 142
Scattering
    square barrier, 140–142
    step potential, 135–138
Schrödinger equation
    solutions
        cubic box, 178
        Dirac delta function, 132
        double well, 135
        harmonic oscillator, 125
        hydrogen atom, 189
        infinite potential well, 93
        Kronig–Penny, 261
        qualitative features, 119–123
        square barrier, 142
        step potential, 138
    time-dependent, 62
    time-independent, 90, 105
Schrödinger's cat, 171–173
Schrödinger, E., 51, 62
Schrieffer, R., 274
Seaborg, G., 307
Selection rule, 195, 207
Semiconductor laser, 272
Semiconductors
    extrinsic, 268–272
        n-type, 269
        p-type, 269
    intrinsic, 266–267
    p-n junction, 270
        depletion zone, 271
Semiempirical mass formula, 282
    asymmetry term, 285–287
    Coulomb term, 284
    pairing term, 287
    surface term, 283
    volume term, 283
Separation of variables, 89
    in Cartesian coordinates, 178
    in multiparticle systems, 211
    in spherical coordinates, 180

Shell model, 282, 287–288
Simulaneity, 373–375
Single-photon source, 18
Singlet
    color, 337, 339
    spin, 215, 334, 338
SLAC, 344–346, 375
Slater determinant, 214
Snell's law, 43
Solar cell, 272
Solar neutrino problem, 347–349
Space-like separation, 321
Special relativity, 367–388
Spectroscopic notation, 216
Spherical coordinates, 179
Spherical harmonics, 183
Spin angular momentum, 196–200
Spontaneous symmetry breaking, 341, 361
Standard Model, 358–363
Stationary state, 91
Stefan–Boltzmann constant, 239
Stefan–Boltzmann law, 239
Step potential, 135–139
Stern, O., 201
Stern–Gerlach experiment, 201–203
Stimulated emission of radiation, 249
Strange quark, 327, 332
Strassmann, F., 306
Strong interactions, 281, 289, 296, 304, 330–340
Super-Kamiokande, 348, 362, 363
Superconductivity, 273–275
Supernova 1987A, 306
Surface term, 283
Symmetry, 351–357
    inversion of coordinates, 354
    phase invariance, 359
    rotational, 352
    translational, 352
Szilard, L., 306

Teller, E., 310
Thermocouple, 265
Thermonuclear weapon, 310
Thin-film interference, 31–32
Thomson, G. P., 61
Time dilation, 369–371, 377
Top quark, 327
Transistor, 272
Translational invariance, 352
Transmission coefficient, 138
Triplet, 216, 334
Tunneling
    non-square barrier, 143
    square barrier, 140–142
Twin paradox, 389

Uhlenbeck, G., 197
Ulam, S., 308
Ultraviolet catastrophe, 254
Uncertainty, 78
Uncertainty relations, 163–164
    energy–time, 167–169, 330–332
    Heisenberg, 163
    orbital angular momentum, 186
Up quark, 327, 333
Uranium hexafluoride, 307

van der Meer, S., 342
van der Waals force, 340
Vector potential, 358
Velocity transformation
    nonrelativisitc, 367
    relativistic, 380–381
Virtual particle, 320
Volume term, 283
von Laue, M., 59
von Neumann, J., 308

W boson, 340–342
Wave equation
    light
        in a medium, 3
        in vacuum, 3
    Schrödinger, 62
Wave function, 62
    physical significance, 64–65
Wave number, 2

Wave packet, 68
    and scattering, 138–139
Wavelength, 2
Waves, 1–3
Weak interaction, 296, 305, 333, 340–342
    nonconservation of parity, 354–355
    nonconservation of strangeness, 333
    powering the sun, 346
Weinberg angle, 341
Wheeler, J., 27, 306
White dwarf star, 225
    Sirius B, 225
Wieman, C., 245
Wien's law, 239
Wigner, E., 359
Wilson, R., 240
Work function, 11
Wu, C-S., 354

Yang, C. N., 354
Young, T., 32, 43
Yukawa, H., 330

Z boson, 340–342
Zeeman effect, 194–195
    anomalous, 197
Zeeman, P., 195
Zero-point energy
    harmonic oscillator, 128–129
    particle in a box, 94

Constants and Conversion Factors

Planck's constant	$h = 6.626 \times 10^{-34}$ J·s $= 4.136 \times 10^{-15}$ eV·s
hbar	$\hbar = h/2\pi = 1.055 \times 10^{-34}$ J·s $= 6.582 \times 10^{-16}$ eV·s
Speed of light	$c = 2.998 \times 10^{8}$ m/s
Elementary charge	$e = 1.602 \times 10^{-19}$ C
Fine-structure constant	$\alpha = e^2/4\pi\epsilon_0 \hbar c = 7.297 \times 10^{-3} = 1/137.036$
Boltzmann constant	$k_B = 1.381 \times 10^{-23}$ J/K $= 8.617 \times 10^{-5}$ eV/K
Avogadro constant	$N_A = 6.022 \times 10^{23}$ particles/mole
Electron mass	$m_e = 9.109 \times 10^{-31}$ kg $= 0.5110$ MeV/c^2
Proton mass	$m_p = 1.673 \times 10^{-27}$ kg $= 938.3$ MeV/c^2
Neutron mass	$m_n = 1.675 \times 10^{-27}$ kg $= 939.6$ MeV/c^2
Bohr radius	$a_0 = 4\pi\epsilon_0 \hbar^2/m_e e^2 = 0.5292 \times 10^{-10}$ m
Rydberg energy	$hcR_\infty = m_e c^2 \alpha^2/2 = 13.61$ eV
Bohr magneton	$\mu_B = e\hbar/2m_e = 5.788 \times 10^{-5}$ eV/T

1 keV $= 10^3$ eV   1 MeV $= 10^6$ eV   1 GeV $= 10^9$ eV   1 TeV $= 10^{12}$ eV

1 μm $= 10^{-6}$ m   1 nm $= 10^{-9}$ m   1 pm $= 10^{-12}$ m   1 fm $= 10^{-15}$ m

1 eV $= 1.602 \times 10^{-19}$ J   1 Å $= 0.1$ nm $= 10^{-10}$ m